New Frontiers in Virology

New Frontiers in Virology

Editor: Harvey O'Brien

R CALLISTO
REFERENCE

www.callistoreference.com

Callisto Reference,
118-35 Queens Blvd., Suite 400,
Forest Hills, NY 11375, USA

Visit us on the World Wide Web at:
www.callistoreference.com

ISBN: 978-1-64116-253-1 (Hardback)

Cataloging-in-Publication Data

New frontiers in virology / edited by Harvey O'Brien.
    p. cm.
Includes bibliographical references and index.
ISBN 978-1-64116-253-1
1. Virology. 2. Microbiology. I. O'Brien, Harvey.
QR360 .N49 2020
579.2--dc23

Table of Contents

Preface

I am honored to present to you this unique book which encompasses the most up-to-date data in the field. I was extremely pleased to get this opportunity of editing the work of experts from across the globe. I have also written papers in this field and researched the various aspects revolving around the progress of the discipline. I have tried to unify my knowledge along with that of stalwarts from every corner of the world, to produce a text which not only benefits the readers but also facilitates the growth of the field.

Virology is the science concerned with the study of viruses and virus-like agents. Study of viruses encompasses an examination of their structure, evolution, classification, interactions and the diseases they cause. Virology also lays due emphasis on the techniques of isolating and culturing viruses, and investigating ways to use them for research and therapy. Virology is a subfield of microbiology and medicine. A major dimension of study in virology is virus classification. Viruses may be classified relative to the host cells that they infect, the shape of their capsid, the viral structure, the nucleic acid used as the genetic material or the viral replication method they employ. The structure and shape of viruses can be studied by NMR spectroscopy, electron microscopy and X-ray crystallography. Viruses cause several infectious diseases, such as measles, common cold, influenza, rabies, hepatitis, AIDS, polio, small pox, etc. Certain viruses may contribute to the development of some cancers. The use of viruses as gene vectors is actively being pursued for gene therapy. They are also being used to introduce new genes in host cells. Virology is an upcoming field of science that has undergone rapid development over the past few decades. This book presents this complex subject in the most comprehensible and easy to understand language. It is a vital tool for all researching and studying this field.

Finally, I would like to thank all the contributing authors for their valuable time and contributions. This book would not have been possible without their efforts. I would also like to thank my friends and family for their constant support.

Editor

Influenza A Virus Entry: Implications in Virulence and Future Therapeutics

Emily Rumschlag-Booms[1] and Lijun Rong[2]

[1] *Department of Biology, Northeastern Illinois University, Chicago, Chicago, IL 60625, USA*
[2] *Department of Microbiology and Immunology, College of Medicine, University of Illinois at Chicago, IL 60612, USA*

Correspondence should be addressed to Emily Rumschlag-Booms; e-booms@neiu.edu

Academic Editor: Hector Aguilar-Carreno

Influenza A viruses have broad host tropism, being able to infect a range of hosts from wild fowl to swine to humans. This broad tropism makes highly pathogenic influenza A strains, such as H5N1, potentially dangerous to humans if they gain the ability to jump from an animal reservoir to humans. How influenza A viruses are able to jump the species barrier is incompletely understood due to the complex genetic nature of the viral surface glycoprotein, hemagglutinin, which mediates entry, combined with the virus's ability to use various receptor linkages. Current therapeutics against influenza A include those that target the uncoating process after entry as well as those that prevent viral budding. While there are therapeutics in development that target entry, currently there are none clinically available. We review here the genetics of influenza A viruses that contribute to entry tropism, how these genetic alterations may contribute to receptor usage and species tropism, as well as how novel therapeutics can be developed that target the major surface glycoprotein, hemagglutinin.

1. Introduction

Influenza viruses belong to the Orthomyxoviridae family, which consists of several genera. The first includes both influenza A and B viruses, while another is comprised of influenza C virus [1]. These classifications are based on the distinct antigenic nature of the internal nucleoprotein and matrix proteins of each virus. Infection with influenza subtypes B and C is mostly restricted to humans [2, 3], while subtype A is able to infect a wide range of hosts including but not limited to humans, swine, horses, domestic and wild birds, fowl, and dogs [4–8]. This broad spectrum of hosts plays a pivotal role in the ability of the virus to reassort, mutate, and spread, all of which contribute to the ever-present global threat of influenza.

Influenza A virus poses the most serious hazard of the three subtypes, causing global economic losses as well as severe health concerns. Influenza A virus is the causative agent of severe respiratory illness infecting nearly 15% of the world's population with upwards of 250,00–500,000 deaths estimated by the World Health Organization. Infections are characterized by upper respiratory distress along with high fever, myalgia, headache and severe malaise, nonproductive cough, sore throat, and rhinitis. Severe illness and death are mainly associated with the young, elderly, and those with compromised immune systems [2].

Influenza viruses have ravaged human and poultry populations around the world for centuries, causing serious illness and death, major economic loss, in addition to instilling fear as the next potential deadly pandemic. During the twentieth century, this virus caused three major pandemics, which resulted in an estimated 20–50 million deaths combined worldwide [9–11]. In the twenty-first century, 2009 Pandemic H1N1 was caused by a reassorted swine strain. The reassortment included influenza viruses of human, avian, and two swine strains [12]. The resultant reassorted swine strain then jumped to humans, spreading around the world within a few weeks [12, 13]. The initial result of this event was more than 22 million reported cases, 13,000 deaths, the blocking of countries' borders, and the closing of numerous schools [14]. A recent study suggests the actual impact may be more than 10 times the initial estimates [15]. While we are currently in the postpandemic phase, this H1N1 strain is the currently circulating endemic influenza strain among human populations. Upwards of 20–40% of the world's population is thought to have immunological protection for the time being, as they have already been exposed to the virus.

Influenza viruses possess several unique characteristics, many of which potentiate the menace posed by this virus. One such feature is the segmented nature of the viral genome [16]. The virus carries eight negative-sense RNA segments. Due to the segmented nature of the viral RNA, if a host cell is infected with two viruses of different influenza strains, the gene segments of one virus can recombine with those of another virus during replication. This reassortment event is referred to as antigenic shift. The newly formed virus can be especially dangerous if a human adapted strain acquires gene(s) which transform it into a highly pathogenic strain, or if a highly pathogenic strain acquires the necessary gene(s) to infect and spread amongst humans. Either scenario is predicted to raise serious threats worldwide, as was the case in 1957 and 1968 [17, 18].

The major influenza pandemics in the twentieth century, along with the 2009 Pandemic H1N1, are thought to have arisen via antigenic shift. The pandemic of 1957, better known as "Asian Influenza" H2N2 virus, was originated in Southern China and spread rapidly to the United States and Europe causing more than 1 million deaths worldwide [19]. Sequence analysis along with biochemical studies suggest that this particular virus was originated from the reassortment or genetic mixing of an avian virus with that of a human virus [19–22]. While the recombinant virus was not particularly virulent, the high level of mortality associated with it is attributed to the immunological naivety of the infected populations. A similar scenario was seen with the pandemic of 1968, the "Hong Kong Influenza." The HA gene of this virus was of the H3 subtype and originated from an avian source along with the PB1 viral polymerase protein [21, 23–25]. These two avian gene segments reassorted with a human virus, creating a new virus with greater virulence and the ability to infect humans. Furthermore, human populations were immunologically naïve to this recombinant virus, making the health impact that much greater. Much devastation and loss are attributed to pandemics arising from antigenic shift and it is antigenic shift that is predicted to be the likely cause of the next pandemic [26]. Furthermore, evidence points to antigenic shift as the perpetrator of the most severe of the influenza pandemics. It is believed that antigenic shift was responsible for the first and most severe pandemic of the 20th century in 1918, killing an estimated 50 million people worldwide [27–32]. Recent sequence analysis of this H1N1 virus, referred to as the "Spanish Flu", strongly suggests that the virus was directly transmitted to humans from an avian source [32].

While antigenic shift is a powerful means of acquiring genetic change, antigenic drift results in more subtle changes in the genome. Antigenic drift in influenza viruses refers to residue substitutions in the virus' coding sequence via point mutations [33]. Due to the lack of a proofreading function in the RNA polymerase, influenza constantly accumulates mutations within its genome during replication. These mutations may be silent or they may alter the virulence and pathogenicity of the virus. For instance, if a highly pathogenic avian virus acquires the necessary mutations that facilitate its ability to efficiently enter and replicate in humans, then the virus can become a serious threat to humans.

The viral surface is studded with two major surface spike glycoproteins, hemagglutinin (HA) and neuraminidase (NA), which differ greatly in genetic variation [8]. In addition, an essential ion channel protein, M2, exists on the virion surface. HA and NA exist on the virion surface in a ratio of approximately $4/5 : 1$, with an estimated 400–600 total spikes. HA is responsible for mediating entry into target cells via the host cell receptor, sialic acid (SA). NA plays a major role during the budding process by releasing progeny virions from the host cell. To date, 16 subtypes of HA and 9 subtypes of NA have been identified [34, 35]. These subtypes have been mainly identified amongst different avian species, as birds are the natural reservoirs of influenza virus [7, 8]. Since entry is the first requirement for infection, it is crucial that we understand its role in host tropism, pathogenesis, as well as the role of differences between HA subtypes and species-specific viruses. Furthermore, HA has garnered recent attention as a target for broad-spectrum neutralizing antibodies [36, 37].

HA has been shown to be an important determinant for influenza virus virulence and pathogenesis. Genomic studies of the 1957 (H2N2) and 1968 (H3N2) pandemics revealed that a major contribution to virulence was due to the exchange of the HA segments between human and avian strains [24]. Sequence comparison of the 1918 (H1N1) virus to other influenza A viruses from various species reveals that the entire 1918 virus is more closely related to avian influenza A viruses than with any other species, namely humans, suggesting that accumulated mutations in the avian HA gene allowed it to better adapt to the human host. The critical role of mutations within the avian virus genome underlies the importance of studying mutations within the H5N1 virus genome that may be critical to sustain infection in and among humans although no sustained human-to-human transmission has been reported yet [38, 39].

HA exists on the virion surface as a trimer of HA_1 and HA_2 subunits linked by disulfide bonding. This surface glycoprotein is first synthesized as a single polypeptide (HA_0) of approximately 550 amino acids, which is highly N-glycosylated. HA assembles into a trimer in the rough endoplasmic reticulum (ER) before passing through the Golgi complex on its way to the cell membrane. For the virus to be infectious, the HA_0 precursor protein must be cleaved into its subunits, HA_1 and HA_2 [40, 41]. If HA is not cleaved, fusion of the viral envelope with the endosomal membrane cannot occur, thus the genomic contents cannot be released within the target cell. At the structural level, cleavage of HA_0 is important because it reveals the hydrophobic N-terminus of HA_2, the fusion peptide, which is inserted into the host membrane during HA-mediated viral/host membrane fusion and viral entry. Upon endocytosis, the acidic pH (5–5.5) environment triggers HA_2 to undergo the irreversible [42–45] conformational changes necessary for fusion to occur, allowing the viral and host membranes to fuse, releasing the viral genomic contents into the cytoplasm to begin replication [46].

Two known classes of proteases are involved in HA_0 cleavage [47–49]. The first known protease class recognizes a single arginine at the cleavage site. HA with such a cleavage site is processed at the cellular surface during the budding process or on released viral particles by secretory proteases, such as tryptase Clara, a trypsin-like protease found in the alveolar fluid of rat lungs, plasmin, and bacterial proteases [50]. This particular set of trypsin-like enzymes is found either in specialized cells or within specific organs, thus viruses carrying such an HA have more restricted activation, infection capability, and therefore limited replication and spread [40, 41, 50]. Due to this restriction, trypsin-like activated viruses are generally thought to be less pathogenic. It is interesting to note that while low pathogenicity is generally associated with a virus whose HA has the trypsin-like cleavage site, the most highly pathogenic human influenza virus was restricted to trypsin-like enzyme cleavage [51]. The enzyme-limited, restricted sites of replication correspond with sites of natural infection for humans and birds, that being limited to the upper-respiratory tract (humans) or gastrointestinal tracts (birds) [20].

The other variant of HA contains a polybasic consensus sequence cleavage site, R-X-K/R-R, which is recognized by the subtilisin-like endoproteases, furin, and PC5/6 [52, 53]. This protease is expressed in the trans-Golgi network, therefore the HA is activated during the exocytic route during virus maturation [53, 54]. As this protease is nearly ubiquitously expressed, a virus with this cleavage site can replicate throughout a host and cause extensive systemic spread. Due to this characteristic, viruses with the polybasic site are generally considered highly pathogenic strains, such as the H5 and H7 strains [53, 55]. A comparative analysis at the entry level between a low pathogenic H5N2 strain and a highly pathogenic H5N1 strain revealed that the major restriction to entry was the cleave site sequence. Replacement of the monobasic cleavage site into the polybasic cleavage site on HA enhanced the HA-mediated entry [56]. However, the worst human influenza epidemic recorded, the 1918 Spanish Flu, did not have the polybasic cleavage site as previously discussed. Instead, it had a single arginine, highlighting the complicated nature of influenza viruses and their virulence and pathogenicity [51].

In addition to the aforementioned two classes of proteases, the following proteolytic enzymes are implicated in HA activation: type II transmembrane serine proteases (TTSPs) TMPRSS2, TMPRSS4, and human airway trypsin-like protease (HAT) [47]. While less is known about these proteases, their activity may be additionally associated with viral pathogenicity and tropism.

Influenza entry tropism is mainly determined by the binding preference of HA_1 to its receptor, SA. SA has long been believed to be the sole receptor for the influenza virus. It was discovered nearly 70 years ago that upon addition of influenza virus to chicken erythrocytes, the cells would agglutinate at $4°C$ [57]. A shift in temperature to $37°C$ would cause the virus to elute, while addition of new influenza virus no longer caused agglutination. This phenomenon suggested that, in addition to binding a surface molecule on the erythrocytes, the virus carries a receptor-destroying enzyme. It was later discovered that the cellular component removed by the virus was SA and that treatment of erythrocytes with purified sialidase from *Vibrio cholerae* prevented agglutination [58–60]. This finding was the first demonstration that SA acts as a receptor for influenza A viruses.

SA encompasses a large family of sugar molecules. The most prevalent member of this family is N-acetylneuraminic acid (NeuAc). It primarily exists as a six-carbon ring with several unique components extending from the ring. The most important feature of SA, with regards to influenza, is the manner in which the free sugar is attached to the host cell surface. Host cells carry various surface glycoproteins and glycolipids, many of which are highly modified. These surface proteins that are modified with a terminal SA play a crucial role in influenza entry, serving as the viral attachment and entry receptor. SA can be attached to the underlying glycocalyx in one of three main linkage patterns, either $\alpha2,3$, $\alpha2,6$, or $\alpha2,8$ [61]. While other linkages exist, these three are the most prevalent in mammalian cells [62].

In addition to viral entry, SA plays an equally important role in determining host tropism. Influenza tropism is highly influenced by the linkage of SA, with avian and human viruses preferentially utilizing different linkages. Avian viruses have been classified as predominantly $\alpha2,3$ specific, while human viruses tend to favor the $\alpha2,6$ linkage [63–66]. These preferences have been established from studies examining SA distribution within a specific host, binding of the virus to SA at the protein level, as well as studies analyzing replication efficiency in the presence of these specific linkages. Increased prevalence of the $\alpha2,3$-linked SA in the avian gastrointestinal tract correlates with the ability of the virus to enter and replicate here, as it is the site of natural infection for birds [23, 61]. As the natural carrier of the virus, avian populations usually display few disease symptoms [7]. Due to the high prevalence of receptors within the gastrointestinal tract, the virus replicates most efficiently there and is excreted through waste products. On the other hand, human influenza viruses predominantly utilize SA in $\alpha2,6$ conformation which is highly prevalent in the upper respiratory tract [23, 65, 67, 68]. SA linkages within the human respiratory tract are present on nasal mucosal epithelial cells, paranasal sinuses, pharynx, trachea, and bronchi, all carrying $\alpha2,6$ SA while $\alpha2,3$ SA is found on nonciliated cuboidal bronchi and alveolus, as well as type II cells within the alveolar wall [69]. Based on these data, it is the low levels of $\alpha2,3$ linkages in the upper tract that have been postulated as being a block for efficient infection and replication of $\alpha2,3$ preferring viruses in humans. On the other hand, flow cytometry studies using lectins specific for $\alpha2,3$ and $\alpha2,6$ SA showed that both linkages were present on human bronchial epithelial cells with $\alpha2,6$ SA being the vast majority [70, 71], suggesting another layer in the complexity of influenza tropism beyond SA preference. In addition, it still remains unclear if all subtypes of influenza virus target the same subset of respiratory cells and/or have the same affinity and avidity for α2,3 and α2,6 SA linkages. Besides terminal sialic-acid linkages, specificity is also influenced by internal linkages along with modification of inner oligosaccharides including sulfation, fucosylation, and sialylation [72, 73]. Overcoming this binding restriction may be one step needed for avian influenza to more efficiently spread from human to human.

The linkage of SA and its influence on influenza entry has been extensively characterized to determine its role in tropism; however in early 2008 an additional level of complexity was revealed. Chandrasekaran et al. reported that the crucial determinant for influenza tropism is the structure of the underlying glycocalyx, not the terminal SA linkage [74]. Using a series of analyses, it was reported that avian viruses prefer SA in a cone-like topology. This shape is adopted by SAs, both α2,3 and α2,6, with short underlying glycan(s) and allows HA to contact Neu5Ac and galactose sugars in a trisaccharide motif. Human viruses are reported to prefer an umbrella-like glycan topology, which is unique to α2,6 SAs with long underlying glycans. This report also concluded that α2,6 alone is insufficient for human transmission, as avian viruses can utilize α2,6 SA on a short sugar chain, suggesting that the virus must adopt the ability to utilize α2,6 SA on a long sugar chain. It was concluded that it is the glycan composition, and thus the SA topology that may influence influenza tropism, not just the SA linkage present.

The region of HA that is responsible for binding SA is referred to as the receptor-binding domain (RBD). This pocket-shaped depression is located at the membrane-distal tip of each monomer within the trimeric HA structure and is comprised of the 190 helix (residues 190–198), the 130 loop (residues 135–138), and the 220 loop (residues 221–228) based on H3 numbering [25]. Several key conserved residues including tyrosine 98, tryptophan 153, and histidine 183 form the base of the binding pocket and are crucial for maintaining the structural basis of the binding pocket as well as in forming interactions with SA [75]. Sequence analysis from several strains of influenza along with structural modeling has given great insight into the residues that play a pivotal role in SA binding preference [25, 76].

Structural studies of HA suggest that avian and human influenza viruses appear to be distinct in their RBDs at positions 226 and 228 [77, 78]. Avian HAs tend to have glutamine and glycine at the respective positions while human HAs carry leucine and serine [78–81]. The avian HA with these residues forms a narrow binding pocket for the α2,3 receptor while the change in residues for human HAs results in a broader pocket for the α2,6 receptor. Neither position 226 nor 228 plays a direct role in binding, rather they seem to influence the contour of the pocket [70, 82]. These differences in pocket shape correlate well with the glycan topology studies. Additionally, it was shown that a lab-adapted strain which prefers α2,6 binding was able to switch preferences to α2,3 when grown in the presence of α-2 macroglobulin, a α2,6 glycoprotein found in high concentrations in horse sera [65]. Based on these and other studies, it seems that residue 226 and 228 have an important indirect role in SA binding compatibility and therefore preference.

Interestingly, the 1918 H1N1 virus HA carried glutamine 226 and glycine 228 corresponding to the avian α2,3 receptor, however the virus is able to bind the α2,6 receptor, demonstrating that changes at these positions are not necessary for altered receptor binding [30]. Further analysis of the 1918 HA revealed that a single change from asparagine to glutamate at position 190 was responsible for the altered binding phenotype [30]. In addition, a change from asparagine to glycine at 225 in combination with the change at 190 increased respiratory transmission of the virus in the ferret model, further highlighting the importance of the RBD residues.

More extensive studies focusing on the H3 subtype revealed that human viruses with α2,6 preference have leucine at 226 and serine at 228, while avian viruses with α2,3 preference have glutamine at 226 and glycine at 228 [63, 82]. Residues 193 and 218 have also been implicated as important determinants for receptor preference in the H3 subtype, however specific residue changes have not been fully established [83].

Similarly, studies focusing on seasonally circulating H1 viruses found avian and human viruses differ at two positions in their binding preference. A proline at 186 and a glycine at 225 correspond with α2,3 type binding, while serine 186 and asparagine 225 are favored by α2,6 type binding [25]. Interestingly, HA proteins of both avian and human viruses have glutamine at 226 and glycine at 228. The discrepancies highlighted by the H3 and H1 studies in combination with the studies on the H1N1 Spanish Flu illustrate the complex nature of receptor binding, that is, not all influenza A viruses behave in a similar way, not even among avian strains nor human strains.

Since the initial H5N1 influenza outbreak that began in China in 1997, several studies have focused on the RBD of this particular viral strain. These studies include the use of sialoglycoconjugates, crystal HA structures, and simulated computer-based binding assays [55, 69, 72, 74, 84–87]. Residues that have been implicated in human receptor-type binding include asparagine 227, asparagine 159, lysine 182, and arginine 192. It is proposed that residues 159, 182, and 192 influence binding by stabilizing the HA binding pocket and maintaining structural integrity, as these residues are not in direct contact with SA. A switch of glutamine to leucine at position 226 and a switch of glycine to serine at position 228 equates a shift from avian receptor to human receptor specificity, as seen in the H3 virus as well [86, 88]. Further studies specifically targeting the HA of the A/Vietnam/1203/2004 H5N1 virus demonstrate the importance of mutation E190D which reduced the binding to 2,3 linkages, as well as the double mutant Q226L/G228S [86]. In addition the H5N1 HA surface residue tyrosine 161 has recently been implicated in altering glycoconjugate recognition and cell-type dependent entry. Substitution of tyrosine 161 to alanine switched the binding preference from N-acetylneuraminic acid to N-glycolylneuraminic acid [87]. It is important to note that different strains of the H5N1 virus from different time points during the virus outbreak were used in these studies.

To better understand the role of naturally acquired mutations and their ability to potentiate sustained transmission, two recent studies showed that as few as four to five amino acid substitutions along with gene reassortment may be sufficient for respiratory droplet transmission between ferrets [89, 90]. Herfst et al. identified (based on H5 numbering) glutamine to leucine at 222 and glycine to serine at 224 as critical residues to alter sialic acid specificity from the avian $\alpha 2,3$ SA to the human $\alpha 2,6$ SA. Two additional HA substitutions, threonine to alanine at 156 and histidine to tyrosine at 103 play a role in disrupting N-linked glycosylation and monomer interaction, respectively. Lastly, a switch from glutamate to lysine at 627 in the polymerase PB2 subunit was identified, which is a common change seen in mammalian adapted influenza strains. Imai et al. identified four substitutions, all within HA. Similar to the results of Herfst et al., it was found that substitutions of asparagine to lysine at 220 and glutamine to leucine at 222 can alter SA specificity from the avian $\alpha 2,3$ SA to the human $\alpha 2,6$ SA. Again, a disruption in N-linked glycosylation via asparagine to aspartate at position 154 was identified along with a change in the stalk region corresponding to threonine to isoleucine at 315.

While the linkage of SA and the residues within the RBD plays a significant role in viral entry, the glycoconjugate to which SA is attached appears to be an additional important factor. Chu and Whittaker determined that cells deficient in the GnT1 gene lack terminal N-linked glycosylation, rendering them deficient in influenza A viral entry [91]. The GnT1 gene encodes the enzyme N-acetylglucosaminyltransferase, which is involved in modification of N-linked glycans in the Golgi apparatus. These cells lack N-glycoproteins, but still possess surface $\alpha 2,3$ SA and $\alpha 2,6$ SA. While considering this phenotype, this mutant Chinese Hamster Ovary (CHO) cell line has the capacity to bind HA and expresses sufficient levels of both SA linkages on the cell surface for attachment. In this mutant CHO cell line, complete entry was blocked, as virions were not endocytosed. These results suggest a specific role for N-linked glycoproteins in influenza A entry, perhaps acting as a cofactor in mediating entry.

There is evidence suggesting that SA is not necessary for infection by influenza. The HA of a human H1N1 strain was shown to bind glycoconjugates other than SA [92]. In addition, Stray et al. demonstrated the ability of desialylated Madin Darby canine kidney (MDCK) cells to be infected by influenza A virus [93]. A study by Nicholls et al. demonstrated the ability of the H5N1 virus to infect upper and lower respiratory tract cells in the presence or absence of $\alpha 2,3$ SA which is believed to be the entry receptor for this virus [94]. In addition, the levels of SA on susceptible and nonsusceptible cells do not correlate with either $\alpha 2,3$ SA or $\alpha 2,6$ SA for an H5N1 strain [95]. Taken together, it is possible that the barrier to efficient human infection and human-to-human transmission of the H5N1 virus is not due to SA linkages, but rather it is due to inefficient use or expression of a yet unidentified entry mediator.

To fight the spread of influenza, prophylactic therapeutics and vaccines continue to be vital methods of control. Vaccines are the primary means of controlling the spread of the virus and are based on inactivated viruses, live attenuated viruses, or purified viral protein(s) that illicit a strong neutralizing antibody response [96]. Of these vaccines, most target the globular head region of HA, containing the receptor-binding domain, thus preventing attachment of the virus to susceptible cells. These neutralizing antibodies are rarely immunoresponsive to an alternate influenza strain, often losing their potency as their corresponding strain acquires mutations during circulation. Furthermore, due to the virus ability to constantly acquire genetic changes, it is difficult to predict what the circulating strain(s) for the upcoming year will be. A mismatch of vaccine strain with circulating strain(s) will offer little to no protection.

In addition to vaccines, anti-flu therapeutics have been developed which can be divided into two classes, anti-NA and anti-M2 [97]. The first class of inhibitors specifically targets the NA protein of the virus, halting the spread of progeny virions [98]. During the budding process of influenza, newly produced progeny virions are tethered to the host cell surface via HA proteins interacting with SA molecules. NA functions in recognizing this HA-SA interaction and cleaves the SA moiety, releasing the viral particle [99]. The currently approved therapeutics for influenza infection include NA inhibitors which block this step, thus preventing release and further spread of the virus, both within the infected host and consequently to others. Included in this category of antivirals are two of the most commonly used therapies, Zanamivir (trade name Relenza) and oseltamivir phosphate (trade name Tamiflu) for the treatment and prevention of influenza A and B viruses. Zanamivir and oseltamivir phosphate are competitive inhibitors for the active site of the NA enzyme [100]. While Zanamivir and oseltamivir phosphate can be highly effective in both treating influenza and in outbreak control, in the 2008-2009 flu season, nearly 100% of H1N1 samples tested by the Center for Disease Control (CDC) were shown to be resistant to oseltamivir phosphate [101, 102]. This high level of resistance highlights the need to develop new antivirals against influenza.

Another class of influenza inhibitors, the adamantanes, is those which block the M2 ion channel on the virion surface. The M2 ion channel is embedded within the virion lipid bilayer and facilitates hydrogen ion transport, ultimately leading to virion uncoating and replication during the entry process [103]. The adamantane derivatives, amantadine and rimantadine, were approved for the treatment and prevention of influenza A only, as only influenza A class viruses have the M2 protein [103]. The CDC and WHO report that greater than 99% of circulating strains are resistant to M2 inhibitors and have recommended their use to be discontinued. Therefore these inhibitors are only used when a specific nonresistant strain is thought to be the causative infectious agent [101]. These antivirals, along with the NA inhibitors, provide a treatment option after infection, however since not all influenza A viruses respond to these treatments, these drugs may be ineffective. In addition, resistance to these treatments over the course of an outbreak or influenza season underlies the urgency to develop new antiviral therapies.

Since there is a lack of clinically available inhibitor(s) which targets the major viral surface protein, HA, recent clinical trials are pushing forward to expand the repertoire of influenza therapeutics, including drugs that target HA, which is the target of most neutralizing antibodies [104]. These new drugs include cyanovirin-N and thiazolides. Cyanovirin-N targets the high-mannose oligosaccharides, neutralizing the viral particle [104, 105]. The thiazolides work to block the maturation of HA posttranslationally [104, 106]. Recently, several groups have identified broad-spectrum neutralizing antibodies that are protective against group 1 (which includes HA subtypes 1, 2, 5, 6, 8, 9, 11, 12, 13, and 16) and group 2 (which includes HA subtypes 3, 4, 7, 10, 14, and 15) influenza A viruses [36, 107–109]. These broad-spectrum neutralizing antibodies (referred to as CR6261, F10, and F16) are further distinct in that they target the stem region of the HA molecule. The proposed model is that by targeting the stem region, more specifically the fusion peptide, these antibodies are able to prevent exposure of the fusion peptide under acid conditions. This trapping mechanism prevents the viral membrane from fusing with the host endosomal membrane, locking the genetic contents within the viral particle. An additional stem-directed neutralizing antibody, CR8020, is active against group 2 influenza A viruses. While not as broad in activity as the aforementioned antibodies, CR8020 could be utilized in conjunction with a group 1 neutralizing antibody, providing a one-two punch against both HA subunits.

Influenza viruses will continue to circulate among animal and human populations, acquiring and exchanging genetic components as they do. Due to the constant change in their genetic profiles, influenza viruses will continue to pose a serious threat, from both an economic and public health point of view. Continued study and surveillance of influenza viruses is further highlighted by our inability to maintain effective influenza vaccines and prophylactic therapeutics. The prospective of new antivirals, especially those that provide broad-spectrum protection, provide renewed efforts in our ability to control influenza infection and spread.

References

[1] F. A. Murphy, "Virus taxonomy," in *Virology*, B. N. Fields, D. M. Knipe, and P. M. Howley, Eds., pp. 15–57, Lippincott-Raven, Philadelphia, Pa, USA, 1996.

[2] D. M. Fleming, M. Zambon, and A. I. M. Bartelds, "Population estimates of persons presenting to general practitioners with influenza-like illness, 1987–96: a study of the demography of influenza-like illness in sentinel practice networks in England and Wales, and in The Netherlands," *Epidemiology & Infection*, vol. 124, no. 2, pp. 245–253, 2000.

[3] K. M. Sullivan and A. S. Monto, "Acute respiratory illness in the community. Frequency of illness and the agents involved," *Epidemiology & Infection*, vol. 110, no. 1, pp. 145–160, 1999.

[4] G. Yuanji, J. Fengen, W. Ping et al., "Isolation of influenza C virus from pigs and experimental infection of pigs with influenza C virus," *Journal of General Virology*, vol. 64, no. 1, pp. 177–182, 1983.

[5] A. Hay, "The virus genome and its replication," in *Textbook of Influenza*, K. G. Nicholson, R. G. Webster, and A. J. Hay, Eds., pp. 43–53, 1998.

[6] A. D. M. E. Osterhaus, B. E. E. Martina, G. F. Rimmelzwaan, T. M. Bestebroer, and R. A. M. Fouchier, "Influenza B virus in seals," *Science*, vol. 288, no. 5468, pp. 1051–1053, 2000.

[7] R. G. Webster, M. Yakhno, V. S. Hinshaw et al., "Intestinal influenza: replication and characterization of influenza viruses in ducks," *Virology*, vol. 84, no. 2, pp. 268–278, 1978.

[8] R. G. Webster, W. J. Bean, O. T. Gorman, T. M. Chambers, and Y. Kawaoka, "Evolution and ecology of influenza A viruses," *Microbiological Reviews*, vol. 56, no. 1, pp. 152–179, 1992.

[9] N. P. Johnson and J. Mueller, "Updating the accounts: global mortality of the 1918–1920 "Spanish" influenza pandemic," *Bulletin of the History of Medicine*, vol. 76, no. 1, pp. 105–115, 2002.

[10] A. W. Crosby, *America's Forgotten Pandemic*, Cambridge University Press, 2003.

[11] K. D. Patterson and G. F. Pyle, "The geography and mortality of the 1918 influenza pandemic," *Bulletin of the History of Medicine*, vol. 65, no. 1, pp. 4–21, 1991.

[12] G. J. D. Smith, D. Vijaykrishna, J. Bahl et al., "Origins and evolutionary genomics of the 2009 swine-origin H1N1 influenza A epidemic," *Nature*, vol. 459, no. 7250, pp. 1122–1125, 2009.

[13] M. Nelson, M. Gramer, A. Vincent, and E. C. Holmes, "Global transmission of influenza viruses from humans to swine," *Journal of General Virology*, vol. 93, part 10, pp. 2195–2203, 2012.

[14] CDC H1N1 flu. Center for Disease Control and Prevention Website.

[15] F. S. Dawood, A. D. Iuliano, C. Reed, M. I. Meltzer, D. K. Shay et al., "Estimated global mortality associated with the first 12 months of 2009 pandemic influenza A H1N1 virus circulation: a modelling study," *The Lancet Infectious Diseases*, vol. 12, no. 9, pp. 687–695, 2012.

[16] R. Lamb and R. Krug, "Orthomyxoviridae: the viruses and their replication," in *Field's Virology*, pp. 1487–1532, 2001.

[17] P. Palese, "Influenza: old and new threats," *Nature Medicine*, vol. 10, no. 12, pp. S82–S87, 2004.

[18] E. D. Kilbourne, "Influenza pandemics: can we prepare for the unpredictable?" *Viral Immunology*, vol. 17, no. 3, pp. 350–357, 2004.

[19] E. D. Kilbourne, "Influenza pandemics of the 20th century," *Emerging Infectious Diseases*, vol. 12, no. 1, pp. 9–14, 2006.

[20] M. C. Zambon, "The pathogenesis of influenza in humans," *Reviews in Medical Virology*, vol. 11, no. 4, pp. 227–241, 2001.

[21] Y.-C. Hsieh, T. Z. Wu, D. P. Liu et al., "Influenza pandemics: past, present and future," *Journal of the Formosan Medical Association*, vol. 105, no. 1, pp. 1–6, 2006.

[22] T. Horimoto and Y. Kawaoka, "Influenza: lessons from past pandemics, warnings from current incidents," *Nature Reviews Microbiology*, vol. 3, no. 8, pp. 591–600, 2005.

[23] T. Ito, J. N. S. S. Couceiro, S. Kelm et al., "Molecular basis for the generation in pigs of influenza A viruses with pandemic potential," *Journal of Virology*, vol. 72, no. 9, pp. 7367–7373, 1998.

[24] Y. Kawaoka, S. Krauss, and R. G. Webster, "Avian-to-human transmission of the PB1 gene of influenza A viruses in the 1957 and 1968 pandemics," *Journal of Virology*, vol. 63, no. 11, pp. 4603–4608, 1989.

[25] J. J. Skehel and D. C. Wiley, "Receptor binding and membrane fusion in virus entry: the influenza hemagglutinin," *Annual Review of Biochemistry*, vol. 69, pp. 531–569, 2000.

[26] N. M. Ferguson, C. Fraser, C. A. Donnelly, A. C. Ghani, and R. M. Anderson, "Public health risk from the avian H5N1 influenza epidemic," *Science*, vol. 304, no. 5673, pp. 968–969, 2004.

[27] C. F. Basler and P. V. Aguilar, "Progress in identifying virulence determinants of the 1918 H1N1 and the Southeast Asian H5N1 influenza A viruses," *Antiviral Research*, vol. 79, no. 3, pp. 166–178, 2008.

[28] R. B. Belshe, "The origins of pandemic influenza—lessons from the 1918 virus," *The New England Journal of Medicine*, vol. 353, no. 21, pp. 2209–2211, 2005.

[29] L. Glaser, J. Stevens, D. Zamarin et al., "A single amino acid substitution in 1918 influenza virus hemagglutinin changes receptor binding specificity," *Journal of Virology*, vol. 79, no. 17, pp. 11533–11536, 2005.

[30] A. H. Reid, T. G. Fanning, J. V. Hultin, and J. K. Taubenberger, "Origin and evolution of the 1918 "Spanish" influenza virus hemagglutinin gene," *Proceedings of the National Academy of Sciences of the United States of America*, vol. 96, no. 4, pp. 1651–1656, 1999.

[31] T. M. Tumpey, C. F. Basler, P. V. Aguilar et al., "Characterization of the reconstructed 1918 Spanish influenza pandemic virus," *Science*, vol. 310, no. 5745, pp. 77–80, 2005.

[32] S. J. Gamblin, L. F. Haire, R. J. Russell et al., "The structure and receptor binding properties of the 1918 influenza hemagglutinin," *Science*, vol. 303, no. 5665, pp. 1838–1842, 2004.

[33] I. A. Wilson and N. J. Cox, "Structural basis of immune recognition of influenza virus hemagglutinin," *Annual Review of Immunology*, vol. 8, pp. 737–787, 1990.

[34] R. A. M. Fouchier, V. Munster, A. Wallensten et al., "Characterization of a novel influenza A virus hemagglutinin subtype (H16) obtained from black-headed gulls," *Journal of Virology*, vol. 79, no. 5, pp. 2814–2822, 2005.

[35] D. Kaye and C. R. Pringle, "Avian influenza viruses and their implication for human health," *Clinical Infectious Diseases*, vol. 40, no. 1, pp. 108–112, 2005.

[36] D. Corti, J. Voss, S. J. Gamblin, G. Codoni, A. Macagno et al., "Neutralizing antibody selected from plasma cells that binds to group 1 and group 2 influenza A hemagglutinins," *Science*, vol. 333, no. 6044, pp. 850–856, 2011.

[37] P. Leyssen, E. de Clercq, and J. Neyts, "Molecular strategies to inhibit the replication of RNA viruses," *Antiviral Research*, vol. 78, no. 1, pp. 9–25, 2008.

[38] T. Y. Aditama, G. Samaan, R. Kusriastuti, O. D. Sampurno, W. Purba et al., "Avian influenza H5N1 transmission in households, Indonesia," *PLoS ONE*, vol. 7, no. 1, Article ID e29971, 2012.

[39] Y. Yang, M. E. Halloran, J. D. Sugimoto, and I. M. Longini Jr., "Detecting human-to-human transmission of avian influenza A (H5N1)," *Emerging Infectious Diseases*, vol. 13, no. 9, pp. 1348–1353, 2007.

[40] H.-D. Klenk, R. Rott, M. Orlich, and J. Blodorn, "Activation of influenza A viruses by trypsin treatment," *Virology*, vol. 68, no. 2, pp. 426–439, 1975.

[41] S. G. Lazarowitz and P. W. Choppin, "Enhancement of the infectivity of influenza A and B viruses by proteolytic cleavage

[42] F. X. Bosch, W. Garten, H.-D. Klenk, and R. Rott, "Proteolytic cleavage of influenza virus hemagglutinins: primary structure of the connecting peptide between HA1 and HA2 determines proteolytic cleavability and pathogenicity of avian influenza viruses," *Virology*, vol. 113, no. 2, pp. 725–735, 1981.

[43] R. T. C. Huang, K. Wahn, H.-D. Klenk, and R. Rott, "Fusion between cell membrane and liposomes containing the glycoproteins of influenza virus," *Virology*, vol. 104, no. 2, pp. 294–302, 1980.

[44] T. Maeda and S.-I. Ohnishi, "Activation of influenza virus by acidic media causes hemolysis and fusion of erythrocytes," *FEBS Letters*, vol. 122, no. 2, pp. 283–287, 1980.

[45] J. White, K. Matlin, and A. Helenius, "Cell fusion by Semliki Forest, influenza, and vesicular stomatitis viruses," *The Journal of Cell Biology*, vol. 89, no. 3, pp. 674–679, 1981.

[46] P. A. Bullough, F. M. Hughson, J. J. Skehel, and D. C. Wiley, "Structure of influenza haemagglutinin at the pH of membrane fusion," *Nature*, vol. 371, no. 6492, pp. 37–43, 1994.

[47] S. Bertram, I. Glowacka, I. Steffen, A. Kühl, and S. Pöhlmann, "Novel insights into proteolytic cleavage of influenza virus hemagglutinin," *Reviews in Medical Virology*, vol. 20, no. 5, pp. 298–310, 2010.

[48] E. Böttcher, T. Matrosovich, M. Beyerle, H.-D. Klenk, W. Garten, and M. Matrosovich, "Proteolytic activation of influenza viruses by serine proteases TMPRSS2 and HAT from human airway epithelium," *Journal of Virology*, vol. 80, no. 19, pp. 9896–9898, 2006.

[49] C. Chaipan, D. Kobasa, S. Bertram et al., "Proteolytic activation of the 1918 influenza virus hemagglutinin," *Journal of Virology*, vol. 83, no. 7, pp. 3200–3211, 2009.

[50] H. Kido, Y. Yokogoshi, K. Sakai et al., "Isolation and characterization of a novel trypsin-like protease found in rat bronchiolar epithelial Clara cells. A possible activator of the viral fusion glycoprotein," *The Journal of Biological Chemistry*, vol. 267, no. 19, pp. 13573–13579, 1992.

[51] T. M. Tumpey, A. García-Sastre, J. K. Taubenberger et al., "Pathogenicity of influenza viruses with genes from the 1918 pandemic virus: functional roles of alveolar macrophages and neutrophils in limiting virus replication and mortality in mice," *Journal of Virology*, vol. 79, no. 23, pp. 14933–14944, 2005.

[52] G. Thomas, "Furin at the cutting edge: from protein traffic to embryogenesis and disease," *Nature Reviews Molecular Cell Biology*, vol. 3, no. 10, pp. 753–766, 2002.

[53] A. Stieneke-Grober, M. Vey, H. Angliker et al., "Influenza virus hemagglutinin with multibasic cleavage site is activated by furin, a subtilisin-like endoprotease," *The EMBO Journal*, vol. 11, no. 7, pp. 2407–2414, 1992.

[54] D. J. Krysan, N. C. Rockwell, and R. S. Fuller, "Quantitative characterization of furin specificity: energetics of substrate discrimination using an internally consistent set of hexapeptidyl methylcoumarinamides," *The Journal of Biological Chemistry*, vol. 274, no. 33, pp. 23229–23234, 1999.

[55] D. J. Hulse, R. G. Webster, R. J. Russell, and D. R. Perez, "Molecular determinants within the surface proteins involved in the pathogenicity of H5N1 influenza viruses in chickens," *Journal of Virology*, vol. 78, no. 18, pp. 9954–9964, 2004.

of the hemagglutinin polypeptide," *Virology*, vol. 68, no. 2, pp. 440–454, 1975.

[56] E. Rumschlag-Booms, Y. Guo, J. Wang, M. Caffrey, and L. Rong, "Comparative analysis between a low pathogenic and a high pathogenic influenza H5 hemagglutinin in cell entry," *Virology Journal*, vol. 6, article 76, 2009.

[57] G. K. Hirst, "Adsorption of influenza hemagglutinins and virus by red blood cells," *The Journal of Experimental Medicine*, vol. 76, no. 2, pp. 195–209, 1942.

[58] F. Burnet and J. D. Stone, "The receptor-destroying enzyme of V. cholerae," *Australian Journal of Experimental Biology & Medical Science*, vol. 25, no. 3, pp. 227–233, 1947.

[59] A. Gottschalk, "Neuraminidase: the specific enzyme of influenza virus and Vibrio cholerae," *Biochimica et Biophysica Acta*, vol. 23, pp. 645–646, 1957.

[60] A. Gottschalk, *The Chemistry and Biology of Sialic Acids and Related Substances*, University Press, Cambridge, Mass, USA, 1960.

[61] Y. Suzuki, "Sialobiology of influenza molecular mechanism of host range variation of influenza viruses," *Biological and Pharmaceutical Bulletin*, vol. 28, no. 3, pp. 399–408, 2005.

[62] C. F. Brewer and T. K. Dam, "*Essentials of Glycobiology*, Edited by A. Varki, R. Cummings, J. Esko, H. Freeze, G. Hart, and J. Marth, Cold Spring Harbor Laboratory Press, Cold Spring Harbor, New York, 1999, 653 pp.," *Carbohydrate Research*, vol. 325, no. 3, pp. 233–234, 2000.

[63] M. N. Matrosovich, A. S. Gambaryan, S. Teneberg et al., "Avian influenza A viruses differ from human viruses by recognition of sialyloligosaccharides and gangliosides and by a higher conservation of the HA receptor-binding site," *Virology*, vol. 233, no. 1, pp. 224–234, 1997.

[64] C. R. Parrish and Y. Kawaoka, "The origins of new pandemic viruses: the acquisition of new host ranges by canine parvovirus and influenza A viruses," *Annual Review of Microbiology*, vol. 59, pp. 553–586, 2005.

[65] G. N. Rogers, T. J. Pritchett, J. L. Lane, and J. C. Paulson, "Differential sensitivity of human, avian, and equine influenza A viruses to a glycoprotein inhibitor of infection: selection of receptor specific variants," *Virology*, vol. 131, no. 2, pp. 394–408, 1983.

[66] G. N. Rogers and J. C. Paulson, "Receptor determinants of human and animal influenza virus isolates: differences in receptor specificity of the H3 hemagglutinin based on species of origin," *Virology*, vol. 127, no. 2, pp. 361–373, 1983.

[67] G. N. Rogers, J. C. Paulson, R. S. Daniels et al., "Single amino acid substitutions in influenza haemagglutinin change receptor binding specificity," *Nature*, vol. 304, no. 5921, pp. 76–78, 1983.

[68] J. N. S. S. Couceiro, J. C. Paulson, and L. G. Baum, "Influenza virus strains selectively recognize sialyloligosaccharides on human respiratory epithelium; the role of the host cell in selection of hemagglutinin receptor specificity," *Virus Research*, vol. 29, no. 2, pp. 155–165, 1993.

[69] K. Shinya, M. Ebina, S. Yamada, M. Ono, N. Kasai, and Y. Kawaoka, "Avian flu: influenza virus receptors in the human airway," *Nature*, vol. 440, no. 7083, pp. 435–436, 2006.

[70] M. Matrosovich, N. Zhou, Y. Kawaoka, and R. Webster, "The surface glycoproteins of H5 influenza viruses isolated from humans, chickens, and wild aquatic birds have distinguishable properties," *Journal of Virology*, vol. 73, no. 2, pp. 1146–1155, 1999.

[71] M. N. Matrosovich, T. Y. Matrosovich, T. Gray, N. A. Roberts, and H.-D. Klenk, "Human and avian influenza viruses target different cell types in cultures of human airway epithelium," *Proceedings of the National Academy of Sciences of the United States of America*, vol. 101, no. 13, pp. 4620–4624, 2004.

[72] J. Stevens, O. Blixt, T. M. Tumpey, J. K. Taubenberger, J. C. Paulson, and I. A. Wilson, "Structure and receptor specificity of the hemagglutinin from an H5N1 influenza virus," *Science*, vol. 312, no. 5772, pp. 404–410, 2006.

[73] J. Stevens, O. Blixt, L. Glaser et al., "Glycan microarray analysis of the hemagglutinins from modern and pandemic influenza viruses reveals different receptor specificities," *Journal of Molecular Biology*, vol. 355, no. 5, pp. 1143–1155, 2006.

[74] A. Chandrasekaran, A. Srinivasan, R. Raman et al., "Glycan topology determines human adaptation of avian H5N1 virus hemagglutinin," *Nature Biotechnology*, vol. 26, no. 1, pp. 107–113, 2008.

[75] N. K. Sauter, J. E. Hanson, G. D. Glick et al., "Binding of influenza virus hemagglutinin to analogs of its cell-surface receptor, sialic acid: analysis by proton nuclear magnetic resonance spectroscopy and X-ray crystallography," *Biochemistry*, vol. 31, no. 40, pp. 9609–9621, 1992.

[76] T. Suzuki, A. Portner, R. A. Scroggs et al., "Receptor specificities of human respiroviruses," *Journal of Virology*, vol. 75, no. 10, pp. 4604–4613, 2001.

[77] S. Chutinimitkul, S. Herfst, J. Steel et al., "Virulence-associated substitution D222G in the hemagglutinin of 2009 pandemic influenza A(H1N1) virus affects receptor binding," *Journal of Virology*, vol. 84, no. 22, pp. 11802–11813, 2010.

[78] A. Vines, K. Wells, M. Matrosovich, M. R. Castrucci, T. Ito, and Y. Kawaoka, "The role of influenza A virus hemagglutinin residues 226 and 228 in receptor specificity and host range restriction," *Journal of Virology*, vol. 72, no. 9, pp. 7626–7631, 1998.

[79] R. J. Connor, Y. Kawaoka, R. G. Webster, and J. C. Paulson, "Receptor specificity in human, avian, and equine H2 and H3 influenza virus isolates," *Virology*, vol. 205, no. 1, pp. 17–23, 1994.

[80] C. W. Naeve, V. S. Hinshaw, and R. G. Webster, "Mutations in the hemagglutinin receptor-binding site can change the biological properties of an influenza virus," *Journal of Virology*, vol. 51, no. 2, pp. 567–569, 1984.

[81] E. Nobusawa, T. Aoyama, H. Kato, Y. Suzuki, Y. Tateno, and K. Nakajima, "Comparison of complete amino acid sequences and receptor-binding properties among 13 serotypes of hemagglutinins of influenza A viruses," *Virology*, vol. 182, no. 2, pp. 475–485, 1991.

[82] C. T. Hardy, S. A. Young, R. G. Webster, C. W. Naeve, and R. J. Owens, "Egg fluids and cells of the chorioallantoic membrane of embryonated chicken eggs can select different variants of influenza A (H3N2) viruses," *Virology*, vol. 211, no. 1, pp. 302–306, 1995.

[83] P. S. Daniels, S. Jeffries, P. Yates et al., "The receptor-binding and membrane-fusion properties of influenza virus variants selected using anti-haemagglutinin monoclonal antibodies," *The EMBO Journal*, vol. 6, no. 5, pp. 1459–1465, 1987.

[84] E. C. J. Claas, J. C. de Jong, R. van Beek, G. F. Rimmelzwaan, and A. D. M. E. Osterhaus, "Human influenza virus A/HongKong/156/97 (H5N1) infection," *Vaccine*, vol. 16, no. 9-10, pp. 977–978, 1998.

[85] J. Stevens, O. Blixt, T. M. Tumpey, J. K. Taubenberger, J. C. Paulson, and I. A. Wilson, "Structure and receptor specificity of the hemagglutinin from an H5N1 influenza virus," *Science*, vol. 312, no. 5772, pp. 404–410, 2006.

[86] G. Ayora-Talavera, H. Shelton, M. A. Scull et al., "Mutations in H5N1 influenza virus hemagglutinin that confer binding to human tracheal airway epithelium," *PLoS ONE*, vol. 4, no. 11, Article ID e7836, 2009.

[87] M. Wang, D. M. Tscherne, C. McCullough, M. Caffrey, A. Garcia-Sastre et al., "Residue Y161 of influenza virus hemagglutinin is involved in viral recognition of sialylated complexes from different hosts," *Journal of Virology*, vol. 86, no. 8, pp. 4455–4462, 2012.

[88] S. Yamada, Y. Suzuki, T. Suzuki et al., "Haemagglutinin mutations responsible for the binding of H5N1 influenza A viruses to human-type receptors," *Nature*, vol. 444, no. 7117, pp. 378–382, 2006.

[89] M. Imai, T. Watanabe, M. Hatta, S. C. Das, M. Ozawa et al., "Experimental adaptation of an influenza H5 HA confers respiratory droplet transmission to a reassortant H5 HA/H1N1 virus in ferrets," *Nature*, vol. 486, pp. 420–428, 2012.

[90] S. Herfst, E. J. A. Schrauwen, M. Linster, S. Chutinimitkul, E. de Wit et al., "Airborne transmission of influenza A/H5N1 virus between ferrets," *Science*, vol. 336, no. 6088, pp. 1534–1541, 2012.

[91] V. C. Chu and G. R. Whittaker, "Influenza virus entry and infection require host cell N-linked glycoprotein," *Proceedings of the National Academy of Sciences of the United States of America*, vol. 101, no. 52, pp. 18153–18158, 2004.

[92] E. M. Rapoport, L. V. Mochalova, H.-J. Gabius, J. Romanova, and N. V. Bovin, "Search for additional influenza virus to cell interactions," *Glycoconjugate Journal*, vol. 23, no. 1 2, pp. 115–125, 2006.

[93] S. J. Stray, R. D. Cummings, and G. M. Air, "Influenza virus infection of desialylated cells," *Glycobiology*, vol. 10, no. 7, pp. 649–658, 2000.

[94] J. M. Nicholls, M. C. W. Chan, W. Y. Chan et al., "Tropism of avian influenza A (H5N1) in the upper and lower respiratory tract," *Nature Medicine*, vol. 13, no. 2, pp. 147–149, 2007.

[95] Y. Guo, E. Rumschlag-Booms, J. Wang et al., "Analysis of hemagglutinin-mediated entry tropism of H5N1 avian influenza," *Virology Journal*, vol. 6, article 39, 2009.

[96] R. Rappuoli and P. R. Dormitzer, "Influenza: options to improve pandemic preparation," *Science*, vol. 336, no. 6088, pp. 1531–1533, 2012.

[97] J. Cinatl, M. Michaelis, and H. W. Doerr, "The threat of avian influenza A (H5N1). Part III: antiviral therapy," *Medical Microbiology and Immunology*, vol. 196, no. 4, pp. 203–212, 2007.

[98] A. Moscona, "Neuraminidase inhibitors for influenza," *The New England Journal of Medicine*, vol. 353, no. 13, pp. 1363–1373, 2005.

[99] J. S. Rossman and R. A. Lamb, "Influenza virus assembly and budding," *Virology*, vol. 411, no. 2, pp. 229–236, 2011.

[100] J. Magano, "Synthetic approaches to the neuraminidase inhibitors zanamivir (Relenza) and oseltamivir phosphate (Tamiflu) for the treatment of influenza," *Chemical Reviews*, vol. 109, no. 9, pp. 4398–4438, 2009.

[101] H. T. Nguyen, A. M. Fry, and L. V. Gubareva, "Neuraminidase inhibitor resistance in influenza viruses and laboratory testing methods," *Antiviral Therapy*, vol. 17, pp. 159–173, 2012.

[102] N. J. Dharan, L. V. Gubareva, J. J. Meyer et al., "Infections with oseltamivir-resistant influenza A(H1N1) virus in the United States," *Journal of the American Medical Association*, vol. 301, no. 10, pp. 1034–1041, 2009.

[103] S. D. Cady, W. Luo, F. Hu, and M. Hong, "Structure and function of the influenza A M2 proton channel," *Biochemistry*, vol. 48, no. 31, pp. 7356–7364, 2009.

[104] D. A. Boltz, J. R. Aldridge, R. G. Webster, and E. A. Govorkova, "Drugs in development for influenza," *Drugs*, vol. 70, no. 11, pp. 1349–1362, 2010.

[105] B. R. O'Keefe, D. F. Smee, J. A. Turpin et al., "Potent anti-influenza activity of cyanovirin-N and interactions with viral hemagglutinin," *Antimicrobial Agents and Chemotherapy*, vol. 47, no. 8, pp. 2518–2525, 2003.

[106] J. F. Rossignol, S. La Frazia, L. Chiappa, A. Ciucci, and M. G. Santoro, "Thiazolides, a new class of anti-influenza molecules targeting viral hemagglutinin at the post-translational level," *The Journal of Biological Chemistry*, vol. 284, no. 43, pp. 29798–29808, 2009.

[107] J. Sui, W. C. Hwang, S. Perez et al., "Structural and functional bases for broad-spectrum neutralization of avian and human influenza A viruses," *Nature Structural and Molecular Biology*, vol. 16, no. 3, pp. 265–273, 2009.

[108] D. C. Ekiert and I. A. Wilson, "Broadly neutralizing antibodies against influenza virus and prospects for universal therapies," *Current Opinion in Virology*, vol. 2, no. 2, pp. 134–141, 2012.

[109] M. Throsby, E. van den Brink, M. Jongeneelen et al., "Heterosubtypic neutralizing monoclonal antibodies cross-protective against H5N1 and H1N1 recovered from human IgM+ memory B cells," *PLoS ONE*, vol. 3, no. 12, Article ID e3942, 2008.

Viruses as Modulators of Mitochondrial Functions

Sanjeev K. Anand[1,2] and Suresh K. Tikoo[1,2,3]

[1] *Vaccine & Infection Disease Organization-International Vaccine Center (VIDO-InterVac),*
  University of Saskatchewan, 120 Veterinary Road, Saskatoon, SK, Canada S7E 5E3

[2] *Veterinary Microbiology, University of Saskatchewan, 120 Veterinary Road, Saskatoon, SK, Canada S7E 5E3*

[3] *School of Public Health, University of Saskatchewan, 120 Veterinary Road, Saskatoon, SK, Canada S7E 5E3*

Correspondence should be addressed to Suresh K. Tikoo; suresh.tik@usask.ca

Academic Editor: Michael Bukrinsky

Mitochondria are multifunctional organelles with diverse roles including energy production and distribution, apoptosis, eliciting host immune response, and causing diseases and aging. Mitochondria-mediated immune responses might be an evolutionary adaptation by which mitochondria might have prevented the entry of invading microorganisms thus establishing them as an integral part of the cell. This makes them a target for all the invading pathogens including viruses. Viruses either induce or inhibit various mitochondrial processes in a highly specific manner so that they can replicate and produce progeny. Some viruses encode the Bcl2 homologues to counter the proapoptotic functions of the cellular and mitochondrial proteins. Others modulate the permeability transition pore and either prevent or induce the release of the apoptotic proteins from the mitochondria. Viruses like Herpes simplex virus 1 deplete the host mitochondrial DNA and some, like human immunodeficiency virus, hijack the host mitochondrial proteins to function fully inside the host cell. All these processes involve the participation of cellular proteins, mitochondrial proteins, and virus specific proteins. This review will summarize the strategies employed by viruses to utilize cellular mitochondria for successful multiplication and production of progeny virus.

1. Introduction

1.1. Mitochondria. Mitochondria are cellular organelles found in the cytoplasm of almost all eukaryotic cells. One of their important functions is to produce and provide energy to the cell in the form of ATP, which help in proper maintenance of the cellular processes, thus making them indispensable for the cell. Besides acting as a powerhouse for the cell, they act as a common platform for the execution of a variety of cellular functions in normal or microorganism infected cells. Mitochondria have been implicated in aging [1, 2], apoptosis [3–7], the regulation of cell metabolism [4, 8], cell-cycle control [9–11], development of the cell [12–14], antiviral responses [15], signal transduction [16], and diseases [17–20].

Although all mitochondria have the same architecture, they vary greatly in shape and size. The mitochondria are composed of outer mitochondrial membrane, inner mitochondrial membrane, intermembrane space (space between outer and inner membrane), and matrix (space within inner mitochondrial membrane). The outer membrane is a smooth phospholipid bilayer, with different types of proteins imbedded in it [21]. The most important of them are the porins, which freely allow the transport (export and import) of the molecules (proteins, ions, nutrients, and ATP) less than 10 kDa across the membranes. The outer membrane surrounds the inner membrane creating an intermembrane space that contains molecules such as Cyt-C, SMAC/Diablo, and endonuclease G. It also acts as a buffer zone between the outer membrane and the inner membrane of mitochondria. The inner membrane is highly convoluted into structures called cristae, which increases the surface area of the membrane and are the seats of respiratory complexes. The inner membrane of mitochondria allows the free transport of oxygen and carbon dioxide. The movement of water through membranes is suggested to be controlled by aquaporins channel protein [22, 23] though a report suggested otherwise [24]. The matrix contains enzymes for the aerobic respiration, dissolved oxygen, water, carbon dioxide, and the recyclable intermediates that serve as energy shuttles and perform other functions.

Mitochondria contain a single 16 kb circular DNA genome, which codes for 13 proteins (mostly subunits of respiratory chains I, II, IV, and V), 22 mitochondrial tRNAs and 2 rRNAs [25, 26]. The mitochondrial genome is not enveloped (like nuclear envelop), contains few introns, and does not follow universal genetic code [27]. Although the majority of the mitochondrial proteins are encoded by nuclear DNA and imported into the mitochondria (reviewed by [21, 28–31]), mitochondria synthesize few proteins that are essential for their respiratory function [1, 27].

Proteins destined to mitochondria have either internally localized [28] or amino terminal localized [21] presequences known as mitochondria/matrix localization signals (MLS), which can be 10–80 amino acid long with predominantly positively charged amino acids. The combination of these presequences with adjacent regions determines the localization of a protein in respective mitochondrial compartments. The outer mitochondrial membrane contains two major translocators, namely, (a) the translocase of outer membrane (TOM) 40, which functions as an entry gate for most mitochondrial proteins with MLS and (b) sorting and assembly machinery (SAM) or translocase of β-barrel (TOB) protein, which is a specialized insertion machinery for beta-barrel membrane proteins [32]. Once proteins pass through the outer membrane, they are recruited by presequence translocase-associated motor (PAM) to the translocase of the inner mitochondrial membrane (TIM) 23 complexes, which mediates the import of proteins to the matrix. Finally, the presequences are cleaved in matrix and proteins are modified to their tertiary structure, and rendered functional [30].

1.2. Viruses. Viruses are acellular obligate intracellular microorganisms that infect the living cells/organisms and are the only exception to cell theory proposed by Schleiden and Schwann in 1838/1839 [33]. The viruses have an outer protein capsid and a nucleic acid core. Usually, the viral nucleic acids can be either DNA (double or single stranded) or RNA (+ or − sense single stranded or double stranded RNA). Some of the viruses are covered with an envelope embedded with glycoproteins. The viruses have long been associated with the living organisms, and it was in the later part of the century that their relationship with various cellular organelles was studied in detail. In order to survive and replicate in the cell, viruses need to take control of the various cellular organelles involved in defense and immune processes. They also require energy to replicate and escape from the cell. Once inside the host cell, they modulate various cellular signal pathways and organelles, including mitochondria, and use them for their own survival and replication. This review summarizes the functions of mitochondria and how viruses modulate them (Figure 1).

2. Viruses Regulate Ca^{2+} Homeostasis in Host Cells

2.1. Ca^{2+} Homeostasis. Ca^{2+} is one of the most abundant and versatile elements in the cell and acts as a second messenger to regulate many cellular processes [34]. Earlier, outer membrane of mitochondria was thought to be permeable to Ca^{2+}, but recent studies suggest that the outer membrane contains voltage-dependent anion channels (VDAC) having Ca^{2+} binding domains, which regulate the entry of Ca^{2+} into the mitochondrial intermembrane space [35–37]. The influx of Ca^{2+} through the inner membrane is regulated by the mitochondrial Ca^{2+} uniporter (MCU), which is a highly selective Ca^{2+} channel that regulates the Ca^{2+} uptake based on mitochondrial membrane potential (MMP). The net movement of charge due to Ca^{2+} uptake is directly proportional to the decrease of MMP [38]. A second mechanism that helps in Ca^{2+} movement across the mitochondria membrane is called "rapid mode" uptake mechanism (RaM) [39]. In this process, Ca^{2+} transports across the mitochondrial membrane by exchange with Na$^+$, which in turn depends upon its exchange with H$^+$ ion and thus MMP. This ion exchange across the mitochondrial membrane decreases MMP and is dependent on electron transport chain (ETC) for its maintenance. A third mechanism involves IP$_3$R, a Ca^{2+} channel in endoplasmic reticulum. IP$_3$R is connected to mitochondrial VDAC through a glucose regulating protein 75 (GRP75). This junction regulates/facilitates Ca^{2+} exchange from IP$_3$R to VDAC [40].

Ca^{2+} efflux mechanism is regulated by the permeability transition pore (PTP). The PTP is assembled in the mitochondrial inner and outer membranes [41, 42], with Ca^{2+} binding sites on the matrix side of the inner membrane. The PTP regulates the mitochondrial Ca^{2+} release by a highly regulated "Flickering" mechanism that controls the opening and closing of the pore [43]. RaM works in sync with ryanodine receptor (RyR) isoform 1, which is another very important calcium release channel [44]. Both RyR and RaM regulate the phenomenon of excitation-metabolism coupling in which cytosolic Ca^{2+} induced contraction is matched by mitochondrial Ca^{2+} stimulation of ox-phos [45]. However, mitochondrial Ca^{2+} overload can result in prolonged opening of the pore leading to pathology [46]. Although Ca^{2+} is involved in the activation of many cellular processes including stimulation of the ATP synthase [47, 48], allosteric activation of Krebs cycle enzymes [49, 50], and the adenine nucleotide translocase (ANT) [51], the primary role of mitochondrial Ca^{2+} is in the stimulation of ox-phos [52–54]. Thus, the elevated mitochondrial Ca^{2+} results in up regulation of the entire ox-phos machinery, which then results in faster respiratory chain activity and higher ATP output, which can then meet the cellular ATP demand. Ca^{2+} also upregulates other mitochondrial functions including activation of N-acetylglutamine synthetase to generate N-acetylglutamine [55], potent allosteric activation of carbamoyl-phosphate synthetase, and the urea cycle [56]. Thus, any perturbation in mitochondrial or cytosolic Ca^{2+} homeostasis has profound implications for cell function. Moreover, mitochondrial Ca^{2+}, particularly at high concentrations experienced in pathology appears to have several negative effects on mitochondrial functions [57].

2.2. Regulation by Viruses. A number of viruses alter the Ca^{2+} regulatory activity of the cell for their survival. Herpes

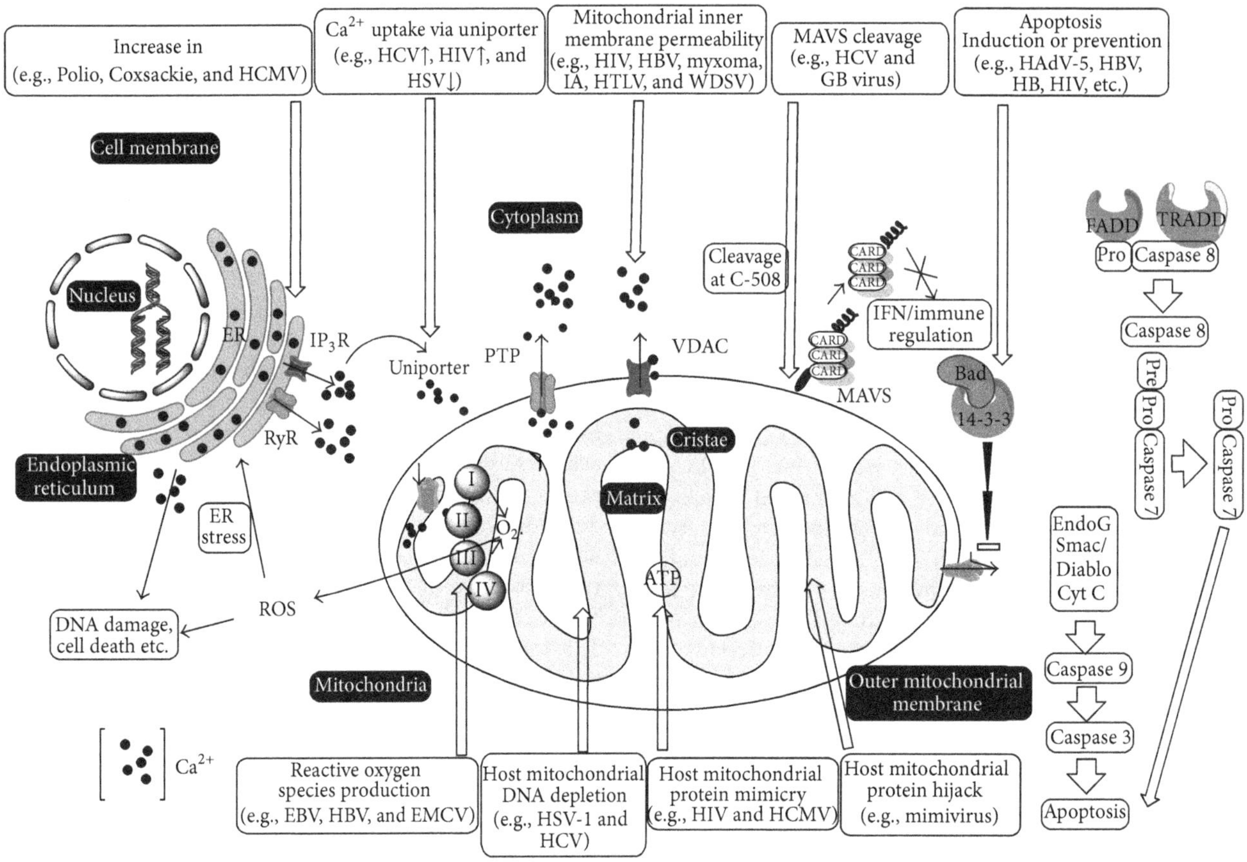

FIGURE 1: Schematic diagram of cell showing mitochondria, nucleus endoplasmic reticulum (ER) and cell membrane. iCa^{2+}: intracellular calcium, FADD: Fas-associated protein with death domain, TRADD: tumor necrosis factor receptor type 1-associated death domain protein, PTP: permeability transition pore, VDAC: voltage-dependent anion channel, IP$_3$R: inositol 1,4,5-trisphosphate receptor, RyR: ryanodine receptor, MAVS: mitochondrial antiviral signaling, I, II, III, and IV are complex I to IV of electron transport chain. O$_2^-$: Superoxide radical, Bad, Bcl-2-associated death promoter, ROS: reactive oxygen species, IFN: interferon, HCMV: human cytomegalovirus, HIV: human immunodeficiency virus, HSV: herpes simplex virus, HBV: hepatitis B virus, HTLV: human T-lymphotropic virus, IA: influenza A virus, WDSV: Walleye dermal sarcoma virus, HCV: hepatitis C virus, HAdV: human adenovirus-5, EBV: Epstein-Barr virus, and EMCV: encephalomyocarditis virus.

simplex type (HSV) 1 virus causes a gradual decline (65%) in mitochondrial Ca^{2+} uptake at 12 hrs lytic cycle [58], which helps in virus replication. Although mitochondrial Ca^{2+} uptake keeps fluctuating throughout the course of a measles virus infection of cells, the total amount of cellular Ca^{2+} remains the same [58] indicating the tight control that the virus exerts over the cellular processes during its life cycle.

The core protein of hepatitis C virus (HCV) targets mitochondria and increases Ca^{2+} [59, 60]. The NS5A protein of HCV causes alterations in Ca^{2+} homeostasis [61–63]. Both of these proteins may be responsible for the pathogenesis of liver disorders associated with HCV infection. Even in the cells coinfected with HCV and human immunodeficiency virus (HIV), these viruses enhance the MCU activity causing cellular stress and apoptosis [59, 64]. The p7 protein of HCV forms porin-like structures [65] and causes Ca^{2+} influx to cytoplasm from storage organelles [66]. These HCV proteins disturb the Ca^{2+} homeostasis at different stages of the infection and thus help to enhance the survival of the cell. Interestingly, interaction of protein X of hepatitis B virus

(HBV) with VDAC causes the release of Ca^{2+} from storage organelles mitochondria/endoplasmic reticulum (ER)/golgi into the cytoplasmic compartment, which appears to help virus replication [67, 68].

The *Nef* protein of HIV interacts with IP$_3$R [69] and induces an increase in cytosolic Ca^{2+} through promotion on T cell receptor-independent activation of the NFAT pathway [70]. Activated NFAT, in turn, causes the low-amplitude intracellular Ca^{2+} oscillation, promoting the viral gene transcription and replication [71].

Ca^{2+} is an important factor for different stages of rotavirus lifecycle and for stability to rotavirus virion [72]. The NSP4 protein of rotavirus increases the cytosolic Ca^{2+} concentration by activation of phospholipase C (PLC) and the resultant ER Ca^{2+} depletion through IP$_3$R [73, 74]. This alteration in Ca^{2+} homeostasis has been attributed to an increase in the permeability of cell membrane [75]. A decrease in cellular Ca^{2+} concentrations toward the end of the life cycle has been reported to enable rotavirus release from the cell [76].

The 2BC protein of poliovirus increases the intracellular Ca^{2+} concentrations in the cells 4 hrs. After infection, which is necessary for viral gene expression [77, 78]. Toward the end of the virus life cycle, the release of Ca^{2+} from the lumen of ER through IP_3R and RyR channels causes accumulation of Ca^{2+} in mitochondria through uniporter and VDAC resulting in mitochondrial dysfunction and apoptosis [79]. On the contrary, the 2B protein of Coxsackie virus decreases the membrane permeability by decreasing Ca^{2+} concentrations in infected cells [80, 81] due to its porin-like activity that results in Ca^{2+} efflux from the organelles. Reduced protein trafficking and low Ca^{2+} concentration in golgi and ER favor the formation of viral replication complexes, downregulate host antiviral immune response, and inhibit apoptosis [82, 83].

Enteroviruses orchestrate the apoptotic process during their life cycle to enhance its entry, survival, and release. The perturbation in cytoplasmic Ca^{2+} homeostasis at 2–4 hrs. postinfection coincides with the inhibition of the apoptotic response that can be attributed to decrease in cytotoxic levels of Ca^{2+} in the cell and the mitochondria. This also provides the virus with optimum conditions for the replication and protein synthesis. Finally, a decrease in mitochondrial and other storage organelles (ER and golgi) Ca^{2+} levels causes an increase in cytosolic Ca^{2+} concentration, leading to the formation of vesicles and cell death, thus assisting in virus release [81, 84, 85].

The pUL37 $\times$ 1 protein of human cytomegalovirus (HCMV) localizes to mitochondria [86] and causes the trafficking of Ca^{2+} from the ER to mitochondria at 4–6 hrs. After infection [87]. Active Ca^{2+} uptake by mitochondrion induces the production of ATP and other Ca^{2+} dependent enzymes accelerating virus replication, and a decrease in Ca^{2+} levels in the ER has antiapoptotic effects [88].

The 6.7K protein encoded by E3 region of HAdV-2 localizes to ER and helps maintain ER Ca^{2+} homeostasis in transfected cells, thus inhibiting apoptosis [89].

3. Viruses Cause Oxidative Stress in Host Cells

3.1. Electron Transport Chain. The mitochondrial respiratory chain is the main and most significant source of reactive oxygen species (RO) in the cell. Superoxide ($O_2^{-\bullet}$) is the primary ROS produced by mitochondria. In the normal state, there is little or no leakage of electrons between the complexes of the electron transport chain (ETC). However, during stress conditions, a small fraction of electrons leave complex III and reach complex IV [90]. This premature electron leakage to oxygen results in the formation of two types of superoxides, namely, O_2^{-}, in its anionic form, and HO_2^{-} in its protonated form.

Leakage of electrons takes place mainly from Q_O sites of complex III, which are situated immediately next to the intermembrane space resulting in the release of superoxides in either the matrix or the innermembrane space of the mitochondria [91–94]. About 25–75% of the total electron leak through Complex III could account for the net extramitochondrial superoxide release [95–97]. Thus, the main source of $O_2^{-\bullet}$ in mitochondria is the ubisemiquinone

radical intermediate ($QH^{\bullet}$) formed during the Q cycle at the Q_O site of complex III [98–100]. Complex I is also a source of ROS, but the mechanism of ROS generation is less clear. Recent reports suggest that glutathionylation [101] or PKA mediated phosphorylation [101–103] of complex I can elevate ROS generation. Backward flow of electron from complex I to complex II can also result in the production of ROS [99].

A variety of cellular defense mechanisms maintain the steady state concentration of these oxidants at nontoxic levels. This delicate balance between ROS generation and metabolism may be disrupted by various xenobiotics including viral proteins. The main reason for generation of ROS in virus-infected cells is to limit the virus multiplication. However, ROS also acts as a signal for various cellular pathways, and the virus utilizes the chaos generated inside the cell for its replication.

3.2. Viruses Induce Reactive Oxygen Species. A number of viruses cause oxidative stress to the host cells, which directly or indirectly helps them to survive. Human-Adenovirus-(HAdV-) 5 has been reported to induce the rupture of endosomal membrane upon infection resulting in the release of lysosomal cathepsins, which prompt the production of ROS. Cathepsins also induce the disruption of mitochondrial membrane leading to the release of ROS from mitochondria thus causing the oxidative stress [104].

The core protein of HCV causes oxidative stress in the cell and alters apoptotic pathways [64, 105–107]. The E1, E2, NS3, and core protein of HCV are potent ROS inducers and can cause host DNA damage independently [107, 108] or mediated by nitric oxide (NO) thus aiding in virus replication.

The ROS is generated during HIV infection [64, 109–111]. H_2O_2, an ROS generated during HIV infection strongly induces HIV long terminal repeat (LTR) via NF-kappa B activation. Impaired LTR activity ablates the LTR activation in response to ROS thus aiding in virus replication [112]. HIV also causes extensive cellular damage due to increased ROS production and decreased cytosolic antioxidant production [113]. Coinfection of HIV and HCV causes the hepatic fibrosis, the progression of which is regulated through the generation of ROS in an NF-κB dependent manner [113].

Epstein-Barr virus (EBV) causes increased oxidative stress in the host cells within 48 hrs. During the lytic cycle indicating the role of ROS in virus release [114]. Oxidative stress activates the EBV early gene BZLF-1, which causes the reactivation of EBV lytic cycle [114]. This has been proposed to play an important role in the pathogenesis of EBV-associated diseases including malignant transformations [115, 116].

Interestingly, HBV causes both an increase and a decrease in oxidative stress to enhance its survival in the host cells [117, 118]. HBV induces strong activation of Nrf2/ARE-regulated genes *in vitro* and *in vivo* through the activation of c-Raf and MEK by HBV protein X thus protecting the cells from HBV induced oxidative stress and promoting establishment of the infection [119]. The protein X of HBV also induces the ROS mediated upregulation of Forkhead box class O4 (Foxo4), enhancing resistance to oxidative stress-induced cell death

[120]. However, reports also suggest that upon exposure to oxidative stress, HBV protein X accelerates the loss of Mcl-1 protein via caspase-3 cascade thus inducing pro apoptotic effects [118]. Coinfection of HCV also causes the genotoxic effects in peripheral blood lymphocytes due to increased oxidative damage and decreased MMP [121]. It is possible that contradictory functions of protein X of HBV cold occur at different stages of virus replication.

Encephalomyocarditis virus (EMCV) causes oxidative stress in the cells during infection damaging the neurons, which is an important process in the pathogenesis of EMCV infection [122].

4. Viruses Regulate Mitochondrial Membrane Potential in Host Cells

4.1. Mitochondrial Membrane Potential. Membrane potential (MP) is the difference in voltage or electrical potential between the interior and the exterior of a membrane. The membrane potential is generated either by electrical force (mutual attraction or repulsion between both positive or negative) and/or by diffusion of particles from high to low concentrations. The mitochondrial membrane potential (MMP) is an MP ($\cong$ 180 mV) across the inner membrane of mitochondria, which provides energy for the synthesis of ATP. Movement of protons from complex I to V of electron transport chain (ETC) located in the inner mitochondrial membrane creates an electric potential across the inner membrane, which is important for proper maintenance of ETC and ATP production. Reported MMP values for mitochondria (*in vivo*) differ from species to species and from one organ to another depending upon the mitochondria function, protein composition, and the amount of oxidative phosphorylation activity required in that part of the body [43].

The voltage dependent anionic channels (VDACs) also known as mitochondrial porins form channels in the outer mitochondrial membranes and act as primary pathway for the movement of metabolites across the outer membrane [37, 96, 123–125]. In addition, a number of factors including oxidative stress, calcium overload, and ATP depletion induce the formation of nonspecific mitochondrial permeability transition pores (MPTP) in the inner mitochondrial membrane, which is also responsible for the maintenance of MMP [36, 37, 126]. The outer membrane VDACs, inner membrane adenine nucleotide translocase (ANT) [127], and cyclophilin D (CyP-D) in matrix are the structural elements of the mitochondrial permeability transition pore (MPTP).

When open, MPTP increases the permeability of the inner mitochondrial membrane to ions and solutes up to 1.5 kDa, which causes dissipation of the MMP and diffusion of solutes down their concentration gradients, by a process known as the permeability transition [128, 129]. The MPTP opening is followed by osmotic water flux, passive swelling, outer membrane rupture, and release of proapoptotic factors leading to the cell death [42, 130]. Because of the consequent depletion of ATP and Ca^{2+} deregulation, opening of the MPTP had been proposed to be a key element in determining the fate of the cell before a role for mitochondria in apoptosis was proposed [129].

The MMP can be altered by a variety of stimuli including sudden burst of ROS [43, 107], Ca^{2+} overload in the mitochondria or the cell [48, 57, 131], and/or by proteins of invading viruses [109, 132, 133]. In general, an increase or decrease in MMP is related to the induction or prevention of apoptosis, respectively. Prevention of apoptosis during early stages of virus infection is a usual strategy employed by viruses to prevent host immune response and promote their replication. On the contrary, induction of apoptosis during later stages of virus infection is a strategy used by viruses to release the progeny virions for dissemination to the surrounding cells.

4.2. Regulation by Viruses. Many viral proteins alter mitochondrial ion permeability and/or membrane potential for their survival in the cell. The p7, a hydrophobic integral membrane [134] viroprotein [135] of HCV, localizes to mitochondria [66] and controls membrane permeability to cations [66, 136] promoting cell survival for virus replication [135].

The R (Vpr) protein of HIV, a small accessory protein, localizes to the mitochondria, interacts with ANT, modulates MPTP, and induces loss of MMP promoting release of Cyto C [137] leading to cell death [138, 139]. The Tat protein of HIV also modulates MPTP leading to the accumulation of Tat in mitochondria and induction of loss of MMP resulting in caspase dependent apoptosis [140].

The M11L protein of myxoma poxvirus localizes to the mitochondria, interacts with the mitochondrial peripheral benzodiazepine receptor (PBR), and regulates MPTP [141] inhibiting MMP loss [142] and thus inhibiting induction of apoptosis during viral infection [143]. The FIL protein of vaccinia virus downregulates proapoptotic Bcl-2 family protein Bak and, inhibits the loss of the MMP and the release of Cyt-C [144, 145]. The crmA/Spi-2 protein of vaccinia virus, a caspase 8 inhibitor, modulates MPTP thus preventing apoptosis [146].

The PB1-F2 protein of influenza A viruses localizes to the mitochondria [147–150] and interacts with VDAC1 and ANT3 [151] resulting in decreased MMP, which induces the release of proapoptotic proteins causing cell death. Recent evidence shows that PB1-F2 is also able to form nonselective protein channel pores resulting in the alteration of mitochondrial morphology, dissipation of MMP, and cell death [150]. The M2 protein of influenza virus, a viroprotein, causes the alteration of mitochondrial morphology, dissipation of MMP, and cell death (reviewed by [135]).

The p13II, an accessory protein encoded by x-II ORF of human T-lymphotropic virus (HTLV), a new member of the viroprotein family [152], localizes to the mitochondria of infected cells and increases the MMP leading to apoptosis [153] and mitochondrial swelling [153–155].

The Orf C protein of Walleye dermal sarcoma virus (WDSV) localizes to the mitochondria [156] and induces perinuclear clustering of mitochondria and loss of MMP [156] leading to the release of proapoptotic factors thus causing apoptosis.

The 2B protein of Coxsackie virus decreases MMP by decreasing the Ca^{2+} concentrations in infected cells [80, 81].

5. Viruses Regulate Apoptosis

5.1. Apoptosis. During the coevolution of viruses with their hosts, viruses have developed several strategies to manipulate the host cell machinery for their survival, replication, and release from the cell. Viruses target the cellular apoptotic machinery at critical stages of viral replication to meet their ends [157, 158]. Depending upon the need, a virus may inhibit [159] or induce [160] apoptosis for the obvious purpose of replication and spread, respectively [158, 159]. Interference in mitochondrial function can cause either cell death due to deregulation of the Ca^{2+} signaling pathways and ATP depletion or apoptosis due to regulation of Bcl-2 family proteins. Apoptosis is a programmed cell death [161] characterized by membrane blebbing, condensation of the nucleus and cytoplasm, and endonucleosomal DNA cleavage. The process starts as soon as the cell senses physiological or stress stimuli, which disturbs the homeostasis of the cell [162, 163]. Apoptotic cell death can be considered as an innate response to limit the growth of microorganisms including viruses attacking the cell.

Two major pathways, namely, the extrinsic and the intrinsic are involved in triggering apoptosis [163, 164]. The extrinsic pathway is mediated by signaling through death receptors like tumor necrosis factor or Fas ligand receptor causing the assembly of death inducing signaling complex (DISC) with the recruitment of proteins like caspases leading to the mitochondrial membrane permeabilization. In the intrinsic pathway, the signals act directly on the mitochondria leading to mitochondrial membrane permeabilization before caspases are activated causing the release of Cyt-C [165, 166], which recruits APAF1 [167, 168] resulting in direct activation of caspase 9 [35, 169]. Both the extrinsic and the intrinsic processes congregate at the activation of downstream effector caspases, (i.e., caspase-3) [170] which is responsible for inducing the morphological changes observed in an apoptotic cell. In addition to Cyt-C, Smac/DIABLO as well as caspase independent death effectors inducing factor (AIF) and endonuclease G [171–173] acts as an activator of the caspase.

The B cell lymphoma- (Bcl-) 2 family of proteins tightly regulate the apoptotic events involving the mitochondria [174, 175]. More than 20 mammalian Bcl-2 family proteins have been described to date [176, 177]. They have been classified by the presence of Bcl-2 homology (BH) domains arranged in the order BH4-BH3-BH2-BH1 and the C-terminal hydrophobic transmembrane (TM) domain, which anchors them to the outer mitochondrial membrane [178]. The highly conserved BH1 and BH2 domains are responsible for antiapoptotic activity and multimerization of Bcl-2 family proteins. The BH3 domain is mainly responsible for proapoptotic activity and the less conserved BH4 domain is required for the antiapoptotic activities of Bcl-2 and Bcl-X_L proteins [174, 178]. Most of the antiapoptotic proteins are multidomain proteins, which contain all four BH domains (BH1 to BH4) and a TM domain. In contrast, proapoptotic proteins are either multidomain proteins, which contain three BH domains (BH1 to BH3) or single domain proteins, which contain one domain (BH3) [158]. The Bcl-2 proteins regulate the MMP depending upon whether they belong to the pro- or antiapoptotic branch of the family, respectively. The MMP marks the dead end of apoptosis beyond which cells are destined to die [125, 166, 179–183].

5.2. Regulation by Viruses. Viruses encode homologs of Bcl-2 (vBcl-2) proteins, which can induce (pro-apototic) or prevent (antiapoptotic) apoptosis thus helping viruses to complete their life cycle in the host cells [117, 163, 175]. While the vBcl-2s and the cellular Bcl-2s share limited sequence homology, their secondary structures are predicted to be quite similar [158, 174, 184]. During primary infection, interplay between vBcl-2 and other proteins enhances the lifespan of the host cells resulting in efficient production of viral progeny and ultimately spread of infection to the new cells. It also favors viral persistence in the cells by enabling the latently infecting viruses to make the transition to productive infection. The pathways and strategies used by viruses to induce/inhibit apoptosis have been reviewed earlier [185].

Many viruses encode for the homologs of antiapoptotic Bcl-2 proteins, which preferentially localize to the mitochondria and may interact with the other proapoptotic Bax homologues. The E1B19K encoded by human-adenovirus- (HAdV-) 5 contains BH1 and BH3-like domains and blocks TNF-alpha-mediated death signaling by inhibiting a form of Bax that interrupts the caspase activation downstream of caspase-8 and upstream of caspase-9 [186, 187]. Like HAdV-5 E1B19K [186], some viruses encode Bcl-2 homologues lacking BH4 domain, which are thought to act by inhibiting proapoptotic members of Bcl-2 family proteins. The FPV309 protein encoded by fowl pox virus contains highly conserved BH1 and BH2-like domains, and a cryptic BH3 domain, interacts with Bax protein and inhibits apoptosis [188]. The A179L protein encoded by African swine fever virus (ASFV) contains BH1 and BH2 domains and, interacts with Bax-Bak proteins and inhibits apoptosis [189, 190]. The Bcl-2 homolog (vBcl-2) encoded by Herpesvirus saimiri (HVS) contains BH3 and BH4-like domains and interacts with Bax, thus stabilizing mitochondria against a variety of apoptotic stimuli preventing the cell death [191]. The E4 ORF encoded by equine Herpesvirus-3 contains BH1 and BH2 domains [192], which may interact with Bax and be essential for antiapoptotic activity [193].

Viruses also encode homologs of proapoptotic Bcl-2 proteins. The HBV encodes protein X, a vBcl-2 protein containing BH3, which localizes to the mitochondria and interacts with VDACs inducing the loss of the MMP leading to apoptosis [117, 121, 194, 195] or interacts with Hsp60 and induces apoptosis [196]. In contrast, another study revealed the protective effects of HB-X in response to proapoptotic stimuli (Fas, TNF, and serum withdrawal) but not from chemical apoptotic stimuli [197]. The protein X of HBV is known to stimulate NFκB [198, 199], SAPK [200, 201], and PI3K/PKB [202] to prevent apoptosis. It is possible that the diverse functions of HBV protein X occur at different times of virus replication cycle in the infected cells. The BALF1 protein encoded by EBV contains BH1 and BH4 domains [203], which interacts with the Bax-Bak proteins [192] and inhibits the antiapoptotic activity of the EBV BHRF1 and the Kaposi Sarcoma virus (KSV) Bcl-2 protein, both of which contain BH1 and BH2 domains [204] and interact with BH3 only proteins [205].

The effects of viral Bcl-2 homologues are thus apparently centered around mitochondria and include prevention or induction of MMP loss. The induction of MMP loss leads to the release of Cyto C and other proapoptotic signals into the cytosol and activation of downstream caspases leading to the cell death and dissemination of viruses to neighbouring cells for further infection.

Viruses encode pro/anti apoptotic proteins, which show no homology to Bcl-2 proteins [158]. The E6 protein of human papilloma virus (HPV) downregulates Bax signal upstream of mitochondria [206, 207] and prevents the release of Cyto C, AIF, and Omi, thus preventing apoptosis [208]. This E6 activity towards another Bcl2 family proapoptotic protein Bak is a key factor promoting the survival of HPV-infected cells, which in turn facilitates the completion of viral life cycle [207]. Enterovirus (EV) 71 induces conformational changes in Bax and increases its expression in cells following infection and induces the activation of caspases 3, 8, and PARP causing caspase dependent apoptosis [209]. On the contrary, Rubella viral capsid binds to Bax, forms oligoheteromers, and prevents the formation of pores on mitochondrial membrane thus preventing Bax induced apoptosis [210].

Viruses also encode proteins, which act as viral mitochondrial inhibitors of apoptosis (vMIA) thus protecting the cells. A splice variant of UL37 of HCMV acts as vMIA and protects the cells from apoptosis [211] thereby helping viruses to complete their replication cycle. It localizes to mitochondria and interacts with ANT [211] and Bax [212, 213]. HCMV vMIA has an N-terminal mitochondrial localization domain and a C-terminal antiapoptotic domain [211], which recruits Bax to mitochondria and prevents loss of MMP. It protects the cells against CD95 ligation [211] and oxidative stress-induced cell death [214, 215] and prevents mitochondrial fusion [216] thus promoting cell survival.

vMIA does not inhibit the apoptotic events upstream of mitochondria but can influence events like preservation of ATP generation, inhibition of Cyto C release, and caspase 9 activation, following induction of apoptosis. However, the exact mechanisms of the events around vMIA still remain a question.

6. Viruses Modulate Mitochondrial Antiviral Immunity

6.1. Mitochondrial Antiviral Immunity. Cells respond to virus attack by activating a variety of signal transduction pathways leading to the production of interferons [217], which limit or eliminate the invading virus. The presence of viruses inside the cell is first sensed by pattern recognition receptors (PRRs) that recognize the pathogen associated molecular patterns (PAMPs). PRRs include toll-like receptors (TLRs), nucleotide oligomerization domain (NOD) like receptors (NLRs), and retinoic acid-inducible gene I (RIG-I) like receptors (RLRs). Mitochondria have been associated with RLRs, which include retinoic acid-inducible gene I (RIG-I) [218] and melanoma differentiation-associated gene 5 (Mda-5) [219]. Both are cytoplasm-located RNA helicases that recognize dsRNA. The

N-terminus of RIG-1 has caspase activation and recruitment domains (CARDs) whereas C-terminus has RNA helicase activity [218], which recognizes and binds to uncapped and unmodified RNA generated by viral polymerases in ATPase dependent manner. This causes conformational changes and exposes its CARD domains to bind and activate downstream effectors leading to the formation of enhanceosome [220] triggering NFκB production. RLRs have recently been reviewed in detail [221–223].

A CARD domain containing protein named mitochondrial antiviral signaling (MAVS) [15, 224], virus-induced signaling adaptor (VISA) [225], IFN-β promoter stimulator 1 (IPS-1) [226], or CARD adaptor inducing IFN-β (CARDIF) protein [227] acts downstream of the RIG-I. Besides the presence of N-terminal CARD domain, MAVS contains a proline-rich region and a C-terminal hydrophobic transmembrane (TM) region, which targets the protein to the mitochondrial outer membrane and is critical for its activity [15]. The TM region of the MAVS resembles the TM domains of many C-terminal tail-anchored proteins on the outer membrane of the mitochondria including Bcl-2 and Bcl-xL [15]. Recent reports indicate that MAVS has an important role in inducing the antiviral defenses in the cell. Overexpression of MAVS leads to the activation of NFκB and IRF-3, leading to the induction of type I interferon response, which is abrogated in the absence of MAVS [15] thus indicating the specific role of MAVS in inducing antiviral response. MAVS has also been shown to prevent apoptosis by its interaction with VDAC [228] and preventing the opening of MPTP.

6.2. Regulation by Viruses. Some viruses induce cleavage of MAVs from outer membranes of mitochondria [227, 229] thus greatly reducing their ability to induce interferon response. HCV persists in the host by lowering the host cell immune response including inhibiting the production of IFN-β by RIG-I pathway [230–232]. The NS3/4A protein of HCV colocalizes with mitochondrial MAVS [227, 229] leading to the cleavage of MAVS at amino acid 508. Since free form of the MAVS is not functional, the dislodging of MAV from the mitochondria inactivates MAVS [227] thus helping in paralyzing the host defense against HCV. Interestingly, another member of family *Flaviviridae* GB virus B shares 28% amino acid homology with HCV over the lengths of their open-reading frames [233]. The NS3/4A protein of GB virus also cleaves MAVS in a manner similar to HCV, thus effectively compromising the host immune response by preventing the production of interferons [234]. Other viruses like influenza A translocate RIG-I/MAVS components to the mitochondria of infected human primary macrophages and regulate the antiviral/apoptotic signals increasing the viral survivability [235].

7. Viruses Hijack Host Mitochondrial Proteins

Over the years, viruses have perfected different strategies to establish complex relationships with their host with the sole purpose of preserving their existence. One such strategy involves the hijacking of the host cell mitochondrial proteins.

The p32, a mitochondria-associated cellular protein, is a member of a complex involved in the import of cytosolic proteins to the nucleus. Upon entry into the cell, adenovirus hijacks this protein and piggybacks it to transport its genome to the nucleus [236], thereby increasing its chances of survival and establishment in the host cell. During HIV-1 assembly, tRNA^{Lys} iso-acceptors are selectively incorporated into virions, and tRNA$_3^{Lys}$ binds to HIV genome and is used as the primer for reverse transcription [237]. In humans, a single gene produces both cytoplasmic and mitochondrial Lys tRNA synthetases (LysRSs) by alternative splicing [238]. The mitochondrial LysRS is produced as a preprotein, which is transported into the mitochondria. The premitochondrial or mitochondrial LysRS is specifically packaged into HIV [239] and acts as a primer to initiate the replication of HIV-I RNA genome, which then binds to a site complementary to the 3'-end 18 nucleotides of tRNA$_3^{Lys}$. It is proposed that HIV viral protein R (Vpr) alters the permeability of the mitochondria [138] leading to the release of premito- or mito-LysRS, which then interacts with Vpr [240] and gets packed into the progeny virions.

Viperin, an interferon inducible protein, is induced in the cells in response to viral infection [241]. This protein has been shown to prevent the release of influenza virus particles from the cells by trapping them in lipid rafts inside the cells thereby preventing its dissemination [242]. During infection, HCMV induces IFN independent expression of viperin, which interacts with HCMV encoded vMIA protein resulting in relocation of viperin from ER to mitochondria. In mitochondria, viperin interacts with mitochondrial tri-functional protein and decreases ATP generation by disrupting oxidation of fatty acids, which results in disrupting actin cytoskeleton of the cells and enhancing the viral infectivity [243].

8. Viruses Alter Intracellular Distribution of Mitochondria

Viruses alter the intracellular distribution of mitochondria either by concentrating the mitochondria near the viral factories to meet energy requirements during viral replication or by cordoning off the mitochondria within cytoplasm to prevent the release of mediators of apoptosis. The protein X of HBV causes microtubule mediated perinuclear clustering of the mitochondria by p38 mitogen-activated protein kinase (MAPK) mediated dynein activity [244]. HCV nonstructural protein 4A (NS4A) either alone or together with NS3, (in the form of the NS3/4A polyprotein) accumulates on mitochondria and changes their intracellular distribution [245]. HIV-1 infection causes clustering of the mitochondria in the infected cells [246]. Interestingly, ASFV causes the microtubule-mediated clustering of the mitochondria around virus factories in the cell providing energy for virus release [247]. Similar changes were observed in the chick embryo fibroblasts infected with frog virus 3, where degenerate mitochondria surrounding virus factories were found [248].

9. Viruses Mimic the Host Mitochondrial Proteins

Molecular mimicry is "the theoretical possibility that sequence similarities between foreign and self-peptides are sufficient to result in the cross-activation of autoreactive T or B cells by pathogen-derived peptides" [249, 250]. Since structure follows the function, viruses, during their coevolution with hosts have evolved to mimic the host proteins to meet their ends during progression of their life cycle inside the cell. Mimicking aids the viruses to gain access to host cellular machinery and greatly helps in their survival in the hostile host environment.

Mimivirus, a member of the newly created virus family *Mimiviridae*, encodes a eukaryotic mitochondria carrier protein (VMC-I) [251], which mimics the host cell's mitochondrial carrier protein and thus controls the mitochondrial transport machinery in infected cells. It helps to transport ADP, dADP, TTP, dTTP, and UTP in exchange for dATP, thus exploiting the host for energy requirements during replication of its A+T rich genome [251]. Besides VMC-I, mimivirus encodes several other proteins (L359, L572, R776, R596, R740, R824 L81, R151, R900, and L908) with putative mitochondria localization signals, which suggest that mimivirus has evolved a strategy to take over the host mitochondria and exploited its physiology to compensate for its energy requirements and biogenesis [251]. Viral Bcl-2 homologues (vBcl-2) are other groups of viral proteins that mimic the host cell Bcl-2s and have been described elsewhere in this review.

10. Viruses Cause Host Mitochondrial DNA Depletion

Mammalian mitochondria contain a small circular genome, which synthesizes enzymes for oxidative phosphorylation and mitochondrial RNAs (mtRNAs) [27]. To increase the chance of survival, some viruses appear to have adopted the strategy of damaging the host cell mitochondrial DNA. Since mitochondria act as a source of energy and play an important role in antiviral immunity as well, it is possible that damage to mitochondrial DNA may help in evading mitochondrial antiviral immune responses [252].

During productive infection of mammalian cells *in vitro*, HSV-1 induces the rapid and complete degradation of host mitochondrial DNA [252]. The UL12.5 protein of HSV-1 localizes to the mitochondria and induces DNA depletion in the absence of other viral gene products [252, 253]. The immediate early Zta protein of EBV interacts with mitochondrial single stranded DNA binding protein resulting in reduced mitochondrial DNA (mtDNA) replication and enhanced viral DNA replication [254]. HCV causes the reactive oxygen species and nitrous oxide mediated DNA damage in host mtDNA [107, 255]. Interestingly, depletion of mtDNA has also been observed in HIV/HCV coinfected humans [256].

11. Conclusions

Though progress has been made in understanding the interaction of viruses with mitochondria-mediated pathways, the pathways linking the detection of viral infection by PRRs (or exact mechanism by which PRRs recognize the PAMPs) and their link to mitochondria-mediated cell death remain poorly understood. Role of the mitochondria in immunity and viral mechanisms to evade them highlights the fact that even after billions of years of coevolution, the fight for the survival is still going on. Both the host and the viruses are evolving, finding new ways to survive. It may be interesting to note that mitochondria mediated apoptosis might be an evolutionary adaptation by which they might have effectively prevented the entry of other microorganisms trying to gain entry into the host cell and thus effectively establishing themselves as an integral part of the cell.

Acknowledgments

The authors thank Dr. Vikram Misra, Veterinary Microbiology, University of Saskatchewan, for his vision and advice. They thank Sherry Hueser for carefully proofreading the paper. The paper is published with the permission of Director VIDO as VIDO article no. 617. Suresh K. Tikoo is funded by grants from Natural Sciences and Engineering Research Council of Canada.

References

[1] D. C. Wallace, "A mitochondrial paradigm of metabolic and degenerative diseases, aging, and cancer: a dawn for evolutionary medicine," *Annual Review of Genetics*, vol. 39, pp. 359–407, 2005.

[2] D. C. Chan, "Mitochondria: dynamic organelles in disease, aging, and development," *Cell*, vol. 125, no. 7, pp. 1241–1252, 2006.

[3] A. Antignani and R. J. Youle, "How do Bax and Bak lead to permeabilization of the outer mitochondrial membrane?" *Current Opinion in Cell Biology*, vol. 18, no. 6, pp. 685–689, 2006.

[4] H. Chen and D. C. Chan, "Emerging functions of mammalian mitochondrial fusion and fission," *Human Molecular Genetics*, vol. 14, no. 2, pp. R283–R289, 2005.

[5] I. Gradzka, "Mechanisms and regulation of the programmed cell death," *Postepy Biochemii*, vol. 52, no. 2, pp. 157–165, 2006.

[6] H. M. McBride, M. Neuspiel, and S. Wasiak, "Mitochondria: more than just a powerhouse," *Current Biology*, vol. 16, no. 14, pp. R551–R560, 2006.

[7] G. Kroemer, L. Galluzzi, and C. Brenner, "Mitochondrial membrane permeabilization in cell death," *Physiological Reviews*, vol. 87, no. 1, pp. 99–163, 2007.

[8] C. A. Mannella, "Structure and dynamics of the mitochondrial inner membrane cristae," *Biochimica et Biophysica Acta*, vol. 1763, no. 5-6, pp. 542–548, 2006.

[9] D. G. Hardie, J. W. Scott, D. A. Pan, and E. R. Hudson, "Management of cellular energy by the AMP-activated protein kinase system," *The FEBS Letters*, vol. 546, no. 1, pp. 113–120, 2003.

[10] R. G. Jones, D. R. Plas, S. Kubek et al., "AMP-activated protein kinase induces a p53-dependent metabolic checkpoint," *Molecular Cell*, vol. 18, no. 3, pp. 283–293, 2005.

[11] S. Mandal, P. Guptan, E. Owusu-Ansah, and U. Banerjee, "Mitochondrial regulation of cell cycle progression during development as revealed by the tenured mutation in Drosophila," *Developmental Cell*, vol. 9, no. 6, pp. 843–854, 2005.

[12] L. E. Bakeeva, Y. S. Chentsov, and V. P. Skulachev, "Mitochondrial framework (reticulum mitochondriale) in rat diaphragm muscle," *Biochimica et Biophysica Acta*, vol. 501, no. 3, pp. 349–369, 1978.

[13] L. E. Bakeeva, Y. S. Chentsov, and V. P. Shulachev, "Intermitochondrial contacts in myocardiocytes," *Journal of Molecular and Cellular Cardiology*, vol. 15, no. 7, pp. 413–420, 1983.

[14] S. Honda and S. Hirose, "Stage-specific enhanced expression of mitochondrial fusion and fission factors during spermatogenesis in rat testis," *Biochemical and Biophysical Research Communications*, vol. 311, no. 2, pp. 424–432, 2003.

[15] R. B. Seth, L. Sun, C. K. Ea, and Z. J. Chen, "Identification and characterization of MAVS, a mitochondrial antiviral signaling protein that activates NF-κB and IRF3," *Cell*, vol. 122, no. 5, pp. 669–682, 2005.

[16] E. Bossy-Wetzel, M. J. Barsoum, A. Godzik, R. Schwarzenbacher, and S. A. Lipton, "Mitochondrial fission in apoptosis, neurodegeneration and aging," *Current Opinion in Cell Biology*, vol. 15, no. 6, pp. 706–716, 2003.

[17] C. W. Olanow and W. G. Tatton, "Etiology and pathogenesis of Parkinson's disease," *Annual Review of Neuroscience*, vol. 22, pp. 123–144, 1999.

[18] S. K. van den Eeden, C. M. Tanner, A. L. Bernstein et al., "Incidence of Parkinson's disease: variation by age, gender, and race/ethnicity," *The American Journal of Epidemiology*, vol. 157, no. 11, pp. 1015–1022, 2003.

[19] L. J. Martin, "Mitochondriopathy in Parkinson disease and amyotrophic lateral sclerosis," *Journal of Neuropathology and Experimental Neurology*, vol. 65, no. 12, pp. 1103–1110, 2006.

[20] R. McFarland, R. W. Taylor, and D. M. Turnbull, "Mitochondrial disease—its impact, etiology, and pathology," in *Current Topics in Developmental Biology*, J. C. St John, Ed., pp. 113–155, Academic Press, New York, NY, USA, 2007.

[21] D. Rapaport, "Finding the right organelle. Targeting signals in mitochondrial outer-membrane proteins," *EMBO Reports*, vol. 4, no. 10, pp. 948–952, 2003.

[22] M. Amiry-Moghaddam, H. Lindland, S. Zelenin et al., "Brain mitochondria contain aquaporin water channels: evidence for the expression of a short AQP9 isoform in the inner mitochondrial membrane," *FASEB Journal*, vol. 19, no. 11, pp. 1459–1467, 2005.

[23] G. Calamita, D. Ferri, P. Gena et al., "The inner mitochondrial membrane has aquaporin-8 water channels and is highly permeable to water," *The Journal of Biological Chemistry*, vol. 280, no. 17, pp. 17149–17153, 2005.

[24] B. Yang, D. Zhao, and A. S. Verkman, "Evidence against functionally significant aquaporin expression in mitochondria," *The Journal of Biological Chemistry*, vol. 281, no. 24, pp. 16202–16206, 2006.

[25] G. S. Shadel and D. A. Clayton, "Mitochondrial DNA maintenance in vertebrates," *Annual Review of Biochemistry*, vol. 66, pp. 409–435, 1997.

[26] E. A. Shoubridge, "The ABcs of mitochondrial transcription," *Nature Genetics*, vol. 31, no. 3, pp. 227–228, 2002.

[27] G. Burger, M. W. Gray, and B. F. Lang, "Mitochondrial genomes: anything goes," *Trends in Genetics*, vol. 19, no. 12, pp. 709–716, 2003.

[28] W. Neupert and J. M. Herrmann, "Translocation of proteins into mitochondria," *Annual Review of Biochemistry*, vol. 76, pp. 723–749, 2007.

[29] A. Chacinska, C. M. Koehler, D. Milenkovic, T. Lithgow, and N. Pfanner, "Importing mitochondrial proteins: machineries and mechanisms," *Cell*, vol. 138, no. 4, pp. 628–644, 2009.

[30] O. Schmidt, N. Pfanner, and C. Meisinger, "Mitochondrial protein import: from proteomics to functional mechanisms," *Nature Reviews Molecular Cell Biology*, vol. 11, no. 9, pp. 655–667, 2010.

[31] M. van der Laan, D. P. Hutu, and P. Rehling, "On the mechanism of preprotein import by the mitochondrial presequence translocase," *Biochimica et Biophysica Acta*, vol. 1803, no. 6, pp. 732–739, 2010.

[32] S. J. Habib, T. Waizenegger, M. Lech, W. Neupert, and D. Rapaport, "Assembly of the TOB complex of mitochondria," *The Journal of Biological Chemistry*, vol. 280, no. 8, pp. 6434–6440, 2005.

[33] T. Schwann, "Microscopical researches into the accordance in the structure and growth of animals and plants," in *Contributions to Phytogenesis*, M. J. Schleiden, Ed., Sydenham Society, London, UK, 1847.

[34] M. J. Berridge, M. D. Bootman, and P. Lipp, "Calcium—a life and death signal," *Nature*, vol. 395, no. 6703, pp. 645–648, 1998.

[35] D. R. Green and J. C. Reed, "Mitochondria and apoptosis," *Science*, vol. 281, no. 5381, pp. 1309–1312, 1998.

[36] S. V. Chorna, V. I. Dosenko, N. A. Strutyns'ka, H. L. Vavilova, and V. F. Sahach, "Increased expression of voltage-dependent anion channel and adenine nucleotide translocase and the sensitivity of calcium-induced mitochondrial permeability transition opening pore in the old rat heart," *Fiziolohichnyi Zhurnal*, vol. 56, no. 4, pp. 19–25, 2010.

[37] Y. Liu, L. Gao, Q. Xue et al., "Voltage-dependent anion channel involved in the mitochondrial calcium cycle of cell lines carrying the mitochondrial DNA A4263G mutation," *Biochemical and Biophysical Research Communications*, vol. 404, no. 1, pp. 364–369, 2011.

[38] Y. Kirichok, G. Krapivinsky, and D. E. Clapham, "The mitochondrial calcium uniporter is a highly selective ion channel," *Nature*, vol. 427, no. 6972, pp. 360–364, 2004.

[39] T. E. Gunter and K. K. Gunter, "Uptake of calcium by mitochondria: transport and possible function," *IUBMB Life*, vol. 52, no. 3–5, pp. 197–204, 2002.

[40] G. Szabadkai, K. Bianchi, P. Várnai et al., "Chaperone-mediated coupling of endoplasmic reticulum and mitochondrial Ca^{2+} channels," *Journal of Cell Biology*, vol. 175, no. 6, pp. 901–911, 2006.

[41] A. P. Halestrap, "What is the mitochondrial permeability transition pore?" *Journal of Molecular and Cellular Cardiology*, vol. 46, no. 6, pp. 821–831, 2009.

[42] A. P. Halestrap, "A pore way to die: the role of mitochondria in reperfusion injury and cardioprotection," *Biochemical Society Transactions*, vol. 38, no. 4, pp. 841–860, 2010.

[43] M. Hüttemann, I. Lee, A. Pecinova, P. Pecina, K. Przyklenk, and J. W. Doan, "Regulation of oxidative phosphorylation, the mitochondrial membrane potential, and their role in human disease," *Journal of Bioenergetics and Biomembranes*, vol. 40, no. 5, pp. 445–456, 2008.

[44] V. Petronilli, B. Persson, M. Zoratti, J. Rydstrom, and G. F. Azzone, "Flow-force relationships during energy transfer between mitochondrial proton pumps," *Biochimica et Biophysica Acta*, vol. 1058, no. 2, pp. 297–303, 1991.

[45] W. Xia, Y. Shen, H. Xie, and S. Zheng, "Involvement of endoplasmic reticulum in hepatitis B virus replication," *Virus Research*, vol. 121, no. 2, pp. 116–121, 2006.

[46] W. J. H. Koopman, L. G. J. Nijtmans, C. E. J. Dieteren et al., "Mammalian mitochondrial complex I: biogenesis, regulation, and reactive oxygen species generation," *Antioxidants and Redox Signaling*, vol. 12, no. 12, pp. 1431–1470, 2010.

[47] S. A. Susin, H. K. Lorenzo, N. Zamzami et al., "Molecular characterization of mitochodrial apoptosis-inducing factor," *Nature*, vol. 397, no. 6718, pp. 441–446, 1999.

[48] R. S. Balaban, "The role of Ca^{2+} signaling in the coordination of mitochondrial ATP production with cardiac work," *Biochimica et Biophysica Acta*, vol. 1787, no. 11, pp. 1334–1341, 2009.

[49] M. E. Wernette, R. S. Ochs, and H. A. Lardy, "Ca^{2+} stimulation of rat liver mitochondrial glycerophosphate dehydrogenase," *The Journal of Biological Chemistry*, vol. 256, no. 24, pp. 12767–12771, 1981.

[50] J. G. McCormack and R. M. Denton, "Mitochondrial Ca^{2+} transport and the role of intramitochondrial Ca^{2+} in the regulation of energy metabolism," *Developmental Neuroscience*, vol. 15, no. 3–5, pp. 165–173, 1993.

[51] V. Mildaziene, R. Baniene, Z. Nauciene et al., "Calcium indirectly increases the control exerted by the adenine nucleotide translocator over 2-oxoglutarate oxidation in rat heart mitochondria," *Archives of Biochemistry and Biophysics*, vol. 324, no. 1, pp. 130–134, 1995.

[52] R. A. Haworth, D. R. Hunter, and H. A. Berkoff, "Contracture in isolated adult rat heart cells. Role of Ca^{2+}, ATP, and compartmentation," *Circulation Research*, vol. 49, no. 5, pp. 1119–1128, 1981.

[53] J. A. Copello, S. Barg, A. Sonnleitner et al., "Differential activation by Ca^{2+}, ATP and caffeine of cardiac and skeletal muscle ryanodine receptors after block by Mg^{2+}," *Journal of Membrane Biology*, vol. 187, no. 1, pp. 51–64, 2002.

[54] P. Nasr, H. I. Gursahani, Z. Pang et al., "Influence of cytosolic and mitochondrial Ca^{2+}, ATP, mitochondrial membrane potential, and calpain activity on the mechanism of neuron death induced by 3-nitropropionic acid," *Neurochemistry International*, vol. 43, no. 2, pp. 89–99, 2003.

[55] J. D. Johnston and M. D. Brand, "The mechanism of Ca^{2+} stimulation of citrulline and N-acetylglutamate synthesis by mitochondria," *Biochimica et Biophysica Acta*, vol. 1033, no. 1, pp. 85–90, 1990.

[56] J. D. McGivan, N. M. Bradford, and J. Mendes-Mourão, "The regulation of carbamoyl phosphate synthase activity in rat liver mitochondria," *Biochemical Journal*, vol. 154, no. 2, pp. 415–421, 1976.

[57] T. I. Peng and M. J. Jou, "Oxidative stress caused by mitochondrial calcium overload," *Annals of the New York Academy of Sciences*, vol. 1201, pp. 183–188, 2010.

[58] K. Lund and B. Ziola, "Cell sonicates used in the analysis of how measles and herpes simplex type 1 virus infections influence Vero cell mitochondrial calcium uptake," *Canadian Journal of Biochemistry and Cell Biology*, vol. 63, no. 11, pp. 1194–1197, 1985.

[59] Y. Li, D. F. Boehning, T. Qian, V. L. Popov, and S. A. Weinman, "Hepatitis C virus core protein increases mitochondrial ROS production by stimulation of Ca^{2+} uniporter activity," *FASEB Journal*, vol. 21, no. 10, pp. 2474–2485, 2007.

[60] R. V. Campbell, Y. Yang, T. Wang et al., "Effects of hepatitis C core protein on mitochondrial electron transport and production of reactive oxygen species," *Methods in Enzymology*, vol. 456, pp. 363–380, 2009.

[61] G. Gong, G. Waris, R. Tanveer, and A. Siddiqui, "Human hepatitis C virus NS5A protein alters intracellular calcium levels, induces oxidative stress, and activates STAT-3 and NF-κB," *Proceedings of the National Academy of Sciences of the United States of America*, vol. 98, no. 17, pp. 9599–9604, 2001.

[62] M. Kalamvoki and P. Mavromara, "Calcium-dependent calpain proteases are implicated in processing of the hepatitis C virus NS5A protein," *Journal of Virology*, vol. 78, no. 21, pp. 11865–11878, 2004.

[63] N. Dionisio, M. V. Garcia-Mediavilla, S. Sanchez-Campos et al., "Hepatitis C virus NS5A and core proteins induce oxidative stress-mediated calcium signalling alterations in hepatocytes," *Journal of Hepatology*, vol. 50, no. 5, pp. 872–882, 2009.

[64] M. K. Baum, S. Sales, D. T. Jayaweera et al., "Coinfection with hepatitis C virus, oxidative stress and antioxidant status in HIV-positive drug users in Miami," *HIV Medicine*, vol. 12, no. 2, pp. 78–86, 2011.

[65] G. A. Cook and S. J. Opella, "NMR studies of p7 protein from hepatitis C virus," *European Biophysics Journal*, vol. 39, no. 7, pp. 1097–1104, 2010.

[66] S. D. C. Griffin, R. Harvey, D. S. Clarke, W. S. Barclay, M. Harris, and D. J. Rowlands, "A conserved basic loop in hepatitis C virus p7 protein is required for amantadine-sensitive ion channel activity in mammalian cells but is dispensable for localization to mitochondria," *Journal of General Virology*, vol. 85, no. 2, pp. 451–461, 2004.

[67] M. J. Bouchard, L. H. Wang, and R. J. Schneider, "Calcium signaling by HBx protein in hepatitis B virus DNA replication," *Science*, vol. 294, no. 5550, pp. 2376–2378, 2001.

[68] Y. Choi, S. G. Park, J. H. Yoo, and G. Jung, "Calcium ions affect the hepatitis B virus core assembly," *Virology*, vol. 332, no. 1, pp. 454–463, 2005.

[69] M. Foti, L. Cartier, V. Piguet et al., "The HIV Nef protein alters Ca²⁺ signaling in myelomonocytic cells through SH3-mediated protein-protein interactions," *The Journal of Biological Chemistry*, vol. 274, no. 49, pp. 34765–34772, 1999.

[70] A. Manninen and K. Saksela, "HIV-1 Nef interacts with inositol trisphosphate receptor to activate calcium signaling in T cells," *Journal of Experimental Medicine*, vol. 195, no. 8, pp. 1023–1032, 2002.

[71] S. Kinoshita, L. Su, M. Amano, L. A. Timmerman, H. Kaneshima, and G. P. Nolan, "The T cell activation factor NF-ATc positively regulates HIV-1 replication and gene expression in T cells," *Immunity*, vol. 6, no. 3, pp. 235–244, 1997.

[72] M. C. Ruiz, J. Cohen, and F. Michelangeli, "Role of Ca²⁺ in the replication and pathogenesis of rotavirus and other viral infections," *Cell Calcium*, vol. 28, no. 3, pp. 137–149, 2000.

[73] P. Tian, M. K. Estes, Y. Hu, J. M. Ball, C. Q. Zeng, and W. P. Schilling, "The rotavirus nonstructural glycoprotein NSP4 mobilizes Ca²⁺ from the endoplasmic reticulum," *Journal of Virology*, vol. 69, no. 9, pp. 5763–5772, 1995.

[74] Y. Díaz, M. E. Chemello, R. Peña et al., "Expression of nonstructural rotavirus protein NSP4 mimics Ca²⁺ homeostasis changes induced by rotavirus infection in cultured cells," *Journal of Virology*, vol. 82, no. 22, pp. 11331–11343, 2008.

[75] J. L. Zambrano, Y. Díaz, F. Peña et al., "Silencing of rotavirus NSP4 or VP7 expression reduces alterations in Ca²⁺ homeostasis induced by infection of cultured cells," *Journal of Virology*, vol. 82, no. 12, pp. 5815–5824, 2008.

[76] M. C. Ruiz, O. C. Aristimuño, Y. Díaz et al., "Intracellular disassembly of infectious rotavirus particles by depletion of

[77] A. Irurzun, J. Arroyo, A. Alvarez, and L. Carrasco, "Enhanced intracellular calcium concentration during poliovirus infection," *Journal of Virology*, vol. 69, no. 8, pp. 5142–5146, 1995.

[78] R. Aldabe, A. Irurzun, and L. Carrasco, "Poliovirus protein 2BC increases cytosolic free calcium concentrations," *Journal of Virology*, vol. 71, no. 8, pp. 6214–6217, 1997.

[79] C. Brisac, F. Téoulé, A. Autret et al., "Calcium flux between the endoplasmic reticulum and mitochondrion contributes to poliovirus-induced apoptosis," *Journal of Virology*, vol. 84, no. 23, pp. 12226–12235, 2010.

[80] J. L. Nieva, A. Agirre, S. Nir, and L. Carrasco, "Mechanisms of membrane permeabilization by picornavirus 2B viroporin," *The FEBS Letters*, vol. 552, no. 1, pp. 68–73, 2003.

[81] F. J. M. van Kuppeveld, A. S. de Jong, W. J. G. Melchers, and P. H. G. M. Willems, "Enterovirus protein 2B po(u)res out the calcium: a viral strategy to survive?" *Trends in Microbiology*, vol. 13, no. 2, pp. 41–44, 2005.

[82] A. S. de Jong, H. J. Visch, F. de Mattia et al., "The coxsackievirus 2B protein increases efflux of ions from the endoplasmic reticulum and Golgi, thereby inhibiting protein trafficking through the Golgi," *The Journal of Biological Chemistry*, vol. 281, no. 20, pp. 14144–14150, 2006.

[83] A. S. de Jong, F. de Mattia, M. M. van Dommelen et al., "Functional analysis of picornavirus 2B proteins: effects on calcium homeostasis and intracellular protein trafficking," *Journal of Virology*, vol. 82, no. 7, pp. 3782–3790, 2008.

[84] F. J. M. van Kuppeveld, J. G. J. Hoenderop, R. L. L. Smeets et al., "Coxsackievirus protein 2B modifies endoplasmic reticulum membrane and plasma membrane permeability and facilitates virus release," *EMBO Journal*, vol. 16, no. 12, pp. 3519–3532, 1997.

[85] M. Campanella, A. S. de Jong, K. W. H. Lanke et al., "The coxsackievirus 2B protein suppresses apoptotic host cell responses by manipulating intracellular Ca²⁺ homeostasis," *The Journal of Biological Chemistry*, vol. 279, no. 18, pp. 18440–18450, 2004.

[86] P. Bozidis, C. D. Williamson, D. S. Wong, and A. M. Colberg-Poley, "Trafficking of UL37 proteins into mitochondrion-associated membranes during permissive human cytomegalovirus infection," *Journal of Virology*, vol. 84, no. 15, pp. 7898–7903, 2010.

[87] R. Sharon-Friling, J. Goodhouse, A. M. Colberg-Poley, and T. Shenk, "Human cytomegalovirus pUL37x1 induces the release of endoplasmic reticulum calcium stores," *Proceedings of the National Academy of Sciences of the United States of America*, vol. 103, no. 50, pp. 19117–19122, 2006.

[88] P. Pinton, D. Ferrari, E. Rapizzi, F. Di Virgilio, T. Pozzan, and R. Rizzuto, "The Ca²⁺ concentration of the endoplasmic reticulum is a key determinant of ceramide-induced apoptosis: significance for the molecular mechanism of Bcl-2 action," *EMBO Journal*, vol. 20, no. 11, pp. 2690–2701, 2001.

[89] A. R. Moise, J. R. Grant, T. Z. Vitalis, and W. A. Jefferies, "Adenovirus E3-6.7K maintains calcium homeostasis and prevents apoptosis and arachidonic acid release," *Journal of Virology*, vol. 76, no. 4, pp. 1578–1587, 2002.

[90] P. H. Chan, K. Niizuma, and H. Endo, "Oxidative stress and mitochondrial dysfunction as determinants of ischemic neuronal death and survival," *Journal of Neurochemistry*, vol. 109, no. 1, pp. 133–138, 2009.

[91] F. Muller, A. R. Crofts, and D. M. Kramer, "Multiple Q-cycle bypass reactions at the Qo site of the cytochrome bc1 complex," *Biochemistry*, vol. 41, no. 25, pp. 7866–7874, 2002.

[92] F. L. Muller, A. G. Roberts, M. K. Bowman, and D. M. Kramer, "Architecture of the Q-o site of the cytochrome bc1 complex probed by superoxide production," *Biochemistry*, vol. 42, no. 21, pp. 6493–6499, 2003.

[93] F. L. Muller, Y. Liu, and H. van Remmen, "Complex III releases superoxide to both sides of the inner mitochondrial membrane," *The Journal of Biological Chemistry*, vol. 279, no. 47, pp. 49064–49073, 2004.

[94] V. P. Skulachev, "Bioenergetic aspects of apoptosis, necrosis and mitoptosis," *Apoptosis*, vol. 11, no. 4, pp. 473–485, 2006.

[95] J. St-Pierre, J. A. Buckingham, S. J. Roebuck, and M. D. Brand, "Topology of superoxide production from different sites in the mitochondrial electron transport chain," *The Journal of Biological Chemistry*, vol. 277, no. 47, pp. 44784–44790, 2002.

[96] D. Han, F. Antunes, R. Canali, D. Rettori, and E. Cadenas, "Voltage-dependent anion channels control the release of the superoxide anion from mitochondria to cytosol," *The Journal of Biological Chemistry*, vol. 278, no. 8, pp. 5557–5563, 2003.

[97] S. Miwa, J. St-Pierre, L. Partridge, and M. D. Brand, "Superoxide and hydrogen peroxide production by Drosophila mitochondria," *Free Radical Biology and Medicine*, vol. 35, no. 8, pp. 938–948, 2003.

[98] H. Tsutsui, T. Ide, and S. Kinugawa, "Mitochondrial oxidative stress, DNA damage, and heart failure," *Antioxidants and Redox Signaling*, vol. 8, no. 9-10, pp. 1737–1744, 2006.

[99] D. F. Stowe and A. K. S. Camara, "Mitochondrial reactive oxygen species production in excitable cells: modulators of mitochondrial and cell function," *Antioxidants and Redox Signaling*, vol. 11, no. 6, pp. 1373–1414, 2009.

[100] H. Tsutsui, S. Kinugawa, and S. Matsushima, "Mitochondrial oxidative stress and dysfunction in myocardial remodelling," *Cardiovascular Research*, vol. 81, no. 3, pp. 449–456, 2009.

[101] J. M. Taylor, D. Quilty, L. Banadyga, and M. Barry, "The vaccinia virus protein F1L interacts with Bim and inhibits activation of the pro-apoptotic protein Bax," *The Journal of Biological Chemistry*, vol. 281, no. 51, pp. 39728–39739, 2006.

[102] M. Ott, J. D. Robertson, V. Gogvadze, B. Zhivotovsky, and S. Orrenius, "Cytochrome c release from mitochondria proceeds by a two-step process," *Proceedings of the National Academy of Sciences of the United States of America*, vol. 99, no. 3, pp. 1259–1263, 2002.

[103] S. Raha, A. T. Myint, L. Johnstone, and B. H. Robinson, "Control of oxygen free radical formation from mitochondrial complex I: roles for protein kinase A and pyruvate dehydrogenase kinase," *Free Radical Biology and Medicine*, vol. 32, no. 5, pp. 421–430, 2002.

[104] K. A. McGuire, A. U. Barlan, T. M. Griffin, and C. M. Wiethoff, "Adenovirus type 5 rupture of lysosomes leads to cathepsin B-dependent mitochondrial stress and production of reactive oxygen species," *Journal of Virology*, vol. 85, no. 20, pp. 10806–10813, 2011.

[105] S. Nishina, K. Hino, M. Korenaga et al., "Hepatitis C virus-induced reactive oxygen species raise hepatic iron level in mice by reducing hepcidin transcription," *Gastroenterology*, vol. 134, no. 1, pp. 226–238, 2008.

[106] N. S. R. de Mochel, S. Seronello, S. H. Wang et al., "Hepatocyte NAD(P)H oxidases as an endogenous source of reactive oxygen species during hepatitis C virus infection," *Hepatology*, vol. 52, no. 1, pp. 47–59, 2010.

[107] M. J. Hsieh, Y. S. Hsieh, T. Y. Chen, and H. L. Chiou, "Hepatitis C virus E2 protein induce reactive oxygen species (ROS)-related fibrogenesis in the HSC-T6 hepatic stellate cell line," *Journal of Cellular Biochemistry*, vol. 112, no. 1, pp. 233–243, 2010.

[108] K. Machida, G. Mcnamara, K. T. Cheng et al., "Hepatitis C virus inhibits DNA damage repair through reactive oxygen and nitrogen species and by interfering with the ATM-NBS1/Mre11/Rad50 DNA repair pathway in monocytes and hepatocytes," *Journal of Immunology*, vol. 185, no. 11, pp. 6985–6998, 2010.

[109] I. I. Kruman, A. Nath, and M. P. Mattson, "HIV-1 protein tat induces apoptosis of hippocampal neurons by a mechanism involving caspase activation, calcium overload, and oxidative stress," *Experimental Neurology*, vol. 154, no. 2, pp. 276–288, 1998.

[110] M. A. Baugh, "HIV: reactive oxygen species, enveloped viruses and hyperbaric oxygen," *Medical Hypotheses*, vol. 55, no. 3, pp. 232–238, 2000.

[111] L. Gil, A. Tarinas, D. Hernandez et al., "Altered oxidative stress indexes related to disease progression marker in human immunodeficiency virus infected patients with antiretroviral therapy," *Biomedicine and Aging Pathology*, vol. 1, no. 1, pp. 8–15, 2011.

[112] C. W. Pyo, Y. L. Yang, N. K. Yoo, and S. Y. Choi, "Reactive oxygen species activate HIV long terminal repeat via post-translational control of NF-κB," *Biochemical and Biophysical Research Communications*, vol. 376, no. 1, pp. 180–185, 2008.

[113] W. Lin, G. Wu, S. Li et al., "HIV and HCV cooperatively promote hepatic fibrogenesis via induction of reactive oxygen species and NF κB," *The Journal of Biological Chemistry*, vol. 286, no. 4, pp. 2665–2674, 2011.

[114] S. Lassoued, B. Gargouri, A. E. F. El Feki, H. Attia, and J. van Pelt, "Transcription of the epstein-barr virus lytic cycle activator BZLF-1 during oxidative stress induction," *Biological Trace Element Research*, vol. 137, no. 1, pp. 13–22, 2010.

[115] S. Lassoued, R. B. Ameur, W. Ayadi, B. Gargouri, R. B. Mansour, and H. Attia, "Epstein-Barr virus induces an oxidative stress during the early stages of infection in B lymphocytes, epithelial, and lymphoblastoid cell lines," *Molecular and Cellular Biochemistry*, vol. 313, no. 1-2, pp. 179–186, 2008.

[116] B. Gargouri, J. van Pelt, A. E. F. El Feki, H. Attia, and S. Lassoued, "Induction of Epstein-Barr virus (EBV) lytic cycle in vitro causes oxidative stress in lymphoblastoid B cell lines," *Molecular and Cellular Biochemistry*, vol. 324, no. 1-2, pp. 55–63, 2009.

[117] Y. J. Kim, J. K. Jung, S. Y. Lee, and K. L. Jang, "Hepatitis B virus X protein overcomes stress-induced premature senescence by repressing p16INK4a expression via DNA methylation," *Cancer Letters*, vol. 288, no. 2, pp. 226–235, 2010.

[118] L. Hu, L. Chen, G. Yang et al., "HBx sensitizes cells to oxidative stress-induced apoptosis by accelerating the loss of Mcl-1 protein via caspase-3 cascade," *Molecular Cancer*, vol. 10, article 43, 2011.

[119] S. Schaedler, J. Krause, K. Himmelsbach et al., "Hepatitis B virus induces expression of antioxidant response element-regulated genes by activation of Nrf2," *The Journal of Biological Chemistry*, vol. 285, no. 52, pp. 41074–41086, 2010.

[120] R. Srisuttee, S. S. Koh, E. H. Park et al., "Up-regulation of Foxo4 mediated by hepatitis B virus X protein confers resistance to oxidative stress-induced cell death," *International Journal of Molecular Medicine*, vol. 28, no. 2, pp. 255–260, 2011.

[121] A. Bhargava, S. Khan, H. Panwar et al., "Occult hepatitis B virus infection with low viremia induces DNA damage, apoptosis

and oxidative stress in peripheral blood lymphocytes," *Virus Research*, vol. 153, no. 1, pp. 143–150, 2010.

[122] Y. Ano, A. Sakudo, T. Kimata, R. Uraki, K. Sugiura, and T. Onodera, "Oxidative damage to neurons caused by the induction of microglial NADPH oxidase in encephalomyocarditis virus infection," *Neuroscience Letters*, vol. 469, no. 1, pp. 39–43, 2010.

[123] M. Colombini, E. Blachly-Dyson, and M. Forte, "VDAC, a channel in the outer mitochondrial membrane," *Ion channels*, vol. 4, pp. 169–202, 1996.

[124] M. Forte, E. Blachly-Dyson, and M. Colombini, "Structure and function of the yeast outer mitochondrial membrane channel, VDAC," *Society of General Physiologists Series*, vol. 51, pp. 145–154, 1996.

[125] S. Villinger, R. Briones, K. Giller et al., "Functional dynamics in the voltage-dependent anion channel," *Proceedings of the National Academy of Sciences of the United States of America*, vol. 107, no. 52, pp. 22546–22551, 2010.

[126] E. Pebay-Peyroula, C. Dahout-Gonzalez, R. Kahn, V. Trézéguet, G. J. Lauquin, and G. Brandolin, "Structure of mitochondrial ADP/ATP carrier in complex with carboxyatractyloside," *Nature*, vol. 426, no. 6962, pp. 39–44, 2003.

[127] D. R. Hunter and R. A. Haworth, "The Ca^{2+}-induced membrane transition in mitochondria. The protective mechanisms," *Archives of Biochemistry and Biophysics*, vol. 195, no. 2, pp. 453–459, 1979.

[128] K. D. Garlid, X. Sun, P. Paucek, and G. Woldegiorgis, "Mitochondrial cation transport systems," *Methods in Enzymology*, vol. 260, pp. 331–348, 1995.

[129] P. Bernardi, "Mitochondrial transport of cations: channels, exchangers, and permeability transition," *Physiological Reviews*, vol. 79, no. 4, pp. 1127–1155, 1999.

[130] A. P. Halestrap, "Calcium, mitochondria and reperfusion injury: a pore way to die," *Biochemical Society Transactions*, vol. 34, no. 2, pp. 232–237, 2006.

[131] K. Szydlowska and M. Tymianski, "Calcium, ischemia and excitotoxicity," *Cell Calcium*, vol. 47, no. 2, pp. 122–129, 2010.

[132] C. Piccoli, R. Scrima, G. Quarato et al., "Hepatitis C virus protein expression causes calcium-mediated mitochondrial bioenergetic dysfunction and nitro-oxidative stress," *Hepatology*, vol. 46, no. 1, pp. 58–65, 2007.

[133] M. Gac, J. Bigda, and T. W. Vahlenkamp, "Increased mitochondrial superoxide dismutase expression and lowered production of reactive oxygen species during rotavirus infection," *Virology*, vol. 404, no. 2, pp. 293–303, 2010.

[134] S. Carrère-Kremer, C. Montpellier-Pala, L. Cocquerel, C. Wychowski, F. Penin, and J. Dubuisson, "Subcellular localization and topology of the p7 polypeptide of hepatitis C virus," *Journal of Virology*, vol. 76, no. 8, pp. 3720–3730, 2002.

[135] M. E. Gonzalez and L. Carrasco, "Viroporins," *The FEBS Letters*, vol. 552, no. 1, pp. 28–34, 2003.

[136] D. Pavlovic, D. C. A. Neville, O. Argaud et al., "The hepatitis C virus p7 protein forms an ion channel that is inhibited by long-alkyl-chain iminosugar derivatives," *Proceedings of the National Academy of Sciences of the United States of America*, vol. 100, no. 10, pp. 6104–6108, 2003.

[137] A. Azuma, A. Matsuo, T. Suzuki, T. Kurosawa, X. Zhang, and Y. Aida, "Human immunodeficiency virus type 1 Vpr induces cell cycle arrest at the G1 phase and apoptosis via disruption of mitochondrial function in rodent cells," *Microbes and Infection*, vol. 8, no. 3, pp. 670–679, 2006.

[138] E. Jacotot, L. Ravagnan, M. Loeffler et al., "The HIV-1 viral protein R induces apoptosis via a direct effect on the mitochondrial permeability transition pore," *Journal of Experimental Medicine*, vol. 191, no. 1, pp. 33–46, 2000.

[139] A. Deniaud, C. Brenner, and G. Kroemer, "Mitochondrial membrane permeabilization by HIV-1 Vpr," *Mitochondrion*, vol. 4, no. 2-3, pp. 223–233, 2004.

[140] A. Macho, M. A. Calzado, L. Jiménez-Reina, E. Ceballos, J. León, and E. Muñoz, "Susceptibility of HIV-1-TAT transfected cells to undergo apoptosis. Biochemical mechanisms," *Oncogene*, vol. 18, no. 52, pp. 7543–7551, 1999.

[141] H. Everett, M. Barry, X. Sun et al., "The myxoma poxvirus protein, M11L, prevents apoptosis by direct interaction with the mitochondrial permeability transition pore," *Journal of Experimental Medicine*, vol. 196, no. 9, pp. 1127–1139, 2002.

[142] H. Everett, M. Barry, S. F. Lee et al., "M11L: a novel mitochondria-localized protein of myxoma virus that blocks apoptosis of infected leukocytes," *Journal of Experimental Medicine*, vol. 191, no. 9, pp. 1487–1498, 2000.

[143] J. L. Macen, K. A. Graham, S. F. Lee, M. Schreiber, L. K. Boshkov, and G. McFadden, "Expression of the myxoma virus tumor necrosis factor receptor homologue and M11L genes is required to prevent virus-induced apoptosis in infected rabbit T lymphocytes," *Virology*, vol. 218, no. 1, pp. 232–237, 1996.

[144] S. T. Wasilenko, T. L. Stewart, A. F. A. Meyers, and M. Barry, "Vaccinia virus encodes a previously uncharacterized mitochondrial-associated inhibitor of apoptosis," *Proceedings of the National Academy of Sciences of the United States of America*, vol. 100, no. 2, pp. 14345–14350, 2003.

[145] S. T. Wasilenko, L. Banadyga, D. Bond, and M. Barry, "The vaccinia virus F1L protein interacts with the proapoptotic protein Bak and inhibits Bak activation," *Journal of Virology*, vol. 79, no. 22, pp. 14031–14043, 2005.

[146] S. T. Wasilenko, A. F. A. Meyers, K. V. Helm, and M. Barry, "Vaccinia virus infection disarms the mitochondrion-mediated pathway of the apoptotic cascade by modulating the permeability transition pore," *Journal of Virology*, vol. 75, no. 23, pp. 11437–11448, 2001.

[147] K. Bruns, N. Studtrucker, A. Sharma et al., "Structural characterization and oligomerization of PB1-F2, a proapoptotic influenza A virus protein," *The Journal of Biological Chemistry*, vol. 282, no. 1, pp. 353–363, 2007.

[148] W. Chen, P. A. Calvo, D. Malide et al., "A novel influenza A virus mitochondrial protein that induces cell death," *Nature Medicine*, vol. 7, no. 12, pp. 1306–1312, 2001.

[149] J. S. Gibbs, D. Malide, F. Hornung, J. R. Bennink, and J. W. Yewdell, "The influenza A virus PB1-F2 protein targets the inner mitochondrial membrane via a predicted basic amphipathic helix that disrupts mitochondrial function," *Journal of Virology*, vol. 77, no. 13, pp. 7214–7224, 2003.

[150] M. Henkel, D. Mitzner, P. Henklein et al., "Proapoptotic influenza A virus protein PB1-F2 forms a nonselective ion channel," *PLoS ONE*, vol. 5, no. 6, Article ID e11112, 2010.

[151] M. Danishuddin, S. N. Khan, and A. U. Khan, "Molecular interactions between mitochondrial membrane proteins and the C-terminal domain of PB1-F2: an in silico approach," *Journal of Molecular Modeling*, vol. 16, no. 3, pp. 535–541, 2010.

[152] M. Silic-Benussi, O. Marin, R. Biasiotto, D. M. D'Agostino, and V. Ciminale, "Effects of human T-cell leukemia virus type 1 (HTLV-1) p13 on mitochondrial K$^+$ permeability: a new member of the viroporin family?" *The FEBS Letters*, vol. 584, no. 10, pp. 2070–2075, 2010.

[153] V. Ciminale, L. Zotti, D. M. D'Agostino et al., "Mitochondrial targeting of the p13(II) protein coded by the x-II ORF of human T-cell leukemia/lymphotropic virus type I (HTLV-I)," *Oncogene*, vol. 18, no. 31, pp. 4505–4514, 1999.

[154] R. Biasiotto, P. Aguiari, R. Rizzuto, P. Pinton, D. M. D'Agostino, and V. Ciminale, "The p13 protein of human T cell leukemia virus type 1 (HTLV-1) modulates mitochondrial membrane potential and calcium uptake," *Biochimica et Biophysica Acta*, vol. 1797, no. 6-7, pp. 945–951, 2010.

[155] M. Silic-Benussi, I. Cavallari, T. Zorzan et al., "Suppression of tumor growth and cell proliferation by p13II, a mitochondrial protein of human T cell leukemia virus type 1," *Proceedings of the National Academy of Sciences of the United States of America*, vol. 101, no. 17, pp. 6629–6634, 2004.

[156] W. A. Nudson, J. Rovnak, M. Buechner, and S. L. Quackenbush, "Walleye dermal sarcoma virus Orf C is targeted to the mitochondria," *Journal of General Virology*, vol. 84, no. 2, pp. 375–381, 2003.

[157] E. White, "Mechanisms of apoptosis regulation by viral oncogenes in infection and tumorigenesis," *Cell Death and Differentiation*, vol. 13, no. 8, pp. 1371–1377, 2006.

[158] L. Galluzzi, C. Brenner, E. Morselli, Z. Touat, and G. Kroemer, "Viral control of mitochondrial apoptosis," *PLoS Pathogens*, vol. 4, no. 5, Article ID e1000018, 2008.

[159] C. A. Benedict, P. S. Norris, and C. F. Ware, "To kill or be killed: viral evasion of apoptosis," *Nature Immunology*, vol. 3, no. 11, pp. 1013–1018, 2002.

[160] S. Hay and G. Kannourakis, "A time to kill: viral manipulation of the cell death program," *Journal of General Virology*, vol. 83, no. 7, pp. 1547–1564, 2002.

[161] J. F. Kerr, A. H. Wyllie, and A. R. Currie, "Apoptosis: a basic biological phenomenon with wide-ranging implications in tissue kinetics," *The British Journal of Cancer*, vol. 26, no. 4, pp. 239–257, 1972.

[162] E. Gulbins, S. Dreschers, and J. Bock, "Role of mitochondria in apoptosis," *Experimental Physiology*, vol. 88, no. 1, pp. 85–90, 2003.

[163] V. Borutaite, "Mitochondria as decision-makers in cell death," *Environmental and Molecular Mutagenesis*, vol. 51, no. 5, pp. 406–416, 2010.

[164] C. M. Sanfilippo and J. A. Blaho, "The facts of death," *International Reviews of Immunology*, vol. 22, no. 5-6, pp. 327–340, 2003.

[165] X. Liu, C. N. Kim, J. Yang, R. Jemmerson, and X. Wang, "Induction of apoptotic program in cell-free extracts: requirement for dATP and cytochrome c," *Cell*, vol. 86, no. 1, pp. 147–157, 1996.

[166] C. Castanier and D. Arnoult, "Mitochondrial dynamics during apoptosis," *Medecine/Sciences*, vol. 26, no. 10, pp. 830–835, 2010.

[167] H. Zou, W. J. Henzel, X. Liu, A. Lutschg, and X. Wang, "Apaf-1, a human protein homologous to C. elegans CED-4, participates in cytochrome c-dependent activation of caspase-3," *Cell*, vol. 90, no. 3, pp. 405–413, 1997.

[168] M. Karbowski, "Mitochondria on guard: role of mitochondrial fusion and fission in the regulation of apoptosis," *Advances in Experimental Medicine and Biology*, vol. 687, pp. 131–142, 2010.

[169] X. M. Sun, M. MacFarlane, J. Zhuang, B. B. Wolf, D. R. Green, and G. M. Cohen, "Distinct caspase cascades are initiated in receptor-mediated and chemical-induced apoptosis," *The Journal of Biological Chemistry*, vol. 274, no. 8, pp. 5053–5060, 1999.

[170] A. Ashkenazi and V. M. Dixit, "Death receptors: signaling and modulation," *Science*, vol. 281, no. 5381, pp. 1305–1308, 1998.

[171] K. F. Ferri and G. Kroemer, "Organelle-specific initiation of cell death pathways," *Nature Cell Biology*, vol. 3, no. 11, pp. E255–E263, 2001.

[172] L. Ravagnan, T. Roumier, and G. Kroemer, "Mitochondria, the killer organelles and their weapons," *Journal of Cellular Physiology*, vol. 192, no. 2, pp. 131–137, 2002.

[173] S. Ohta, "A multi-functional organelle mitochondrion is involved in cell death, proliferation and disease," *Current Medicinal Chemistry*, vol. 10, no. 23, pp. 2485–2494, 2003.

[174] N. N. Danial, A. Gimenez-Cassina, and D. Tondera, "Homeostatic functions of BCL-2 proteins beyond apoptosis," *Advances in Experimental Medicine and Biology*, vol. 687, pp. 1–32, 2010.

[175] M. E. Soriano and L. Scorrano, "The interplay between BCL-2 family proteins and mitochondrial morphology in the regulation of apoptosis," *Advances in Experimental Medicine and Biology*, vol. 687, pp. 97–114, 2010.

[176] S. Krishna, I. C. C. Low, and S. Pervaiz, "Regulation of mitochondrial metabolism: yet another facet in the biology of the oncoprotein Bcl-2," *Biochemical Journal*, vol. 435, no. 3, pp. 545–551, 2011.

[177] F. Llambi and D. R. Green, "Apoptosis and oncogenesis: give and take in the BCL-2 family," *Current Opinion in Genetics and Development*, vol. 21, no. 1, pp. 12–20, 2011.

[178] L. Scorrano and S. J. Korsmeyer, "Mechanisms of cytochrome c release by proapoptotic BCL-2 family members," *Biochemical and Biophysical Research Communications*, vol. 304, no. 3, pp. 437–444, 2003.

[179] M. Crompton, "Bax, Bid and the permeabilization of the mitochondrial outer membrane in apoptosis," *Current Opinion in Cell Biology*, vol. 12, no. 4, pp. 414–419, 2000.

[180] N. J. Waterhouse, J. E. Ricci, and D. R. Green, "And all of a sudden it's over: mitochondrial outer-membrane permeabilization in apoptosis," *Biochimie*, vol. 84, no. 2-3, pp. 113–121, 2002.

[181] A. S. Belzacq, H. L. A. Vieira, F. Verrier et al., "Bcl-2 and Bax modulate adenine nucleotide translocase activity," *Cancer Research*, vol. 63, no. 2, pp. 541–546, 2003.

[182] N. Zamzami and G. Kroemer, "Apoptosis: mitochondrial membrane permeabilization—the (w)hole story?" *Current Biology*, vol. 13, no. 2, pp. R71–R73, 2003.

[183] G. Paradies, G. Petrosillo, V. Paradies, and F. M. Ruggiero, "Role of cardiolipin peroxidation and Ca^{2+} in mitochondrial dysfunction and disease," *Cell Calcium*, vol. 45, no. 6, pp. 643–650, 2009.

[184] A. Cuconati and E. White, "Viral homologs of BCL-2: role of apoptosis in the regulation of virus infection," *Genes and Development*, vol. 16, no. 19, pp. 2465–2478, 2002.

[185] B. J. Thomson, "Viruses and apoptosis," *International Journal of Experimental Pathology*, vol. 82, no. 2, pp. 65–76, 2001.

[186] D. Perez and E. White, "TNF-α signals apoptosis through a bid-dependent conformational change in Bax that is inhibited by E1B 19K," *Molecular Cell*, vol. 6, no. 1, pp. 53–63, 2000.

[187] B. M. Pützer, T. Stiewe, K. Parssanedjad, S. Rega, and H. Esche, "E1A is sufficient by itself to induce apoptosis independent of p53 and other adenoviral gene products," *Cell Death and Differentiation*, vol. 7, no. 2, pp. 177–188, 2000.

[188] L. Banadyga, J. Gerig, T. Stewart, and M. Barry, "Fowlpox virus encodes a Bcl-2 homologue that protects cells from apoptotic death through interaction with the proapoptotic protein bak," *Journal of Virology*, vol. 81, no. 20, pp. 11032–11045, 2007.

[189] A. Brun, C. Rivas, M. Esteban, J. M. Escribano, and C. Alonso, "African swine fever virus gene A179L, a viral homologue of bcl-2, protects cells from programmed cell death," *Virology*, vol. 225, no. 1, pp. 227–230, 1996.

[190] Y. Revilla, A. Cebrián, E. Baixerás, C. Martínez, E. Viñuela, and M. L. Salas, "Inhibition of apoptosis by the African swine fever virus Bcl-2 homologue: role of the BH1 domain," *Virology*, vol. 228, no. 2, pp. 400–404, 1997.

[191] T. Derfuss, H. Fickenscher, M. S. Kraft et al., "Antiapoptotic activity of the herpesvirus saimiri-encoded Bcl-2 homolog: stabilization of mitochondria and inhibition of caspase-3-like activity," *Journal of Virology*, vol. 72, no. 7, pp. 5897–5904, 1998.

[192] W. L. Marshall, C. Yim, E. Gustafson et al., "Epstein-Barr virus encodes a novel homolog of the bcl-2 oncogene that inhibits apoptosis and associates with Bax and Bak," *Journal of Virology*, vol. 73, no. 6, pp. 5181–5185, 1999.

[193] X. M. Yin, Z. N. Oltvai, and S. J. Korsmeyer, "BH1 and BH2 domains of Bcl-2 are required for inhibition of apoptosis and heterodimerization with Bax," *Nature*, vol. 369, no. 6478, pp. 321–323, 1994.

[194] Z. Rahmani, K. W. Huh, R. Lasher, and A. Siddiqui, "Hepatitis B virus X protein colocalizes to mitochondria with a human voltage-dependent anion channel, HVDAC3, and alters its transmembrane potential," *Journal of Virology*, vol. 74, no. 6, pp. 2840–2846, 2000.

[195] Y. W. Lu and W. N. Chen, "Human hepatitis B virus X protein induces apoptosis in HepG2 cells: role of BH3 domain," *Biochemical and Biophysical Research Communications*, vol. 338, no. 3, pp. 1551–1556, 2005.

[196] Y. Tanaka, F. Kanai, T. Kawakami et al., "Interaction of the hepatitis B virus X protein (HBx) with heat shock protein 60 enhances HBx-mediated apoptosis," *Biochemical and Biophysical Research Communications*, vol. 318, no. 2, pp. 461–469, 2004.

[197] J. Diao, A. A. Khine, F. Sarangi et al., "X protein of hepatitis B virus inhibits Fas-mediated apoptosis and is associated with up-regulation of the SAPK/JNK pathway," *The Journal of Biological Chemistry*, vol. 276, no. 11, pp. 8328–8340, 2001.

[198] A. S. Kekule, U. Lauer, L. Weiss, B. Luber, and P. H. Hofschneider, "Hepatitis B virus transactivator HBx uses a tumour promoter signalling pathway," *Nature*, vol. 361, no. 6414, pp. 742–745, 1993.

[199] F. Su and R. J. Schneider, "Hepatitis B virus HBx protein activates transcription factor NF-κB by acting on multiple cytoplasmic inhibitors of rel-related proteins," *Journal of Virology*, vol. 70, no. 7, pp. 4558–4566, 1996.

[200] J. Benn, F. Su, M. Doria, and R. J. Schneider, "Hepatitis B virus HBx protein induces transcription factor AP-1 by activation of extracellular signal-regulated and c-Jun N-terminal mitogen-activated protein kinases," *Journal of Virology*, vol. 70, no. 8, pp. 4978–4985, 1996.

[201] F. Henkler, A. R. Lopes, M. Jones, and R. Koshy, "Erk-independent partial activation of AP-1 sites by the hepatitis B virus HBx protein," *Journal of General Virology*, vol. 79, no. 11, pp. 2737–2742, 1998.

[202] W. L. Shih, M. L. Kuo, S. E. Chuang, A. L. Cheng, and S. L. Doong, "Hepatitis b virus x protein inhibits transforming growth factor-β-induced apoptosis through the activation of phosphatidylinositol 3-kinase pathway," *The Journal of Biological Chemistry*, vol. 275, no. 33, pp. 25858–25864, 2000.

[203] J. Komano, M. Sugiura, and K. Takada, "Epstein-barr virus contributes to the malignant phenotype and to apoptosis resistance in Burkitt's lymphoma cell line Akata," *Journal of Virology*, vol. 72, no. 11, pp. 9150–9156, 1998.

[204] D. S. Bellows, M. Howell, C. Pearson, S. A. Hazlewood, and J. M. Hardwick, "Epstein-Barr virus BALF1 is a BCL-2-like antagonist of the herpesvirus antiapoptotic BCL-2 proteins," *Journal of Virology*, vol. 76, no. 5, pp. 2469–2479, 2002.

[205] A. M. Flanagan and A. Letai, "BH3 domains define selective inhibitory interactions with BHRF-1 and KSHV BCL-2," *Cell Death and Differentiation*, vol. 15, no. 3, pp. 580–588, 2008.

[206] M. Thomas and L. Banks, "Human papillomavirus (HPV) E6 interactions with Bak are conserved amongst E6 proteins from high and low risk HPV types," *Journal of General Virology*, vol. 80, no. 6, pp. 1513–1517, 1999.

[207] S. Jackson, C. Harwood, M. Thomas, L. Banks, and A. Storey, "Role of Bak in UV-induced apoptosis in skin cancer and abrogation by HPV E6 proteins," *Genes and Development*, vol. 14, no. 23, pp. 3065–3073, 2000.

[208] S. Leverrier, D. Bergamaschi, L. Ghali et al., "Role of HPV E6 proteins in preventing UVB-induced release of pro-apoptotic factors from the mitochondria," *Apoptosis*, vol. 12, no. 3, pp. 549–560, 2007.

[209] Z. M. Sun, Y. Xiao, L. L. Ren, X. B. Lei, and J. W. Wang, "Enterovirus 71 induces apoptosis in a Bax dependent manner," *Zhonghua Shi Yan He Lin Chuang Bing Du Xue Za Zhi*, vol. 25, no. 1, pp. 49–52, 2011.

[210] C. S. Ilkow, I. S. Goping, and T. C. Hobman, "The rubella virus capsid is an anti-apoptotic protein that attenuates the pore-forming ability of Bax," *PLoS Pathogens*, vol. 7, no. 2, Article ID e1001291, 2011.

[211] V. S. Goldmacher, L. M. Bartle, A. Skaletskaya et al., "A cytomegalovirus-encoded mitochondria-localized inhibitor of apoptosis structurally unrelated to Bcl-2," *Proceedings of the National Academy of Sciences of the United States of America*, vol. 96, no. 22, pp. 12536–12541, 1999.

[212] D. Arnoult, L. M. Bartle, A. Skaletskaya et al., "Cytomegalovirus cell death suppressor vMIA blocks Bax- but not Bak-mediated apoptosis by binding and sequestering Bax at mitochondria," *Proceedings of the National Academy of Sciences of the United States of America*, vol. 101, no. 21, pp. 7988–7993, 2004.

[213] D. Poncet, N. Larochette, A. Pauleau et al., "An anti-apoptotic viral protein that recruits Bax to mitochondria," *The Journal of Biological Chemistry*, vol. 279, no. 21, pp. 22605–22614, 2004.

[214] H. L. A. Vieira, A. S. Belzacq, D. Haouzi et al., "The adenine nucleotide translocator: a target of nitric oxide, peroxynitrite, and 4-hydroxynonenal," *Oncogene*, vol. 20, no. 32, pp. 4305–4316, 2001.

[215] P. Boya, M. C. Morales, R. Gonzalez-Polo et al., "The chemopreventive agent N-(4-hydroxyphenyl)retinamide induces apoptosis through a mitochondrial pathway regulated by proteins from the Bcl-2 family," *Oncogene*, vol. 22, no. 40, pp. 6220–6230, 2003.

[216] A. L. McCormick, V. L. Smith, D. Chow, and E. S. Mocarski, "Disruption of mitochondrial networks by the human cytomegalovirus UL37 gene product viral mitochondrion-localized inhibitor of apoptosis," *Journal of Virology*, vol. 77, no. 1, pp. 631–641, 2003.

[217] M. G. Katze, Y. He, and M. Gale Jr., "Viruses and interferon: a fight for supremacy," *Nature Reviews Immunology*, vol. 2, no. 9, pp. 675–687, 2002.

[218] M. Yoneyama, M. Kikuchi, T. Natsukawa et al., "The RNA helicase RIG-I has an essential function in double-stranded RNA-induced innate antiviral responses," *Nature Immunology*, vol. 5, no. 7, pp. 730–737, 2004.

[219] J. Andrejeva, K. S. Childs, D. F. Young et al., "The V proteins of paramyxoviruses bind the IFN-inducible RNA helicase, mda-5, and inhibit its activation of the IFN-β promoter," *Proceedings of the National Academy of Sciences of the United States of America*, vol. 101, no. 49, pp. 17264–17269, 2004.

[220] T. Maniatis, J. V. Falvo, T. H. Kim et al., "Structure and function of the interferon-β enhanceosome," *Cold Spring Harbor Symposia on Quantitative Biology*, vol. 63, pp. 609–620, 1998.

[221] I. Scott, "The role of mitochondria in the mammalian antiviral defense system," *Mitochondrion*, vol. 10, no. 4, pp. 316–320, 2010.

[222] C. Castanier and D. Arnoult, "Mitochondrial localization of viral proteins as a means to subvert host defense," *Biochimica et Biophysica Acta*, vol. 1813, no. 4, pp. 575–583, 2011.

[223] C. Wang, X. Liu, and B. Wei, "Mitochondrion: an emerging platform critical for host antiviral signaling," *Expert Opinion on Therapeutic Targets*, vol. 15, no. 5, pp. 647–665, 2011.

[224] R. B. Seth, L. Sun, and Z. J. Chen, "Antiviral innate immunity pathways," *Cell Research*, vol. 16, no. 2, pp. 141–147, 2006.

[225] L. G. Xu, Y. Y. Wang, K. J. Han, L. Y. Li, Z. Zhai, and H. B. Shu, "VISA is an adapter protein required for virus-triggered IFN-β signaling," *Molecular Cell*, vol. 19, no. 6, pp. 727–740, 2005.

[226] T. Kawai, K. Takahashi, S. Sato et al., "IPS-1, an adaptor triggering RIG-I- and Mda5-mediated type I interferon induction," *Nature Immunology*, vol. 6, no. 10, pp. 981–988, 2005.

[227] E. Meylan, J. Curran, K. Hofmann et al., "Cardif is an adaptor protein in the RIG-I antiviral pathway and is targeted by hepatitis C virus," *Nature*, vol. 437, no. 7062, pp. 1167–1172, 2005.

[228] Y. Xu, H. Zhong, and W. Shi, "MAVS protects cells from apoptosis by negatively regulating VDAC1," *Molecular and Cellular Biochemistry*, vol. 375, no. 1-2, p. 219, 2010.

[229] X. D. Li, L. Sun, R. B. Seth, G. Pineda, and Z. J. Chen, "Hepatitis C virus protease NS3/4A cleaves mitochondrial antiviral signaling protein off the mitochondria to evade innate immunity," *Proceedings of the National Academy of Sciences of the United States of America*, vol. 102, no. 49, pp. 17717–17722, 2005.

[230] E. Foy, K. Li, C. Wang et al., "Regulation of interferon regulatory factor-3 by the hepatitis C virus serine protease," *Science*, vol. 300, no. 5622, pp. 1145–1148, 2003.

[231] A. Breiman, N. Grandvaux, R. Lin et al., "Inhibition of RIG-I-dependent signaling to the interferon pathway during hepatitis C virus expression and restoration of signaling by IKKε," *Journal of Virology*, vol. 79, no. 7, pp. 3969–3978, 2005.

[232] E. Foy, K. Li, R. Sumpter Jr. et al., "Control of antiviral defenses through hepatitis C virus disruption of retinoic acid-inducible gene-I signaling," *Proceedings of the National Academy of Sciences of the United States of America*, vol. 102, no. 8, pp. 2986–2991, 2005.

[233] B. Beames, D. Chavez, and R. E. Lanford, "GB virus B as a model for hepatitis C virus," *ILAR Journal*, vol. 42, no. 2, pp. 152–160, 2001.

[234] Z. Chen, Y. Benureau, R. Rijnbrand et al., "GB virus B disrupts RIG-I signaling by NS3/4A-mediated cleavage of the adaptor protein MAVS," *Journal of Virology*, vol. 81, no. 2, pp. 964–976, 2007.

[235] T. Öhman, J. Rintahaka, N. Kalkkinen, S. Matikainen, and T. A. Nyman, "Actin and RIG-I/MAVS signaling components translocate to mitochondria upon influenza a virus infection of human primary macrophages," *Journal of Immunology*, vol. 182, no. 9, pp. 5682–5692, 2009.

[236] D. A. Matthews and W. C. Russell, "Adenovirus core protein V interacts with p32—a protein which is associated with both the mitochondria and the nucleus," *Journal of General Virology*, vol. 79, no. 7, pp. 1677–1685, 1998.

[237] S. Cen, A. Khorchid, H. Javanbakht et al., "Incorporation of lysyl-tRNA synthetase into human immunodeficiency virus type 1," *Journal of Virology*, vol. 75, no. 11, pp. 5043–5048, 2001.

[238] E. Tolkunova, H. Park, J. Xia, M. P. King, and E. Davidson, "The human lysyl-tRNA synthetase gene encodes both the cytoplasmic and mitochondrial enzymes by means of an unusual: alternative splicing of the primary transcript," *The Journal of Biological Chemistry*, vol. 275, no. 45, pp. 35063–35069, 2000.

[239] M. Kaminska, V. Shalak, M. Francin, and M. Mirande, "Viral hijacking of mitochondrial lysyl-tRNA synthetase," *Journal of Virology*, vol. 81, no. 1, pp. 68–73, 2007.

[240] L. A. Stark and R. T. Hay, "Human immunodeficiency virus type 1 (HIV-1) viral protein R (Vpr) interacts with Lys-tRNA synthetase: implications for priming of HIV-1 reverse transcription," *Journal of Virology*, vol. 72, no. 4, pp. 3037–3044, 1998.

[241] L. Q. Qiu, P. Cresswell, and K. C. Chin, "Viperin is required for optimal Th2 responses and T-cell receptor-mediated activation of NF-κB and AP-1," *Blood*, vol. 113, no. 15, pp. 3520–3529, 2009.

[242] X. Wang, E. R. Hinson, and P. Cresswell, "The interferon-inducible protein viperin inhibits influenza virus release by perturbing lipid rafts," *Cell Host and Microbe*, vol. 2, no. 2, pp. 96–105, 2007.

[243] J. Y. Seo, R. Yaneva, E. R. Hinson, and P. Cresswell, "Human cytomegalovirus directly induces the antiviral protein viperin to enhance infectivity," *Science*, vol. 332, no. 6033, pp. 1093–1097, 2011.

[244] S. Kim, H. Y. Kim, S. Lee et al., "Hepatitis B virus X protein induces perinuclear mitochondrial clustering in microtubule-and dynein-dependent manners," *Journal of Virology*, vol. 81, no. 4, pp. 1714–1726, 2007.

[245] Y. Nomura-Takigawa, M. Nagano-Fujii, L. Deng et al., "Nonstructural protein 4A of Hepatitis C virus accumulates on mitochondria and renders the cells prone to undergoing mitochondria-mediated apoptosis," *Journal of General Virology*, vol. 87, no. 7, pp. 1935–1945, 2006.

[246] J. S. Radovanović, V. Todorović, I. Boričić, M. Janković-Hladni, and A. Korać, "Comparative ultrastructural studies on mitochondrial pathology in the liver of AIDS patients: clusters of mitochondria, protuberances, "minimitochondria," vacuoles, and virus-like particles," *Ultrastructural Pathology*, vol. 23, no. 1, pp. 19–24, 1999.

[247] G. Rojo, M. Chamorro, M. L. Salas, E. Vinuela, J. M. Cuezva, and J. Salas, "Migration of mitochondria to viral assembly sites in African swine fever virus-infected cells," *Journal of Virology*, vol. 72, no. 9, pp. 7583–7588, 1998.

[248] D. C. Kelly, "Frog virus 3 replication: electron microscope observations on the sequence of infection in chick embryo fibroblasts," *Journal of General Virology*, vol. 26, no. 1, pp. 71–86, 1975.

[249] R. S. Fujinami and M. B. A. Oldstone, "Amino acid homology between the encephalitogenic site of myelin basic protein and virus: mechanism for autoimmunity," *Science*, vol. 230, no. 4729, pp. 1043–1045, 1985.

[250] A. P. Kohm, K. G. Fuller, and S. D. Miller, "Mimicking the way to autoimmunity: an evolving theory of sequence and structural homology," *Trends in Microbiology*, vol. 11, no. 3, pp. 101–105, 2003.

[251] M. Monné, A. J. Robinson, C. Boes, M. E. Harbour, I. M. Fearnley, and E. R. S. Kunji, "The mimivirus genome encodes a mitochondrial carrier that transports dATP and dTTP," *Journal of Virology*, vol. 81, no. 7, pp. 3181–3186, 2007.

[252] H. A. Saffran, J. M. Pare, J. A. Corcoran, S. K. Weller, and J. R. Smiley, "Herpes simplex virus eliminates host mitochondrial DNA," *EMBO Reports*, vol. 8, no. 2, pp. 188–193, 2007.

[253] J. A. Corcoran, H. A. Saffran, B. A. Duguay, and J. R. Smiley, "Herpes simplex virus UL12.5 targets mitochondria through a mitochondrial localization sequence proximal to the N terminus," *Journal of Virology*, vol. 83, no. 6, pp. 2601–2610, 2009.

[254] A. Wiedmer, P. Wang, J. Zhou et al., "Epstein-Barr virus immediate-early protein Zta co-opts mitochondrial single-stranded DNA binding protein to promote viral and inhibit mitochondrial DNA replication," *Journal of Virology*, vol. 82, no. 9, pp. 4647–4655, 2008.

[255] K. Machida, K. T. Cheng, C. K. Lai, K. S. Jeng, V. M. Sung, and M. M. C. Lai, "Hepatitis C virus triggers mitochondrial permeability transition with production of reactive oxygen species, leading to DNA damage and STATS activation," *Journal of Virology*, vol. 80, no. 14, pp. 7199–7207, 2006.

[256] C. de Mendoza, L. Martin-Carbonero, P. Barreiro et al., "Mitochondrial DNA depletion in HIV-infected patients with chronic hepatitis C and effect of pegylated interferon plus ribavirin therapy," *AIDS*, vol. 21, no. 5, pp. 583–588, 2007.

3

Cellular Factors Implicated in Filovirus Entry

Suchita Bhattacharyya,[1] **and Thomas J. Hope**[2]

[1] *University of Mumbai and Department of Atomic Energy-Centre for Excellence in Basic Sciences, Health Centre Building, Vidyanagari, Kalina, Santacruz East, Mumbai 400098, India*
[2] *Department of Cell and Molecular Biology, Feinberg School of Medicine, Northwestern University, 303 East Superior Avenue, Chicago, IL 60611, USA*

Correspondence should be addressed to Thomas J. Hope; thope@northwestern.edu

Academic Editor: Amiya K. Banerjee

Although filoviral infections are still occurring in different parts of the world, there are no effective preventive or treatment strategies currently available against them. Not only do filoviruses cause a deadly infection, but they also have the potential of being used as biological weapons. This makes it imperative to comprehensively study these viruses in order to devise effective strategies to prevent the occurrence of these infections. Entry is the foremost step in the filoviral replication cycle and different studies have reported the involvement of a myriad of cellular factors including plasma membrane components, cytoskeletal proteins, endosomal components, and cytosolic factors in this process. Signaling molecules such as the TAM family of receptor tyrosine kinases comprising of Tyro3, Axl, and Mer have also been implicated as putative entry factors. Additionally, filoviruses are suggested to bind to a common receptor and recent studies have proposed T-cell immunoglobulin and mucin domain 1 (TIM-1) and Niemann-Pick C1 (NPC1) as potential receptor candidates. This paper summarizes the existing literature on filoviral entry with a special focus on cellular factors involved in this process and also highlights some fundamental questions. Future research aimed at answering these questions could be very useful in designing novel antiviral therapeutics.

1. Introduction

The *Filoviridae* family comprises of three genera: *Ebolavirus*, *Marburgvirus*, and *Cuevavirus* (tentative). These enveloped viruses are nonsegmented with negative-sense RNA and produce filamentous virions, which are pleomorphic in shape [1]. *Ebolavirus* has five known species: Zaire (EBOV), Sudan (SUDV), Reston (RESTV), Tai Forest (TAFV), and Bundibugyo (BDBV) while *Marburgvirus* has only one species: Marburg virus (MARV) [2–9]. EBOV and MARV and are known to be serologically, biochemically, and genetically distinct [10, 11].

The filoviral genome encodes seven structural proteins: envelope glycoprotein (GP), major matrix protein (VP40), nucleoprotein (NP), polymerase cofactor (VP35), replication/transcription protein (VP30), minor matrix protein (VP24), and RNA dependent DNA polymerase (L). In addition to this, EBOV also expresses a small, secreted, nonstructural glycoprotein (sGP) (see [12] for a comprehensive review).

Filoviruses are transmitted through contact with infected blood or body fluids [13] and can infect many cell types across different host species with lymphocytes being the notable exception [14, 15]. Although filoviruses are known to be pantropic, their preferred target cells include hepatocytes, dendritic cells, endothelial cells, macrophages, and monocytes (see [16] for a detailed review). Several species of fruit bats are suggested to act reservoirs for these viruses [17–21] and destroying these reservoirs could help to curtail the spread of these viruses. EBOV and MARV cause a fatal form of hemorrhagic fever [2, 6, 9, 12] and there are no vaccines or drugs currently available against them. Moreover, the US Centers for Disease Control and Prevention (CDC) has classified filoviruses as possible weapons for bioterrorism [22]. Therefore, these viruses need to be studied under Biosafety Level 4 conditions, which restricts the number of research laboratories that can work with these infectious viruses.

Entry is the earliest step in the viral replication cycle and the filoviral entry process broadly involves the following steps: binding of the virus to its receptor(s)/attachment factors on the cell surface; uptake of the virus; intracellular trafficking of the virus in endosomes via clathrin, macropinocytic and/or caveolae-mediated endocytic pathways; viral fusion and release of the nucleocapsid into the cytoplasm. Earlier reports investigating various steps of the entry process have yielded conflicting results with various studies implicating or refuting the involvement of different cellular factors and endocytic pathways in this process. This paper summarizes the key findings underlying the various steps involved in filoviral entry with a special focus on the cellular factors implicated in this process and also discusses some unresolved issues in this field.

2. The Filoviral GP Mediates Entry into Target Cells

The filoviral GP is the only protein present on the virus surface and facilitates receptor binding as well as fusion of the virus envelope with the host cell membrane [23]. GP is a type I transmembrane glycoprotein encoded by the fourth gene from the 3′ end of the genome [24], and is expressed as homotrimers, which form spikes on the surface of virus. Folding and assembly of EBOV GP trimers occurs independently of other viral proteins [25].

GP is expressed as a precursor protein that is post-translationally cleaved by a cellular proprotein convertase furin into GP1 (140kD) and GP2 (26kD) [26], which are linked by disulfide bonds. The GP1 subunit contains the receptor binding site and a heavily glycosylated mucin-like region (MLR), which facilitates viral attachment to target cells but is not required for viral entry *in vitro* [27]. The MLR also contains several epitopes, which are recognized by anti-GP antibodies to facilitate antibody-dependent enhancement of filoviral infection *in vitro* [28–30]. Furthermore, the crystal structure of EBOV GP demonstrates that the receptor binding site of GP1 is masked by a glycan cap and the MLR and therefore, removal of these regions could perhaps expose additional sites required for receptor/cofactor binding [31, 32]. The GP2 subunit contains two heptad repeat regions, which facilitate assembly of GP into trimers, a transmembrane anchor sequence, and the fusion loop [25, 33]. In MARV GP, the putative fusion domain is located 91 amino acids from the furin cleavage site [34]. The carboxy (C) terminus region of EBOV GP and MARV GP is very homologous and contains seven highly conserved cysteine residues, is high in proline content and has a short hydrophilic tail [24].

Despite the extensive homology at the C terminus, EBOV GP and MARV GP also exhibit several important distinctions. EBOV GP and MARV GP only share 31% identity in their amino acid sequence [35] and do not cross-react serologically [5]. Also, MARV GP is synthesized as a 170kD protein, which is encoded by a single open reading frame (ORF) [24, 36]; while EBOV GP is encoded in two ORFs

and expression of the full-length GP occurs by transcriptional RNA editing [4].

3. Cellular Plasma Membrane Components Involved in Attachment and Uptake of Filoviruses

Given the broad tissue tropism and host range of filoviruses, it was believed that the receptors of these viruses are ubiquitously expressed in most cells. Subsequently, beta 1 integrins [37] and several lectins such as DC-SIGN, DC-SIGNR, L-SIGN, and hMGL were shown to be involved in filovirus entry [38–42]. Matsuno and colleagues have demonstrated that the efficiency of C-type lectin mediated MARV entry differs between different strains [43] and that these lectins are not functional receptors for filoviral entry [44]. The role of another ubiquitous cellular factor folate receptor alpha in filoviral entry has been implicated and refuted by different groups [45, 46]. Two reports have suggested that the TAM family of receptor tyrosine kinases comprising of Tyro3, Axl and Mer are employed by filoviruses for entry [47, 48]. A more extensive analysis by Brindley and coworkers demonstrated that while Axl facilitated viral attachment and macropinocytic uptake of EBOV in several cell lines and primary cells, it did not bind to GP directly and hence is not a receptor for EBOV [49].

EBOV GP and MARV GP were initially suggested to bind to distinct cell surface residues for entry [14] and were also speculated to use different receptors for internalization into diverse cell types [50]. However, it is now known that these viruses bind to a common receptor [51–53].

Recently, T-cell immunoglobulin and mucin domain 1 (TIM-1) was reported to be a common receptor for EBOV and MARV [54]. TIM-1 is also known to bind to phosphatidyl serine, which is exposed on the surface of apoptotic cells and thereby facilitates phagocytosis of these cells [55]. Since viruses such as influenza are known to trigger expression of phosphatidyl serine on the surface of infected cells [56], it is possible that filoviruses also trigger expression of phosphatidyl serine on the surface of infected cells, which could then bind to TIM-1 and thereby facilitate viral uptake. Interestingly, TIM-1 is not expressed in macrophages and dendritic cells [57], which are the primary target cells of filoviral infection. Therefore, it is also possible that TIM-1 is merely one of many cellular factors that facilitate filoviral entry. The detailed mechanisms governing the interactions between filoviral GP and these cellular factors remain to be understood.

4. Cytoskeletal Components Involved in Filoviral Entry

The involvement of cytoskeletal proteins in EBOV entry has been widely reported. Using pseudotyped virus, Yonezawa and coworkers showed that microtubules and microfilaments are required for EBOV entry [58]. Similarly, Ruthel and colleagues demonstrated that the EBOV matrix protein VP40 directly associates with microtubules [59].

Several studies have also demonstrated the involvement of actin and actin regulatory factors in EBOV entry [60–62]. Using fluorescently labeled EBOV, Saeed and coworkers showed that phosphoinositide-3 kinase (PI3K), Akt, and Rac1 are required for entry [63]. Using WT Zaire EBOV, Kolokoltsov and colleagues demonstrated a requirement of calcium/calmodulin kinase (CAMK2) in entry [64]. All these studies also support the role of macropinocytosis in EBOV entry.

5. Involvement of Clathrin, Macropinocytosis, and Caveolae Endocytic Pathways in Filoviral Entry

Using chemical inhibitors to block endocytosis, several groups have shown that filoviruses are endocytosed in a pH-dependent manner [14, 65–67]. Clathrin, macropinocytic, and caveolae-mediated endocytic pathways have all been implicated to be involved in filoviral entry. A few studies have also reported the concomitant use of multiple endocytic pathways in filoviral entry. However, the relative contribution of each of these endocytic pathways in filoviral entry into different cell types is still unclear.

Using wild type as well as pseudotyped viruses, we and others have shown that filoviruses use clathrin-mediated endocytosis as an entry pathway [66–69]. We also performed a comprehensive analysis of the clathrin pathway using HIV pseudotyped with EBOV GP or MARV GP and found that filoviruses have a common requirement for several cellular factors of this pathway including clathrin heavy chain (CHC), phosphatidylinositol binding clathrin assembly protein (PICALM), epsin1, intersectin 1, dynamin 2, NUMB, low density lipoprotein receptor adaptor protein 1 (LDL-RAP1), inositol polyphosphate phosphatase-like 1 (INPPL1), RALBP1-associated Eps domain containing 1 (REPS1), and RALBP1-associated Eps domain containing 2 (REPS2). Interestingly, while EBOV GP mediated entry was found to require Eps15, AP-2, and DAB2; MARV GP mediated entry was independent of these cellular factors and instead required Arrestin, beta 1 (ARRB1) [68]. This differential requirement for key components of the clathrin pathway in EBOV GP versus MARV GP mediated entry could perhaps be attributed to the differences in the composition of the GPs of these two viruses or the usage of additional cellular factors/coreceptors during entry.

Numerous groups have also described a role of macropinocytosis in EBOV entry [47, 60, 61, 69, 70]. Using biologically contained virions and virus-like particles (VLPs), Nanbo and colleagues showed that EBOV virions co-localize with sorting nexin (SNX) 5, which is a constituent of macropinosomes [70]. Hunt and colleagues demonstrated that Axl enhances macropinocytic uptake of EBOV [47]. Other cellular factors that were implicated in EBOV entry via macropinocytosis include p21-activated kinase (Pak1), ADP-ribosylation factor 6 (Arf6), C-terminal-binding protein 1 (CtBP1), Protein kinase C (PKC), and Phospholipase C (PLC) [47, 60, 61]. The role of dynamin 2 in macropinocytic entry of filoviruses was implicated and refuted by different groups

[60, 61, 70]. Also, macropinocytic uptake of EBOV was shown to be independent of viral morphology [61]. However, the role of filoviral morphology in entry by clathrin and caveolae pathways has not yet been established.

A few reports have also demonstrated that filoviruses can simultaneously use multiple endocytic pathways for entry [47, 66, 69] and it was suggested that perhaps the virus preferentially chooses one pathway over another based on the type of target cells or receptors used [50].

Although lipid rafts and membrane cholesterol were shown to be required for filoviral entry [58, 71], there are conflicting reports on the role of caveolae that are composed of membrane cholesterol, in filoviral entry with different studies implicating [47, 66, 72] or refuting [45] the involvement of caveolae in filoviral entry.

Therefore, future studies examining the relative contribution and preference of each of these endocytic pathways in filoviral entry into target cells would prove insightful.

6. Endosomal Constituents Involved in Filoviral Entry

Studies examining the trafficking of filoviruses have revealed that after entry, the virus traffics from early to late endosomes/lysosomes. Using GFP-labeled virions and VLPs, Saeed and colleagues have demonstrated that EBOV traffics from EEA1 and Rab5-positive early endosomes to Rab7-positive late endosomes in HEK293T and Vero cells [60]. Similarly, Nanbo and coworkers have shown using biologically contained virions and VLPs that EBOV localizes to Rab7-positive late endosomes in Vero cells [70].

Several studies in Vero and Jurkat cell lines as well as mouse embryonic fibroblasts (MEFs) from cathepsins B and L deficient mice have demonstrated that proteolytic cleavage of EBOV GP by these lysosomal cysteine proteases removes the glycan cap and MLR of GP1 to produce a stable GP intermediate, which is necessary for infection [66, 73–75]. In contrast, Martinez and colleagues have reported that cathepsin L is not required for EBOV entry into human dendritic cells, which are one of the primary target cells of filoviral infection [76]. Moreover, a recent study by Misasi and coworkers showed using Vero and MEF cell lines that Zaire and Tai Forest species of EBOV require cathepsin B, while Sudan and Reston species as well as MARV do not [77]. Hence, the role of cathepsins B and L in filoviral entry into different cell types is not completely understood.

Recent reports have demonstrated that the endosomal membrane protein Niemann-Pick C1 (NPC1) can directly bind to EBOV GP and is an intracellular receptor for filoviruses [78–80]. These studies point towards the interesting possibility that cell surface receptors such as TIM-1 and endosomal receptors such as NPC1 perhaps act in concert with each other to facilitate filoviral entry.

7. Cytosolic Cellular Factors Involved in Filoviral Entry

Several cytosolic factors were shown to be required for filoviral entry. Using EBOV GP pseudotyped virus, Yonezawa

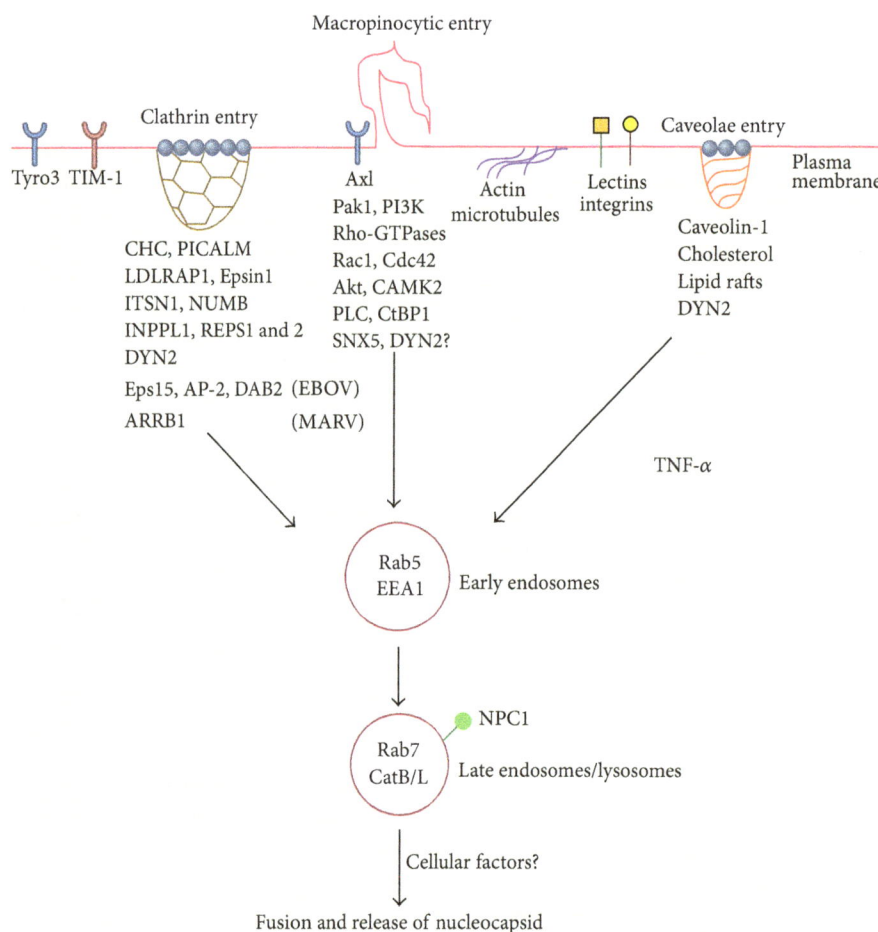

FIGURE 1: Schematic representation of cellular endocytic pathways and factors implicated in filovirus entry.

and colleagues have demonstrated that TNF-α enhances viral entry and fusion [58]. Similarly, using EBOV GP pseudo-typed virus, Quinn and coworkers have showed that Rho B and C are required for EBOV entry [62]. Future studies investigating the involvement of additional cytosolic factors and the signaling pathways triggered by them to facilitate filoviral entry would be very useful.

8. Intracellular Factors Involved in Fusion and Release of Filoviral Nucleocapsid into the Cytoplasm

After GP1 is cleaved by the host cysteine cathepsins, the cleaved GP binds to NPC1 [78–80] and undergoes a series of conformational changes resulting in the refolding of GP2 into a six-helix bundle and insertion of its fusion loop into the host membrane. Viral membrane fusion results in the release of NP, VP35, VP30, L, and the RNA genome into the host cell cytoplasm. The cellular factors and molecular mechanisms governing the different steps of the fusion process are not clearly understood.

9. Implications of Using a Common Cellular Receptor for Entry

Since filoviruses are suggested to bind to a common cellular receptor and yet can enter via multiple endocytic pathways, it is possible that these viruses require different corecep-tor(s) and/or different processed or modified forms of the same primary receptor or coreceptor(s) for entry. Also, the involvement of cell surface receptors such as TIM-1 as well as endosomal receptors such as NPC1 in filoviral entry suggests that filoviruses utilize multiple receptors at various stages of the entry process. Future studies dissecting the interaction of filoviral GP with these receptors could be very insightful.

10. Therapeutic Implications

Small molecule inhibitors of NPC1 were shown to inhibit EBOV infection [78]. Hence, future research aimed at designing small molecule inhibitors of TIM-1 could be very useful for therapeutic purposes. Since TIM-1 can facilitate phagocytosis [55], specific inhibitors of phagocytosis can also be explored as potential therapeutic candidates. Additionally, several receptor tyrosine kinase inhibitors are already being used for treatment of numerous cancers [81–84] and therefore, specific inhibitors of TAM receptors could also be developed as anti-filoviral drug candidates.

11. Future Directions

The mechanisms governing filoviral entry are not completely understood although recent studies have identified several cellular factors, which play critical roles in this process. Figure 1 summarizes our existing knowledge on filoviral entry and the key cellular factors implicated in this process. However, there are several important pending questions the answers to which will greatly enhance our understanding of this field and also promote development of new avenues of therapy.

Understanding the detailed interactions of filoviruses with their cellular receptors/entry factors would be very useful in designing effective strategies to block these interactions. Since cellular factors are fixed targets, they are ideal candidates for development of effective broad-spectrum antiviral therapeutics. Therefore, it would be important to investigate the following broad issues.

(i) How do endosomal receptors such as NPC1 interact with cell surface receptors such as TIM-1 to facilitate viral entry? What are the molecular mechanisms governing the interactions of filoviral receptors with each other? Do the same residues of Filoviral GP bind to all the putative filoviral receptors? How do the filoviral receptors interact with key components of endocytic pathways to participate in filoviral entry?

(ii) Do EBOV and MARV require any additional receptors and coreceptor(s) for entry? If so, are these receptors and coreceptor(s) conserved between the two viruses? Is the requirement for receptors and/or coreceptors cell type specific? Does differential expression of receptors/coreceptors on different cell types play a role in determining the preference for one endocytic pathway over the other? What are the molecular mechanisms governing the interactions of filoviral GP with its cellular receptors/coreceptors that enables filoviruses to exhibit broad tissue tropism and host range?

(iii) What are the molecular mechanisms governing the induction of conformational changes in GP downstream of GP-NPC1 binding, to drive membrane fusion and release of the viral nucleocapsid into the cytoplasm? Which cellular factors play a role in this process?

Future research should be aimed at answering the abovementioned issues, which could help to reveal as well as characterize the many intricacies involved in receptor binding, uptake, and entry of filovirus particles into target cells.

Acknowledgments

The authors regret the inadvertent omission of any relevant studies due to space constraints. T. J. Hope is supported by National Institutes of Health (NIH) Grant AI052051 and is also an Elizabeth Glaser Scientist.

References

[1] A. Sanchez, T. W. Geisbert, and H. Feldmann, "Filoviridae: marburg and Ebola viruses," in *Fields Virology*, D. Knipe, Ed., pp. 1409–1448, 5th edition, 2007.

[2] J. H. Kuhn, S. Becker, H. Ebihara et al., "Proposal for a revised taxonomy of the family Filoviridae: classification, names of taxa and viruses, and virus abbreviations," *Archives of Virology*, vol. 155, no. 12, pp. 2083–2103, 2010.

[3] S. Mahanty and M. Bray, "Pathogenesis of filoviral haemorrhagic fevers," *Lancet Infectious Diseases*, vol. 4, no. 8, pp. 487–498, 2004.

[4] A. Sanchez, S. G. Trappier, B. W. J. Mahy, C. J. Peters, and S. T. Nichol, "The virion glycoproteins of Ebola viruses are encoded in two reading frames and are expressed through transcriptional editing," *Proceedings of the National Academy of Sciences of the United States of America*, vol. 93, no. 8, pp. 3602–3607, 1996.

[5] H. Feldmann, S. T. Nichol, H. D. Klenk, C. J. Peters, and A. Sanchez, "Characterization of filoviruses based on differences in structure and antigenicity of the virion glycoprotein," *Virology*, vol. 199, no. 2, pp. 469–473, 1994.

[6] H. Feldmann, H. D. Klenk, and A. Sanchez, "Molecular biology and evolution of filoviruses," *Archives of Virology*, vol. 7, pp. 81–100, 1993.

[7] M. E. G. Miranda, M. E. White, M. M. Dayrit, C. G. Hayes, T. G. Ksiazek, and J. P. Burans, "Seroepidemiological study of filovirus related to Ebola in the Philippines," *The Lancet*, vol. 337, no. 8738, pp. 425–426, 1991.

[8] P. B. Jahrling, T. W. Geisbert, D. W. Dalgard et al., "Preliminary report: isolation of Ebola virus from monkeys imported to USA," *The Lancet*, vol. 335, no. 8688, pp. 502–505, 1990.

[9] M. P. Kiley, E. T. W. Bowen, and G. A. Eddy, "Filoviridae: a taxonomic home for Marburg and Ebola viruses?" *Intervirology*, vol. 18, no. 1-2, pp. 24–32, 1982.

[10] H. Feldmann and H. D. Klenk, "Marburg and Ebola viruses," *Advances in virus research*, vol. 47, pp. 1–52, 1996.

[11] K. M. Johnson, J. V. Lange, P. A. Webb, and F. A. Murphy, "Isolation and partial characterisation of a new virus causing acute haemorrhagic fever in Zaire," *The Lancet*, vol. 1, no. 8011, pp. 569–571, 1977.

[12] H. Feldmann, S. Jones, H. D. Klenk, and H. J. Schnittler, "Ebola virus: from discovery to vaccine," *Nature Reviews Immunology*, vol. 3, no. 8, pp. 677–685, 2003.

[13] S. F. Dowell, R. Mukunu, T. G. Ksiazek, A. S. Khan, P. E. Rollin, and C. J. Peters, "Transmission of Ebola hemorrhagic fever: a study of risk factors in family members, Kikwit, Democratic Republic of the Congo, 1995. Commission de Lutte contre les Epidemies a Kikwit," *Journal of Infectious Diseases*, vol. 179, supplement 1, pp. S87–S91, 1999.

[14] S. Y. Chan, R. F. Speck, M. C. Ma, and M. A. Goldsmith, "Distinct mechanisms of entry by envelope glycoproteins of Marburg and Ebola (Zaire) viruses," *Journal of Virology*, vol. 74, no. 10, pp. 4933–4937, 2000.

[15] R. J. Wool-Lewis and P. Bates, "Characterization of Ebola virus entry by using pseudotyped viruses: identification of receptor-deficient cell lines," *Journal of Virology*, vol. 72, no. 4, pp. 3155–3160, 1998.

[16] P. Aleksandrowicz, K. Wolf, D. Falzarano, H. Feldmann, J. Seebach, and H. J. Schnittler, "Viral haemorrhagic fever and vascular alterations," *Hamostaseologie*, vol. 28, no. 1-2, pp. 77–84, 2008.

[17] A. Groseth, H. Feldmann, and J. E. Strong, "The ecology of Ebola virus," *Trends in Microbiology*, vol. 15, no. 9, pp. 408–416, 2007.

[18] E. M. Leroy, B. Kumulungui, X. Pourrut et al., "Fruit bats as reservoirs of Ebola virus," *Nature*, vol. 438, no. 7068, pp. 575–576, 2005.

[19] X. Pourrut, M. Souris, J. S. Towner et al., "Large serological survey showing cocirculation of Ebola and Marburg viruses in Gabonese bat populations, and a high seroprevalence of both viruses in *Rousettus aegyptiacus*," *BMC Infectious Diseases*, vol. 9, article 1471, p. 159, 2009.

[20] R. Swanepoel, S. B. Smit, P. E. Rollin et al., "Studies of reservoir hosts for Marburg virus," *Emerging Infectious Diseases*, vol. 13, no. 12, pp. 1847–1851, 2007.

[21] J. S. Towner, X. Pourrut, C. G. Albariño et al., "Marburg virus infection detected in a common African bat," *PloS One*, vol. 2, no. 1, p. e764, 2007.

[22] L. Borio, T. Inglesby, C. J. Peters et al., "Hemorrhagic fever viruses as biological weapons: medical and public health management," *Journal of the American Medical Association*, vol. 287, no. 18, pp. 2391–2405, 2002.

[23] H. Feldmann and M. P. Kiley, "Classification, structure, and replication of filoviruses," *Current Topics in Microbiology and Immunology*, vol. 235, pp. 1–21, 1999.

[24] C. Will, E. Muhlberger, D. Linder, W. Slenczka, H. D. Klenk, and H. Feldmann, "Marburg virus gene 4 encodes the virion membrane protein, a type I transmembrane glycoprotein," *Journal of Virology*, vol. 67, no. 3, pp. 1203–1210, 1993.

[25] A. Sanchez, Z. Y. Yang, L. Xu, G. J. Nabel, T. Crews, and C. J. Peters, "Biochemical analysis of the secreted and virion glycoproteins of Ebola virus," *Journal of Virology*, vol. 72, no. 8, pp. 6442–6447, 1998.

[26] V. E. Volchkov, H. Feldmann, V. A. Volchkova, and H. D. Klenk, "Processing of the Ebola virus glycoprotein by the proprotein convertase furin," *Proceedings of the National Academy of Sciences of the United States of America*, vol. 95, no. 10, pp. 5762–5767, 1998.

[27] G. Simmons, R. J. Wool-Lewis, F. Baribaud, R. C. Netter, and P. Bates, "Ebola virus glycoproteins induce global surface protein down-modulation and loss of cell adherence," *Journal of Virology*, vol. 76, no. 5, pp. 2518–2528, 2002.

[28] A. Takada, "Filovirus tropism: cellular molecules for viral entry," *Frontiers in Microbiology*, vol. 3, article 34, 2012.

[29] E. Nakayama, D. Tomabechi, K. Matsuno et al., "Antibody-dependent enhancement of marburg virus infection," *Journal of Infectious Diseases*, vol. 204, supplement 3, pp. S978–S985, 2011.

[30] A. Takada, H. Ebihara, H. Feldmann, T. W. Geisbert, and Y. Kawaoka, "Epitopes required for antibody-dependent enhancement of Ebola virus infection," *Journal of Infectious Diseases*, vol. 196, supplement 2, pp. S347–S356, 2007.

[31] J. E. Lee and E. O. Saphire, "Ebolavirus glycoprotein structure and mechanism of entry," *Future Virology*, vol. 4, no. 6, pp. 621–635, 2009.

[32] J. E. Lee, M. L. Fusco, A. J. Hessell, W. B. Oswald, D. R. Burton, and E. O. Saphire, "Structure of the Ebola virus glycoprotein bound to an antibody from a human survivor," *Nature*, vol. 454, no. 7201, pp. 177–182, 2008.

[33] V. N. Malashkevich, B. J. Schneider, M. L. McNally, M. A. Milhollen, J. X. Pang, and P. S. Kim, "Core structure of the envelope glycoprotein GP2 from Ebola virus at 1.9-Å resolution," *Proceedings of the National Academy of Sciences of the United States of America*, vol. 96, no. 6, pp. 2662–2667, 1999.

[34] V. E. Volchkov, V. A. Volchkova, U. Ströher et al., "Proteolytic processing of Marburg virus glycoprotein," *Virology*, vol. 268, no. 1, pp. 1–6, 2000.

[35] A. Sanchez, S. G. Trappier, U. Ströher, S. T. Nichol, M. D. Bowen, and H. Feldmann, "Variation in the glycoprotein and VP35 genes of Marburg virus strains," *Virology*, vol. 240, no. 1, pp. 138–146, 1998.

[36] H. Feldmann, C. Will, M. Schikore, W. Slenczka, and H. D. Klenk, "Glycosylation and oligomerization of the spike protein of Marburg virus," *Virology*, vol. 182, no. 1, pp. 353–356, 1991.

[37] A. Takada, S. Watanabe, H. Ito, K. Okazaki, H. Kida, and Y. Kawaoka, "Downregulation of $\beta 1$ integrins by Ebola virus glycoprotein: implication for virus entry," *Virology*, vol. 278, no. 1, pp. 20–26, 2000.

[38] A. Marzi, A. Akhavan, G. Simmons et al., "The signal peptide of the ebolavirus glycoprotein influences interaction with the cellular lectins DC-SIGN and DC-SIGNR," *Journal of Virology*, vol. 80, no. 13, pp. 6305–6317, 2006.

[39] X. Ji, G. G. Olinger, S. Aris, Y. Chen, H. Gewurz, and G. T. Spear, "Mannose-binding lectin binds to Ebola and Marburg envelope glycoproteins, resulting in blocking of virus interaction with DC-SIGN and complement-mediated virus neutralization," *Journal of General Virology*, vol. 86, no. 9, pp. 2535–2542, 2005.

[40] A. Takada, K. Fujioka, M. Tsuiji et al., "Human macrophage C-type lectin specific for galactose and N-acetylgalactosamine promotes filovirus entry," *Journal of Virology*, vol. 78, no. 6, pp. 2943–2947, 2004.

[41] G. Simmons, J. D. Reeves, C. C. Grogan et al., "DC-SIGN and DC-SIGNR bind Ebola glycoproteins and enhance infection of macrophages and endothelial cells," *Virology*, vol. 305, no. 1, pp. 115–123, 2003.

[42] C. P. Alvarez, F. Lasala, J. Carrillo, O. Muñiz, A. L. Corbí, and R. Delgado, "C-type lectins DC-SIGN and L-SIGN mediate cellular entry by Ebola virus in cis and in trans," *Journal of Virology*, vol. 76, no. 13, pp. 6841–6844, 2002.

[43] K. Matsuno, N. Kishida, K. Usami et al., "Different potential of C-type lectin-mediated entry between Marburg virus strains," *Journal of Virology*, vol. 84, no. 10, pp. 5140–5147, 2010.

[44] K. Matsuno, E. Nakayama, O. Noyori et al., "C-type lectins do not act as functional receptors for filovirus entry into cells," *Biochemical and Biophysical Research Communications*, vol. 403, no. 1, pp. 144–148, 2010.

[45] G. Simmons, A. J. Rennekamp, N. Chai, L. H. Vandenberghe, J. L. Riley, and P. Bates, "Folate receptor alpha and caveolae are not required for Ebola virus glycoprotein-mediated viral infection," *Journal of Virology*, vol. 77, no. 24, pp. 13433–13438, 2003.

[46] S. Y. Chan, C. J. Empig, F. J. Welte et al., "Folate receptor-α is a cofactor for cellular entry by Marburg and Ebola viruses," *Cell*, vol. 106, no. 1, pp. 117–126, 2001.

[47] C. L. Hunt, A. A. Kolokoltsov, R. A. Davey, and W. Maury, "The Tyro3 receptor kinase Axl enhances macropinocytosis of *Zaire ebolavirus*," *Journal of Virology*, vol. 85, no. 1, pp. 334–347, 2011.

[48] M. Shimojima, A. Takada, H. Ebihara et al., "Tyro3 family-mediated cell entry of Ebola and Marburg viruses," *Journal of Virology*, vol. 80, no. 20, pp. 10109–10116, 2006.

[49] M. A. Brindley, C. L. Hunt, A. S. Kondratowicz et al., "Tyrosine kinase receptor Axl enhances entry of *Zaire ebolavirus* without direct interactions with the viral glycoprotein," *Virology*, vol. 415, no. 2, pp. 83–94, 2011.

[50] T. Hoenen, A. Groseth, D. Falzarano, and H. Feldmann, "Ebola virus: unravelling pathogenesis to combat a deadly disease," *Trends in Molecular Medicine*, vol. 12, no. 5, pp. 206–215, 2006.

[51] J. H. Kuhn, S. R. Radoshitzky, A. C. Guth et al., "Conserved receptor-binding domains of *Lake Victoria marburgvirus* and *Zaire ebolavirus* bind a common receptor," *Journal of Biological Chemistry*, vol. 281, no. 23, pp. 15951–15958, 2006.

[52] B. Manicassamy, J. Wang, E. Rumschlag et al., "Characterization of Marburg virus glycoprotein in viral entry," *Virology*, vol. 358, no. 1, pp. 79–88, 2007.

[53] J. Wang, B. Manicassamy, M. Caffrey, and L. Rong, "Characterization of the receptor-binding domain of ebola glycoprotein in viral entry," *Virologica Sinica*, vol. 26, no. 3, pp. 156–170, 2011.

[54] A. S. Kondratowicz, N. J. Lennemann, P. L. Sinn et al., "T-cell immunoglobulin and mucin domain 1 (TIM-1) is a receptor for *Zaire ebolavirus* and *Lake Victoria marburgvirus*," *Proceedings of the National Academy of Sciences of the United States of America*, vol. 108, no. 20, pp. 8426–8431, 2011.

[55] N. Kobayashi, P. Karisola, V. Peña-Cruz et al., "TIM-1 and TIM-4 glycoproteins bind phosphatidylserine and mediate uptake of apoptotic cells," *Immunity*, vol. 27, no. 6, pp. 927–940, 2007.

[56] A. Shiratsuchi, M. Kaido, T. Takizawa, and Y. Nakanishi, "Phosphatidylserine-mediated phagocytosis of influenza A virus-infected cells by mouse peritoneal macrophages," *Journal of Virology*, vol. 74, no. 19, pp. 9240–9244, 2000.

[57] T. Ichimura, E. J. P. V. Asseldonk, B. D. Humphreys, L. Gunaratnam, J. S. Duffield, and J. V. Bonventre, "Kidney injury molecule-1 is a phosphatidylserine receptor that confers a phagocytic phenotype on epithelial cells," *Journal of Clinical Investigation*, vol. 118, no. 5, pp. 1657–1668, 2008.

[58] A. Yonezawa, M. Cavrois, and W. C. Greene, "Studies of Ebola virus glycoprotein-mediated entry and fusion by using pseudotyped human immunodeficiency virus type 1 virions: involvement of cytoskeletal proteins and enhancement by tumor necrosis factor alpha," *Journal of Virology*, vol. 79, no. 2, pp. 918–926, 2005.

[59] G. Ruthel, G. L. Demmin, G. Kallstrom et al., "Association of Ebola virus matrix protein Vp40 with microtubules," *Journal of Virology*, vol. 79, no. 8, pp. 4709–4719, 2005.

[60] M. F. Saeed, A. A. Kolokoltsov, T. Albrecht, and R. A. Davey, "Cellular entry of Ebola virus involves uptake by a macropinocytosis-like mechanism and subsequent trafficking through early and late endosomes," *PLoS Pathogens*, vol. 6, no. 9, Article ID e01110, 2010.

[61] N. Mulherkar, M. Raaben, J. C. de la Torre, S. P. Whelan, and K. Chandran, "The Ebola virus glycoprotein mediates entry via a non-classical dynamin-dependent macropinocytic pathway," *Virology*, vol. 419, no. 2, pp. 72–83, 2011.

[62] K. Quinn, M. A. Brindley, M. L. Weller et al., "Rho GTPases modulate entry of Ebola virus and vesicular stomatitis virus pseudotyped vectors," *Journal of Virology*, vol. 83, no. 19, pp. 10176–10186, 2009.

[63] M. F. Saeed, A. A. Kolokoltsov, A. N. Freiberg, M. R. Holbrook, and R. A. Davey, "Phosphoinositide-3 kinase-akt pathway controls cellular entry of Ebola virus," *PLoS Pathogens*, vol. 4, no. 8, Article ID e1000141, 2008.

[64] A. A. Kolokoltsov, M. F. Saeed, A. N. Freiberg, M. R. Holbrook, and R. A. Davey, "Identification of novel cellular targets for therapeutic intervention against Ebola virus infection by siRNA screening," *Drug Development Research*, vol. 70, no. 4, pp. 255–265, 2009.

[65] N. Chazal, G. Singer, C. Aiken, M. L. Hammarskjöld, and D. Rekosh, "Human immunodeficiency virus type 1 particles pseudotyped with envelope proteins that fuse at low pH no longer require Nef for optimal infectivity," *Journal of Virology*, vol. 75, no. 8, pp. 4014–4018, 2001.

[66] A. Sanchez, "Analysis of filovirus entry into vero E6 cells, using inhibitors of endocytosis, endosomal acidification, structural integrity, and cathepsin (B and L) activity," *Journal of Infectious Diseases*, vol. 196, supplement 2, pp. S251–S258, 2007.

[67] S. Bhattacharyya, K. L. Warfield, G. Ruthel, S. Bavari, M. J. Aman, and T. J. Hope, "Ebola virus uses clathrin-mediated endocytosis as an entry pathway," *Virology*, vol. 401, no. 1, pp. 18–28, 2010.

[68] S. Bhattacharyya, T. J. Hope, and J. A. T. Young, "Differential requirements for clathrin endocytic pathway components in cellular entry by Ebola and Marburg glycoprotein pseudovirions," *Virology*, vol. 419, no. 1, pp. 1–9, 2011.

[69] P. Aleksandrowicz, A. Marzi, N. Biedenkopf et al., "Ebola virus enters host cells by macropinocytosis and clathrin-mediated endocytosis," *Journal of Infectious Diseases*, vol. 204, supplement 3, pp. S957–S967, 2011.

[70] A. Nanbo, M. Imai, S. Watanabe et al., "Ebolavirus is internalized into host cells via macropinocytosis in a viral glycoprotein-dependent manner," *PLoS Pathogens*, vol. 6, no. 9, Article ID e01121, 2010.

[71] S. Bavari, C. M. Bosio, E. Wiegand et al., "Lipid raft microdomains: a gateway for compartmentalized trafficking of Ebola and Marburg viruses," *Journal of Experimental Medicine*, vol. 195, no. 5, pp. 593–602, 2002.

[72] C. J. Empig and M. A. Goldsmith, "Association of the caveola vesicular system with cellular entry by filoviruses," *Journal of Virology*, vol. 76, no. 10, pp. 5266–5270, 2002.

[73] R. L. Kaletsky, G. Simmons, and P. Bates, "Proteolysis of the Ebola virus glycoproteins enhances virus binding and infectivity," *Journal of Virology*, vol. 81, no. 24, pp. 13378–13384, 2007.

[74] K. Schornberg, S. Matsuyama, K. Kabsch, S. Delos, A. Bouton, and J. White, "Role of endosomal cathepsins in entry mediated by the Ebola virus glycoprotein," *Journal of Virology*, vol. 80, no. 8, pp. 4174–4178, 2006.

[75] K. Chandran, N. J. Sullivan, U. Felbor, S. P. Whelan, and J. M. Cunningham, "Virology: endosomal proteolysis of the Ebola virus glycoprotein is necessary for infection," *Science*, vol. 308, no. 5728, pp. 1643–1645, 2005.

[76] O. Martinez, J. Johnson, B. Manicassamy et al., "Zaire Ebola virus entry into human dendritic cells is insensitive to cathepsin L inhibition," *Cellular Microbiology*, vol. 12, no. 2, pp. 148–157, 2010.

[77] J. Misasi, K. Chandran, J.-Y. Yang et al., "Filoviruses require endosomal cysteine proteases for entry but exhibit distinct protease preferences," *Journal of Virology*, vol. 86, no. 6, pp. 3284–3292, 2012.

[78] M. Côté, J. Misasi, T. Ren et al., "Small molecule inhibitors reveal Niemann-Pick C1 is essential for Ebola virus infection," *Nature*, vol. 477, no. 7364, pp. 344–348, 2011.

[79] J. E. Carette, M. Raaben, A. C. Wong et al., "Ebola virus entry requires the cholesterol transporter Niemann-Pick C1," *Nature*, vol. 477, no. 7364, pp. 340–343, 2011.

[80] E. H. Miller, G. Obernosterer, M. Raaben et al., "Ebola virus entry requires the host-programmed recognition of an intracellular receptor," *EMBO Journal*, vol. 31, no. 8, pp. 1947–1960, 2012.

[81] M. Sahade, F. Caparelli, and P. M. Hoff, "Cediranib: a VEGF receptor tyrosine kinase inhibitor," *Future Oncology*, vol. 8, no. 7, pp. 775–781, 2012.

[82] Z. Gao, B. Han, H. Wang, C. Shi, L. Xiong, and A. Gu, "Clinical observation of gefitinib as a first-line therapy in sixty-eight patients with advanced NSCLC," *Oncology Letters*, vol. 3, no. 5, pp. 1064–1068, 2012.

[83] C. Carmichael, C. Lau, D. Y. Josephson, and S. K. Pal, "Comprehensive overview of axitinib development in solid malignancies: focus on metastatic renal cell carcinoma," *Clinical Advances in Hematology and Oncology*, vol. 10, no. 5, pp. 307–314, 2012.

[84] X. Wu, Y. Jin, I. H. Cui et al., "Addition of vandetanib to chemotherapy in advanced solid cancers: a meta-analysis," *Anti-Cancer Drugs*, vol. 23, no. 7, pp. 731–738, 2012.

4

Genotyping of HCV RNA Reveals that 3a is the most Prevalent Genotype in Mardan, Pakistan

Sajid Ali,[1,2] **Ayaz Ahmad,**[1] **Raham Sher Khan,**[1] **Sanaullah Khan,**[3] **Muhammad Hamayun,**[4] **Sumera Afzal Khan,**[2] **Amjad Iqbal,**[5] **Abid Ali Khan,**[2] **Abdul Wadood,**[6] **Taj Ur Rahman,**[7] **and Ali Hydar Baig**[3]

[1] *Department of Biotechnology, Abdul Wali Khan University Mardan, Mardan 23200, Pakistan*
[2] *Center of Biotechnology & Microbiology, University of Peshawar, Peshawar 25120, Pakistan*
[3] *Department of Biotechnology, KUST, Kohat 26000, Pakistan*
[4] *Department of Botany, Abdul Wali Khan University Mardan, Mardan 23200, Pakistan*
[5] *Department of Agriculture, Abdul Wali Khan University Mardan, Mardan 23200, Pakistan*
[6] *Department of Biochemistry, Abdul Wali Khan University Mardan, Mardan 23200, Pakistan*
[7] *Department of Chemistry, Abdul Wali Khan University Mardan, Mardan 23200, Pakistan*

Correspondence should be addressed to Ayaz Ahmad; ahdayazb5@awkum.edu.pk

Academic Editor: Jay C. Brown

The clinical outcomes of patients infected with hepatitis C virus (HCV) range from acute resolving hepatitis to chronic liver diseases such as liver cirrhosis or hepatocellular carcinoma. Identification of the infecting virus genotype is indispensable for the exploration of many aspects of HCV infection, including epidemiology, pathogenesis, and response to antiviral therapy. 1419 individuals were screened for anti-HCV in this study, of which 166 (11.7%) were found reactive by ICT (Immunochromatographic test). These 166 anti-HCV positive and 26 normal individuals were further analyzed. RNA was extracted from serum and reverse-transcribed to cDNA and the core region of HCV genome was targeted and amplified by multiplex PCR. HCV RNA was detected in 121 individuals, of which 87 were male and 34 were female. Genotype 3a was the most prevalent among all the genotypes observed followed by 3b. Genotypes 1a, 2a, and 2b were found in 10.89%, 13.22%, and 6.61% patients, respectively. 25.41% of the HCV RNA positive samples were not typed. 6.05% of patients were found having mixed genotypes. These findings will not only help the physicians to prescribe more appropriate treatment for the HCV infection but will also draw the attention of health-related policy makers to devise strategies to curb the disease more effectively.

1. Introduction

Hepatitis C virus (HCV) is the most frequent cause of chronic viral hepatitis worldwide. In the recent years, infection with HCV has emerged as one of the most common causes of acute and chronic liver diseases all over the world [1]. HCV is a member of the *Flaviviridae* family that bears approximately 10 kb long positive sense single-stranded RNA (ssRNA) genome. Since anti-HCV testing alone cannot differentiate between acute, chronic, or resolved infection, a supplementary test must also be carried out, involving measurement of anti-HCV immunoglobulin G activity index [2] or antibody reactivities to specific HCV structural and nonstructural proteins [3], to confirm a positive anti-HCV result [1]. HCV is known to have high rate of genetic heterogeneity [4]. This has allowed HCV strains to be classified into a number of genetically distinct groups, known as genotypes, subtypes, isolates, and quasispecies [5]. The genetic variability among HCV strains is 65.8%–68.7% nucleotide sequence identities of full-length sequences for types, 76.9%–80.1% nucleotide sequence identities of full-length sequences for subtypes, and 90.8%–99% nucleotide sequence identities of full-length sequences for isolates and quasispecies [6]. Six major genotypes, that is, 1 through 6 and more than 50 subtypes, have been identified so far. These genotypes differ by 31 to 34% in their nucleotide sequences whereas the subtypes differ by 20 to 23% in their full-length genomic sequences. This extensive genetic heterogeneity, as well as the tendency for mutation, has hindered vaccine development against this virus [5]. As patients infected with different genotypes respond differently to antiviral drug therapy, identification of the infecting genotype is inevitable to guide the correct dose and duration of current combination therapy (pegylated alpha interferon plus ribavirin) [7].

Hepatitis C is also common in Pakistan but accurate epidemiological information is quite limited. In the outer edges of the cities and in remote areas, unqualified medical and dental practitioners, lady health visitors, midwives, and barbers often use unsterilized instruments which are major potential sources of spreading HCV infection in the urban and rural population of Pakistan [8]. Although the exact ratio of this chronic disease is not known, various studies have shown that 3–7% of population of Pakistan is infected [9]. An earlier study reported that 8.9% of population is infected with HCV in Mardan [10], which is the second largest city in Khyber Pakhtunkhwa province of Pakistan. It is significantly prominent to inspect the degree and distribution of HCV genotypes in district Mardan, Khyber Pakhtunkhwa, where detection and genotype determination preceding therapy are sporadic. Therefore, appropriate information to make individual treatment is required in order to maximize the chance of successful treatment outcome for each individual patient, rendering HCV genotyping assays important and useful tools to optimize treatment type, duration, and dose. This study was designed to determine the active HCV RNA infection as well as to know about the HCV genotypes circulating in the study area.

2. Materials and Method

All the blood samples were collected from different areas (hospitals and clinical laboratories) of Mardan. A 5 cc of blood was taken in a Vacutainer tube and serum was isolated or whole blood was stored at $-80°C$ for further analysis at the Department of Biotechnology, KUST. While collecting blood samples, a proforma was filled to collect medical/clinical information from all the individuals and obtain their consent. Patient's history, liver function tests (LFTs), jaundice, blood transfusion, and other different tests such as HBs Ag and anti-HCV, if carried out, were noted. This study was conducted with the approval of ethics committee of Kohat University of Science and Technology (KUST) Kohat, Pakistan.

2.1. Immunochromatographic Test (ICT). Initially all the samples were screened for HCV antibodies using ICT device kit (Accurate Diagnostic, Canada), as described according to the manufacturer instructions.

2.2. RNA Extraction and cDNA Synthesis (for 5′UTR Detection). RNA was extracted from $300 \mu L$ of blood sample by using RNA extraction kit (RNA purification kit, Ultrascript,

Anagen Technologies, Inc., USA). cDNA was synthesized from the 5′UTR region of extracted RNA, using primers.

2.3. Regular PCR (for 5′UTR Detection). cDNA synthesized was then amplified in next round of regular PCR using a sense and an antisense primer specific for 5′UTR [11]. cDNA was used from the previous round and was run in thermocycler. The cycling conditions for regular PCR were consisting of 25 cycles in three steps I, II, and III.

2.4. Nested PCR (for HCV 5′UTR Detection). After regular PCR, nested PCR was carried out using the next pair of primers internal to the first one [11]. A mixture was prepared in the same way as for regular PCR except the primers. The PCR cycling conditions were the same as for regular PCR.

2.5. Electrophoresis. $12 \mu L$ of the amplified cDNA was resolved on 2% agarose gel. cDNA bands (230 bp) were identified by comparing with a DNA ladder marker of 100 bp (Fermentas, USA) and visualized under UV illumination using gel documentation system.

2.6. HCV Genotyping

2.6.1. cDNA Synthesis for Core Region. cDNA was synthesized from the core region of extracted RNA with specific primers [11]. The reaction was carried out in thermocycler (Techne Inc., USA) using the same program as for 5′UTR.

2.6.2. HCV 1st Round PCR for Genotyping. Two primers (one reverse and one forward) [11] specific for core region of HCV were used to amplify cDNA in a thermocycler (Techne Inc., USA) with *Taq* DNA polymerase (Fermentas, USA).

2.6.3. HCV Genotype-Specific PCR. Genotyping with type-specific primers [11] for the core region of HCV genome was performed for the nine most common subtypes and types of HCV (1a, 1b, 2a, 2b, 3a, 3b, 4, 5a, and 6a). For the discrimination of different products of the HCV genotypes amplified, the type-specific PCR mix was divided into mix A and mix B. For mix B all the reagents were the same except the antisense primers. The HCV genotype-specific PCR/multiplex PCR was performed using the same program as for 1st round of PCR.

2.6.4. Electrophoresis. PCR products were resolved on 2% agarose gel, prepared in 0.5X TBE buffer, and visualized under ultraviolet light. HCV genotypes were determined by comparing the amplified product (cDNA bands) of a specific genotype with 100 bp DNA ladder marker (Fermentas, USA).

3. Results

A total of 1419 subjects were screened for anti-HCV, of which 166 were found positive for HCV antibodies, comprising 11.7% of the screened population (Figure 1 and Table 1). Of the total screened individuals, 192 were selected for further

(a)

Mix-I

Mix-II

(b)

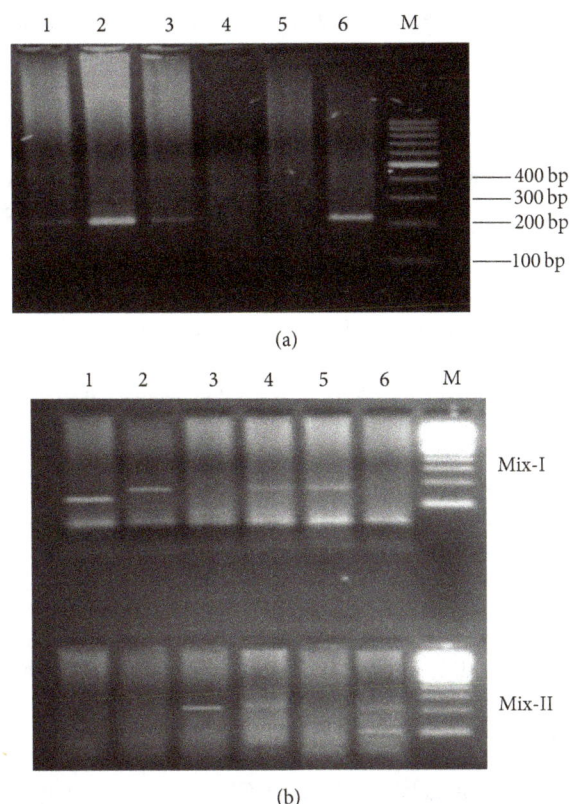

FIGURE 1: (a) Screening of samples for HCV RNA by amplifying 5′UTR. Lanes 1, 2, and 3: positive samples (232 bp band of 5′UTR), Lane 4: negative sample, Lane 5: negative control, Lane 6: positive control, and Lane M: 100 bp DNA ladder. (b) Amplified products of different HCV genotypes. Lane 1 = 139 bp (HCV genotype 2a), Lane 2 = 176 bp HCV genotype 3b, Lane 3 = 232 bp (HCV genotype 3a), Lane 4 = 176 and 232 bp (HCV genotypes 3b and 3a), Lane 5 = 176 bp HCV genotype 3b, Lane 6 = 232 and 99 bp (HCV genotypes 3a and 4), and Lane M = 100 bp DNA ladder.

FIGURE 2: Percentage of HCV genotypes.

TABLE 1: Screening general population of Mardan for anti-HCV and HCV RNA.

	Male	Female	Total (male + female)
Total subjects	757	662	1419
Anti-HCV +Ve	103	63	166
HCV RNA +Ve	87	34	121

study, such that 166 were anti-HCV positive and 26 were from anti-HCV negative (normal individuals), as control, with no apparent history or symptoms of hepatitis C. 103 males and 63 females were anti-HCV positive whereas in general population 17 males and 9 females were anti-HCV negative (Table 1).

All subjects were analyzed with PCR for HCV RNA detection. Of the total 192 individuals, 121 were found to be HCV RNA positive while 71 were negative. Of these 121 HCV RNA positive individuals, 87 were males and 34 were females. HCV RNA was not detected in 16 males and 29 females which were anti-HCV positive. In all 26 anti-HCV negative individuals (general population), HCV RNA was not

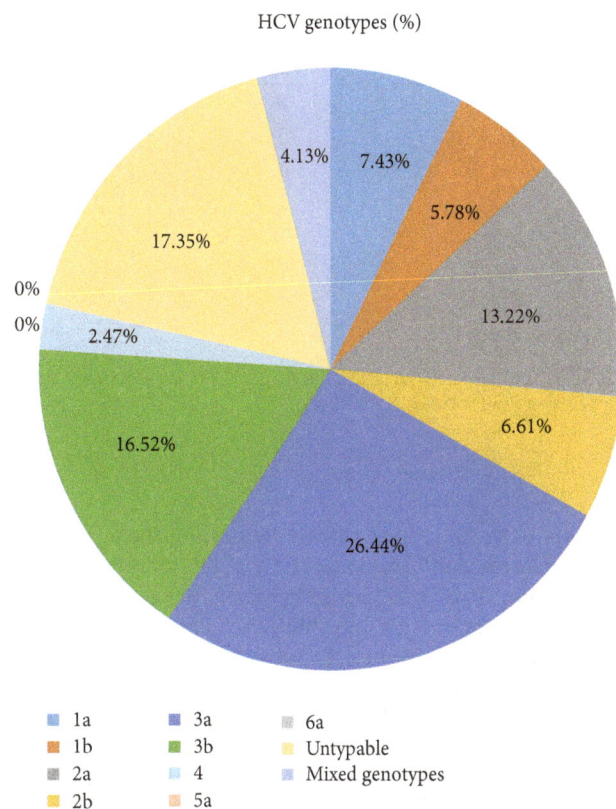

observed after amplification with nested PCR which indicates true negative controls.

The prevalence of different genotypes in the subject population was analyzed by type-specific PCR targeting the core region of HCV genome. HCV genotype 3a was detected in 26.44% of the anti-HCV positive individuals, the most prevalent genotype in the studied area. HCV genotype 3b was observed in 16.52% followed by genotype 2a, 13.22%, and genotype 1a was 7.43%. HCV genotypes 5a and 6a were not detected in any of the HCV RNA positive patients. 17.35% of samples were such that where no genotype was observed and were declared as untypable by the method used. Mixed infection of HCV genotypes was found in 4.13% samples (Figure 2).

The chances of HCV infection in males are relatively higher than females because of more frequent visits of males to high risk areas like barber, and so forth, in this region. The genotype 3a was found to infect males more frequently followed by 3b and 2a, respectively, whereas genotypes 2b and 1a were found to be the major causes of infections in females. Data was analyzed using Roy's Largest Root for correlation of HCV genotype to gender and was found nonsignificant (RLR = 0.834, $P > 0.05$ NS) (Table 2).

4. Discussion

Hepatitis C virus (HCV) has variable clinical outcomes in different infected patients ranging from acute resolving

TABLE 2: Gender-wise distribution of HCV genotypes.

Genotypes	Male	Female
1a	6	3
1b	5	2
2a	11	5
2b	2	6
3a	25	7
3b	16	4
4	2	1
Mixed genotypes	4	1
Untypable	17	4

Roy's Largest Root = 0.834, $P > 0.05$ NS.

hepatitis to chronic liver diseases including liver cirrhosis and Hepatocellular Carcinoma (HCC) [12]. Identification of the infecting virus genotype is important for the exploration of many aspects of HCV infection, including epidemiology, pathogenesis, and response to antiviral therapy [13, 14]. A suitable and reliable HCV genotyping method is inevitable for large-scale epidemiological and experimental studies [11]. A number of laboratory procedures aimed at identifying the HCV genotypes have been described [15]. HCV genotype determination in full-length genomic sequence analysis followed by phylogenetic analysis is still the golden standard. Though this system is expensive and time consuming and cannot be adapted to clinical studies or extensive standard use [15]. PCR has been broadly used for genotyping [13], which is based upon the amplification of virus sequences in clinical specimens, using type-specific primers that specifically amplify different genotypes [11, 16]. Information obtained from various parts of the world has focused on the increasing implication of HCV genotyping and stressed the need of easy, reliable, cost effective, and fast techniques for mass screening.

The precise and sensitive measurement of HCV RNA is important for medical management of infected patients and to know about the biology of HCV and HCV infection in research [17]. The clinical causes, risk factors, and severity of HCV infections and HCV genotypes are still poorly defined in district Mardan, like other districts of Khyber Pakhtunkhwa. Antibodies against HCV were observed in 11.70% of the studied population, which shows high prevalence in comparison to 4.57% reported by Muhammad and Jan in district Buner [18].

Our data showed that genotype 3a (26.44%) followed by genotype 3b (16.52%) is prevalent in district Mardan and the study of Idrees and Riazuddin [19], where genotypes 3a and 3b were dominant among the studied population [19]. Also in another study by Idrees and Riazuddin, HCV genotype 3a was found predominant in general population [19]. Regarding occurrence of HCV genotypes in other countries like Argentina, genotype 2 is more prevalent [20]. Similarly, Alfonso et al. (2001) observed genotype 1 in 61.4% and genotype 3 in 23.7% infected patients in Latin American region [21]. In a recent review we reported that genotype 3a accounts for 68.94% of HCV infections in Punjab, 76.88% in Sindh, 58% in KPK, and 60.71% in Balochistan provinces of Pakistan [22].

Analysis of epidemiological data showed marked differences between patients with single infections and those with apparently mixed infections illustrated by the type-specific PCR [13]. Mixed infection was observed in our study in 4.13% of samples (Figure 2). The mixed genotypes may be due to infection by different HCV types by repeated exposure to HCV [23]. In addition, 17.35% of the individuals which were confirmed by nested PCR for HCV RNA were not typed by type-specific PCR. That is not in agreement with our previous findings that showed 20.16% of the HCV positive individuals in KPK were untypable [22].

HCV is known as silent killer because in majority of the cases no proper signs or symptoms are visible in the early stages of infection and when symptoms appear then the treatment is difficult. Secondly, in most of the developing countries diagnosis is not proper due to lack of facilities. The people are screened for anti-HCV through strips or by anti-HCV ELISA test for HCV infection but these methods are more erroneous and lack sensitivity [24]. Therefore, molecular detection of HCV by PCR based methods is inevitable due to higher levels of sensitivity and specificity than the serological methods. HCV infection is a typical example of diseases in which direct detection of the virus is essential for a correct diagnosis. In comparison to other existing in vitro assays, RT-PCR has extra prospective for its diagnosis as it offers an ultimate detection of HCV [25].

The present study shows that the distribution of HCV genotype 3 in Mardan is similar to that in other areas of Pakistan. HCV types 2 and 3 are prevalent in this area which can better respond to interferon therapy but types 1 and 4 were also circulating which need longer treatment. Proper epidemiological studies and treatment strategies should be initiated in this area. Appropriate preventive measures should be also taken into consideration to control the spread of this dreadful disease.

Acknowledgments

The authors are thankful to Abdul Wali Khan University Mardan (AWKUM), Kohat University of Science and Technology (KUST), University of Peshawar, and the Higher Education Commission (HEC) of Pakistan for funding the project.

References

[1] D. B. Strader, T. Wright, D. L. Thomas, and L. B. Seeff, "Diagnosis, management, and treatment of hepatitis C," *Hepatology*, vol. 39, no. 4, pp. 1147–1171, 2004.

[2] S. Klimashevskaya, A. Obriadina, T. Ulanova et al., "Distinguishing acute from chronic and resolved hepatitis C virus (HCV) infections by measurement of anti-HCV immunoglob-

ulin G avidity index," *Journal of Clinical Microbiology*, vol. 45, no. 10, pp. 3400–3403, 2007.

[3] A. C. Araujo, I. V. Astrakhantseva, H. A. Fields, and S. Kamili, "Distinguishing acute from chronic hepatitis C virus (HCV) infection based on antibody reactivities to specific HCV structural and nonstructural proteins," *Journal of Clinical Microbiology*, vol. 49, no. 1, pp. 54–57, 2011.

[4] L. Stuyver, W. van Arnhem, A. Wyseur, F. Hernandez, E. Delaporte, and G. Maertens, "Classification of hepatitis C viruses based on phylogenetic analysis of the envelope 1 and nonstructural 5B regions and identification of five additional subtypes," *Proceedings of the National Academy of Sciences of the United States of America*, vol. 91, no. 21, pp. 10134–10138, 1994.

[5] J. Bukh, R. H. Miller, and R. H. Purcell, "Genetic heterogeneity of hepatitis C virus: quasispecies and genotypes," *Seminars in Liver Disease*, vol. 15, no. 1, pp. 41–63, 1995.

[6] T. Sy and M. M. Jamal, "Epidemiology of hepatitis C virus (HCV) infection," *International Journal of Medical Sciences*, vol. 3, no. 2, pp. 41–46, 2006.

[7] J. Pawlotsky, "Mechanisms of antiviral treatment efficacy and failure in chronic hepatitis C," *Antiviral Research*, vol. 59, no. 1, pp. 1–11, 2003.

[8] M. Nafees, M. S. Bhatti, and I. U. Haq, "Sero-prevalence of HCV Antibodies in population attending Madina Teaching hospital, Faisalabad," *Annals of King Edward Medical University*, vol. 13, no. 4, pp. 57–62, 2010.

[9] S. Akhtar, T. Moatter, S. I. Azam, M. H. Rahbar, and S. Adil, "Prevalence and risk factors for intrafamilial transmission of hepatitis C virus in Karachi, Pakistan," *Journal of Viral Hepatitis*, vol. 9, no. 4, pp. 309–314, 2002.

[10] N. S. Ali, K. Jamal, and R. Qureshi, "Hepatitis B vaccination status and identification of risk factors for hepatitis B in health care workers," *Journal of the College of Physicians and Surgeons. Pakistan*, vol. 15, no. 5, pp. 257–260, 2005.

[11] T. Ohno, M. Mizokami, R. Wu et al., "New hepatitis C virus (HCV) genotyping system that allows for identification of HCV genotypes 1a, 1b, 2a, 2b, 3a, 3b, 4, 5a, and 6a," *Journal of Clinical Microbiology*, vol. 35, no. 1, pp. 201–207, 1997.

[12] Y. S. Lee, S. K. Yoon, E. S. Chung et al., "The relationship of histologic activity to serum ALT, HCV genotype and HCV RNA titers in chronic hepatitis C," *Journal of Korean Medical Science*, vol. 16, no. 5, pp. 585–591, 2001.

[13] X. Forns, M. D. Maluenda, F. X. López-Labrador et al., "Comparative study of three methods for genotyping hepatitis C virus strains in samples from Spanish patients," *Journal of Clinical Microbiology*, vol. 34, no. 10, pp. 2516–2521, 1996.

[14] K. Nagayama, M. Kurosaki, N. Enomoto, Y. Miyasaka, F. Marumo, and C. Sato, "Characteristics of hepatitis C viral genome associated with disease progression," *Hepatology*, vol. 31, no. 3, pp. 745–750, 2000.

[15] J. Pawlotsky, L. Prescott, P. Simmonds et al., "Serological determination of hepatitis C virus genotype: comparison with a standardized genotyping assay," *Journal of Clinical Microbiology*, vol. 35, no. 7, pp. 1734–1739, 1997.

[16] H. Okamoto, Y. Sugiyama, S. Okada et al., "Typing hepatitis C virus by polymerase chain reaction with type-specific primers: application to clinical surveys and tracing infectious sources," *Journal of General Virology*, vol. 73, part 3, pp. 673–679, 1992.

[17] M. T. Pyne, E. Q. Konnick, A. Phansalkar, and D. R. Hillyard, "Evaluation of the abbott investigational use only RealTime hepatitis C virus (HCV) assay and comparison to the roche TaqMan HCV analyte-specific reagent assay," *Journal of Clinical Microbiology*, vol. 47, no. 9, pp. 2872–2878, 2009.

[18] N. Muhammad and M. A. Jan, "Epidemiology of hepatitis C virus (HCV) infection," *Journal of the College of Physicians and Surgeons. Pakistan*, vol. 15, no. 1, p. 11, 2005.

[19] M. Idrees and S. Riazuddin, "Frequency distribution of hepatitis C virus genotypes in different geographical regions of Pakistan and their possible routes of transmission," *BMC Infectious Diseases*, vol. 8, article 69, 2008.

[20] J. R. Oubiña, J. F. Quarleri, M. A. Sawicki et al., "Hepatitis C virus and GBV-C/hepatitis G virus in Argentine patients with porphyria cutanea tarda," *Intervirology*, vol. 44, no. 4, pp. 215–218, 2001.

[21] V. Alfonso, D. Flichman, S. Sookoian, V. A. Mbayed, and R. H. Campos, "Phylogenetic characterization of genotype 4 hepatitis C virus isolates from Argentina," *Journal of Clinical Microbiology*, vol. 39, no. 5, pp. 1989–1992, 2001.

[22] S. Attaullah, S. Khan, and I. Ali, "Hepatitis C virus genotypes in Pakistan: a systemic review," *Virology Journal*, vol. 8, article 433, 2011.

[23] Y.-W. Hu, E. Balaskas, M. Furione et al., "Comparison and application of a novel genotyping method, semiautomated primer-specific and mispair extension analysis, and four other genotyping assays for detection of hepatitis C virus mixed-genotype infections," *Journal of Clinical Microbiology*, vol. 38, no. 8, pp. 2807–2813, 2000.

[24] J.-S. Li, L. Vitvitski, S.-P. Tong, and C. Trepo, "Identification of the third major genotype of hepatitis C virus in France," *Biochemical and Biophysical Research Communications*, vol. 199, no. 3, pp. 1474–1481, 1994.

[25] J. Albadalejo, R. Alonso, R. Antinozzi et al., "Multicenter evaluation of the COBAS AMPLICOR HCV assay, an integrated PCR system for rapid detection of hepatitis C virus RNA in the diagnostic laboratory," *Journal of Clinical Microbiology*, vol. 36, no. 4, pp. 862–865, 1998.

5

Nelfinavir Impairs Glycosylation of Herpes Simplex Virus 1 Envelope Proteins and Blocks Virus Maturation

Soren Gantt,[1,2,3] Eliora Gachelet,[4] Jacquelyn Carlsson,[5] Serge Barcy,[1] Corey Casper,[3,5,6,7] and Michael Lagunoff[4]

[1]Seattle Children's Research Institute, University of Washington, Seattle, WA 98101, USA
[2]Department of Pediatrics, University of Washington, Seattle, WA 98105, USA
[3]Department of Global Health, University of Washington, Seattle, WA 98195, USA
[4]Department of Microbiology, University of Washington, Seattle, WA 98195, USA
[5]Department of Medicine, University of Washington, Seattle, WA 98195, USA
[6]Department of Epidemiology, University of Washington, Seattle, WA 98195, USA
[7]Fred Hutchinson Cancer Research Center, Seattle, WA 98109, USA

Correspondence should be addressed to Soren Gantt; sgantt@cfri.ca

Academic Editor: Gary S. Hayward

Nelfinavir (NFV) is an HIV-1 aspartyl protease inhibitor that has numerous effects on human cells, which impart attractive antitumor properties. NFV has also been shown to have *in vitro* inhibitory activity against human herpesviruses (HHVs). Given the apparent absence of an aspartyl protease encoded by HHVs, we investigated the mechanism of action of NFV herpes simplex virus type 1 (HSV-1) in cultured cells. Selection of HSV-1 resistance to NFV was not achieved despite multiple passages under drug pressure. NFV did not significantly affect the level of expression of late HSV-1 gene products. Normal numbers of viral particles appeared to be produced in NFV-treated cells by electron microscopy but remain within the cytoplasm more often than controls. NFV did not inhibit the activity of the HSV-1 serine protease nor could its antiviral activity be attributed to inhibition of Akt phosphorylation. NFV was found to decrease glycosylation of viral glycoproteins B and C and resulted in aberrant subcellular localization, consistent with induction of endoplasmic reticulum stress and the unfolded protein response by NFV. These results demonstrate that NFV causes alterations in HSV-1 glycoprotein maturation and egress and likely acts on one or more host cell functions that are important for HHV replication.

1. Introduction

Human herpesvirus (HHV) infections are ubiquitous and are responsible for substantial morbidity and mortality worldwide, particularly among people infected with human immunodeficiency virus (HIV). Herpes simplex virus (HSV) and cytomegalovirus (CMV) infections can be recurrent and difficult to treat in HIV coinfected individuals [1]. Moreover, genital HSV infection has been associated with greater risks of HIV acquisition, transmission, and progression of disease [2]. HHV-8 and Epstein-Barr virus infections cause the most common AIDS-defining malignancies, Kaposi sarcoma and non-Hodgkin lymphoma, respectively [3]. Although greatly reduced by effective antiretroviral therapy (ART), complications of HHV infections remain among the most common medical problems in people infected with HIV worldwide [3–7].

Currently available antiviral drugs to treat or prevent complications of HHV infections all directly or indirectly target the viral polymerase [8]. Each of these drugs has one or more important limitations, including selection of drug-resistant viral mutants, significant toxicities, and/or poor bioavailability requiring intravenous administration. For example, treatment of acyclovir-resistant HSV or ganciclovir-resistant CMV infections requires the use of intravenous foscarnet or cidofovir, both of which are associated with nephrotoxicity. As such, new agents that are effective for HHV infections are needed that are safe, orally bioavailable and have a high barrier to resistance.

Nelfinavir (NFV) is a first-generation HIV aspartyl protease inhibitor recently found to block production of multiple HHVs [9]. Furthermore, because it also has potent antitumor and antiangiogenic properties, clinical trials are ongoing to evaluate NFV for the treatment of several cancers [10–15]. The mechanisms by which NFV acts on tumor cells are multifactorial and include inhibition of cellular proteases, Akt activation, and NFκ-B signaling, as well as induction of the endoplasmic reticulum (ER) stress, the unfolded protein response (UPR), and autophagy [11, 16, 17].

In contrast to the aspartyl protease required for HIV maturation, HHVs utilize a serine protease, which for HSV-1 is the gene product of UL26 open reading frame (ORF) [18]. As such, we hypothesized that NFV does not inhibit HHV replication by acting on the viral protease. Furthermore, we speculated that NFV acting on a nonprotease viral target would be improbable and that its antiviral activity is more likely due to one or more of its effects on host cells, which could impair efficient viral replication. We therefore investigated mechanism of action of NFV on HSV-1, by attempting to identify host cell or viral targets of the drug *in vitro*.

2. Methods

2.1. Cells and Virus. Human fibroblasts (HF) were cultivated in Dulbecco's modified Eagle's medium (DMEM; Gibco) containing 10% fetal bovine serum (FBS) and 100 units per mL penicillin G and 100 μg per mL streptomycin (Pen-Strep) and maintained at 37°C in a humidified 5% CO_2 atmosphere, as previously described [9]. HEK293-T cells were maintained in DMEM supplemented with 10% FBS and Pen-Strep. HSV-1 (strain F) was a gift from Keith Jerome (Fred Hutchinson Cancer Research Center).

2.2. Drugs. NFV and indinavir (IDV) were obtained through the AIDS Research and Reference Reagent Program, Division of AIDS, NIAID, NIH. NFV was solubilized in dimethyl sulfoxide (DMSO). IDV and acyclovir (Sigma-Aldrich) were solubilized in water.

2.3. Antibodies. The primary antibodies used were as follows: mouse monoclonal anti-HSV gB (Virusys), mouse monoclonal anti-HSV1 gC clone 3G9 (Abcam), rabbit polyclonal anti-LC3B (Cell Signaling Technology), mouse monoclonal anti-β-actin, anti-FLAG M2 monoclonal, and anti-HA monoclonal antibody, clone HA-7 (Sigma-Aldrich). For Western blotting, the secondary antibodies used were peroxidase-conjugated AffiniPure F(ab′)2 fragment of goat anti-mouse IgG(H + L) or peroxidase-conjugated AffiniPure F(ab′)2 fragment of goat anti-rabbit IgG(H + L) (Jackson ImmunoResearch). Secondary antibodies for immunofluorescence were AlexaFluor 594 F(ab′)2 fragment of goat anti-mouse IgG(H + L) or AlexaFluor 488 F(ab′)2 fragment of goat anti-rabbit IgG(H + L) (Life Technologies).

2.4. Lectins and Eastern Blotting. Biotinylated lectins used in the Eastern blots (Vector Laboratories, Inc.) were as follows: peanut agglutinin (PNA), *Ricinus communis* agglutinin I (RCA I), wheat germ agglutinin (WGA), and concanavalin A (ConA). Total cellular proteins (0.75–1.0 μg loaded per lane) were separated by SDS-PAGE and transferred to a PVDF membrane. The membranes were blocked with 5% BSA in Tris-buffered saline with 0.3% Tween-20 (TBS-T), incubated with 1-2 μg lectin/mL in blocking buffer for 1-2 hours, washed three times in TBS-T, and then incubated 1 hour with horseradish-peroxidase-conjugated avidin D (Vector Labs) at 1-2 μg/mL in TBS-T. The blots were washed three times in TBS-T following the avidin incubation. Detection was performed using an enhanced chemiluminescence method (Pierce ECL Plus).

2.5. Selection of Resistance. HF were infected with HSV-1 at an MOI of 0.1 for 1 hour at 37°C and then incubated at 37°C with either 5 μM NFV, 1 μM ACV, or no drug for 3–5 days until most cells were lysed. Four passages were made for each group. Virus was harvested with 3 freeze-thaw cycles followed by the addition of nonfat milk. Virus titers were determined by plaque assay previously described [9].

2.6. HSV-1 Protease Activity Assay. The effect of NFV on HSV-1 maturational protease activity was assayed by transfecting HEK293-T cells with vectors expressing the protease (VP24; N-terminal 247 amino acids of the UL26 ORF gene product), as well as two substrates: the HSV-1 capsid scaffold protein (amino acids 307–635 of the UL26 ORF gene product; full-length product of UL26.5; ICP35) and a catalytically inactive mutant (S129A) of the full-length protease-scaffold protein transcript (amino acids 1–635 of the UL26 ORF gene product) [18]. The protease coding sequence contained an N-terminal HA tag and each substrate contained an N-terminal FLAG tag; all vectors were synthesized by Blue Heron Biotech, LLC. Each substrate was expressed alone or in combination with the protease, in the presence of NFV or a vehicle control. Expression was under the control of the CMV6 promoter. The plasmid pEGFP-N2, which expresses a GFP variant from the CMV immediate-early promoter, was used to monitor transfection efficiencies. Transfection was performed using Mirus *Trans*IT-293 reagent (MIR 2700) according to the manufacturer's instructions. NFV (10 μM final concentration) or DMSO (0.1% final concentration) was added to the cells concurrently with the DNA and transfection reagent. Cultures were incubated 24 hours with no change of medium, at which time the cells were harvested on ice by scraping into RIPA buffer containing protease inhibitors (Roche Complete Mini EDTA-Free Protease Inhibitor Cocktail tablets). Protein concentrations were determined by BCA assay (Pierce). Extracted proteins (10 μg/lane) were separated by SDS-PAGE and transferred to PVDF membranes by tank transfer overnight. The blots were blocked with 5% nonfat dry milk and probed with anti-FLAG M2 mouse monoclonal antibody (Sigma), recognizing the tag on the scaffold protein, anti-HA monoclonal antibody, clone HA-7 (Sigma), and recognizing the tag on VP24.

2.7. Transmission Electron Microscopy. Human foreskin fibro-blasts (HFFs) were mock-infected or infected with HSV-1 (strain F) at an MOI of 10 or 50. After 1 hour, the inoculum was replaced with medium containing 0.1% DMSO or 10 μM NFV. At 16 or 20 hpi the medium was replaced with a 1:1 mixture of DMEM and 1/2 Karnovsky's fixative [19] and the culture was returned to the incubator for 10 minutes. The mixture was replaced by 1/2 Karnovsky's, and the cells were incubated a further 30–60 minutes at room temperature. The cells were scraped, transferred to a microcentrifuge tube, and pelleted at 200 ×g. The pellet was resuspended in 1 mL of fixative and then dehydrated, embedded, sectioned, and affixed to grids according to standard methods. The grids were examined on a JEOL 1230 or a JEOL JEM 1400 trans-mission electron microscope at the Electron Microscopy Lab at the Fred Hutchinson Cancer Research Center.

2.8. Immunofluorescence Microscopy. HFFs were seeded in 4-well chamber slides to a confluence of approximately 70%. They were mock-infected or infected with HSV-1 at an MOI of 3. After 1 h, the inoculum was replaced by medium with 0.1% DMSO or 10 μM NFV, and the cultures were returned to the incubator. Sixteen hours after infection, the cells were fixed in 4% paraformaldehyde in phosphate-buffered saline. Cells were then permeabilized with 0.2% Tween-20. Endogenous peroxidase activity was inhibited with 3% H_2O_2. The cells were then incubated with 10% nonfat dry milk containing 1% normal goat serum (blotto/NGS; Jackson) to block nonspe-cific binding. Primary antibody binding was revealed with anti-mouse IgG-HRP (Jackson) followed by a 10-minute TSA amplification with TSA-594 (Life Technologies). The nuclei were stained with TO-PRO 3. Slides were mounted after addition of SlowFade (Life Technologies) and analyzed by confocal microscopy. Confocal images were generated on an LSM 5 Pascal system (Zeiss).

3. Results

3.1. NFV Inhibits HSV-1 Replication by a Mechanism Distinct from That of Acyclovir. We passaged HSV-1 in the presence of NFV to determine if resistant mutants could be selected in an attempt to elucidate the mechanism of action of NFV on HHV production. As expected, [20] multistep passage under drug pressure was readily selected for high-level resistance to acyclovir (Figure 1). In contrast, no significant change in sus-ceptibility to NFV was observed despite parallel passaging in the presence of NFV. Of note, the inhibitory activity of NFV was not different between acyclovir-resistant and wild type isolates, further suggesting a distinct antiviral mechanism.

3.2. NFV Does Not Inhibit HSV-1 Protease Activity. To deter-mine if NFV, an aspartyl protease inhibitor, could act on the essential HSV-1 UL26 serine protease, we tested whether NFV could inhibit the activity of the HSV-1 protease on its scaffold protein substrates using a cotransfection assay. At NFV con-centrations that potently block production of infectious HSV-1, there was no effect on the activity of the HSV-1 protease expressed in HEK293T cells (Figure 2).

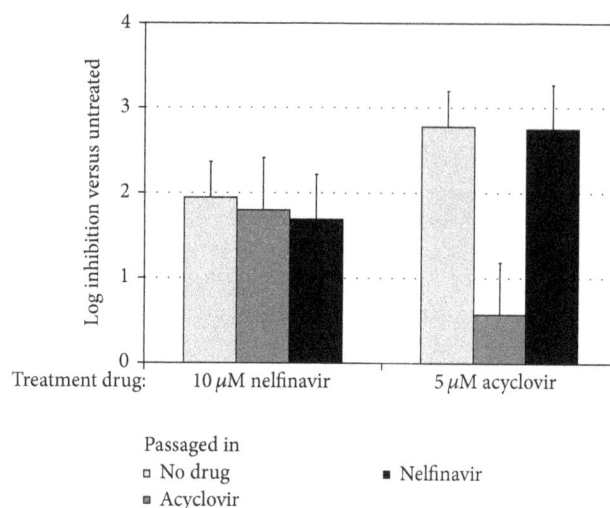

FIGURE 1: NFV inhibits production of infectious HSV-1 through a mechanism distinct from that of acyclovir and does not readily select for antiviral resistance *in vitro.* HSV-1 was passaged four times in HF in the presence of either 5 μM NFV or 1 μM acyclovir, the approximate IC_{50} of each drug. Virus passaged in acyclovir showed increased resistance to inhibition by acyclovir but was inhibited by NFV similarly to the control HSV-1 isolate. In contrast, passage in the presence of NFV did not result in a significant change in sus-ceptibility to either drug. Shown are the results from three separate experiments.

3.3. Inhibition of HSV-1 Replication by NFV Cannot Be Attrib-uted to a Decrease in Akt Activation. One of the prominent effects of NFV on human cells that has been described is inhibition of the Akt signaling pathway by reducing the phosphorylation of Akt by phosphatidylinositol-3 kinase [21–23]. HSV-1 infection results in an increase in the level of Akt phosphorylation (Figure 3), as has been previously described [24]. The Akt inhibitor LY294002 completely suppressed Akt phosphorylation in HSV-1 infected cells, but NFV did not reduce the levels of phosphorylated Akt even at drug con-centrations that potently block virus production (Figure 3). Furthermore, LY294002 treatment of HSV-1 infected HF cells did not reduce the production of infectious virus by plaque assay (not shown), consistent with published data [24]. Therefore, it is unlikely that the documented inhibition of NFV on AKT activation plays a role in the drug's inhibition of HSV-1.

3.4. Ultrastructure of NFV-Treated Cells Infected with HSV-1. As shown by Kalu et al. [25] and supported by our preliminary experiments (data not shown), late HSV-1 gene expression in HF cells was not reduced by NFV treatment. We therefore explored the effects of NFV on HSV-1 infected HF cells by transmission electron microscopy. Normal numbers of virus particles appeared to be produced in NFV-treated cells. Similar numbers of capsids were observed in the nucleus of untreated and NFV-treated cells (approximately 39 and 44, resp., Figures 4(a) and 4(b)). However, compared with untreated cells, capsids in NFV-treated cells appeared to be disproportionately retained within the cytoplasm (~14 versus

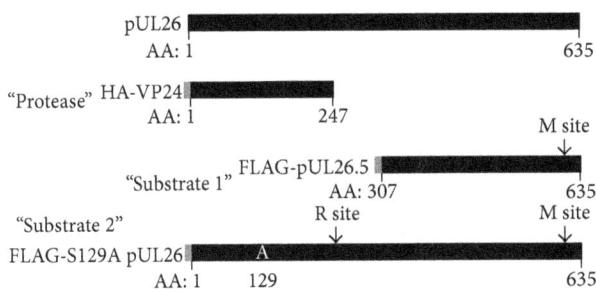

(a)

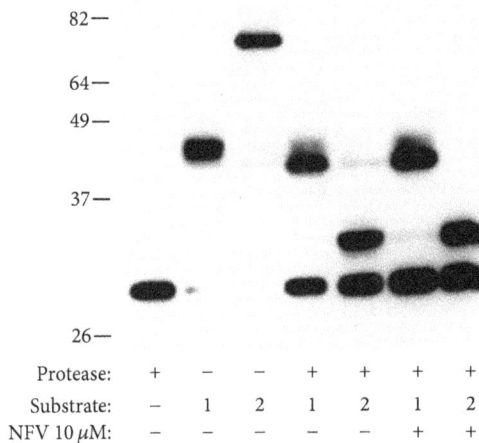

(b)

FIGURE 2: NFV does not inhibit activity of the HSV-1 protease. The coding sequence of the full-length protein product of UL26 ORF (pUL26) and the expression proteins employed are diagrammed in panel (a). Transfection vectors were constructed to express the VP24 protease with an N-terminal HA tag ("protease"), the pUL26.5 with an N-terminal FLAG tag (Substrate 1), or the S129A pUL26 mutant inactive protease-scaffold protein with an N-terminal FLAG tag (Substrate 2). Tags are depicted in grey at the left end of each construct. The release (R) and maturation (M) cleavage sites are shown. After transfection into HEK293-T cells, construct expression and substrate cleavage were evaluated by Western blot using anti-HA and FLAG antibodies. Each protein expressed alone was of the expected size (28.8, 37.9, and 70.6 kD, resp., including tag). Similarly, coexpression of protease with each of the 2 substrates resulted in the N-terminal cleavage products of the expected sizes. The generation of substrate cleavage products was not affected by the addition of NFV at concentrations that inhibit HSV-1 replication *in vitro*.

79, resp.), in which virus particles were more often located outside the plasma membrane (~51 versus 3), suggesting a block in virus maturation or egress. In addition, in contrast to those in the cytoplasm of untreated cells, cytoplasmic capsids in NFV-treated cells were rarely observed to be enveloped (~9 versus 0 in the fields shown; Figures 4(c) and 4(d)). Interestingly, although endoplasmic reticulum (ER) stress, the unfolded protein response (UPR), and autophagy are well known effects of NFV [16, 26–29], neither ER dilation nor the abundance of double-membrane bound vesicles consistent with autophagosomes appeared consistently different between NVF-treated and untreated HSV-1-infected cells.

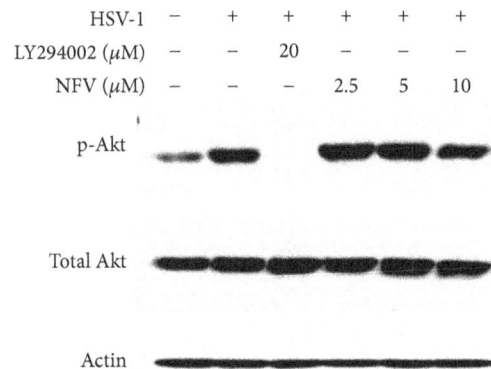

FIGURE 3: Decreased Akt activation does not account for the ability of NFV to inhibit HSV-1 viral replication. HSV-1 infection of Vero cells resulted in an increase in levels of phosphorylated (p)-Akt at 2 hours compared to uninfected cells. Although the Akt inhibitor LY294002 could completely suppress Akt phosphorylation in HSV-1 infected cells, NFV did not substantially reduce the levels of p-Akt even at drugs concentrations that potently block virus production.

3.5. Glycoprotein Posttranslational Processing and Localization Are Impaired by NFV. During evaluation of HSV-1 glycoproteins gB and gC expression by Western blotting, it was apparent that NFV treatment of infected cells resulted in increased electrophoretic mobility (Figure 5(a)) compared to untreated controls or IDV-treated cells. The apparent change in molecular weight of these viral proteins was estimated to be consistent with the reduction in glycosylation [30–32]. To assess the effect of NFV on protein glycosylation in HSV-1-infected cells, Eastern blotting was performed using lectins PNA, RCA-I, WGA, and ConA (Figure 5(a)). Compared to untreated or IDV-treated cells, NFV resulted in a marked reduction in staining by PNA, RCA-I, and WGA indicating decreased addition of galactose, N-acetyl-D-galactosamine, and N-acetyl-D-glucosamine [33, 34]. In contrast, ConA staining was not appreciably reduced, suggesting relatively normal levels of oligomannose-type N-glycans. NFV resulted in a clear alteration in the subcellular localization of HSV-1 gB (Figure 5(b)) by immunofluorescent antibody staining. Compared with control treatments in which viral envelope glycoprotein staining uniformly delineated the plasma membrane of HSV-1-infected HF cells, with NFV treatment staining appeared predominantly perinuclear, suggesting improper trafficking to the cell surface. Consistent with the electron microscopy results, using immunofluorescence, no increase in LC3-II staining, a marker of autophagy, was apparent (data not shown).

4. Discussion

NFV is an HIV aspartyl protease inhibitor that, in addition to its antiretroviral activity, has complex effects on numerous human cellular functions, including on Akt signaling, inhibition of cellular proteases, and induction of ER stress, many of which might contribute to its ability to broadly inhibit tumor cell growth as well as replication of several non-HIV viruses [9, 11, 16, 17, 35, 36]. In this study, we specifically explored

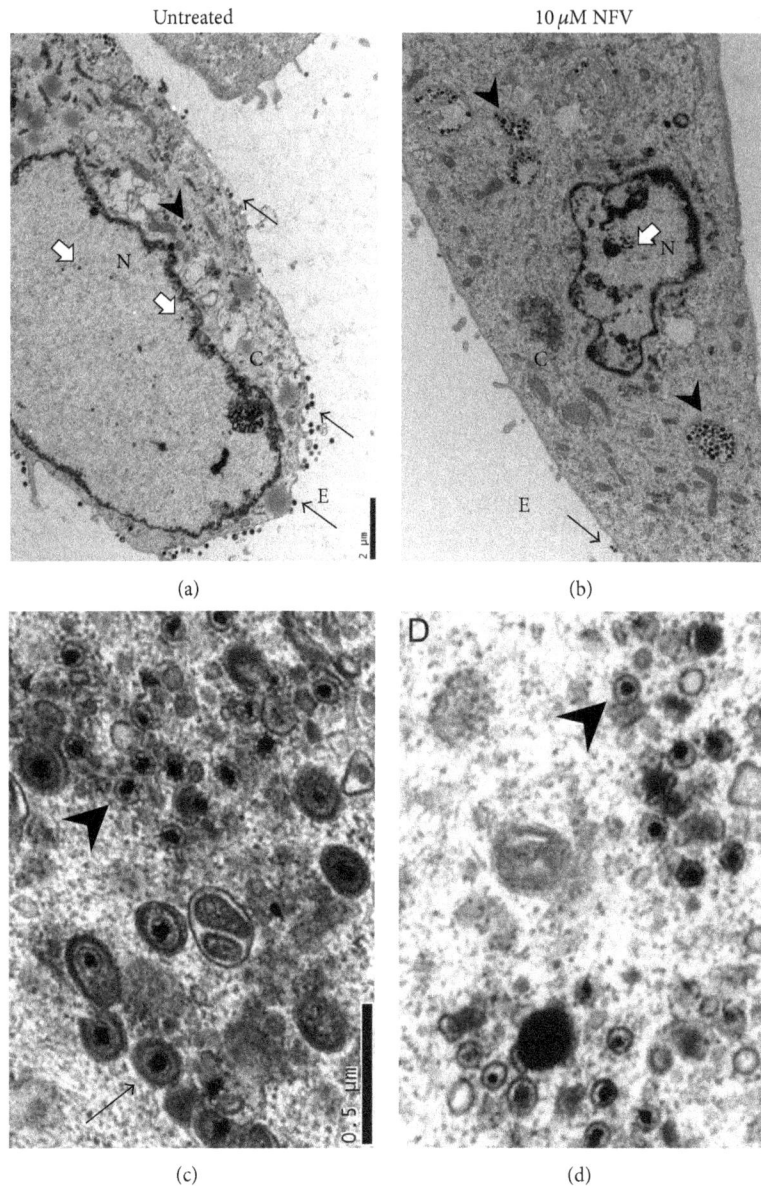

FIGURE 4: NFV treatment impairs HSV-1 maturation and egress. Shown are representative transmission election micrographs, performed 20 hours after HSV-1 infection of HF that were either untreated (panels (a) and (c)) or treated with 10 μM NFV (panels (b) and (d)). Low power micrographs are shown in panels (a) and (b); bar = 2 μm. White arrows indicate HSV-1 capsids in the nucleus (N), black arrowheads indicate capsids in the cytoplasm (C), and black arrows indicate extracellular (E) virions. Similar numbers of capsids were observed in the nucleus of untreated and NFV-treated cells (approximately 39 and 44, resp.). However, untreated cells were observed to have fewer capsids in the cytoplasm (~14 versus 79) and substantially more extracellular virus particles (~51 versus 3). Cytoplasmic virus particles are shown in panels (c) and (d); bar = 0.5 μm. Enveloped virus particles (indicated in panel (c) by a black arrow) were commonly observed in untreated cells but rarely in NFV-treated cells (approximately 9 versus 0, resp., in the fields shown). Capsids in the cytoplasm of NFV-treated cells were almost exclusively nonenveloped (indicated by arrowheads; approximately 16 versus 23 in panels (c) and (d), resp.).

the mechanism(s) by which NFV might inhibit production of infectious HSV-1 *in vitro*. We found that NFV acts on HSV-1 late in virus production, without a detectable effect on late viral gene expression. Furthermore, abundant virus particles were observed within infected cells, though envelopment and release of virus appeared to be substantially diminished. This is consistent with a recently published study by Kalu et al., which also reported that NFV blocked HSV-1 maturation and egress [25].

NFV showed no activity on the VP24 protease, which was expected given that it is a serine protease that lacks structural or functional similarity with the HIV (aspartyl) protease. This was shown using a transfection system to examine enzymatic activity in trans and supports the finding that scaffold protein cleavage appears unaffected in HSV-1-infected cells treated with NFV [25]. No resistance to NFV could be selected under conditions that readily resulted in acyclovir-resistant HSV-1, which is again consistent with findings by Kalu et al. [25].

(a)

Untreated   10 μM NFV

(b)

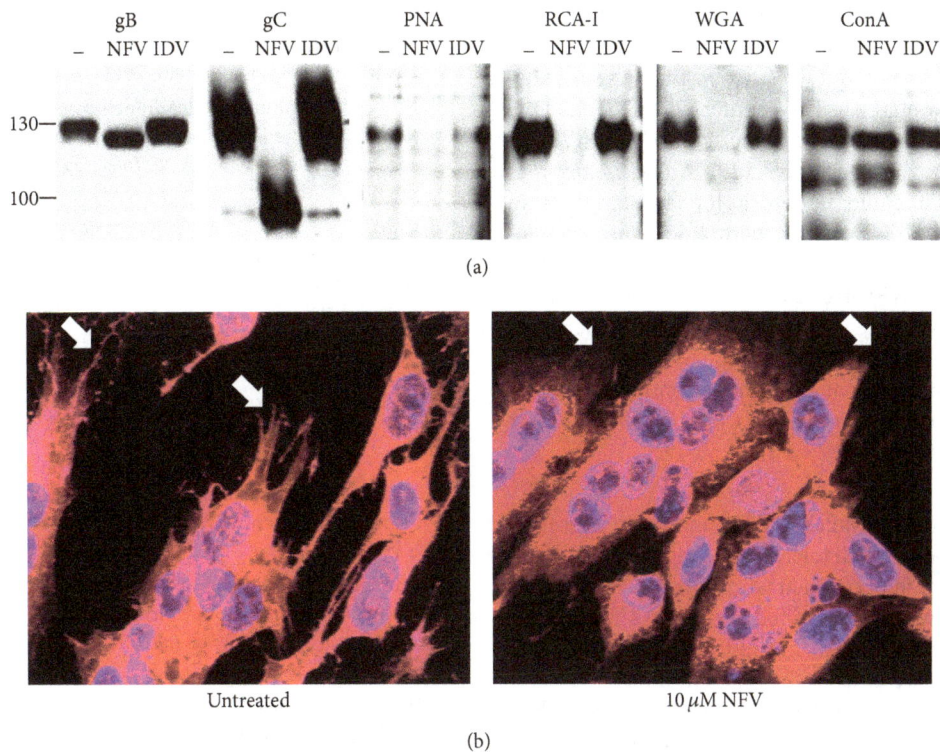

FIGURE 5: NFV affects the glycosylation and subcellular localization of viral gene products but does not change the level of expression. (a) Western blots of HSV-1 infected HF cells show increased mobility in gB and gC with NFV treatment, compared to cells that were untreated or treated with indinavir (IDV). Eastern blots show that altered staining by lectins (peanut agglutinin, PNA; ricinus communis agglutinin I, RCA-I; wheat germ agglutinin, WGA; and concanavalin A, ConA) demonstrates reduced overall glycosylation in NFV-treated cells. (b) HSV-1 gB localization (red staining) is altered in NFV-treated HF cells, showing decreased delineation of plasma membrane processes (indicated by white arrows) by IFA compared to untreated cells. Nuclei are stained blue.

Though NFV-resistant HSV-1 may well be isolated using other conditions, this finding suggests a relatively high barrier to resistance *in vitro* and suggests a mechanism of action on a host cell function required for virus production, rather than a direct effect on a viral target [37–41]. Indeed, many of the cellular functions affected by NFV have similarly been described to play a role in HSV-1 replication.

NFV inhibits cellular proteases and the proteasome, which leads to accumulation and inefficient removal of misfolded proteins in the ER and Golgi [16, 42, 43]. The finding that NFV resulted in impaired viral protein glycosylation and trafficking is consistent with these processes and again validates the recent findings by Kalu et al. [25]. Of note, based on ConA staining, N-linkage of immature (high mannose) carbohydrates appeared relatively normal [33]. These mannose structures are largely assembled in the cytoplasm, whereas trimming and modification of more complex sugar residues occur in the ER and Golgi. We found that the impairment of viral glycoprotein processing is at least one mechanism by which NFV reduces infectious HSV-1 production. Agents that induce ER stress, such as thapsigargin, similarly interfere with HSV-1 glycoprotein posttranslational processing and production of infectious virus [31]. Numerous studies have reported that tunicamycin, which blocks the synthesis of the N-acetylglucosamine-lipid

intermediates, and other inhibitors of protein glycosylation decrease the infectious yield of HSV-1 *in vitro* [44–46]. Furthermore, tunicamycin does not affect the level of late viral gene product expression, and normal appearing capsids were noted within the cytoplasm, similar to the effects we observed with NFV. It is unclear, however, whether impaired HSV-1 envelope protein glycosylation would block virus egress based on studies using cell lines deficient in N-acetylglucosaminyl transferase activity, in which virus yield was only mildly reduced [47].

This work has several important limitations. The effects of NFV are highly pleiotropic, and we stress that NFV might affect the production of infectious HSV-1 through multiple mechanisms. In addition, based on what is known about NFV's effects on tumor cells [11, 21], the most relevant mechanism(s) of action may differ with respect to individual HHV, cell type, and drug concentration. By necessity, all of the possible mechanisms by which NFV might affect HSV-1 replication were not evaluated.

Autophagy, a catabolic process that maintains cellular homeostasis under conditions of stress, is a prominent effect of NFV [11, 16]. HSV-1 encodes genes to block autophagy in infected cells, including infected cell protein (ICP) 34.5, which is required for neurovirulence [48]. Increased levels of autophagy can reduce production of infectious HSV-1 similarly to NFV, with relatively normal viral gene expression and formation of viral particles that are retained within cells [49, 50]. We were not able to formally show that NFV increased autophagy in HSV-1 infected cells under the conditions used. NFV treatment appeared to result in retention of cytoplasmic virus particles within single membrane-bound vesicles that could be late autophagosomes as in other reports [49, 50]. However, pathognomonic double-membrane structures were not convincingly observed in greater numbers in NFV-treated versus untreated HSV-1-infected cells nor were we able to show that NFV increased LC3-II staining during HSV-1 infection. Autophagy can be difficult to demonstrate [51] and cannot currently be excluded as a potential contributor to the effect of NFV on HSV-1 or other HHVs [48].

NFV is an orally bioavailable, FDA-approved treatment for HIV infection and is well tolerated during long-term use. Since the advent of more potent and convenient options, NFV is no longer widely prescribed for ART [4]. However, because of its activity against a broad range of tumor cell types, there is intense interest in repositioning NFV as a cancer chemotherapeutic agent, perhaps at higher doses than those for ART [11, 42, 52]. NFV could also be beneficial in the treatment or suppression of infection with HSV and other HHVs, particularly in the setting of resistance to first-line antivirals, in patients with cancer and/or HIV. Additional studies are indicated to further elucidate the mechanism(s) of action of NFV on HHV infections and to evaluate its efficacy in clinical trials.

Acknowledgments

This work was supported by grants from the NIH: KL2 RR025015-01, UL1 RR025014, and University of Washington Center for AIDS Research P30 AI027757.

References

[1] Panel on Opportunistic Infections in HIV-Infected Adults and Adolescents, "Guidelines for the prevention and treatment of opportunistic infections in HIV-infected adults and adolescents: recommendations from the Centers for Disease Control and Prevention, the National Institutes of Health, and the HIV Medicine Association of the Infectious Diseases Society of America," September 2014, http://aidsinfo.nih.gov/contentfiles/lvguidelines/adult_oi.pdf.

[2] R. H. Gray, M. J. Wawer, R. Brookmeyer et al., "Probability of HIV-1 transmission per coital act in monogamous, heterosexual, HIV-1-discordant couples in Rakai, Uganda," The Lancet, vol. 357, no. 9263, pp. 1149–1153, 2001.

[3] A. Jemal, F. Bray, M. M. Center, J. Ferlay, E. Ward, and D. Forman, "Global cancer statistics," CA: A Cancer Journal for Clinicians, vol. 61, no. 2, pp. 69–90, 2011.

[4] Adolescents PoAGfAa, Guidelines for the Use of Antiretroviral Agents in HIV-1-Infected Adults and Adolescents, Department of Health and Human Services, 2013.

[5] N. Ford, Z. Shubber, P. Saranchuk et al., "Burden of HIV-Related cytomegalovirus retinitis in resource-limited settings: a systematic review," Clinical Infectious Diseases, vol. 57, no. 9, pp. 1351–1361, 2013.

[6] C. M. Posavad, A. Wald, S. Kuntz et al., "Frequent reactivation of herpes simplex virus among HIV-1-infected patients treated with highly active antiretroviral therapy," Journal of Infectious Diseases, vol. 190, no. 4, pp. 693–696, 2004.

[7] S. Lodi, M. Guiguet, D. Costagliola, M. Fisher, A. de Luca, and K. Porter, "Kaposi sarcoma incidence and survival among HIV-infected homosexual men after HIV seroconversion," Journal of the National Cancer Institute, vol. 102, no. 11, pp. 784–792, 2010.

[8] E. de Clercq, "A cutting-edge view on the current state of antiviral drug development," Medicinal Research Reviews, vol. 33, no. 6, pp. 1249–1277, 2013.

[9] S. Gantt, J. Carlsson, M. Ikoma et al., "The HIV protease inhibitor nelfinavir inhibits Kaposi's sarcoma-associated herpesvirus replication in vitro," Antimicrobial Agents and Chemotherapy, vol. 55, no. 6, pp. 2696–2703, 2011.

[10] W. A. Chow, C. Jiang, and M. Guan, "Anti-HIV drugs for cancer therapeutics: back to the future?" The Lancet Oncology, vol. 10, no. 1, pp. 61–71, 2009.

[11] S. Gantt, C. Casper, and R. F. Ambinder, "Insights into the broad cellular effects of nelfinavir and the HIV protease inhibitors supporting their role in cancer treatment and prevention," Current Opinion in Oncology, vol. 25, no. 5, pp. 495–502, 2013.

[12] T. B. Brunner, M. Geiger, G. G. Grabenbauer et al., "Phase I trial of the human immunodeficiency virus protease inhibitor nelfinavir and chemoradiation for locally advanced pancreatic cancer," Journal of Clinical Oncology, vol. 26, no. 16, pp. 2699–2706, 2008.

[13] J. Buijsen, G. Lammering, R. L. H. Jansen et al., "Phase I trial of the combination of the Akt inhibitor nelfinavir and chemoradiation for locally advanced rectal cancer," Radiotherapy and Oncology, vol. 107, no. 2, pp. 184–188, 2013.

[14] R. Rengan, R. Mick, D. Pryma et al., "A phase I trial of the HIV protease inhibitor nelfinavir with concurrent chemoradiotherapy for unresectable stage IIIA/IIIB non-small cell lung cancer: a report of toxicities and clinical response," Journal of Thoracic Oncology, vol. 7, no. 4, pp. 709–715, 2012.

[15] J. Pan, M. Mott, B. Xi et al., "Phase I study of nelfinavir in liposarcoma," Cancer Chemotherapy and Pharmacology, vol. 70, no. 6, pp. 791–799, 2012.

[16] J. J. Gills, J. LoPiccolo, and P. A. Dennis, "Nelfinavir, a new anticancer drug with pleiotropic effects and many paths to autophagy," Autophagy, vol. 4, no. 1, pp. 107–109, 2008.

[17] L. Xie, T. Evangelidis, and P. E. Bourne, "Drug discovery using chemical systems biology: weak inhibition of multiple kinases may contribute to the anti-cancer effect of nelfinavir," PLoS Computational Biology, vol. 7, no. 4, Article ID e1002037, 2011.

[18] F. Y. Liu and B. Roizman, "The herpes simplex virus 1 gene encoding a protease also contains within its coding domain the gene encoding the more abundant substrate," Journal of Virology, vol. 65, no. 10, pp. 5149–5156, 1991.

[19] M. J. Karnovsk, "A formaldehyde-glutaraldehyde fixative of high osmolality for use in electron microscopy," Journal of Cell Biology, vol. 27, pp. A137–A138, 1965.

[20] L. E. Schnipper and C. S. Crumpacker, "Resistance of herpes simplex virus to acycloguanosine: role of viral thymidine kinase and DNA polymerase loci," *Proceedings of the National Academy of Sciences of the United States of America*, vol. 77, no. 4, pp. 2270–2273, 1980.

[21] J. J. Gills, J. LoPiccolo, J. Tsurutani et al., "Nelfinavir, a lead HIV protease inhibitor, is a broad-spectrum, anticancer agent that induces endoplasmic reticulum stress, autophagy, and apoptosis in vitro and in vivo," *Clinical Cancer Research*, vol. 13, no. 17, pp. 5183–5194, 2007.

[22] A. M. Petrich, V. Leshchenko, P.-Y. Kuo et al., "Akt inhibitors MK-2206 and nelfinavir overcome mTOR inhibitor resistance in diffuse large B-cell lymphoma," *Clinical Cancer Research*, vol. 18, no. 9, pp. 2534–2544, 2012.

[23] M. Kraus, J. Bader, H. Overkleeft, and C. Driessen, "Nelfinavir augments proteasome inhibition by bortezomib in myeloma cells and overcomes bortezomib and carfilzomib resistance," *Blood Cancer Journal*, vol. 3, no. 3, article e103, 2013.

[24] M.-J. Hsu, C.-Y. Wu, H.-H. Chiang, Y.-L. Lai, and S.-L. Hung, "PI3K/Akt signaling mediated apoptosis blockage and viral gene expression in oral epithelial cells during herpes simplex virus infection," *Virus Research*, vol. 153, no. 1, pp. 36–43, 2010.

[25] N. N. Kalu, P. J. Desai, C. M. Shirley, W. Gibson, P. A. Dennis, and R. F. Ambinder, "Nelfinavir inhibits maturation and export of herpes simplex virus 1," *Journal of Virology*, vol. 88, no. 10, pp. 5455–5461, 2014.

[26] M. Guan, K. Fousek, and W. A. Chow, "Nelfinavir inhibits regulated intramembrane proteolysis of sterol regulatory element binding protein-1 and activating transcription factor 6 in castration-resistant prostate cancer," *The FEBS Journal*, vol. 279, no. 13, pp. 2399–2411, 2012.

[27] E. Mahoney, K. Maddocks, J. Flynn et al., "Identification of endoplasmic reticulum stress-inducing agents by antagonizing autophagy: a new potential strategy for identification of anticancer therapeutics in B-cell malignancies," *Leukemia & Lymphoma*, vol. 54, no. 12, pp. 2685–2692, 2013.

[28] S. Thomas, N. Sharma, E. B. Golden et al., "Preferential killing of triple-negative breast cancer cells in vitro and in vivo when pharmacological aggravators of endoplasmic reticulum stress are combined with autophagy inhibitors," *Cancer Letters*, vol. 325, no. 1, pp. 63–71, 2012.

[29] A. Brüning, P. Burger, M. Vogel et al., "Nelfinavir induces the unfolded protein response in ovarian cancer cells, resulting in ER vacuolization, cell cycle retardation and apoptosis," *Cancer Biology and Therapy*, vol. 8, no. 3, pp. 222–228, 2009.

[30] D. C. Johnson and P. G. Spear, "O-linked oligosaccharides are acquired by herpes simplex virus glycoproteins in the Golgi apparatus," *Cell*, vol. 32, no. 3, pp. 987–997, 1983.

[31] S. Chatterjee, S. Nishimuro, and R. J. Whitley, "Expression of HSV-1 glycoproteins in tunicamycin-treated monkey kidney cells," *Biochemical and Biophysical Research Communications*, vol. 167, no. 3, pp. 1139–1145, 1990.

[32] C. E. Isaacs, W. Xu, R. K. Pullarkat, and R. Kascsak, "Retinoic acid reduces the yield of herpes simplex virus in Vero cells and alters the N-glycosylation of viral envelope proteins," *Antiviral Research*, vol. 47, no. 1, pp. 29–40, 2000.

[33] M. E. Taylor and K. Drickamer, *Introduction to Glycobiology*, Oxford University Press, Oxford, UK, 2011.

[34] A. Varki, "NCBI bookshelf," in *Essentials of Glycobiology*, p. 1, Cold Spring Harbor Laboratory Press, New York, NY, USA, 2nd edition, 2009.

[35] N. Yamamoto, R. Yang, Y. Yoshinaka et al., "HIV protease inhibitor nelfinavir inhibits replication of SARS-associated coronavirus," *Biochemical and Biophysical Research Communications*, vol. 318, no. 3, pp. 719–725, 2004.

[36] M. Federico, "HIV-protease inhibitors block the replication of both vesicular stomatitis and influenza viruses at an early post-entry replication step," *Virology*, vol. 417, no. 1, pp. 37–49, 2011.

[37] F. N. Linero, C. S. Sepúlveda, F. Giovannoni et al., "Host cell factors as antiviral targets in arenavirus infection," *Viruses*, vol. 4, no. 9, pp. 1569–1591, 2012.

[38] W. Coley, K. Kehn-Hall, R. van Duyne, and F. Kashanchi, "Novel HIV-1 therapeutics through targeting altered host cell pathways," *Expert Opinion on Biological Therapy*, vol. 9, no. 11, pp. 1369–1382, 2009.

[39] M. A. Khattab, "Targeting host factors: a novel rationale for the management of hepatitis C virus," *World Journal of Gastroenterology*, vol. 15, no. 28, pp. 3472–3479, 2009.

[40] S. A. Krumm, J. M. Ndungu, J.-J. Yoon et al., "Potent host-directed small-molecule inhibitors of myxovirus RNA-dependent RNA-polymerases," *PLoS ONE*, vol. 6, no. 5, Article ID e20069, 2011.

[41] B. Pastorino, A. Nougairède, N. Wurtz, E. Gould, and X. de Lamballerie, "Role of host cell factors in flavivirus infection: implications for pathogenesis and development of antiviral drugs," *Antiviral Research*, vol. 87, no. 3, pp. 281–294, 2010.

[42] A. Brüning, A. Gingelmaier, K. Friese, and I. Mylonas, "New prospects for nelfinavir in non-HIV-related diseases," *Current Molecular Pharmacology*, vol. 3, no. 2, pp. 91–97, 2010.

[43] P. Pyrko, A. Kardosh, W. Wang, W. Xiong, A. H. Schönthal, and T. C. Chen, "HIV-1 protease inhibitors nelfinavir and atazanavir induce malignant glioma death by triggering endoplasmic reticulum stress," *Cancer Research*, vol. 67, no. 22, pp. 10920–10928, 2007.

[44] L. I. Pizer, G. H. Cohen, and R. J. Eisenberg, "Effect of tunicamycin on herpes simplex virus glycoproteins and infectious virus production," *Journal of Virology*, vol. 34, no. 1, pp. 142–153, 1980.

[45] E. Katz, E. Margalith, and D. Duksin, "Antiviral activity of tunicamycin on herpes simplex virus," *Antimicrobial Agents and Chemotherapy*, vol. 17, no. 6, pp. 1014–1022, 1980.

[46] K. G. Kousoulas, D. J. Bzik, N. Deluca, and S. Person, "The effect of ammonium chloride and tunicamycin on the glycoprotein content and infectivity of herpes simplex virus type 1," *Virology*, vol. 125, no. 2, pp. 468–474, 1983.

[47] G. Campadelli-Fiume, L. Poletti, F. Dall'Olio, and F. Serafini-Cessi, "Infectivity and glycoprotein processing of herpes simplex virus type 1 grown in a ricin-resistant cell line deficient in N-acetylglucosaminyl transferase I," *Journal of Virology*, vol. 43, no. 3, pp. 1061–1071, 1982.

[48] G. S. Taylor, J. Mautner, and C. Münz, "Autophagy in herpesvirus immune control and immune escape," *Herpesviridae*, vol. 2, article 2, 2011.

[49] V. le Sage and B. W. Banfield, "Dysregulation of autophagy in murine fibroblasts resistant to HSV-1 infection," *PLoS ONE*, vol. 7, no. 8, Article ID e42636, 2012.

[50] Y. Pei, Z.-P. Chen, H.-Q. Ju et al., "Autophagy is involved in antiviral activity of pentagalloylglucose (PGG) against Herpes simplex virus type 1 infection *in vitro*," *Biochemical and Biophysical Research Communications*, vol. 405, no. 2, pp. 186–191, 2011.

[51] D. J. Klionsky, H. Abeliovich, P. Agostinis et al., "Guidelines for the use and interpretation of assays for monitoring autophagy in higher eukaryotes," *Autophagy*, vol. 4, no. 2, pp. 151–175, 2008.

[52] W. Wu, R. Zhang, and D. R. Salahub, "Nelfinavir: a magic bullet to annihilate cancer cells?" *Cancer Biology and Therapy*, vol. 8, pp. 233–235, 2009.

6

Recombinant Varicella-Zoster Virus Vaccines as Platforms for Expression of Foreign Antigens

Wayne L. Gray

Department of Microbiology and Immunology, University of Arkansas for Medical Sciences, 4301 West Markham Street, Little Rock, AR 72205, USA

Correspondence should be addressed to Wayne L. Gray; graywaynel@uams.edu

Academic Editor: Alain Kohl

Varicella-zoster virus (VZV) vaccines induce immunity against childhood chickenpox and against shingles in older adults. The safety, efficacy, and widespread use of VZV vaccines suggest that they may also be effective as recombinant vaccines against other infectious diseases that affect the young and the elderly. The generation of recombinant VZV vaccines and their evaluation in animal models are reviewed. The potential advantages and limitations of recombinant VZV vaccines are addressed.

1. Introduction

Varicella-zoster virus (VZV) vaccines provide immune protection against diseases that affect both the young and the elderly. The live, attenuated varicella vaccine (VARIVAX) immunizes children against chickenpox, a childhood disease characterized by fever and vesicular skin rash. The VZV Zostavax vaccine protects older adults against herpes zoster (shingles), a vesicular skin disease caused by VZV reactivation from latently infected neural ganglia.

The proven safety and effectiveness of the varicella and shingles vaccines provide support for recombinant VZV (rVZV) vaccines to induce immunity against not only VZV but also against other pathogens. The ability of VZV vaccines to safely induce long-lasting humoral and cellular immune responses provides advantages over other live, attenuated vaccine vectors and over killed and subunit vaccines. This review summarizes research to develop and evaluate VZV vectors as recombinant vaccines against other diseases.

1.1. Varicella and Herpes Zoster. Varicella is a highly contagious disease of children and adolescents [1]. The infectious agent is transmitted through aerosols by coughs and sneezes or by direct contact with rash secretions. Following a 10–14 day incubation period, fever and malaise arise along with the characteristic vesicular skin rash, which occurs on the face, torso, and extremities. Chickenpox is generally a benign disease in otherwise healthy children with symptoms usually completely resolved within two weeks of disease onset and with subsequent life-long immunity against varicella. However, complications may include varicella pneumonia, hepatitis, encephalitis, and secondary bacterial infections. Immunocompromised children, including cancer and AIDS patients, are particularly susceptible to severe, sometimes life-threatening varicella [2, 3]. Adults who escape VZV exposure during childhood have more severe varicella symptoms and complications. VZV infection in pregnant women during the first 28 weeks of gestation may lead to fetal infection and congenital malformations [4]. Later maternal infection may cause neonatal varicella. Prior to routine vaccination, varicella was annually responsible for 100–200 varicella-related deaths and 11,000 hospitalizations in the U.S.

Following resolution of the primary disease, VZV establishes life-long latent infection within neural ganglia [5]. Later in life, the virus may reactivate, travel along sensory nerves, and reinfect the skin. Herpes zoster, which occurs primarily, but not exclusively, in older adults, is characterized by unilateral pain and a vesicular rash that are limited to the dermatome innervated by a single spinal or cranial nerve [1]. Shingles affects approximately 20–30% of individuals with estimated one million cases in the U.S. each year. Postherpetic neuralgia, a common complication of zoster, is characterized by intense, sometimes chronic, pain. Immunosuppressed cancer and AIDS patients are at risk for severe herpes zoster with possible disseminated disease [3].

Acyclovir and related nucleoside analogs may be effective against chickenpox and shingles, particularly if administered early in the course of infection. However, prevention against VZV disease outbreaks has been facilitated by development of VZV vaccines.

1.2. VZV Vaccines. The VZV VARIVAX vaccine induces immunity and protection against chickenpox in children [1]. The vaccine, approved in the U.S. in 1995, is the first human herpesvirus vaccine licensed for clinical use. The live vaccine is derived from a clinical isolate, VZV Oka, and was attenuated by passage in guinea pig embryo and human WI38 cells, although the molecular basis of attenuation remains unknown. An initial immunization is recommended for children 12–15 months of age followed by a booster vaccination at ages 4 to 6. Administered by intradermal inoculation, this two-dose immunization approach reduces the incidence of varicella in children by 90–95%. The varicella vaccine is now recommended as a routine childhood vaccination in the U.S. with over 80% of school-age children immunized each year. Vaccination is also beneficial for susceptible adults, who are at risk for severe varicella. The vaccine is well tolerated with only minor side effects, including soreness at the inoculation site, fever, and mild rash. The VZV vaccine is even safe and effective for some cancer patients and for some immunosuppressed patients, including HIV-infected children with depression of CD4+ T cells [6]. The vaccine is not recommended for pregnant females. The VZV Oka vaccine virus is capable of establishing latent infection in neural ganglia and may reactivate to cause herpes zoster, although the incidence appears to be significantly less than that caused by wild-type VZV [7, 8].

The shingles (Zostavax) vaccine is also composed of live, attenuated VZV Oka but is administered at a >10-fold higher viral dose than the varicella vaccine [9]. The vaccine, administered by subcutaneous injection, reduces the incidence of shingles and postherpetic neuralgia overall by 50% and 67%, respectively, and lasts at least for 5 years [10]. The U.S. Food and Drug Administration (FDA) approved the shingles vaccine in 2006 for individuals aged 50 and older. The Centers for Disease Control (CDC) recommend that adults aged 60 and older receive this vaccine. Vaccine efficacy decreases with advancing age (64% in persons 60–69 years, 41% in persons 70–79, and 18% in persons over 80) [11]. Only minor side effects are observed, including possible irritation at the site of inoculation and headache. The shingles vaccine is not approved for immunocompromised individuals, although it has been shown to be safe and effective in some cancer patients [12].

2. The VZV Oka Vaccine as a Potential Recombinant Vaccine Vector

The varicella and shingles VZV vaccines are candidates as vehicles for delivery of foreign antigens and immunization against other infectious agents. VZV vaccines are highly immunogenic inducing long-lasting neutralizing antibody and cellular immune responses. In addition to their proven safety, efficacy, and widespread use, the vaccines offer advantages as a vector for the expression of heterologous antigens

(1) The size of the VZV genome (125 kb) allows stable insertion of multiple heterologous genes into specific loci without effecting viral replication [28].

(2) VZV replicates in the cell nucleus, so foreign genes may be properly spliced, and viral antigens are expressed, processed, and presented in infected cells as they are in natural infection [25].

(3) The replication competent virus amplifies foreign antigen expression—a potential advantage over replication defective vectors.

(4) A protective Th1 type T-cell immune response against VZV and the foreign antigen is induced [26].

(5) VZV DNA does not contain genetic elements associated with oncogenesis.

(6) The VZV host range is limited to man, limiting uncontrolled environmental spread.

(7) Periodic subclinical reactivation of VZV from latency may provide restimulation of immune responses to VZV and foreign antigens as discussed below [58].

A recombinant varicella or shingles vaccine could be ideally suited for prevention of other diseases that affect children or the elderly, respectively. An rVZV vaccine offers greater safety as a general population vaccine compared to other live viral vectors, such as vaccinia virus, and the varicella vaccine is the only live viral vaccine approved for certain groups of cancer and otherwise immunosuppressed patients, including children with HIV infection [13].

3. Genetic Approaches for Generation of rVZV

Initial approaches to insert foreign genes into the VZV genome involved cotransfection of susceptible human melanoma cells (MeWo) with genomic VZV DNA and a foreign gene flanked by homologous VZV DNA sequences. Genetic recombination yielded rVZV plaques which were detected by immunohistochemistry using antibody to the foreign gene product [14]. rVZV was plaque purified upon serial propagation. While successful, this approach was laborious, due to the cell-associated nature of VZV, making it difficult to clonally isolate rVZV.

Development of a cosmid-based system for manipulation of the VZV genome was an important advance in VZV genetics [15]. The 125 kb VZV genome was cloned into four overlapping 30–45 kb cosmids which are cotransfected into MeWo cells. Recombination of the homologous overlapping DNA sequences yields infectious rVZV and viral plaques within ten days (Figure 1). This approach facilitates creation of rVZV mutants by site-specific mutagenesis or insertion of foreign genes within individual VZV cosmids by RecA-assisted restriction endonuclease (RARE) cleavage [16], followed by transfection of all four VZV cosmids into MeWo

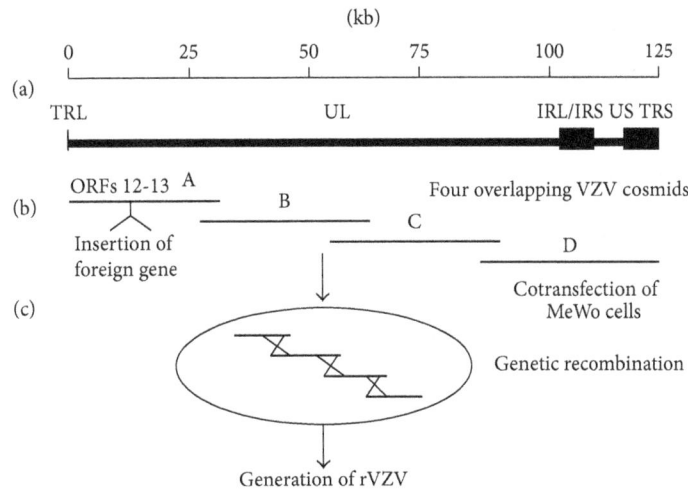

FIGURE 1: Cosmid approach to generate rVZV. (a) The 125 kb VZV genome consists of a unique long (UL) component flanked by internal and terminal inverted repeat sequences (IRL and TRL) covalently linked to a unique short (US) component bracketed by internal and terminal inverted repeats (IRS and TRS) [17]. (b) Four cosmids (30–45 kb), representing the entire VZV genome, are transfected into MeWo cells yields infectious virus. In this example, a foreign gene with a promoter is inserted into cosmid A, within the intergenic region between VZV ORFs 12 and 13.

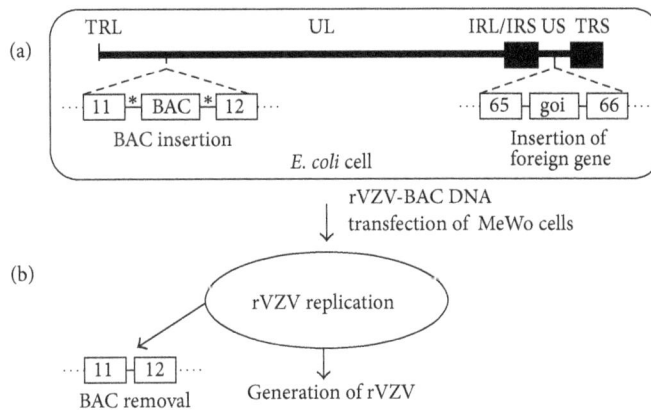

FIGURE 2: VZV-BAC approach to generate rVZV. BAC sequences are inserted into the VZV genome within the intergenic region between ORFs 11 and 12 and maintained within E. coli. A foreign gene of interest (goi) with promoter is inserted between ORFs 65 and 66. rVZV-BAC DNA is transfected into MeWo cells to generate infectious rVZV. BAC sequences flanked by lox P sites (*) may be removed from the VZV genome by Cre recombination.

cells. The effectiveness of this system is limited by the requirement for efficient transfection of all four of the cosmids into the same cell and multiple recombination events.

An improvement over the cosmid system was insertion of the entire VZV genome into a bacterial artificial chromosome (BAC) which allowed stable maintenance of the VZV genome within E. coli (Figure 2) [18–20]. Infectious VZV can be reconstituted upon transfection of rVZV-BAC DNA into MeWo cells. The BAC sequences have been inserted within the intergenic region of VZV open reading frames (ORFs) 11 and 12 [18, 19] or between ORFs 65 and 66 [20] and are generally flanked by loxP sites, permitting removal of the bacterial vector sequences from the VZV genome by Cre recombination. A recent advance permits markerless, self-excision of the BAC sequences upon reconstitution in MeWo cells [20]. Development of a VZV-BAC provides an efficient approach for introduction of site-specific mutations and insertions, including foreign genes, into the VZV genome.

4. Insertion of Foreign Genes within the VZV Genome

The large size (125 kb) of VZV DNA facilitates insertion of one or multiple foreign genes within the VZV genome. VZV encodes several genes which are nonessential for in vitro replication and are targets for insertion of foreign genes. For example, several rVZVs have incorporated the foreign gene within the viral thymidine kinase (TK) gene [14, 21, 22]. These rVZV-TK$^-$ viruses replicate as efficiently as wild-type VZV in cell culture. Generation of rVZV-TK$^-$ is facilitated by selection with nucleoside analogs such as bromodeoxyuridine.

However, rVZV vaccines generated by insertion of foreign genes within the viral TK gene may not be clinically useful as effective antiviral treatment for VZV infections with acyclovir and other nucleoside analogs depends on a functional TK. In other cases, the exogenous genes are inserted within the intergenic regions of VZV ORFs 12 and 13 or between ORFs 65 and 66.

A variety of recombination methods including *RecA+* recombination and *Red* recombination may be used to insert genes of other pathogens into specific sites within VZV cosmids or BACs taking advantage of homologous flanking VZV sequences [20, 23]. Tn7-mediated site-specific transposition has also been used to introduce a foreign gene into VZV-BAC DNA [24].

Foreign genes within rVZV have been expressed from an endogenous VZV promoter, such as the TK or gE promoters [14, 21]. In these rVZVs, the foreign gene is expressed as an immediate early, early, or late gene as during natural infection. In other rVZVs, the foreign gene is expressed from the human cytomegalovirus immediate early gene promoter/enhancer (HCMV-IE) [23]. While this commonly used promoter induces strong gene expression of foreign genes, long-term constitutive expression could result in deleterious accumulation of the recombinant protein.

5. Recombinant VZV Vaccines

Several studies provide support for the VZV vaccine virus as a vector for the expression of foreign genes. Initially, an rVZV expressing the Epstein-Barr virus (EBV) membrane glycoprotein (gp350/220) was generated and plaque purified [14]. The VZV gpI (gE) promoter and initial 35 amino acids were fused in-frame with the EBV gp gene beginning at codon 21. This construct flanked by VZV TK DNA sequences was cotransfected with VZV Oka DNA into MeWo cells. Homologous recombination of TK sequences permitted insertion of the EBV gp into VZV DNA within the viral TK gene, as confirmed by Southern blot hybridization. Approximately, 10% of viral plaques expressed EBV gp antigen as detected by immunohistochemistry using EBV monoclonal antibody. The EBV antigen was glycosylated and presented on cellular plasma membranes in a manner similar to expression in EBV-infected cells [14]. The EBV gp was also demonstrated on rVZV virions by immune electron microscopy. Due to insertional inactivation of the viral TK gene, the rVZV replicated in infected MeWo cells in the presence of bromodeoxyuridine and this property can be used for selection of rVZV.

An rVZV Oka expressing the hepatitis B surface antigen (HBs) was generated by inserting the HBs gene into the viral TK ORF [21, 25]. The HBs gene was expressed from the VZV TK promoter, initiated at the VZV TK ATG initiation codon, and followed by codons for 25 amino acids of the HBs pre-S2 and the entire HBs protein. The construct was cotransfected with VZV Oka DNA into MRC-5 (human lung) cells. rVZV-HBs virus was detected by immunofluorescence using monoclonal antibody to HBs and was plaque purified. The HBs was synthesized as 26K and 30K proteins within

infected cells. The HBs was glycosylated with N- and O-linked glycans and secreted as 30K and 35K proteins in the culture supernatant [25]. Immunization of guinea pigs with rVZV-HBs by subcutaneous inoculation induced antibody titers to HBs comparable to that induced by recombinant HBs generated in yeast. In addition, rVZV-HBs induced Th1-type cell-mediated immune responses including delayed-type hypersensitivity [26].

The herpes simplex type 2 (HSV-2) glycoprotein D (gD) was inserted into the unique short region of the VZV Oka genome within the intergenic regions between ORFs 65 and 66 using the VZV cosmid genetic system [27]. HSV-2 gD expression was driven more efficiently from the endogenous HSV-2 gD promoter than the VZV gE promoter. The HSV-2 gD was expressed within the cell cytoplasm and on the surface of rVZV-HSV2gD infected cells as indicated by immunofluorescence and immunoblot analysis using HSV-2 gD-specific antibody. In addition, immune electron microscopy detected HSV-2 gD on the envelope of rVZV-HSV2gD virions. The HSV-2 gD expression did not affect viral replication as the VZV-HSV2gD replicated as efficiently as wild-type VZV Oka in cell culture. The VZV genome does not encode a gD homolog which may be related to its species specificity for a limited number of human cell lines. However, addition of the HSV-2 gD did not enhance the permissiveness VZV for cells which it does not efficiently replicate. Immunization of guinea pigs with rVZV-HSV2gD induced neutralizing antibodies against HSV-2 and significantly reduced disease severity, including prevention of hindlimb paralysis and mortality following intravaginal HSV-2 challenge. In a subsequent study, an rVZV Oka expressing both the HSV-2 gD and gB was constructed [28]. The HSV-2 gD was inserted within the VZV ORF 13 and expressed from the natural HSV-2 gD promoter, while the HSV-2 gB was inserted within the ORFs 65-66 intergenic region and driven by the SV40 early promoter. Immunization of guinea pigs by subcutaneous inoculation induced humoral and cellular immune responses against HSV-2 gB and gD, including neutralizing antibodies, and reduced viral shedding (>100 fold), disease severity, and mortality following intravaginal HSV-2 challenge. However, the rVZV-HSV2gD and rVZV-HSV2gBgD vaccines did not prevent HSV-2 latency and reactivation.

An rVZV vaccine expressing the HIV-1 env gene was constructed by insertion of the HIV-1 env gene encoding amino acids 296–463 fused to HBs antigen within the VZV TK gene [22]. The rVZV-HIVenv was selected by resistance to sorivudine, a nucleoside analog, and immunofluorescence employing an HIVenv monoclonal antibody. rVZV-HIVenv immunization induced humoral and cellular immune responses against the HIV gp120 antigen in vaccinated guinea pigs.

The mumps virus (MV) hemagglutinin-neuraminidase (HN) gene was inserted within the VZV Oka ORF 13 using the VZV-BAC system [23]. The HN was efficiently expressed in the cytoplasm and on the surface of rVZV-MVHN infected cells but not within the envelope of purified virions. The rVZV-MVHN replicated as efficiently as wild-type VZV, and cell tropism was not altered. Immunization of guinea pigs induced neutralizing antibodies against VZV and MV.

A subsequent study showed that MV fusion (F) protein could be inserted into the rVZV genome, and the F protein was efficiently expressed in rVZV-MVF infected cells [24].

The ability of an rVZV vaccine to induce immunity against simian immunodeficiency virus (SIV) was investigated in a nonhuman primate model [29]. The SIV env gene expressed from the HCMV IE promoter was inserted into the VZV genome between ORFs 65 and 66. Rhesus monkeys, immunized with rVZV-SIVenv by intranasal, intratracheal, and intramuscular inoculation, generated antibodies to VZV and SIV env as indicated by ELISA and immunoprecipitation. However, SIV-specific cytotoxic T lymphocyte or CD8+ T-cell responses were not detected. Following challenge with pathogenic SIVsmE660, rVZV-SIV immunized animals exhibited increased SIV titers, rapid CD4 T-cell proliferation, and enhanced progression to simian AIDS, compared to control monkeys immunized with wild-type VZV. The authors speculated that the limited ability of VZV to replicate in tissues of rhesus monkeys led to inefficient induction of CD8+ T-cell responses and that the CD4 T-cell proliferation resulted in higher SIV replication and enhanced disease progression. Thus, nonhuman primates may not be the optimal animal model to evaluate rVZV-SIV vaccines. The problem can be avoided by using the simian varicella model to evaluate potential recombinant varicella vaccines as described below.

rVZV expressing foreign genes can also be employed to investigate the molecular basis of viral pathogenesis. The VZV cosmid system was used to generate rVZV Oka expressing a fusion protein consisting of the immediate early gene ORF 63 (IE63) and its duplicate gene ORF 70 fused in frame to click beetle luciferase genes [30]. IE63 expression was detected within the cytoplasm of rVZV-63/70-*Luc*-infected melanoma cells by confocal microscopy and the virus replicated *in vitro* as efficiently as wild-type VZV. The rVZV-63/70-*Luc* retained pathogenicity in the SCIDhu mouse model of VZV pathogenesis. *In vivo* IE63 expression over the course of at least 21 days was measured in rVZV-63/70-*Luc*-infected human skin and dorsal root ganglia xenografts by whole animal imaging.

6. Simian Varicella, a Model to Evaluate Recombinant Varicella Vaccines

Studies on VZV pathogenesis, antiviral therapy, and vaccines are hampered because VZV inoculation of laboratory animals, ranging from mice to monkeys, fails to induce a varicella-like disease. However, a natural disease closely resembling human varicella occurs in nonhuman primates with symptoms characterized by fever and vesicular skin rash [31]. Simian varicella epizootics occur sporadically in facilities housing Old World monkeys [32]. In some epizootics, the clinical symptoms are mild, similar to the rather benign varicella seen normally in children. In others, a more severe, disseminated varicella disease is observed with symptoms similar to those seen in some adults and immunosuppressed patients.

The causative agent, simian varicella virus (SVV), is a primate herpesvirus with properties resembling human VZV. The SVV and VZV genomes are similar in size and structure, share 70–75% DNA homology, and are colinear in gene organization [33–35]. SVV and VZV are antigenically related as indicated by the ability of VZV immunization to confer immune protection against acute simian varicella in monkeys following SVV challenge [36, 37]. Like VZV, SVV establishes latent infection in neural ganglia and may reactivate to cause secondary disease similar to herpes zoster [38, 39].

Based upon the genetic and antigenic relatedness of SVV and VZV and the clinical similarities between simian and human varicella, SVV infection of primates is a useful animal model for studying viral pathogenesis and for evaluating the effectiveness of antiviral agents and vaccines [31, 40, 41].

The SVV model offers an approach to evaluate recombinant varicella vaccines. The initial recombinant SVV (rSVV) was generated by insertion of the green fluorescent protein (GFP) gene expressed from the Rous sarcoma virus (RSV) promoter into the intergenic region between SVV ORFs 65 and 66 by homologous recombination [42]. Insertion of the GFP gene into this region of the SVV genome did not inhibit viral growth in cell culture. The rSVV-GFP was pathogenic in African green monkeys inducing necrotizing pneumonitis upon intratracheal inoculation. GFP expression was detected within infected cells derived from lung tissues on day 10 postinfection.

An SVV cosmid genetic system was employed to generate an rSVV expressing SIV antigens. The SIV gag and env genes were inserted within the SVV glycoprotein C (ORF 14) gene, and SIV antigen expression in infected Vero cells was confirmed by immunofluorescence and immunoblot analysis using SIV monoclonal antibodies [43]. The rSVV-SIVenv and rSVV-SIVgag replicated as efficiently in infected Vero cells as wild-type SVV. The rSVV-SIVenv and rSVV-SIVgag were initially evaluated by coinfection in infected African green monkeys inoculated by intratracheal and subcutaneous inoculation. The viruses were attenuated in the monkeys as indicated by reduced skin rash and viremia compared to wild-type SVV infection [43, 44]. The attenuation is likely due to the insertional inactivation of the SVV gC, which has been associated with VZV and herpesvirus virulence [45–47]. The rSVV-SIVgag and SVV-SIVenv vaccines induced humoral and cellular immune responses to the SIV antigens as revealed by ELISA and ELISPOT analyses, respectively. Each of the viruses also established viral latency as indicated by detection of rSVV DNA in neural ganglia by PCR analysis.

A subsequent study evaluated the ability of the rSVV-SIVenv and rSVV-SIVgag vaccines to immunize rhesus macaque monkeys against SIV infection [44]. The rSVVgag/env immunization induced antibody and cellular immune responses against SIV. Six months following the rSVV immunization, the monkeys were challenged with pathogenic SIV strain SIVmac251-CX-1 by intravenous inoculation. The rSVV-SIVgag/env immunization reduced SIV plasma viral loads by 100-fold in rSVV-SIVgag/env immunized monkeys compared to monkeys immunized with a negative control rSVV. Increased CD4+ T-cell proliferation and SIV-specific polyfunctional cytokine responses correlated to the reduced viremia in rSVV-SIVgag/env immunized monkeys [48]. The results of this study suggest that an rVZV-HIV vaccine could be an effective approach for AIDS vaccination.

An rVZV vaccine expressing respiratory syncytial virus (RSV) antigens could be effective for immunization of children or older adults who are most susceptible to RSV-induced respiratory disease. To evaluate a recombinant varicella vaccine, rSVV expressing the RSV glycoprotein G and the M2 matrix protein were constructed employing the SVV cosmid system [49]. Immunization of rhesus monkeys with the rSVV-RSVG and M2 vaccines induced neutralizing antibody responses against RSV.

The 125 kb SVV genome has recently been cloned into a BAC and stably maintained in *E. coli* [50]. The SVV BAC has been used to insert site-specific mutations within the SVV genome using *Red*-mediated recombination. The SVV BAC genetic system will facilitate the generation of rSVV expressing foreign genes and evaluation of candidate recombinant vaccines.

7. Influence of Preexisting VZV Immunity and Viral Latency on rVZV Vaccines

A recombinant varicella vaccine may be effective for VZV seronegative children. However, older children and adults are generally VZV seropositive either by natural VZV infection or by prior VZV vaccination. Preexisting antibodies to VZV could potentially hamper the effectiveness of an rVZV vaccine, possibly by neutralization of the vaccine virus. Indeed, humoral and cellular immune responses to a model antigen expressed by a recombinant HSV-1 were reduced in HSV-1 seropositive mice [51]. However, effective immunization under this constraint may be feasible. Preexisting immunity to the vector may not hamper immune responses to the recombinant antigen when the vaccine is administered by a route different from that of natural infection. For example, subcutaneous or mucosal immunization is not inhibited by prior systemic infection [52, 53]. In addition, the immunogenicity of some recombinant vectors, such as cytomegalovirus, measles, and Sindbis virus, are not influenced by preexisting immunity to the vector [54–56]. The influence of preexisting immunity on the immunogenicity of rVZV vaccines has not been investigated but is an important factor to be addressed to assess vaccine safety and effectiveness.

The VZV vaccine virus, like wild-type VZV, can establish latent infection in neural ganglia and has the potential to reactivate to cause herpes zoster [7, 57]. However, the incidence of reactivation disease for the VZV vaccine virus appears to be significantly less than that caused by wild-type VZV [7]. VZV may undergo sporadic subclinical reactivation resulting in periodic restimulation of VZV immune responses and life-long immunity against VZV [58]. Subclinical reactivation of an rVZV vaccine virus could be an advantage, providing sporadic immune boosts against both the VZV antigens as well as the recombinant antigen.

8. Conclusions

Viral vectors offer novel opportunities as vaccine platforms, particularly against diseases for which traditional vaccine approaches have not been effective. Recombinant viral vaccines against several human diseases are currently in clinical trials [59]. Most of these recombinant vaccines utilize adenovirus or poxvirus as viral vectors. The attenuated, replication-competent VZV Oka vaccine, with widespread use, an established safety profile, and an ability to induce long-lasting antibody and T-cell-mediated immunity provides an opportunity to develop effective recombinant vaccines. rVZV vaccines could be utilized as a primary vaccine or as part of a heterologous prime-boost strategy coupled with a DNA vaccine or adenovirus/poxvirus combination. rVZV varicella and shingles vaccines offer the flexibility to develop immunization strategies against diseases that affect children and the elderly, respectively.

Acknowledgment

This work was supported by Public Health Service Grant AI052373 of the National Institutes of Health.

References

[1] J. I. Cohen, S. E. Straus, and A. M. Arvin, "Varicella-zoster virus replication, pathogenesis, and management," in *Field's Virology*, D. M. Knipe, P. M. Howley, D. E. Griffin et al., Eds., pp. 2774–2840, Lippincott Williams & Wilkins, Philadelphia, Pa, USA, 5th edition, 2007.

[2] S. Feldman, W. T. Hughes, and C. B. Daniel, "Varicella in children with cancer: seventy seven cases," *Pediatrics*, vol. 56, no. 3, pp. 388–397, 1975.

[3] S. M. Wood, S. S. Shah, A. P. Steenhoff, and R. M. Rutstein, "Primary varicella and herpes zoster among HIV-infected children from 1989 to 2006," *Pediatrics*, vol. 121, no. 1, pp. e150–e156, 2008.

[4] A. Shrim, G. Koren, D. Farine et al., "Management of varicella infection (chickenpox) in pregnancy," *Journal of the Obstetrics and Gynaecology Canada*, vol. 34, no. 3, pp. 287–292, 2012.

[5] D. Gilden, R. Mahalingam, M. A. Nagel, S. Pugazhenthi, and R. J. Cohrs, "Review: the neurobiology of varicella zoster virus infection," *Neuropathology and Applied Neurobiology*, vol. 37, no. 5, pp. 441–463, 2011.

[6] M. J. Levin, A. A. Gershon, A. Weinberg, L.-Y. Song, T. Fentin, and B. Nowak, "Administration of live varicella vaccine to HIV-infected children with current or past significant depression of CD4+ T cells," *Journal of Infectious Diseases*, vol. 194, no. 2, pp. 247–255, 2006.

[7] A. A. Gershon, C. Chen, and L. Davis, "Latency of varicella zoster virus in dorsal root, cranial, and enteric ganglia in vaccinated children," *Transactions of the American Clinical and Climatological Association*, vol. 123, pp. 17–35, 2012.

[8] S. S. Chaves, P. Haber, K. Walton et al., "Safety of varicella vaccine after licensure in the United States: experience from reports to the vaccine adverse event reporting system, 1995–2005," *Journal of Infectious Diseases*, vol. 197, no. 2, pp. S170–S177, 2008.

[9] M. N. Oxman, "Zoster vaccine: current status and future prospects," *Clinical Infectious Diseases*, vol. 51, no. 2, pp. 197–213, 2010.

[10] M. N. Oxman, M. J. Levin, G. R. Johnson et al., "A vaccine to prevent herpes zoster and postherpetic neuralgia in older adults," *New England Journal of Medicine*, vol. 352, no. 22, pp. 2271–2365, 2005.

[11] K. E. Schmader, M. J. Levin, J. W. Gnann et al., "Efficacy, safety, and tolerability of herpes zoster vaccine in persons aged 50–59 years," *Clinical Infectious Diseases*, vol. 54, no. 7, pp. 922–928, 2012.

[12] E. Naidus, L. Damon, B. S. Schwartz, C. Breed, and C. Liu, "Experience with use of Zostavax in patients with hematologic malignancy and hematopoietic cell transplant recipients," *American Journal of Hematology*, vol. 87, no. 1, pp. 123–125, 2012.

[13] M. Son, E. D. Shapiro, P. LaRussa et al., "Effectiveness of varicella vaccine in children infected with HIV," *Journal of Infectious Diseases*, vol. 201, no. 12, pp. 1806–1810, 2010.

[14] R. S. Lowe, P. M. Keller, B. J. Keech et al., "Varicella-zoster virus as a live vector for the expression of foreign genes," *Proceedings of the National Academy of Sciences of the United States of America*, vol. 84, no. 11, pp. 3896–3900, 1987.

[15] J. I. Cohen and K. E. Seidel, "Generation of varicella-zoster virus (VZV) and viral mutants from cosmid DNAs: VZV thymidylate synthetase is not essential for replication *in vitro*," *Proceedings of the National Academy of Sciences of the United States of America*, vol. 90, no. 15, pp. 7376–7380, 1993.

[16] L. J. Ferrin and R. D. Camerini-Otero, "Selective cleavage of human DNA: RecA-assisted restriction endonuclease (RARE) cleavage," *Science*, vol. 254, no. 5037, pp. 1494–1497, 1991.

[17] A. J. Davison and J. E. Scott, "The complete DNA sequence of varicella-zoster virus," *Journal of General Virology*, vol. 67, no. 9, pp. 1759–1816, 1986.

[18] K. Nagaike, Y. Mori, Y. Gomi et al., "Cloning of the varicella-zoster virus genome as an infectious bacterial artificial chromosome in *Escherichia coli*," *Vaccine*, vol. 22, no. 29-30, pp. 4069–4074, 2004.

[19] H. Yoshii, P. Somboonthum, M. Takahashi, K. Yamanishi, and Y. Mori, "Cloning of full length genome of varicella-zoster virus vaccine strain into a bacterial artificial chromosome and reconstitution of infectious virus," *Vaccine*, vol. 25, no. 27, pp. 5006–5012, 2007.

[20] B. K. Tischer, B. B. Kaufer, M. Sommer, F. Wussow, A. M. Arvin, and N. Osterrieder, "A self-excisable infectious bacterial artificial chromosome clone of varicella-zoster virus allows analysis of the essential tegument protein encoded by ORF9," *Journal of Virology*, vol. 81, no. 23, pp. 13200–13208, 2007.

[21] K. Shiraki, Y. Hayakawa, H. Mori et al., "Development of immunogenic recombinant Oka varicella vaccine expressing hepatitis B virus surface antigen," *Journal of General Virology*, vol. 72, no. 6, pp. 1393–1399, 1991.

[22] K. Shiraki, H. Sato, Y. Yoshida et al., "Construction of Oka varicella vaccine expressing human imunodeficiency virus env antigen," *Journal of Medical Virology*, vol. 64, no. 2, pp. 89–95, 2001.

[23] P. Somboonthum, H. Yoshii, S. Okamoto et al., "Generation of a recombinant Oka varicella vaccine expressing mumps virus hemagglutinin-neuraminidase protein as a polyvalent live vaccine," *Vaccine*, vol. 25, no. 52, pp. 8741–8755, 2007.

[24] P. Somboonthum, T. Koshizuka, S. Okamoto et al., "Rapid and efficient introduction of a foreign gene into bacterial artificial chromosome-cloned varicella vaccine by Tn7-mediated site-specific transposition," *Virology*, vol. 402, no. 1, pp. 215–221, 2010.

[25] K. Shiraki, H. Ochiai, S. Matsui et al., "Processing of hepatitis B virus surface antigen expressed by recombinant Oka varicella vaccine virus," *Journal of General Virology*, vol. 73, no. 6, pp. 1401–1407, 1992.

[26] T. Kamiyama, H. Sato, T. Takahara, S. Kageyama, and K. Shiraki, "Novel immunogenicity of Oka varicella vaccine vector expressing hepatitis B surface antigen," *Journal of Infectious Diseases*, vol. 181, no. 3, pp. 1158–1161, 2000.

[27] T. C. Heineman, B. L. Connelly, N. Bourne, L. R. Stanberry, and J. Cohen, "Immunization with recombinant Varicella-Zoster virus expressing herpes simplex virus type 2 glycoprotein D reduces the severity of genital herpes in guinea pigs," *Journal of Virology*, vol. 69, no. 12, pp. 8109–8113, 1995.

[28] T. C. Heineman, L. Pesnicak, M. A. Ali, T. Krogmann, N. Krudwig, and J. I. Cohen, "Varicella-zoster virus expressing HSV-2 glycoproteins B and D induces protection against HSV-2 challenge," *Vaccine*, vol. 22, no. 20, pp. 2558–2565, 2004.

[29] S. I. Staprans, A. P. Barry, G. Silvestri et al., "Enhanced SIV replication and accelerated progression to AIDS in macaques primed to mount a CD4 T cell response to the SIV envelope protein," *Proceedings of the National Academy of Sciences of the United States of America*, vol. 101, no. 35, pp. 13026–13031, 2004.

[30] S. L. Oliver, L. Zerboni, M. Sommer, J. Rajamani, and A. M. Arvin, "Development of recombinant varicella-zoster viruses expressing luciferase fusion proteins for live *in vivo* imaging in human skin and dorsal root ganglia xenografts," *Journal of Virological Methods*, vol. 154, no. 1-2, pp. 182–193, 2008.

[31] W. L. Gray, "Simian varicella: a model for human varicella-zoster virus infections," *Reviews in Medical Virology*, vol. 14, no. 6, pp. 363–381, 2004.

[32] W. L. Gray, "Simian varicella in Old World monkeys," *Comparative Medicine*, vol. 58, no. 1, pp. 22–30, 2008.

[33] W. L. Gray, C. Y. Pumphrey, W. T. Ruyechan, and T. M. Fletcher, "The simian varicella virus and varicella zoster virus genomes are similar in size and structure," *Virology*, vol. 186, no. 2, pp. 562–572, 1992.

[34] C. Y. Pumphrey and W. L. Gray, "The genomes of simian varicella virus and varicella zoster virus are colinear," *Virus Research*, vol. 26, no. 3, pp. 255–266, 1992.

[35] W. L. Gray, B. Starnes, M. W. White, and R. Mahalingam, "The DNA sequence of the simian varicella virus genome," *Virology*, vol. 284, no. 1, pp. 123–130, 2001.

[36] A. D. Felsenfeld and N. J. Schmidt, "Antigenic relationships among several simian varicella like viruses and varicella zoster virus," *Infection and Immunity*, vol. 15, no. 3, pp. 807–812, 1977.

[37] A. D. Felsenfeld and N. J. Schmidt, "Varicella-zoster virus immunizes patas monkeys against simian varicella-like disease," *Journal of General Virology*, vol. 42, no. 1, pp. 171–178, 1979.

[38] R. Mahalingam, D. Smith, M. Wellish et al., "Simian varicella virus DNA in dorsal root ganglia," *Proceedings of the National Academy of Sciences of the United States of America*, vol. 88, no. 7, pp. 2750–2752, 1991.

[39] P. G. E. Kennedy, E. Grinfeld, V. Traina-Dorge, D. H. Gilden, and R. Mahalingam, "Neuronal localization of simian varicella virus DNA in ganglia of naturally infected african green monkeys," *Virus Genes*, vol. 28, no. 3, pp. 273–276, 2004.

[40] W. L. Gray, R. J. Williams, R. Chang, and K. F. Soike, "Experimental simian varicella virus infection of St. Kitts vervet

monkeys," *Journal of Medical Primatology*, vol. 27, no. 4, pp. 177–183, 1998.

[41] W. L. Gray, "Pathogenesis of simian varicella virus," *Journal of Medical Virology*, vol. 70, supplement 1, pp. S4–S8, 2003.

[42] R. Mahalingam, M. Wellish, T. White et al., "Infectious simian varicella virus expressing the green fluorescent protein," *Journal of NeuroVirology*, vol. 4, no. 4, pp. 438–444, 1998.

[43] Y. Ou, V. Traina-Dorge, K. A. Davis, and W. L. Gray, "Recombinant simian varicella viruses induce immune responses to simian immunodeficiency virus (SIV) antigens in immunized vervet monkeys," *Virology*, vol. 364, no. 2, pp. 291–300, 2007.

[44] V. Traina-Dorge, B. Pahar, P. Marx et al., "Recombinant varicella vaccines induce neutralizing antibodies and cellular immune responses to SIV and reduce viral loads in immunized rhesus macaques," *Vaccine*, vol. 28, no. 39, pp. 6483–6490, 2010.

[45] P. R. Kinchington, P. Ling, M. Pensiero, B. Moss, W. T. Ruyechan, and J. Hay, "The glycoprotein products of varicella-zoster virus gene 14 and their defective accumulation in a vaccine strain (Oka)," *Journal of Virology*, vol. 64, no. 9, pp. 4540–4548, 1990.

[46] P. R. Kinchington, P. Ling, M. Pensiero, A. Gershon, J. Hay, and W. T. Ruyechan, "A possible role for glycoprotein gpV in the pathogenesis of varicella-zoster virus," in *Immunobiology and Prophylaxis of Human Herpesvirus Infections*, C. Lopez, Ed., pp. 83–91, Plenum Press, New York, NY, USA, 1990.

[47] J. F. Moffat, L. Zerboni, P. R. Kinchington, C. Grose, H. Kaneshima, and A. M. Arvin, "Attenuation of the vaccine Oka strain of varicella-zoster virus and role of glycoprotein C in alphaherpesvirus virulence demonstrated in the SCID-hu mouse," *Journal of Virology*, vol. 72, no. 2, pp. 965–974, 1998.

[48] B. Pahar, W. L. Gray, K. Phelps et al., "Increased cellular immune responses and CD4+ T-cell proliferation correlated with reduced plasma viral load in SIV challenged recombinant simian varicella virus-simian immunodeficiency virus (rSVV-SIV) vaccinated rhesus macaques," *Virology Journal*, vol. 9, pp. 160–168, 2012.

[49] T. M. Ward, V. Traina-Dorge, K. A. Davis, and W. L. Gray, "Recombinant simian varicella viruses expressing respiratory syncytial virus antigens are immunogenic," *Journal of General Virology*, vol. 89, no. 3, pp. 741–750, 2008.

[50] W. L. Gray, F. Zhou, J. Noffke, and B. K. Tischer, "Cloning the simian varicella virus genome in *E. coli* as an infectious bacterial artificial chromosome," *Archives of Virology*, vol. 156, no. 5, pp. 739–746, 2011.

[51] H. Lauterbach, C. Ried, A. L. Epstein, P. Marconi, and T. Brocker, "Reduced immune responses after vaccination with a recombinant herpes simplex virus type 1 vector in the presence of antiviral immunity," *Journal of General Virology*, vol. 86, no. 9, pp. 2401–2410, 2005.

[52] J. Mestecky, M. W. Russell, and C. O. Elson, "Perspectives on mucosal vaccines: is mucosal tolerance a barrier?" *Journal of Immunology*, vol. 179, no. 9, pp. 5633–5638, 2007.

[53] Z. Moldoveanu, A. N. Vzorov, W. Q. Huang, J. Mestecky, and R. W. Compans, "Induction of immune responses to SIV antigens by mucosally administered vaccines," *AIDS Research and Human Retroviruses*, vol. 15, no. 16, pp. 1469–1476, 1999.

[54] C. Moriya, S. Horiba, M. Inoue et al., "Antigen-specific T-cell induction by vaccination with a recombinant Sendai virus vector even in the presence of vector-specific neutralizing antibodies in rhesus macaques," *Biochemical and Biophysical Research Communications*, vol. 371, no. 4, pp. 850–854, 2008.

[55] S. Brandler and F. Tangy, "Recombinant vector derived from live attenuated measles virus: potential for flavivirus vaccines," *Comparative Immunology, Microbiology and Infectious Diseases*, vol. 31, no. 2-3, pp. 271–291, 2008.

[56] S. G. Hansen, C. Vieville, N. Whizin et al., "Effector memory T cell responses are associated with protection of rhesus monkeys from mucosal simian immunodeficiency virus challenge," *Nature Medicine*, vol. 15, no. 3, pp. 293–299, 2009.

[57] S. Iyer, M. K. Mittal, and R. L. Hodinka, "Herpes zoster and meningitis resulting from reactivation of varicella vaccine virus in an immunocompetent child," *Annals of Emergency Medicine*, vol. 53, no. 6, pp. 792–795, 2009.

[58] S. Schünemann, C. Mainka, and M. H. Wolff, "Subclinical reactivation of varicella-zoster virus in immunocompromised and immunocompetent individuals," *Intervirology*, vol. 41, no. 2-3, pp. 98–102, 1998.

[59] C. S. Rollier, A. Reyes-Sandoval, M. G. Cottingham, K. Ewer, and A. V. S. Hill, "Viral vectors as vaccine platforms: deployment in sight," *Current Opinion in Immunology*, vol. 23, no. 3, pp. 377–382, 2011.

7

Production of Platinum Atom Nanoclusters at One End of Helical Plant Viruses

Yuri Drygin,[1] **Olga Kondakova,**[2] **and Joseph Atabekov**[1,2]

[1] *Belozersky Institute of Physico-Chemical Biology, Lomonosov Moscow State University, Vorob'evy gory 1, Building 40, Moscow 119992, Russia*
[2] *Department of Biology, Lomonosov Moscow State University, Vorob'evy gory 1, Building 12, Moscow 119992, Russia*

Correspondence should be addressed to Yuri Drygin; drygin@belozersky.msu.ru

Academic Editor: Jay C. Brown

Platinum atom clusters (Pt nanoparticles, Pt-NPs) were produced selectively at one end of helical plant viruses, tobacco mosaic virus (TMV) and potato virus X (PVX), when platinum coordinate compounds were reduced chemically by borohydrides. Size of the platinum NPs depends on conditions of the electroless deposition of platinum atoms on the virus. Results suggest that the Pt-NPs are bound concurrently to the terminal protein subunits and the $5'$ end of encapsidated TMV RNA. Thus, a special structure of tobacco mosaic virus and potato X virus particles with nanoparticles of platinum, which looks like a push-pin with platinum head and virus needle, was obtained. Similar results were obtained with ultrasonically fragmented TMV particles. By contrast, the Pt-NPs fully filled the central axial hole of *in vitro* assembled RNA-free TMV-like particles. We believe that the results presented here will be valuable in the fundamental understanding of interaction of viral platforms with ionic metals and in a mechanism of nanoparticles formation.

1. Introduction

One current goal in nanobiotechnology is to investigate biological nanoplatforms capable of binding metal atoms with the aim of generating new bioinorganic materials for nanoelectronics and medicine. Helical plant viruses, in particular tobacco mosaic virus (TMV) and potato virus X (PVX), have been widely used as templates and scaffolds in nanotechnology [1–12].

TMV particles are rod-like with a diameter of 18 nm and modal length of 300 nm. They consist of 2130 identical 17.5 kDa protein subunits helically arranged around a cylindrical canal and closely packed into a rigid tube. The two-layer cylindrical substructure, each layer consisting of a ring of 17 molecules of coat protein (CP), is known as a "disk," and 16 1/3 molecules are present in each turn of the assembled helix. RNA is introduced between the CP turns and follows the helix of protein subunits [13, 14]. Apparently, the front stereochemical surface of terminal CP molecules of the helical TMV particle is not equivalent to inner surfaces of other CP subunits.

It is known that in the absence of RNA the viral CP may be *in vitro* assembled into several types of aggregate. In particular, TMV CP can be *in vitro* assembled into virus-like particles (VLPs) that are structurally similar to native virions [15, 16].

The virions of another helical virus, PVX, are flexuous filaments with modal lengths in the range of 470–580 nm and a diameter of approximately 13 nm [17].

In this work, we found that ions of platinum coordinate compounds reduced by borohydrides could form nucleation centers for the very selective growth of nanoparticles on one pole of helical plant viruses (tobacco mosaic virus and potato virus X).

2. Materials and Methods

2.1. Modification of Viruses with (Dien)platinum, (Chloro(diethylenetriamine)platinum(II) Chloride or [Pt(dien)Cl]Cl), and Sodium Borohydride. 100 μL of TMV or PVX suspension (0.4 mg/mL) in TDW (deionized double distilled water) was mixed with 2 μL of 25 mM [Pt(dien)Cl]Cl in TDW. The mixture was incubated for 1 h at 30°C, then borate buffer 0.5 M (pH 8.3) was added up to 0.1 M. The reaction mixture was cooled in an ice bath, and sodium borohydride (5mg/mL) was added portion wise to a final concentration of 3 mM or 10 mM. To do Pt-NPs larger and produce short nanowires, concentration of (dien)platinum was increased two fold. The reaction mixture was diluted 4-fold with TDW and samples were examined with transmission electron microscopes Zeis LEO912 AB OMEGA supplied with the energy filter or with TEM Jeol JEM 1011.

2.2. Modification of TMV with (Dien)platinum and Hypophosphite. 100 μL of suspension of virus (TMV, PVX, 1 mg/mL) was mixed with 4 μL 20 mM (dien)platinum in water (pH~ 7), kept for 60 min at 30°C. Then reducing agent sodium hypophosphite (Fluka) was added to a final concentration of 30 mM, and the mixture was incubated for 20 min at room temperature.

2.3. Electron Microscopy of the Virus RNA and Ribonucleoproteins. 2% collodion solution (Fluka) strengthened by an evaporated carbon was used for preparation of support films specimens for electron microscopy. Usually, 5 μL of TMV suspension was placed on the grid for 1 min, after which the droplet was removed by a filter paper. While samples were examined with transmission electron microscopes Zeis LEO912 AB OMEGA supplied with the energy filter, no staining was done. In some cases (indicated), virus particles were briefly stained with 2% uranyl acetate for additional positive contrasting.

To observe the RNA, samples were prepared by the protein-free monolayer spreading method described in [18]. Concentrations in the spreading solution were 0.1 mg/mL TMV, 3 M urea, 3.7% formaldehyde, and 0.01% benzyldimethylalkylammonium chloride. The hypophase was 0.15 M sodium acetate and 0.1% formaldehyde. The monolayer was picked up on collodion support, and the sample was rotatory shadowed with tungsten under 7°.

3. Results and Discussion

3.1. Polar Electroless Deposition of Platinum onto Helical Viruses. In the present work, the structure of bioinorganic nanocomplexes, produced by platinum atoms reduced on a platform of helical plant viruses (TMV, PVX) and on RNA-free TMV VLPs, was studied.

Previously, to synthesize DNA [19] and RNA [20] probes, we used the platinum coordinate compound chloro(diethylenetriamine)platinum(II) chloride (mw 369.15) as a label. [Pt(dien)Cl]Cl was synthesized according to Watt and Cude [21]. It is known that at a ratio of 1 and less per 10 nucleotide residues of double-stranded DNA, (dien)platinum is associated almost exclusively with guanine residues at position N7 [22].

Because it was possible that the (dien)platinum would react with the TMV protein, it was necessary to determine the working concentrations of the reactants and the optimal (dien)platinum/virus ratio to obtain stable complexes

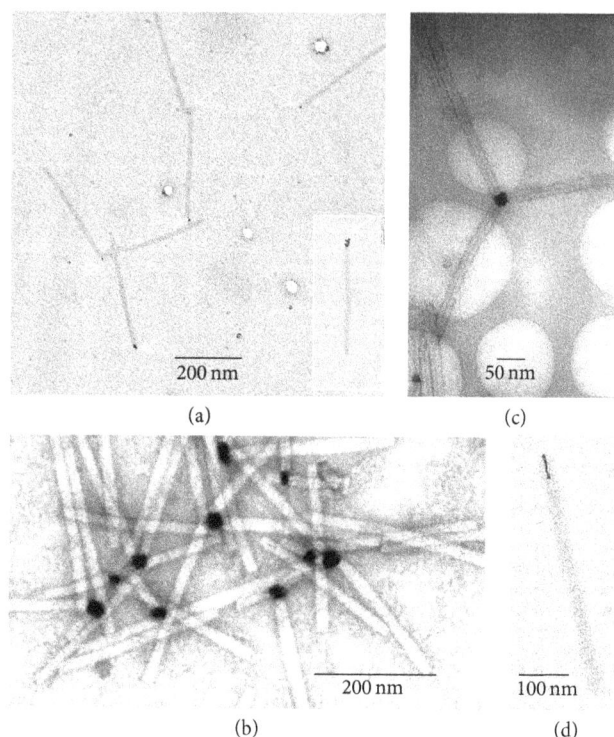

FIGURE 1: Polar growth of the platinum atom clusters on the TMV platform. Platinum nanoparticles were formed after chemical reduction of (dien)platinum by sodium borohydride to a final concentration of 3 mM (a) or 10 mM ((b)–(d)), see Section 2). Samples were examined with transmission electron microscopes Zeis LEO912 AB OMEGA supplied with the energy filter (specimens ((a), (d)) and Jeol JEM 1011 (specimens (b), (c)). No staining of specimens ((a), (d)) was carried out. Samples ((b), (c)) were stained slightly with 2% uranyl acetate. Arrows in (a) indicate the platinum atom clusters bound to the end of the virus particles.

and avoid TMV aggregation. As found by sedimentation analysis in an analytical ultracentrifuge, the optimal ratio of (dien)platinum to the virus (in moles of CP) must be 50 : 1 or less to obtain stable (dien)platinum reacted TMV particles in solution.

Chemical reduction of the complex platinum ions bound to the TMV platform was performed in aqueous sodium borohydride or dimethylamine borane solutions using different reaction times and excesses of the reducing agent. Clusters of platinum atoms and short nanowires formed on the native TMV scaffold were revealed under these experimental conditions (Figure 1) by transmission electron microscopy (TEM).

Figure 1(a) shows that small (~1-2 nm) Pt nanoparticles (Pt-NPs) could be formed on one end of the TMV particles upon exposure to 0.5 mM [Pt(dien)Cl]Cl in 3 mM solution of the reducing agent at pH~8. It is noteworthy that 55% of TMV particles (200 particles were analyzed) contained the Pt nanoparticles on one end, while the number of virus particles carrying Pt-NP clusters on both ends was negligible (less than 0.5%), and no lateral electroless deposition of the platinum on the virus exterior was detected. The formation of Pt-NPs was also observed on one end of the end-to-end aggregates of TMV particles (longer than 300 nm) and TMV fragments shorter than 300 nm. Figure 1(a) (inset) shows that several small discrete Pt-NPs could also be linked to one end of the TMV particle; presumably the primary platinum nanoparticle works here as a nucleation center.

Increasing concentrations of (dien)platinum to 1 mM and borohydride to 10 mM resulted in an increase in the size of Pt-NPs to ≤30 nm as shown in Figures 1(b) and 1(c). Frequently, these TMV particles are aggregated by their terminal "Pt-NP ends" into star-like structures (Figure 1(c)). TMV particles with the central hole partly filled with Pt-nanowire (Figure 1(d)) were also revealed, comprising 20% of the total number of the virus particles. The Pt-nanowires were 10–30 nm in length with a diameter of approximately 3 nm and resembled a small helix.

To prove the platinum origin of the nanoparticles, solution of TMV with bound platinum nanoparticles was concentrated, and diffraction of electron on nanoparticles was examined. Diffraction of electrons for concentrated sample, shown in Figure 1(b) and unstained with uranyl acetate, was obtained by energy-filtering TEM and showed typical picture of face-centered cubic lattice of platinum (Figure 2), with intrinsic interplanar distances within the measurement error.

Production of Pt-NPs by reduction of the (dien)platinum-pretreated TMV with dimethylamine borane (10 mM, pH 7, 60 min, r.t.) was less efficient than that with sodium borohydride. Presumably, this was because of dimethylamine borane's weaker reducing potential [23].

While potassium tetrachloroplatinate (IV) was reduced by sodium borohydride in weakly alkaline conditions, polar growth of clusters of platinum atoms on virus particles was observed. One hundred of the TMV particles were examined, 55% had platinum atom clusters at one end and 3% at both ends (not shown).

To further elucidate this phenomenon, we examined the interaction of (dien)platinum and its reduction by sodium borohydride with another helical virus, the potexvirus. Figure 3 shows that several discrete small Pt-NPs were associated with one end of filamentous PVX, suggesting that the polar binding of Pt-NPs to helical viruses is a common phenomenon.

The (dien)platinum reacts in preference with guanine residues of DNA [22] and, presumably, of RNA [20]. It is known that TMV and PVX RNAs contain only two identical terminal chemical structures, both at the 5′ end, which are guanine residues in the cap group (m^{7}GpppGp) [24, 25].

It appears that the recognition of one end of the TMV (and PVX) particle by Pt atoms is due to identical 5′-end cap group and peculiarities of the polar geometry of the viral helix. By its selectivity this phenomenon is similar to the specific interaction of monoclonal antibodies with antigenically different terminal surfaces at the virus particle ends [26, 27].

There were three possible mechanisms for platinum atom interaction with the 5′ end of the initial TMV and PVX particles: (i) (dien)platinum interacts with the RNA cap structure and Pt-NPs interact strongly with the 5′ end of the virus RNAs; (ii) (dien)platinum atoms and Pt-NPs recognize and interact strongly with the exposed surface exterior of

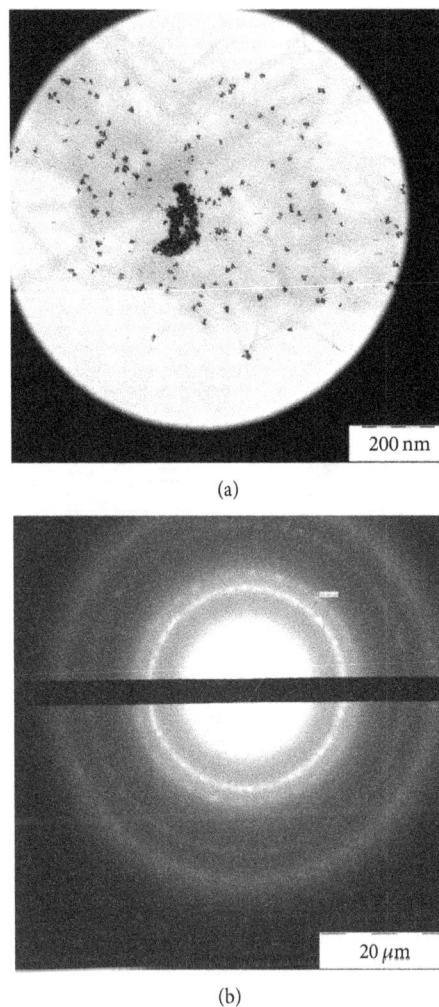

(a)

(b)

FIGURE 2: Electron diffraction of the Pt-NPs-TMV preparation. Large area image of the sample in Figure 1(b) concentrated on the specimen grid and its electron diffraction ((a) and (b), resp.).

the 5′-terminal protein or with a disk face structure; (iii) both interaction types of (dien)platinum and Pt-NPs are realized. To select between these mechanisms, the following experiments were carried out.

3.2. Pt-NPs Are Deposited on the 5′ End of TMV RNA. It has been shown that the stripping of TMV CP subunits from the 5′ end of RNA is induced by urea, DMSO (dimethyl sulfoxide), alkali, and detergent [28, 29]. Polar stripping results in the production of rods with a tail of RNA protruding from one end. The 5′-proximal 69 nucleotides of TMV RNA lack guanine bases and interact more weakly with CP subunits compared to other regions of RNA [30]. As a consequence, polar 5′ to 3′ uncoating of RNA occurs because of the removal of the terminal CP molecules.

Therefore, (dien)platinum was reduced by sodium borohydride on the tobacco mosaic virus, and the Pt-NPs-containing virus particles were then treated with urea. To fix the structure obtained, formaldehyde was added up to 3.7%. Partly stripped TMV rods with a tail of RNA protruding from one end were produced (Figure 4).

FIGURE 3: Polar binding of the platinum atom clusters to PVX particles. Sample preparation described in Section 2. Specimen was examined under TEM Zeis LEO912 AB OMEGA microscope, no contrasting of specimen.

Treatment of the Pt-NPs-TMV complex with urea leads to a partial stripping of the viral capsid proteins adjacent to the nanoparticle and to the protruding free RNA (Figure 4(c)). The Pt-NPs were still associated with the shortened virus rods after removal of RNA tails by RNase A (20 μg/mL, 30 min, 30°C) (data not shown). However, Figure 4 shows that, after treatment with ultrasound and urea, the Pt-NP is located on the inner edge of the truncated viral particles. How does this translocation of nanoparticles take place?

There are two possible explanations.

(1) Since the disassembly and reassembly of TMV are reversible, and the viral proteins have a high propensity toward association, the $5'$-end complex of "Pt-NPs-RNA-CP" does not move out and retain contact with the $5'$ end of the truncated viral particle via protein-protein interactions. Freed bare RNA is looped out, but bound continuously to the $5'$-terminal facing of the viral capsid (Figures 4(a) and 4(b)). It is well known that the purified TMV preparation has some sticky ribonuclease [31]. Free denatured RNA undergoes simple hydrolysis, producing free ends retaining links with the truncated virus particle and leaving Pt-NPs on the edge of the shortened virus rod (Figure 4(c)).

(2) Pt-NPs bind more strongly to the $5'$-terminal end of TMV preeminently because of the more appropriate spatial arrangement of the complex-forming groups of the virus protein and RNA. Therefore, the nanoparticle translocates from the disembodied capsid protein to the truncated (via the urea) end of the viral particle because of linking with bare RNA is weaker (Figure 4(c)).

This observation allows us to hypothesize that the cap groups were not accessible or weakly associated with the Pt-NPs being encapsidated in the virus helix. This also suggests that the Pt-NPs were linked tightly to CP subunits within the $5'$-terminal facing or several exterior surfaces of subunits in the TMV helix. If so, the Pt-NPs were linked to the domain(s) of CP subunits located at the terminal surface and adjacent

to the central hole of the helical particle. This suggestion is also in line with the ability of Pt-NPs to penetrate the internal canal of TMV tubes, producing nanowires.

3.3. Pt-NPs Are Bound to the $5'$-Terminal Protein Facing of the TMV Virus Particle.

Strong binding of the Pt-NPs to the TMV was confirmed by the linkage resistance to the ultrasonication treatment. Results of two series of experiments illustrate this finding.

(i) The virus was first ultrasonicated at 0°C to produce halves of particles and then reacted with (dien)platinum under the same conditions. The size of fragments varied from 25 to 200 nm with the peak (42%) at 100–120 nm. As expected, approximately half (55%, 100 TMV particles were analyzed) of sonication-generated fragments contained Pt-NPs bound to one end (Figures 5(a) and 5(b)).

(ii) TMV was first reacted with (dien)platinum and then ultrasonicated. TMV fragments ranged in size from 25 to 200 nm with two peaks: 160 nm (22%) and 120 nm (17%). Only 23% of the TMV particles (100 TMV particles were analyzed) were associated with Pt-NPs; approximately half of the Pt-NPs were detached from the virus rods (20–30% at least, some aggregated) (Figure 5(b)). One can propose that the essential part of free Pt-NPs is just shaken off from the TMV fragments by ultrasonication or they are associated with a low contrasted and weakly visible filamentary material (RNA?).

To gain further insight into the origin of a low contrasted filamentary material with apparently bound Pt-NPs, the fragments generated by sonication of reacted with (dien)platinum virus were fixed with formaldehyde, and benzylalcylammonium chloride was then added. To spread over RNA molecules, urea was added up to a final concentration of 3 M, and the reaction cocktail was loaded onto a hypophase (0.15 M sodium acetate and 0.1% formaldehyde).

Significantly, this technique revealed unusual structures comprising the disk-like element (apparently consisting of CP subunits) with Pt-NPs bound to this disk and the tail of RNA connecting the disk with the remainder of the stripped TMV particle (Figure 5(c) and inset). We assume also that the Pt-NP, by binding to the $5'$ end of the RNA and to the $5'$-terminal protein (or disk) facing of the TMV virus particle, weakens the connection between this exterior and the rest of the virus particle.

The results obtained suggest that the rigid protein shell is broken into fragments by sonication, while the flexible RNA remains intact (Figures 4 and 5(c)). This finding can be explained by the varying diametrical stiffness of the virion and RNA. It should be stressed that uncoating the TMV modified with (dien)platinum by ultrasonic treatment starts from one end (Figure 4) and is similar to the native TMV uncoating with urea, DMSO, alkali, SDS (sodium dodecyl sulfate), and heat treatment [28, 29]. This confirms, despite its seemingly symmetrical form (see, e.g., the PDB 3D image for TMV), that the virus particle has chemical and physical structural asymmetries at input and output ends. Interestingly, this is logically related in the biological sense to the polarity of TMV RNA-concurrent with uncoating, translation of the parent virion RNA starts at the $5'$ end [30].

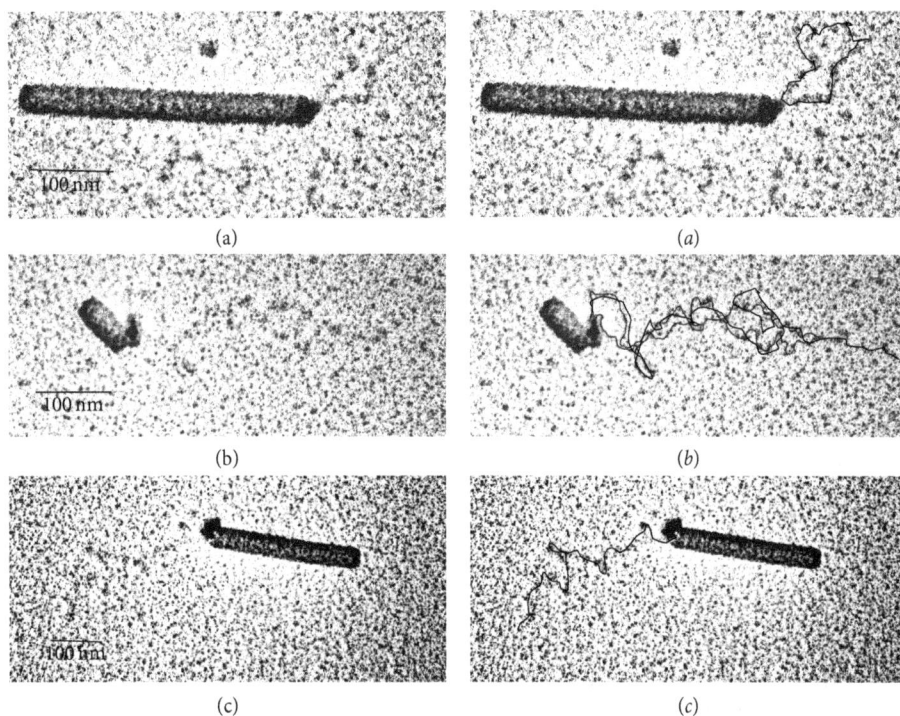

FIGURE 4: Partial uncoating of TMV and Pt-NPs complexes by 6 M urea. Uncoating was allowed to proceed at 0°C for 5 min in 6 M urea and 60 mM sodium phosphate at a virus concentration of 0.5 mg/mL. Shadowed samples (see Section 2) were examined with a Jeol JEM1011 microscope. Analysis was conducted on 130 uncoated partially TMV particles. Left: ((a), (b)) show a loop of the free RNA released from TMV particle retained by Pt nanoparticle and (c) shows the partially hydrolyzed linear RNA. It is important to note that RNA has secondary structure under spreading conditions. Right: ((a)–(c)) merge images with inked over RNA strand.

It appears that the platinum nanoparticle is bound both to the 5′ end of the viral RNA and to the capsid protein, and this association is resistant to the ultrasonic processing and urea treatment. Taken together, these data provide strong evidence that Pt-NPs can be linked directly to the first 5′ surface exterioir of terminal subunits in the TMV CP helix. Concurrent link formation between the Pt-NPs and the 5′ end of encapsidated RNA is also indicated.

It is difficult now to predict with which atoms of amino acid residues of CP and nucleotide residues of RNA at the very 5′-end of the virus RNA interacts (dien)platinum ion without special study of (dien)platinum docking on the pole TMV and PVX exterior surfaces. In connection with this, it would be interesting to add other ligands to a platinum modified virus particle that are able to enter the metal coordinate sphere, using chemical properties of coordinate metals and clusters of their atoms bound to a virus.

3.4. Platinum Modification of TMV-Like Particles.
As mentioned above, RNA-free TMV-like particles can be assembled from viral CP in vitro [15–17].

CP was extracted from TMV using an acetic acid according to Fraenkel-Conrat et al. [32]. Then, a sodium phosphate buffer (100 mM, pH 5.6) was added to the suspension of TMV CP to obtain a final concentration of 1 mg/mL, and the protein was incubated at room temperature for 24 h in order to produce VLP. This solution was stored at 4°C.

It was discovered that (dien)platinum and clusters of the reduced platinum atoms do not bind to the VLP via the polar mode; instead, they fill the central hole of the TMV-like particles throughout their length (Figure 6). This finding supports the conclusion on the high sensitivity of (dien)platinum and/or Pt-NPs to spatial organization of functional chemical groups of CP on the surface of VLP and TMV.

3.5. Reduction of (Dien)platinum with Hypophosphite.
Influence of the reduction agent and the reaction conditions were of much importance to understand the mechanism of the (dien)platinum interaction with TMV. Both, the surface and central axial hole of TMV were stained slightly with tiny atom clusters of platinum when (dien)platinum was reduced by 30 mM hypophosphite at pH 6-7 (Figure 7). Thus, modification of the reaction conditions drastically changed character of (dien)platinum-TMV interaction.

Summing up, one can conclude that platinum coordinate compounds (dien)platinum and tetrachloroplatinate(IV) interact very selectively with spiral viruses TMV and PVX and after chemical reduction with borohydrides form nanoparticles via the polar mode with the 5′ end of the viruses (corresponding to RNA), as demonstrated through the polar uncoating of TMV with urea.

Interestingly Balci et al. [4] described the self-assembling of prepared preliminary gold nanoparticles with a diameter of 6 nm and TMV into a metal-virus nanodumbbells. The gold nanoparticles were selectively bound to both ends of the virus rods and can be enlarged by electroless deposition of gold associated with shortening of the virus particles to yield gold-virus-gold dumbbells.

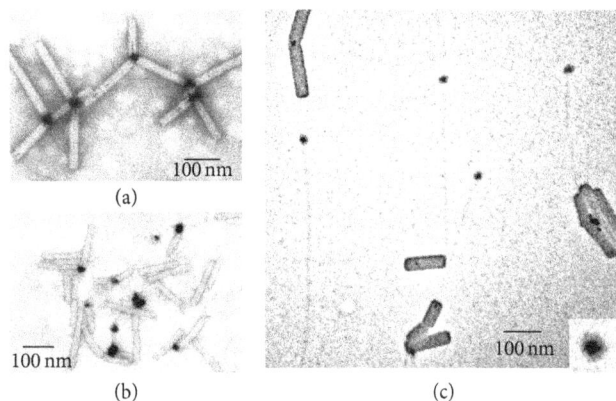

FIGURE 5: Platinum modification of the ultrasonically treated TMV. (a) chilled TMV samples (1 mg/mL, in ice water) were ultrasonicated at 40 W for five seconds six times with intervals of 60 s using the ultrasonic disintegrator MSE Soniprep 150. TMV was then reacted with 1 mM (dien)platinum and reduced by 10 mM sodium borohydride, (pH~8). (b) TMV was first incubated with 1 mM (dien)platinum and reduced with 10 mM sodium borohydride, then, was exposed to 40 W sonication for 5 s six times with intervals of 60 s at 0°C. Samples (a) and (b) were diluted 10-fold with TDW and stained with 2% uranyl acetate. To observe viral RNA (c), a sample prepared as described in (b) was developed via the monolayer technique (see Section 2). Inset shows 3-fold magnification of the Pt nanoparticle apparently included in the terminal TMV protein facing or disk.

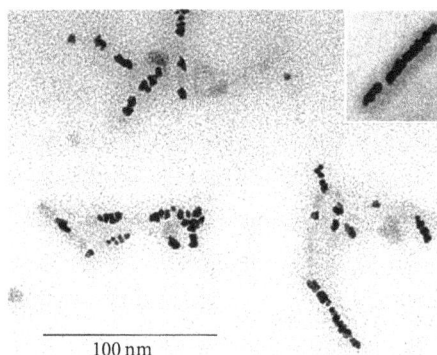

FIGURE 6: Nanoparticles of platinum fill the inner canal of the TMV VLP. TEM images of Pt nanoparticles in the inner canal of the virus-like rods produced from the coat protein of TMV. Incubation of VLP with (dien)platinum followed by electroless deposition of Pt atoms took place under the same conditions as with TMV. Inset in the upper corner indicates clearly the coat protein tube around the Pt nanoparticles. No contrasting of specimen was done.

Complexes of TMV and different ionic metals (Ni, Co, Cu, Fe, Ag, and Au) were examined by us as well. We found that copper could only interact with TMV in a polar mode, 100 TMV particles were examined and found 53% one end- and 1% two end- labeled TMV particles (data not shown) after borohydride reduction.

4. Conclusions and Proposal

Nanoparticles of platinum that grows by polar mode on TMV and PVX helical viruses were produced by the borohydrides

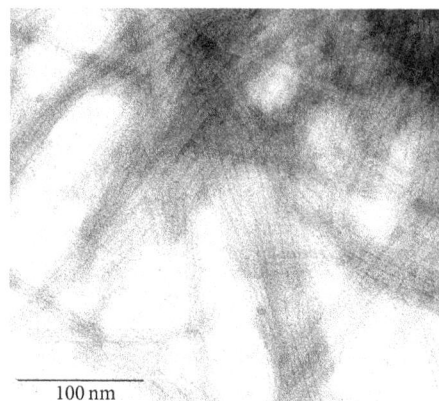

FIGURE 7: Chemical reduction of (dien)platinum on the TMV platform by hypophosphite. No additional staining was carried out.

reduction of platinum coordinate compounds. Size of the platinum NPs depends on conditions of the chemical reduction of platinum atoms on the virus. Electron microscopy examination of polar uncoating of TMV particles showed that the Pt-NPs are bound concurrently to the terminal protein subunits and the $5'$ end of encapsidated TMV RNA.

Above and beyond nanorods, nanowires and nanorings produced earlier on the plant viruses as templates, tobacco mosaic virus, and potato X virus particles in a complex with nanoparticles of platinum, which look like a push-pin with platinum head and virus needle, were obtained. Thus, in this work, a novel structure of tobacco mosaic virus, and potato X virus particles in a complex with nanoparticles of platinum, which demonstrates extraordinary specificity of the coordinate metal interaction with a complex polyvalent biological construction, were obtained.

TMV is established as one of the best-investigated models of macromolecular organization in biology. Moreover, partially and completely TMV can be reconstituted from its RNA and coat protein. The process of the TMV reconstitution was studied kinetically [16]. Thus, population of the TMV nucleoproteins of definite size (length) containing $5'$ pole (in respect to RNA) can be obtained. On the other hand, $3'$-terminal ribose of RNA in these nucleoproteins might be modified (oxidized, e.g.) and fixed covalently on any appropriate surface. One can propose that modified with platinum (or copper) population of these nanosize nucleoproteins of different lengths could serve as comb-like structures, or as strings of different thickness in a stringed instrument that are structurally (and functionally) similar to a musical organ. From our point of view, such wellorganized nanoparticle structures are of high demand in nanoelectronics.

We also cannot exclude that manifold ligands might be linked to the TMV and PVX viruses due to high chemical coordinate potential of platinum atoms.

Acknowledgments

The authors are grateful to Dr. O. V. Karpova for the purified TMV (strain U1) and PVX (Russian strain) and Dr. M.

Arkhipenko for the TMV coat protein preparation. They are thankful to Sergei Ambramchuk and Anatoly Bogdanov for assistance with the electron microscopes operation. They deeply appreciate the help of Dr. Emma Hoyle for English language assistance. This work was supported in parts by the Federal Agency for Science and Innovations contract No. 16.M04.12.0024 from the Russian Federation Ministry of Science and Education and the RBRR grant 12-04-01711-a to Yu.F.D.

References

[1] E. Dujardin, C. Peet, G. Stubbs, J. N. Culver, and S. Mann, "Organization of metallic nanoparticles using tobacco mosaic virus templates," *Nano Letters*, vol. 3, no. 3, pp. 413–417, 2003.

[2] M. Knez, A. Kadri, C. Wege, U. Gösele, H. Jeske, and K. Nielsch, "Atomic layer deposition on biological macromolecules: metal oxide coating of tobacco mosaic virus and ferritin," *Nano Letters*, vol. 6, no. 6, pp. 1172–1177, 2006.

[3] R. J. Tseng, C. Tsai, L. Ma, J. Ouyang, C. S. Ozkan, and Y. Yang, "Digital memory device based on tobacco mosaic virus conjugated with nanoparticles," *Nature Nanotechnology*, vol. 1, no. 1, pp. 72–77, 2006.

[4] S. Balci, K. Noda, A. M. Bittner et al., "Self-assembly of metal-virus nanodumbbells," *Angewandte Chemie*, vol. 46, no. 17, pp. 3149–3151, 2007.

[5] M. Ł. Górzny, A. S. Walton, M. Wnęk, P. G. Stockley, and S. D. Evans, "Four-probe electrical characterization of Pt-coated TMV-based nanostructures," *Nanotechnology*, vol. 19, no. 16, Article ID 165704, 2006.

[6] M. Kobayashi, K. Onodera, Y. Watanabe, and I. Yamashita, "Fabrication of 3-nm platinum wires using a tobacco mosaic virus template," *Chemistry Letters*, vol. 39, no. 6, pp. 616–618, 2010.

[7] X. Chen, K. Gerasopoulos, J. Guo et al., "Virus-enabled silicon anode for lithium-ion batteries," *ACS Nano*, vol. 4, no. 9, pp. 5366–5372, 2010.

[8] A. Mueller, F. J. Eber, C. Azucena et al., "Inducible site-selective bottom-up assembly of virus-derived nanotube arrays on RNA-equipped wafers," *ACS Nano*, vol. 5, no. 6, pp. 4512–4520, 2011.

[9] A. K. Manocchi, S. Seifert, B. Lee, and H. Yi, "In situ small-angle X-ray scattering analysis of palladium nanoparticle growth on tobacco mosaic virus nanotemplates," *Langmuir*, vol. 27, no. 11, pp. 7052–7058, 2011.

[10] J. C. Zhou, C. M. Soto, M. S. Chen et al., "Biotemplating rod-like viruses for the synthesis of copper nanorods and nanowires," *Journal of Nanobiotechnology*, vol. 10, article 18, 2012.

[11] C. Azucena, F. J. Eber, V. Trouillet et al., "New approaches for bottom-up assembly of tobacco mosaic virus-derived nucleo-protein tubes on defined patterns on silica- and polymer-based substrates," *Langmuir*, vol. 28, no. 42, pp. 14867–14877, 2012.

[12] C. Hou, Q. Luo, J. Liu et al., "Construction of GPx active centers on natural protein nanodisk/nanotube: a new way to develop artificial nanoenzyme," *ACS Nano*, vol. 6, no. 10, pp. 8692–8701, 2012.

[13] A. Klug, "The tobacco mosaic virus particle: structure and assembly," *Philosophical Transactions of the Royal Society B*, vol. 354, no. 1383, pp. 531–535, 1999.

[14] F. A. Anderer, "Recent studies on the structure of tobacco mosaic virus," *Advances in Protein Chemistry*, vol. 18, pp. 1–35, 1964.

[15] D. L. D. Caspar, "Assembly and stability of the tobacco mosaic virus particle," *Advances in Protein Chemistry*, vol. 18, pp. 37–121, 1964.

[16] P. J. G. Butler, "Self-assembly of tobacco mosaic virus: the role of an intermediate aggregate in generating both specificity and speed," *Philosophical Transactions of the Royal Society B*, vol. 354, no. 1383, pp. 537–550, 1999.

[17] J. Atabekov, E. Dobrov, O. Karpova, and N. Rodionova, "Potato virus X: structure, disassembly and reconstitution," *Molecular Plant Pathology*, vol. 8, no. 5, pp. 667–675, 2007.

[18] H. J. Vollenweider, J. M. Sogo, and K. T. Koller Th., "A routine method for protein free spreading of double and single stranded nucleic acid molecules," *Proceedings of the National Academy of Sciences of the United States of America*, vol. 72, no. 1, pp. 83–87, 1975.

[19] O. A. Kondakova and Y. F. Drygin, "Diagnostics of potato spindle tubor viroid disease with dien-Pt DNA probe," *Biotechnologya*, vol. 1, pp. 83–90, 1999 (Russian).

[20] Y. F. Drygin, J. G. Atabekov, O. A. Kondakova, S. N. Chirkov, R. A. Zinovkin, and V. I. Kiseleva, "Method for simultaneously detecting a plurality of RNA sequences in a biological sample," Patent WO 123494, 2009.

[21] G. W. Watt and W. A. Cude, "Diethylenetriamine complexes of platinum(II) halides," *Inorganic Chemistry*, vol. 7, no. 2, pp. 335–338, 1968.

[22] N. P. Johnson, J. P. Macquet, J. L. Wiebers, and B. Monsarrat, "Structures of the adducts formed between [Pt(dien)Cl]Cl and DNA in vitro," *Nucleic Acids Research*, vol. 10, no. 17, pp. 5255–5269, 1982.

[23] W. Buchner and H. Niederprünt, "Sodium borohydride and amine-boranes, commercially important reducing agents," *Pure & Applied Chemistry*, vol. 49, no. 6, pp. 733–743, 1977.

[24] D. Zimmern, "The $5'$ end group of tobacco mosaic virus RNA is m7G5′ ppp5′Gp," *Nucleic Acids Research*, vol. 2, no. 7, pp. 1189–1201, 1975.

[25] N. Sonenberg, A. J. Shatkin, and R. P. Ricciardi, "Analysis of terminal structures of RNA from potato virus X," *Nucleic Acids Research*, vol. 5, no. 7, pp. 2501–2512, 1978.

[26] M. H. V. van Regenmortel, D. Altschuh, and G. Zeder-Lutz, "Tobacco mosaic virus: a model antigen to study virus-antibody interactions," *Biochimie*, vol. 75, no. 8, pp. 731–739, 1993.

[27] M. H. V. van Regenmortel, "The antigenicity of tobacco mosaic virus," *Philosophical Transactions of the Royal Society B*, vol. 354, no. 1383, pp. 559–568, 1999.

[28] A. Buzzell, "Action of urea on tobacco mosaic virus. II. The bonds between protein subunits," *Biophysical Journal*, vol. 2, pp. 223–233, 1962.

[29] L. E. Blowers and T. M. A. Wilson, "The effect of urea on tobacco mosaic virus: polarity of diassembly," *Journal of General Virology*, vol. 61, no. 1, pp. 137–141, 1982.

[30] K. W. Mundry, P. A. C. Watkins, T. Ashfield, K. A. Plaskitt, S. Eisele-Walter, and T. M. A. Wilson, "Complete uncoating of the $5'$ leader sequence of tobacco mosaic virus RNA occurs rapidly and is required to initiate cotranslational virus disassembly in vitro," *Journal of General Virology*, vol. 72, no. 4, pp. 769–777, 1991.

[31] H. Fraenkel-Conrat, B. Singer, and A. Tsugita, "Purification of viral RNA by means of bentonite," *Virology*, vol. 14, no. 1, pp. 54–58, 1961.

[32] H. Fraenkel-Conrat, "Degradation of tobacco mosaic virus with acetic acid," *Virology*, vol. 4, no. 1, pp. 1–4, 1957.

Comparison of β-Propiolactone and Formalin Inactivation on Antigenicity and Immune Response of West Nile Virus

Pritom Chowdhury,[1,2] Rashmee Topno,[1] Siraj A. Khan,[1] and Jagadish Mahanta[1]

[1]Arbovirology Group, Entomology and Filariasis Division, Regional Medical Research Centre, ICMR, Northeast Region, Post Box No. 105, Dibrugarh, Assam 786001, India
[2]Department of Biotechnology, Tocklai Tea Research Institute, TRA, Jorhat, Assam 785008, India

Correspondence should be addressed to Siraj A. Khan; sirajkhanicmr@gmail.com

Academic Editor: Trudy Morrison

West Nile Virus (WNV) is a pathogenic arbovirus that belongs to genus *Flavivirus* under family Flaviviridae. Till now there are no approved vaccines against WNV for human use. In this study, the effect of two alkylating agents, formaldehyde and β-PL, generally used for inactivated vaccine preparation, was assessed on the basis of antigenic and immunogenic potential of the inactivated WNV. Lineage 5 WNV isolates were inactivated by both formalin and β-PL treatments. Inactivation was confirmed by repeated passage in BHK-21 cell line and infant mice. Viruses inactivated by both the treatments showed higher antigenicity. Immune response in mice model showed serum anti-WNV antibody titre was moderately higher in formalin inactivated antigen compared to β-PL inactivated antigen. However, no significant differences were observed in neutralization antibody titre. In conclusion, we can state that both formaldehyde and β-PL inactivation processes were found to be equally efficient for inactivation of WNV. However, they need to be compared with other inactivating agents along with study on cell mediated immune response.

1. Introduction

West Nile Virus (WNV) is a pathogenic arbovirus that belongs to genus *Flavivirus*, family Flaviviridae. The natural cycle involves reservoir hosts, namely, wild and domestic birds. Mosquitoes, generally *Culex* species, act as principal vectors and humans act as dead end hosts [1, 2]. Although most of the human WNV infections remain subclinical, febrile illness and neuroinvasive disease develop in ≈20% and <1% of the infected patients, respectively [3]. Initially, the virus was distributed in Africa, Asia, and Europe where it caused infrequent and unpredictable epidemics of mild systemic disease [4]. However, in recent times the virus has spread rapidly to new regions like Romania (1996), United States (1999), and recently Greece and Italy (2010), resulting in hundreds of neurological and fatal cases worldwide [4, 5]. In India, antibodies against WNV were first detected in human sera from Bombay (1952) [6]. Since then, febrile illness in epidemic form and clinically overt encephalitis cases has been observed from southern, central, and western

India [7]. In eastern India, WNV infection causing acute encephalitis syndrome was first reported during 2006 from the state of Assam [8]. WNV has emerged in recent decades as significant burden to public health and its subsequent spread throughout the world has depicted it as a reemerging global pathogen [9]. Currently, there are five inactivated and chimeric vaccines licensed for veterinary use [10]. However, no human vaccine is available although several candidate vaccines are in clinical trial [11]. The economical impact of both clinical and subclinical diseases warrants search for and use of efficient vaccines. Formalin and β-propiolactone (β-PL) are commonly used for inactivation of viruses via chemical reaction with viral capsid proteins and nucleic acids [12]. Both formalin and β-PL have been classified by International Agency for Research in Cancer (IARC) under groups 2A and 2B, respectively [13]. But till now there is no epidemiological data relevant to the evaluation of the carcinogenic risk of the alkylating agent on humans. In general, preserving the integrity of the immunological epitopes is important for vaccine efficacy. Here, we have studied and compared the effect of two alkylating agents, formaldehyde and β-PL generally used for vaccine preparation [10, 14] on the antigenic and immunogenic properties of the WNV.

2. Materials and Methods

2.1. Virus Strain and Cell Propagation. A circulating strain of WNV, WNIRGC07, isolated in 2008 from human AES patient (GeneBank ID: HQ246154) from Assam, India, was used. Phylogenetic analysis placed it on lineage 5 of WNV [15]. Four virus passages in infant *Swiss albino* mice and subsequently two passages in baby hamster kidney (BHK-21) cell line were done to increase adaptability and virus titer. The cell line was obtained from National Centre for Cell Science, Pune, India, and maintained in Eagles Minimal Essential Medium (EMEM, Sigma) supplemented with 10% heat inactivated fetal bovine serum, 7.5% sodium bicarbonate (Sigma), and 2 mM L-glutamine. Tissue culture infectious dose 50 ($TCID_{50}$) of the virus was calculated by cytopathic effect (CPE) method as per protocol of Cui et al. [16].

2.2. Virus Inactivation. The virus infected culture supernatant was clarified by centrifugation at 10,000 rpm for 1 hr at 4°C. β-PL was added to virus suspension at a concentration of 0.1% and kept for 48 hrs at 4°C [17]. Formalin inactivation was done at a concentration of 0.2% and kept for 5 days at 32°C [18]. Both the suspensions were centrifuged at 10,000 rpm for 1 hr at 4°C and the collected supernatant was treated with protamine sulphate (1 μg/mL) and kept at 4°C for 30 min. Subsequently, the suspensions were centrifuged at 5,000 rpm for 20 min and the supernatant was collected and stored at -80°C till further experiment.

2.3. Screening for Virus Infectivity. Inactivated virus preparations were tested for residual infectivity by serial passage into BHK-21 cells and by intracranial inoculation of inactivated virus into suckling mice. For residue of any viral nucleic acid, viral RNA was extracted using QIAamp viral RNA mini kit (Qiagen, Germany), from the brains collected from mice surviving 21 days after inoculation. Reverse transcriptase- (RT-) PCR was done by WNV specific NS5 region using the primer sequence forward 5′-GCTCCGCTGTCCCTGTGA-3′ and reverse 5′- CACTCTCCTCCTGCATGGATG-3′ [19].

2.4. Haemagglutination (HA) Assay. Inactivated virus titers were determined by HA assay as described by Clarke and Casals [20]. Serial twofold dilutions of inactivated virus in bovine albumin phosphate buffer (0.4% BABS) were prepared and incubated in round bottom plate with 0.4% goose erythrocytes in variable antigen diluents (VAD) of pH 6.2, 6.4, 6.6, and 6.8. After incubation at room temperature for 30 min, the hemagglutination titer, expressed as the reciprocal of the highest dilution producing complete hemagglutination was read.

2.5. Inactivated Virus Antigenicity. The antigenicity of inactivated virus preparations was determined by indirect antigen capture ELISA by following standard method [21]. Ninety-six-well microtiter plates (Nunc-MaxiSorp) were coated with

50 μL/well of *Flavivirus* specific monoclonal antibody (HX-2Ab) at a dilution of 1 : 50 in coating buffer. Biotinylated HX-B (courtesy: National Institute of Virology, Pune, India) was used as detector antibody. ELISA reading was taken at 490 OD in triplicate.

2.6. In Vitro Microcytotoxicity Assay/Cell Viability Assay. Cell toxicity assay of inactivated virus preparations was evaluated for cell cytotoxicity in BHK-21 cell lines by using 3-(4,5-dimethylthiazol-2-yl)-2,5-diphenyltetrazolium bromide (MTT) Cell Proliferation Kit (Roche). The percentage of cytotoxicity was calculated as $[(A - B)/A \times 100]$ where A and B are the absorbances of control and treated cells, respectively [22].

2.7. Mice Immunization. For immunogen preparation standard inactivated virus preparations (β-PL and formalin) were mixed with equal volume of Alhydrogel (2% alum, Sigma), incubated at room temperature for 30 min, and stored at 4°C [23]. A total of two groups (n = 8 mice each) of 3- to 4-week-old *Swiss albino* mice were immunized subcutaneously with 50 μL of immunogen. Control group was injected with phosphate buffered saline (PBS). Booster injections with same formulation were given on 14 and 28 days after first immunization.

2.8. Determination of Immune Response. The anti-WNV IgG antibody response in inoculated mice was determined on serum samples collected at 7 days after inoculation by indirect ELISA by following standard procedure. Briefly, a 96-well microtiter plate (Nunc-MaxiSorp) was coated with mouse brain derived WNV antigen at 1 μg/mL in coating buffer (pH 9.6) and kept overnight at 4°C. The plate was blocked with 1% BSA (Sigma) in PBS (pH 7.4). After washing with PBST, wells were incubated with postimmunized sera in triplicate wells (100 μL/well). A serum of control group mice was also included along with standard anti-JEV polyclonal sera raised with JEV strain, P20778, for cross-reactivity screening. Bound antibodies were detected by HRP-labeled goat anti-mouse IgG (Sigma). ortho-Phenylenediamine dihydrochloride (OPD) (Sigma) as chromogen and hydrogen peroxide (H_2O_2) as substrate were used. Colour development was stopped using 1 M H_2SO_4 and plates were read at 490 nm by ELISA reader. To determine the end point of antibody titer, absorbance reading $\geq$ to twice the OD value of negative control serum was considered positive.

To determine neutralization antibody, hyperimmune sera raised against respective viral inactive agent with WNV strain were used. The fourfold diluted sera were mixed with 100 plaque forming units (P.F.U.) of virus strain G22886 and incubated at 37°C for 1 hr. Virus diluents (0.1 mL) mixture was allowed to absorb in preseeded confluent cell monolayer of BHK-21 cell in a 24-well microtiter plate and incubated at 37°C for 1 hr. Then, cells were overlaid with equal amount of 2x MEM and 1.8% carboxymethyl cellulose (CMC) supplemented with 2% FCS. After 60 hrs, medium was discarded and stained with 0.1% amido black staining solution. Neutralizing antibody titers were expressed by 50% of plaque reduction ($PRNT_{50}$).

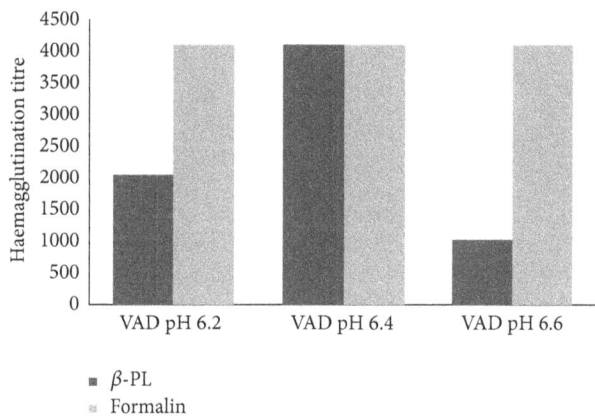

FIGURE 1: Haemagglutination titre of prepared WNV antigen by two different inactivating agents: β-PL and formalin at VAD of pH 6.2, 6.4, and 6.6.

2.9. Statistical Analysis. Results are expressed as mean values for at least three experiments. Comparison between the two groups was made using two sample independent *t*-tests.

3. Result

3.1. Inactivation of WNV. WNV inactivation was confirmed after repeated passage into infant mice and BHK-21 cell monolayer. No mortality was observed in infant mice inoculated intracranially, whereas infants inoculated with live virus died after 3-4 days of inoculation. CPE was not observed in both the inactivated virus infected cell monolayers. PCR amplicons also showed no traces of viral nucleic acid.

3.2. HA Titer and Antigenicity. HA assay showed highest titre of 4096 in both the β-PL and formalin inactivation procedure. However, in β-PL inactivation, antigen titer was found to be dependent on pH values of virus diluent (VAD) where highest titre was obtained at VAD of pH 6.4. However, in antigen prepared with formalin inactivation, uniform high titres were obtained in VADs of pH 6.2, 6.4, and 6.6 (Figure 1).

Antigenicity of formalin and β-PL inactivated virus preparation was checked by indirect ELISA. No significant difference in antigenicity of the virus inactivated by the two inactivating agents was observed (Figure 2).

3.3. Cell Toxicity Evaluation by MTT Assay. Cytotoxicity of inactivated WNV by both β-PL and formalin inactivation on BHK-21 cell lines was quantitatively determined by MTT assay. It was observed that the formalin treated virus induced cell death in 13% of cells and the β-PL treated virus induced cell death in 12% of cells.

3.4. Humoral Immune Response. Humoral antibody response in postimmunized mice sera measured by indirect ELISA showed anti-WNV antibody titre moderately higher in formalin inactivated antigen compared to β-PL inactivated antigen. Statistical analysis between the two groups showed no significant difference in induction of humoral immune

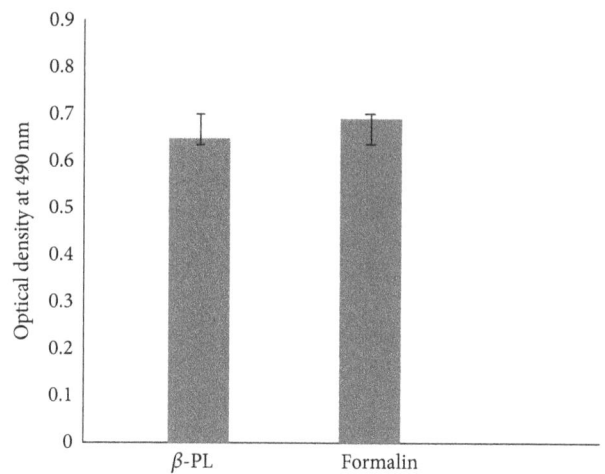

FIGURE 2: Antigenicity of β-PL and formalin inactivated WN preparations evaluated by indirect ELISA.

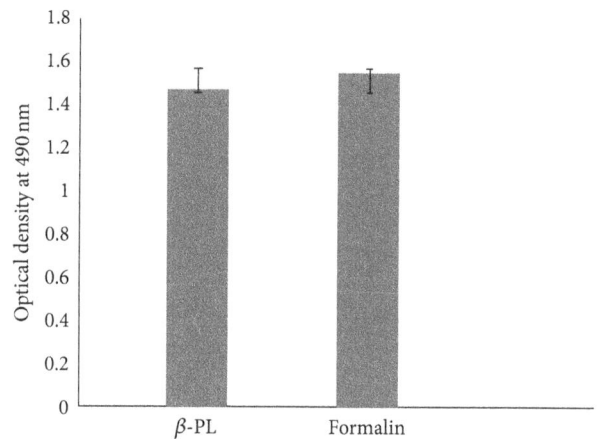

FIGURE 3: Humoral immune response induced in mice model by β-PL and formalin inactivated WN preparations evaluated by indirect ELISA.

response (Figure 3). ELISA titer of JEV specific polyclonal sera is negligible and comes into cut-off of negative value.

The neutralizing antibody response of the antigen obtained by inactivation by either of the methods (β-PL or Formalin) showed higher $PRNT_{50}$ titers ($P < 0.02$) as compared to the control mice group. Although both the experimental groups showed no significant differences in titer levels, but both the inactivation procedures elicited equal protective efficacy. However, the postimmunized WNV specific mice sera showed cross protective neutralizing antibodies of 1 : 25 against JEV.

4. Discussion

WNV invasion continues to expand its geographic distribution. Therefore, studies on safety and efficacy of different vaccine preparation methods are important to develop intervention strategies. The main objective of our study was to evaluate and compare the efficiency of two virus inactivating agents, formalin and β-PL, on WNV. The alkylating agents, formalin and β-PL inactivate viruses *via* chemical reaction with viral capsid proteins and nucleic acids. They were tested in separate experiments at standard usage concentrations [17, 18].

The result obtained for β-PL and formaldehyde inactivation demonstrated that complete inactivation was achieved within 48 to 120 hrs in media containing 0.1% β-PL and 0.2% formaldehyde, respectively. The amount of β-PL required to inactivate the virus differs depending upon the groups of the virus. For arbovirus groups, 0.05 to 0.1% β-PL is sufficient [17]. Formalin was used for inactivation of a hepatitis A virus, and the vaccine was found to be safe and immunogenic in experimental models [18]. It was observed that for formalin inactivation; concentration, pH, and medium composition were critical. A doubled concentration of formalin at 26°C, pH 8.4, for 48 hours was sufficient for inactivation [24]. An incomplete inactivation procedure may prove to be fatal for public health. Outbreak of Venezuelan equine encephalitis was a result of incomplete inactivation of the vaccines prepared by formalin inactivation [25]. β-PL causes structural modification by alkylation and depurination of nucleic acid and is capable of inactivating viruses in 10–15 min at 37°C whereas formalin requires at least 24–96 hrs at 4°C–37°C conditions for inactivation under similar conditions [26, 27]. Although the exact mechanism of RNA degradation through formalin inactivation is not known, it may be due to formalin reaction with viral RNA, forming N-methylol (N-CH2OH), followed by an electrophilic attack to form a methylene bridge between amino groups resulting in cross-linkage between nucleic acids and proteins. This cross-linking inhibits reverse transcription of the extracted RNA and interferes in cDNA synthesis [28].

Studies demonstrated significantly higher degree of antigenicity with β-PL inactivated virus vaccines [29]. In a study comparing β-PL and formalin inactivaton of poliovirus, significant higher antigen recoveries were found in β-PL inactivated antigen compared to formalin antigen [30]. However, in our study, no significant differences were observed in antigenic potential between the β-PL and formalin inactivated WNV. Both showed equal antigenic potency as measured by indirect ELISA using monoclonal antibody against *Flavivirus* (HX-B). Moreover, it may also be supported by high antigen titer obtained in HA assay in both the inactivation processes. In β-PL inactivation, highest titer was obtained with pH 6.6, whereas formalin inactivation showed a high titer with VADs Ph 6.2, 6.6, and 6.8. The pH of the diluent had a critical influence on the HA reaction, affecting not only the titer of the HA preparation but also the appearance of the erythrocyte agglutination pattern. It may so happen that, during β-PL hydrolysis, the pH decreases, resulting in a conformational change in HA as it has been observed in case of influenza virus [31].

Studies carried out in humans and in animal model have indicated the importance of an effective humoral response in preventing *Flavivirus* infection both in the periphery and within the central nervous system [32, 33]. In the present study, both the formaldehyde and β-PL inactivation processes were found to be equally efficient for inactivation

of WNV and were able to elicit humoral immune response against WNV. However, the present study is limited by showing the humoral immune response only till 7 days post inoculation (PI). Studies must be done to screen for immunogenic response every 3 months PI for 1 year. Moreover, other inactivating agents like binary ethylenimine [34] should be further compared and studies assessing influence of different inactivation agent on innate immunity (cell mediated immune response) should be done to formulate the best suitable method for WNV vaccine preparation.

Acknowledgments

This study was supported by Indian Council of Medical Research (ICMR). Pritom Chowdhury and Rashmee Topno are supported by ICMR, Ph.D. Senior Research Fellowship Scheme.

References

[1] G. L. Campbell, A. A. Marfin, R. S. Lanciotti, and D. J. Gubler, "West Nile virus," *Lancet Infectious Diseases*, vol. 2, no. 9, pp. 519–529, 2002.

[2] Z. Hubálek and J. Halouzka, "West Nile fever—a reemerging mosquito-borne viral disease in Europe," *Emerging Infectious Diseases*, vol. 5, no. 5, pp. 643–650, 1999.

[3] E. B. Hayes, J. J. Sejvar, S. R. Zaki, R. S. Lanciotti, A. V. Bode, and G. L. Campbell, "Virology, pathology, and clinical manifestations of West Nile virus disease," *Emerging Infectious Diseases*, vol. 11, no. 8, pp. 1174–1179, 2005.

[4] B. Murgue, H. Zeller, and V. Deubel, "The ecology and epidemiology of West Nile virus in Africa, Europe and Asia," *Current Topics in Microbiology and Immunology*, vol. 267, pp. 195–221, 2002.

[5] A. Papa, P. Perperidou, A. Tzouli, and C. Castilletti, "West Nile virus-neutralizing antibodies in humans in Greece," *Vector-Borne and Zoonotic Diseases*, vol. 10, no. 7, pp. 655–658, 2010.

[6] D. D. Banker, "Preliminary observations on antibody patterns against certain viruses among inhabitants of Bombay city," *Indian Journal of Medical Science*, vol. 6, pp. 733–766, 1952.

[7] S. R. Paramasivan, A. C. Mishra, and D. T. Mourya, "West Nile virus: the Indian scenario," *Indian Journal of Medical Research*, vol. 118, pp. 101–108, 2003.

[8] S. A. Khan, P. Dutta, A. M. Khan et al., "West nile virus infection, Assam, India," *Emerging Infectious Diseases*, vol. 17, no. 5, pp. 947–948, 2011.

[9] L. R. Petersen and J. T. Roehrig, "West Nile virus: a reemerging global pathogen," *Emerging Infectious Diseases*, vol. 7, no. 4, pp. 611–614, 2001.

[10] G. Posadas-Herrera, S. Inoue, I. Fuke et al., "Development and evaluation of a formalin-inactivated West Nile Virus vaccine (WN-VAX) for a human vaccine candidate," *Vaccine*, vol. 28, no. 50, pp. 7939–7946, 2010.

[11] R. Biedenbender, J. Bevilacqua, A. M. Gregg, M. Watson, and G. Dayan, "Phase II, randomized, double-blind, placebo-controlled, multicenter study to investigate the immunogenicity and safety of a West Nile virus vaccine in healthy adults," *Journal of Infectious Diseases*, vol. 203, no. 1, pp. 75–84, 2011.

[12] E. I. Budowsky, E. A. Friedman, and N. V. Zheleznova, "Prin-

ciples of selective inactivation of viral genome. VII. Some peculiarities in determination of viral suspension infectivity during inactivation by chemical agents," *Vaccine*, vol. 9, no. 7, pp. 473–476, 1991.

[13] International Agency for Research on Cancer, *Monographs on the Evaluation of the Carcinogenic Risk of Chemicals to Man*, World Health Organization, Geneva, Switzerland, 2012.

[14] G. A. Logrippo and F. W. Hartman, "Antigenicity of *β*-propiolactone-inactivated virus vaccines," *The Journal of Immunology*, vol. 75, no. 2, pp. 123–128, 1955.

[15] P. Chowdhury, S. A. Khan, P. Dutta, R. Topno, and J. Mahanta, "Characterization of West Nile virus (WNV) isolates from Assam, India: insights into the circulating WNV in north-eastern India," *Comparative Immunology, Microbiology and Infectious Diseases*, vol. 37, no. 1, pp. 39–47, 2014.

[16] J. Cui, D. Counor, D. Shen et al., "Detection of Japanese encephalitis virus antibodies in bats in Southern China," *The American Journal of Tropical Medicine and Hygiene*, vol. 78, no. 6, pp. 1007–1011, 2008.

[17] F. A. Rodrigues and K. H. Pavri, *Technical Manual*, National Institute of Virology, Pune, India, 1972.

[18] V. Pellegrini, N. Fineschi, G. Matteucci et al., "Preparation and immunogenicity of an inactivated hepatitis A vaccine," *Vaccine*, vol. 11, no. 3, pp. 383–387, 1993.

[19] T. Briese, W. G. Glass, and W. I. Lipkin, "Detection of West Nile virus sequences in cerebrospinal fluid," *The Lancet*, vol. 355, no. 9215, pp. 1614–1615, 2000.

[20] D. H. Clarke and J. Casals, "Techniques for hemagglutination and hemagglutination-inhibition with arthropod-borne viruses," *The American Journal of Tropical Medicine and Hygiene*, vol. 7, no. 5, pp. 561–573, 1958.

[21] A. Gajanana, R. Rajendran, V. Thenmozhi, P. P. Samuel, T. F. Tsai, and R. Reuben, "Comparative evaluation of bioassay and ELISA for detection of Japanese encephalitis virus in field collected mosquitos," *The Southeast Asian Journal of Tropical Medicine and Public Health*, vol. 26, no. 1, pp. 91–97, 1995.

[22] C. R. A. Fröhner, R. V. Antonio, T. B. Creczynski-Pasa, C. R. M. Barardi, and C. M. O. Simões, "Cytotoxicity and potential antiviral evaluation of violacein produced by *Chromobacterium violaceum*," *Memorias do Instituto Oswaldo Cruz*, vol. 98, no. 6, pp. 843–848, 2003.

[23] P. Chowdhury, S. A. Khan, R. Topno, P. Dutta, R. N. S. Yadav, and J. Mahanta, "Immunogenic potential and protective efficacy of formalin inactivated circulating Indian strain of West Nile virus," *Asian Pacific Journal of Tropical Medicine*, vol. 7, no. 12, pp. 946–951, 2014.

[24] S. J. Barteling and J. Vreeswijk, "Developments in foot-and-mouth disease vaccines," *Vaccine*, vol. 9, no. 2, pp. 75–88, 1991.

[25] F. Brown, "Review of accidents caused by incomplete inactivation of viruses," *Developments in Biological Standardization*, vol. 81, pp. 103–107, 1993.

[26] S. A. Lawrence, "*β*-propiolactone: viral inactivation in vaccines and plasma," *The PDA Journal of Pharmaceutical Science and Technology*, vol. 54, no. 3, pp. 209–217, 2000.

[27] K. Schaff, "Avian diseases: avian encephalomyelitis immunization with inactivated virus," *American Association of Avian Pathologist*, vol. 3, no. 3, pp. 245–256, 1959.

[28] N. Masuda, T. Ohnishi, S. Kawamoto, M. Monden, and K. Okubo, "Analysis of chemical modification of RNA from formalin-fixed samples and optimization of molecular biology applications for such samples," *Nucleic Acids Research*, vol. 27, no. 22, pp. 4436–4443, 1999.

[29] S. D. Jiang, D. Pye, and J. C. Cox, "Inactivation of poliovirus with *β*-propiolactone," *Journal of Biological Standardization*, vol. 14, no. 2, pp. 103–109, 1986.

[30] H. J. M. Jagt, M. L. E. Bekkers, S. A. J. T. van Bommel, P. van der Marel, and C. C. Schrier, "The influence of the inactivating agent on the antigen content of inactivated Newcastle disease vaccines assessed by the *in vitro* potency test," *Biologicals*, vol. 38, no. 1, pp. 128–134, 2010.

[31] R. W. H. Ruigrok, E. A. Hewat, and R. H. Wade, "Low pH deforms the influenza virus envelope," *Journal of General Virology*, vol. 73, no. 4, pp. 995–998, 1992.

[32] J. T. Roehrig, L. A. Staudinger, A. R. Hunt, J. H. Mathews, and C. D. Blair, "Antibody prophylaxis and therapy for flavivirus encephalitis infections," *Annals of the New York Academy of Sciences*, vol. 951, pp. 286–297, 2001.

[33] M. Ferguson, D. J. Wood, and P. D. Minor, "Antigenic structure of poliovirus in inactivated vaccines," *Journal of General Virology*, vol. 74, no. 4, pp. 685–690, 1993.

[34] M. Habib, I. Hussain, W. H. Fang, Z. I. Rajput, Z. Z. Yang, and H. Irshad, "Inactivation of infectious bursal disease virus by binary ethylenimine and formalin," *Journal of Zhejiang University Science B*, vol. 7, no. 4, pp. 320–323, 2006.

9

Emergence of Hepatitis B Virus Genotype F in Aligarh Region of North India

Hiba Sami,[1] Meher Rizvi,[1] Mohd Azam,[1] Rathindra M. Mukherjee,[2] Indu Shukla,[1] M. R. Ajmal,[3] and Abida Malik[1]

[1] Department of Microbiology, Jawaharlal Nehru Medical College, Aligarh Muslim University, Aligarh 202002, India
[2] Asian Institute of Gastroenterology, Hyderabad 500082, India
[3] Department of Medicine, Jawaharlal Nehru Medical College, Aligarh Muslim University, Aligarh 202002, India

Correspondence should be addressed to Meher Rizvi; rizvimeher@gmail.com

Academic Editor: Trudy Morrison

Introduction. HBV genotypes and subtypes are useful clinical and epidemiological markers. In this study prevalent HBV genotypes were assessed in relation to serological profile and clinical status. *Material & Methods.* 107 cases of HBV were genotyped. Detailed clinical history was elicited from them. HBsAg, HBeAg, anti-HBs, anti-HBe, and anti-HBc-IgM were assessed. HBV genotyping was performed using Kirschberg's type specific primers (TSP-PCR), heminested PCR, and Naito's monoplex PCR. Nucleotide sequencing was performed. *Results.* A total of 97 (91%) were genotyped following the methods of Kirschberg et al./Naito et al. Genotype D was by far the most prevalent genotype 91 (85.04%) in this region. A surprising finding was the detection of genotype F in 5 (4.67%) of our patients. Genotype A strangely was observed only in one case. In 85.7% genotype D was associated with moderate to severe liver disease, 43.9% HBeAg, and 18.7% anti-HBc-IgM positivity. Majority of genotype F (80%) was seen in mild to moderate liver disease. It was strongly associated with HBeAg 60% and 20% anti-HBc-IgM positivity. *Conclusion.* Emergence of genotype F in India merits further study regarding its clinical implications and treatment modalities. Knowledge about HBV genotypes can direct a clinician towards more informed management of HBV patients.

1. Introduction

Hepatitis B virus (HBV) is one of the major public health problems worldwide. About 30% of the world population has serological evidence of current and past infection with HBV [1] and approximately 1 million persons die annually from HBV related chronic liver diseases including severe complications such as liver cirrhosis and hepatocellular carcinoma [2]. Every year there are over 4 million acute clinical cases of HBV and about 25% of carriers. Approximately one million people a year die from chronic active hepatitis, cirrhosis, or primary liver cancer [3].

Genotypically, HBV is divided into 8 groups, A–H. HBV genotypes represent naturally occurring strains of HBV that have evolved over the years in the world. The genotypes and subtypes were identified on the basis of intergroup divergence of 8% or 4% in gene (S) sequence, respectively. They are useful clinical and epidemiological markers [4]. It is also well known

that genotypes vary geographically and correlate strongly with ethnicity [5]. Genotype correlation has been associated with HBV core antigen, HBe antigen seroconversion, activity of liver disease, and treatment response with chronic HBV infections [6, 7]. Type A is prevalent in Europe, Africa, and southeast Asia, including the Philippines. Types B and C are predominant in Asia; type D is common in the Mediterranean area, the Middle East, and India; type E is localized in sub-Saharan Africa; type F (or H) is restricted to Central and South America. Type G has been found in France and Germany. Genotypes A, D, and F are predominant in Brazil and all genotypes occur in the United States with frequencies dependent on ethnicity. The E and F strains appear to have originated in aboriginal populations of Africa and the New World, respectively.

Currently HBV genotypes can be determined by several methods, including direct sequencing [8], restriction fragment length polymorphism analysis (RFLP) [9], line probe assay [10], PCR using type specific primers [11], calorimetric point mutation assay [12], ligase chain reaction assay [13], and enzyme linked immunosorbent assay for genotype specific epitopes [14]. Kirschberg and Naito primers were used to genotype HBV in Aligarh region of North India. The current study was done to determine the prevalence of various HBV genotypes in Aligarh region in patients of acute and chronic hepatitis by using type specific primers and the clinical and demographic characteristics in relation to their genotypes.

2. Material and Methods

2.1. Study Group. Three hundred and thirty patients presenting with the sign and symptoms of liver disease were evaluated on the basis of various investigations such as liver function test (AST, ALT, T. Bilirubin, and ALP), S.creatinine, PT/INR, MELD Score, ultrasound, CT scan, and G.I. endoscopy. Detailed clinical history and rigorous physical examination was conducted on them. 107 confirmed cases of HBV, positive for HBsAg, were included in the study. This study was conducted after obtaining permission from institutional ethics committee of J.N. Medical College and the procedure followed in the study was in accordance with institutional guidelines and informed consent was obtained from all the patients before including them in the study. All patients underwent complete physical examination and detailed clinical history was elicited from them.

2.2. Exclusion Criteria. Patients with autoimmune hepatitis, alcoholic hepatitis, drug induced hepatitis, patients giving history of recent infection, surgery, trauma within the preceding two months, renal insufficiency, or with other acute or chronic inflammatory diseases were excluded from this study. None of the participants had received any antiviral or immunosuppressive therapy before or during the course of this study.

2.3. Serological Investigations. All patients with hepatitis were screened for HAV (hepatitis A virus), HBV (hepatitis B virus), HCV (hepatitis C virus), HEV (hepatitis E virus) and HIV by commercially available ELISA kits: HBsAg, third generation anti-HCV, fourth generation anti-HIV (J. Mitra & Co., Pvt. Ltd., India), anti-HAV IgM, and anti-HEV IgM (DRG International, Inc., USA). HBc lgM antibodies were tested in 107 HBsAg positive samples using DRG Anti-Hepatitis B Core IgM Antigen ELISA kit, (DRG International Inc., USA). The tests were performed according to the manufacturer's instructions. Patients positive for HBV were enrolled for further study. Further on the basis of duration of illness, serum examination, and biochemical examination these patients were regrouped as follows.

(a) Acute Hepatitis B. It is defined as a condition associated with prodromal symptoms preceding the onset of jaundice by 1-2 weeks (e.g., anorexia, nausea, vomiting, and fatigue), fever, an onset of clinical jaundice, diminished constitutional prodromal symptoms, hepatomegaly, jaundice, hyperbilirubinemia (>17 mol/L), serum amino alanine transaminase

(ALT) and aspartate amino transferase (AST) at least fivefold greater than normal, HBsAg (+), anti-HBc-IgM (+), anti-HBc IgG (−), and anti-HBs (−).

(b) Chronic Hepatitis B. It is defined as a condition associated with fatigue, anorexia, jaundice, hepatomegaly, density of the liver harder than normal, splenomegaly, hyperbilirubinemia more than twofold higher than in healthy individuals, serum AST, and ALT twofold higher than in the healthy control group, and HBsAg (+) for longer than 6 months, and anti-HBc type IgG (+), anti-HBc type IgM (−).

(c) Fulminant Hepatic Failure/Hepatic Encephalopathy. FHF was diagnosed if the patients developed hepatic encephalopathy within 4 weeks from the onset of acute hepatitis.

All cases of acute HBV infection were followed up for 6 months to assess seroconversion to anti-HBs (DRG International, Inc., USA). All cases which seroconverted to anti-HBS were included in acute HBV (AHB) group and those who were HBsAg (+) for longer than 6 months were included in chronic HBV (CHB) group. 107 confirmed cases were further screened for HBeAg (DRG International Inc., USA) and anti HBc IgM (DRG International, Inc., USA).

2.3.1. Other Investigations. Liver function tests (LFT) like serum amino alanine transaminase (ALT), serum aspartate amino transferase (AST) and alkaline phosphatase (ALP), bilirubin (direct and indirect), total bilirubin, albumin, globulin, creatinine, and international normalized ratio for prothrombin time were performed.

2.3.2. DNA Extraction. Total DNA from $100 \mu L$ serum was extracted by standard phenol chloroform isoamyl alcohol method [15].

2.3.3. Genotyping and Sequencing of HBV. All 107 cases were subjected to genotyping. HBV genotyping was performed following three pronged approach; initially all samples were subjected to multiplex PCR [16]. HPSF purified oligonucleotides were obtained from MWG-Biotech, Ebersberg, Germany. The sequences of the HBV genotype specific primers were as follows.

(i) *Genotype A*: HBV-GT1-A-s (nt 2331–2360) 5‗-CGG AAA CTA CTG TTG TTA GAC GAC GGG AC-3‗; HBV-GT1-A-as (nt 2701–2665) 5‗-AAT TCC TTT GTC TAA GGG CAA ATA TTT AGT GTG GG-3‗

(ii) *Genotype B*: HBV-GT1-B-s (nt 1470–1491) 5‗-CCG CTT GGG GCT CTA CCG CCC G-3‗; HBV-GT1-B-as (nt 1660–1633) 5‗-CTC TTA TGC AAG ACC TTG GGC AGG TTC C-3‗

(iii) *Genotype C*: HBV-GT1-C-s (nt 2706–2741) 5‗-CCT GAA CAT GCA GTT AAT CAT TAC TTC AAA ACT AGG-3‗; HBV-GT1-C-as (nt 192–165) 5‗-AGC AGG GGT CCT AGG AAT CCT GAT GTT G-3‗

(iv) *Genotype D*: HBV-GT1-D-s (nt 2843–2870) 5‗-ACA GCA TGG GGC AGA ATC TTT CCA CCA G-3‗;

HBV-GT1-D-as (nt 2990—2966) 5_-CCT ACC TTG TTG GCG TCT GGC CAG G-3_

(v) *Genotype E*: HBV-GT1-E-s (nt 2093–2122) 5_-CTA ATG ACT CTA GCT ACC TGG GTG GGT GTA-3_; HBV-GT1-E-as (nt 2880–2853) 5_-CCA TTC GAG AGG GAC CGT CCA AGA AAG C-3_

(vi) *Genotype F*: HBV-GT1-F-s (nt 2843–2871) 5_-ACA GCA TGG GAG CAC CTC TCT CAA CGA CA-3; HBV-GT1-F-as (nt 109–83) 5_-AGA GGC AAT AGT CGG AGC AGG GTT CTG-3_.

Multiplex PCR was carried out in a total volume of 50_1 which contained 25 μL of HotStarTaqTM Master Mix, Qiagen, Hilden, Germany, 1 μL of each sense and antisense primer (10 μmol/L), 8 μL template, and water for a total volume of 50 μL. The thermocycler was programmed to first incubate the samples for 15 min at 95°C, followed by 40 cycles consisting of 94°C for 60 s, 60°C for 60 s, and 72°C for another 2 minutes. After PCR the amplified products were electrophoresed on a 2.5% agarose gel, stained with ethidium bromide, and evaluated under UV light (Biorad, USA). The size of the expected amplified product for each Genotype was: Genotype A: 370 bp, Genotype B: 190 bp, Genotype C: 701 bp, Genotype D: 147 bp, Genotype E: 787 bp, and Genotype F: 481 bp.

Heminested PCR was performed on all the sera which could not be genotyped by the above method. Samples which could not be identified by heminested PCR were subjected to monoplex PCR using the method of Naito et al. [11]. In brief, 10 mL of extracted DNA was subjected to 40 cycles of first round PCR using primers 5'-TCA CCA TAT TCT TGG GAA CAA GA-3' (nt 2823 2845, universal, sense) and 5'-CGA ACC ACT GAA CAA ATG GC-3' (nt 685-704, universal, antisense) amplifying a 1063 bp region of S gene [11]. TSP-PCR was performed in two separate mixes A and B utilizing 1 mL of 1st round PCR product and subjecting it to two rounds of PCR cycles (20 cycles each) as described by Naito et al. [11].

In mix A, primers specific for genotype A (5'-CTC GCG GAG ATT GAC GAG ATG T-3' nt 113–134, type A specific, antisense), and genotype B (5'-CAG GTT GGT GAG TGA CTG GAG A-3' nt 324–345, type B specific, antisense), genotype C (5'-GGT CCT AGG AAT CCT GAT GTT G-3' nt 165–186, type C specific, antisense) and a common universal sense primer (5'-GGC TCA AGT TCA GGA ACA GT-3' nt 67–86, types A to C specific, sense) were used.

In mix B, a common antisense primer (5'-GGA GGC GGA TCT GCT GGC AA-3' nt 3078–3097, specific for types D to F, antisense) along with genotype specific primer D (5'-GCC AAC AAG GTAGGA GCT-3' nt 2979–2996, type D specific, sense), E (5'-CAC CAG AAA TCC AGA TTG GGA CCA-3' nt 2955–2978, type E specific, sense), and F (52-GTT ACG GTC CAG GGT TCA CA-3 nt 3032–3051, type F specific, sense) was used. Mix A allowed for the specific detection of PCR products for types A (68 bp), B (281 bp), and C (122 bp), and mix B allowed for detection of types D (119 bp), E (167 bp), and F (97 bp).

The amplified product of 10 genotype D samples was purified and sequenced by Macrogen, Inc. (Seoul, Korea), using same primers as were used for PCR. Sequencing

reactions were performed in a MJ Research PTC-225 Peltier Thermal Cycler using a ABI PRISM BigDyeTM Terminator Cycle Sequencing Kits with AmpliTaq DNA polymerase (FS enzyme) (Applied Biosystems).

2.3.4. Phylogenetic Analysis. For sequence alignment as well as phylogenetic analysis, we selected the GenBank sequences with the best high scoring matching with HBV reference sequences for each genotype [17]. Sequences, were edited, aligned and analyzed using Clustal W Bioedit software. Genetic distances were calculated using the Kimura two parameter algorithms and phylogenetic trees were constructed by the neighbour joining (NJ) method. To confirm the reliability of the pairwise comparison and phylogenetic tree analysis, bootstrap resampling and reconstruction were carried out 1000 times. Phylogenetic analysis was done using MEGA version MEGA 4 package.

2.3.5. Clinical and Biochemical Grading of Severity of Liver Disease. Disease was classified as mild, moderate, and severe according to the presence or absence of sign and symptoms like icterus, pallor, anorexia, jaundice, nausea, vomiting, splenomegaly, ascites, variceal bleeding, weight loss, and abdominal discomfort. Patients with no sign and symptoms were classified as having mild, those with icterus, pallor, anorexia, jaundice, nausea, vomiting as having moderate, and those with splenomegaly, ascites, variceal bleeding, weight loss, and abdominal discomfort as having severe liver disease. Biochemically the patients in whom all the four parameters (ALT, AST, PT-INR, and MELD) were elevated were classified as having severe, those with elevations in three parameters were classified as having moderate, and those having elevation of only two parameters were classified as having mild liver disease.

2.4. Statistical Analysis. Statistical analysis was performed with the IBM SPSS Statistics 19. Results were expressed as means ± standard deviation or as percentages. Means were compared between groups by using the t-test, ANOVA (one way analysis of variance) and frequency distributions were compared by using the chi-square test.

3. Results

The study group is comprised of 107 HBV infected patients. 52 (48.6%) had acute viral hepatitis (AVH), 32 (30%) had chronic viral hepatitis (CVH), 9 (8.4%) had fulminant hepatic failure (FHF), 13 (12%) were incidentally detected asymptomatic HBsAg positive subjects (IDAHS), and 1 (0.93%) had hepatocellular carcinoma (HCC). The patients (54 men and 36 women) ranged in age from 18 to 70 years. The mean age is 34.58 ± 15.58 years. 48 (44.9%) patients were positive for HBeAg and 21 (19.6%) patients were positive for anti-HBc-IgM antibodies. Of 32 CVH cases, 14 (43.7%) were HBeAg positive, and 18 (56.3%) were HBeAg negative, and 6 (28.6%) were positive for anti-HBc-IgM antibodies.

3.1. Genotype Distribution. In our study we found that HBV genotype D was by far the most prevalent genotype

Table 1: Clinical and biochemical grading of severity of liver disease in different genotypes.

Clinical grading	Genotype D	Genotype F
Mild (%)	13 (14.3)	2 (40)
Moderate (%)	26 (28.6)	2 (40)
Severe (%)	52 (57.1)	1 (20)
Biochemical grading		
Mild (%)	22 (24.1)	1 (20)
Moderate (%)	21 (23.1)	1 (20)
Severe (%)	20 (21.9)	1 (20)

91 (85.04%) in this region. A surprising finding was detection of genotype F in 5 (4.67%) of our patients. Genotype A strangely was observed only in one case. In genotype D patients prevalence of HBeAg was observed in 24 (51%) cases of AVH, 10 (38%) in CVH, 4 (80%) in FHF, and 2 (17%) in IDAHS. Genotype D was found to be associated with moderate to severe liver disease in 78 (85.7%) patients based on clinical grading (Table 1). 17 (18.7%) genotype D patients were positive for anti-HBc-IgM. Of the 18 HBeAg negative cases, HBV DNA was detected in 17 (94.4%) cases pointing to presence of precore mutations.

Using multiplex PCR 53 (49.5%) were genotyped using a modified protocol of Kirschberg et al. [16]. The remaining samples were subjected to heminested PCR [18]. An additional 28 (26%) samples were genotyped by this method. Despite amplification by heminested PCR, 26 samples could not be genotyped. Of these, 20 samples were amplified using monoplex PCR following Naito et al. [11] protocol. 16 (80%) of these samples were genotyped. Thus, a total of 91% were genotyped following methods of Kirschberg et al. [16]/Naito et al. [11]. Genotypes A (1) and F (5) were detected initially by primers of Naito et al. [11] whereas only one of the F genotype was detected by Kirschberg et al. [16] and genotype A was identified only by Naito et al. [11] method.

Genotype F was detected in 1AVH, 1 FHF, and 3 CVH cases and was associated with mild to moderate liver disease in 4 (80%) cases based on clinical and biochemical grading (Table 1). Anti-HBc-IgM positivity was seen in 1 (20%) patient.

The individual with genotype A was associated with high ALT, AST, and PT-INR and was positive for HBeAg and anti-HBc-IgM antibodies (Table 2).

On comparing demographic profile, HBeAg status, and biochemical profile of patients with different HBV genotypes (Table 2), no significant findings were observed in any particular genotype. HBeAg was present in 43.9% in genotype D and 60% in genotype F. Genotype F was associated with mildly elevated levels of ALT while the others were associated with severely deranged levels.

3.2. Genotyping of HBV by Direct Sequencing. DNA sequencing and phylogenetic analysis confirmed genotype D in 10 of our representative samples (Figure 1). A clustering of genotype D in our study was observed with genotype D from

India and Syria, D4 of Italy, D2 of Japan, D5 of Poland, and D6 of Germany [17].

4. Discussion

Different HBV genotypes have their own relatively obvious geographical distribution. HBV genotypes A and D have been well documented from different parts of mainland India [19–22]. In two different studies from northern India, genotypes A and D were found to be equally prevalent [19, 23]. In our study we found that HBV genotype D (85.04%) was by far the most prevalent genotype in the Aligarh region of North India. These results differ from those of two earlier studies in North India where genotypes A and D were found in equal proportions. Contrary to these studies, only one genotype A was identified in our patient group. However, another study from the same region reported genotype D to be predominant (84%) with a low frequency (16%) of genotype A [20]. The pattern of genotype prevalence in this study is in line with studies emanating from Indian subcontinent confirming the high proportion of genotype D in this part of Asia [24]. This study, therefore, highlights the genotypic link between various ethnic groups within the country and people of the neighboring countries.

A surprising finding was detection of genotype F in 4.67% of our patients. Genotype F is considered a new world strain and largely prevalent in central and South America [25, 26]. Genotype F is divided into 4 subgenotypes: F1–F4. Subgenotypes F1 and F2 have been further divided in F1a, F1b, F2a, and F2b [26–29]. In Venezuela subtypes F1, F2, and F3 are found in East and West Amerindians. Among South Amerindians only F3 was found. Subtypes Ia, III, and IV exhibit a restricted geographic distribution (Central America, the North and the South of South America, resp.) while clades Ib and II are found in all the Americas except in the Northern South America and North America, respectively.

One reason for this unexpected finding may be because Aligarh lies in close vicinity to New Delhi which records significant efflux and influx of population to and from different parts of the world. In an era of frequent international travel and human migration, introduction of new HBV genotype to a community might have far reaching effects, including recombination between genotypes or replacement of one genotype by another. However, there is an urgent need to explore other possible reasons for the unusual prevalence of genotype F. In all the five cases the route of transmission appears to be vertical as no other significant factor was identified in any one of them. The patients did not give any history of travel outside India. Prior to this study, Singh et al. [30] have also reported genotype F (3%) from North India. This suggests that genotype F may be indigenous to certain pockets of North India, clustering in and around western UP and Haryana. This unusual finding apparently contradicts the conventional knowledge that HBV genotype closely mirrors ethnic and geographical migration.

In this study Naito primers were more sensitive in detection of HBV genotypes especially genotype F. Naito primers were used when genotyping was unsuccessful using Kirschberg primers. In this way 16 of 20 cases were

TABLE 2: Demographic profile, hepatitis B and antigen (HBeAg) status, and biochemical profile of patients with different HBV genotypes.

Variable	Genotype D (n = 91)	Genotype F (n = 5)	Genotype A (n = 1)	Nontypable (n = 10)
Age, mean years ± SD	34.19 ± 15.01	29.8 ± 12.17	40	38.18 ± 21.87
Ratio of male to female subjects	2.6 : 1	4 : 1	1 : 0	4 : 6
HBeAg positive, no (%)	40 (43.9)	3 (60)	1 (100)	4 (40)
HBeAg negative, no (%)	51 (56.04)	2 (40)	0	6 (60)
Anti-HBc-IgM, no (%)	17 (18.7)	1 (20)	1 (100)	2 (20)
ALT (IU/L) (normal 2–15)	47.61 ± 45.36	23.4 ± 11.54	60	52.90 ± 49.43
AST (IU/L) (normal 2–20)	45.28 ± 44.00	21.8 ± 5.35	90	52.81 ± 41.93
PT-INR	2.40 ± 2.44	1.86 ± 0.79	4.3	2.56 ± 1.79
MELD	19.52 ± 10.57	17 ± 7.41	27	24.18 ± 11.10
Nonnecrotising inflammation (NI)	55 (60.4)	5 (100)	0	6 (60)
Necrotising inflammation (NNI)	36 (39.5)	0	1 (100)	4 (40)

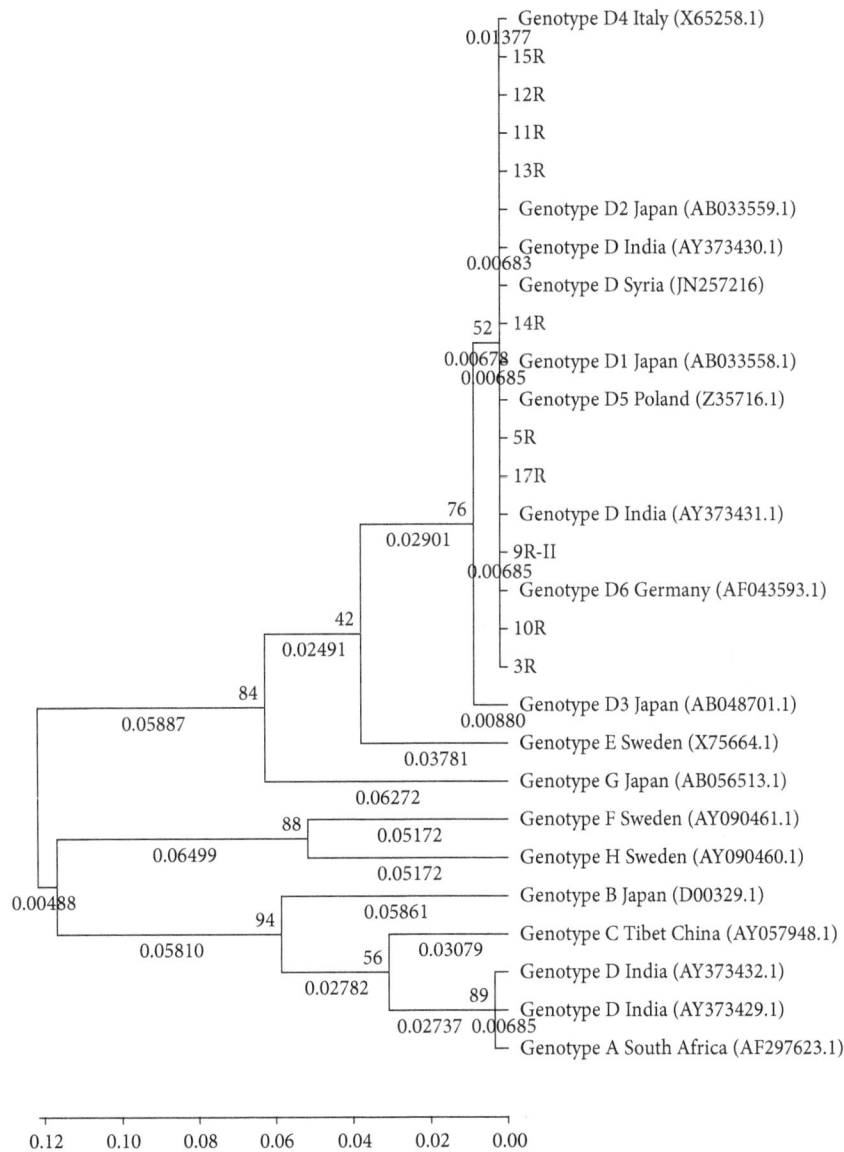

FIGURE 1: Phylogenetic analysis of isolates 3R, 5R, 9R, 10R, 11R, 12R, 13R, 14R, 15R, and 17R (genotype D) from Aligarh region was compared with reference strains from Genbank; phylogenetic analysis was based on comparison of 305 bp of the S-gene and constructed using neighbour joining method (bootstrap test of phylogeny). Robustness of the phylogenetic analysis was evaluated by a 1000 bootstrap replications. Genetic distances were calculated using the Kimura 2-parameter model of nucleotide substitution, MEGA 4 package.

genotyped. We recommend Naito primers for detection of HBV genotypes. The advantage of Kirschberg of course is that it is multiplex, thus saving time. Heminested PCR using Kirschberg primers further increases sensitivity of these primers.

HBV DNA was detected in 90.6% of our patients using primers of Naito et al. [11] and Kirschberg et al. [16]. HBeAg positivity was seen only in 48 (44.9%) cases. Of 32 CVH cases, 14 (43.7%) were HBeAg positive and 18 (56.3%) were HBeAg negative. Of the 18 HBeAg negative cases, DNA was detected in 17 (94.4%) cases (of which 16 were genotype D and 1 was genotype F), suggesting that due to infection for a prolonged duration these strains may have developed mutations in the precore region. Similar precore mutations were observed in a study in Brazil, in which 58% of the HBV patients were HBeAg negative [31]. Of the 18 HBeAg negative cases, majority had severe liver disease: 3 had hepatic decompensation, 12 had cirrhosis, and 3 were stable. Abdo et al. [32] also reported that HBeAg negativity in genotype D was associated with advancing stages of liver disease as compared to HBeAg positive individuals. Further analysis in large-scale longitudinal studies is required to better delineate this relationship.

Based on clinical grading genotype D was associated with moderate to severe liver disease in 85.7% patients. The prevalence of genotype D in different liver disease groups was 90.1% in AVH, 81.2% in CVH, 55.5% in FHF, 92.3% in IDAHS, and 100% in HCC. Several studies have revealed the association of HBV genotypes with the severity of chronic liver disease, but the results are not consistent [33, 34]. Wai et al. [35] and Rodriguez-Frias et al. [36] reported that genotype D is more prevalent in acute disease, and reported a significant association between genotype D and acute liver failure compared to chronic hepatitis. In contrast, in a study from Egypt, Zekri et al. [37] reported that genotype D was significantly more prevalent in chronic active hepatitis than acute hepatitis. In this study genotype D was associated strongly with all stages of liver disease. In our study, while 30.7% patients with genotype D had normal levels of liver enzymes, the number of patients with the same genotype across advancing stages of liver disease comprised 69.23% ($P < 0.001$). These findings suggest that genotype D correlates with advancing liver disease in this region.

On the other hand in 80% cases genotype F was found to be associated with mild to moderate liver disease. In studies by Livingston et al. [38] and Sánchez-Tapias et al. [39], genotype F was associated with severe infection and HCC development in younger population. Genotype F in our study was found to be strongly associated with HBeAg positivity with odds ratio being 1.9 and risk ratio 1.39. Liver enzymes in genotype F in comparison to genotype D were lower. All the cases of genotype F had NNI as against 60% genotype D cases.

In our study we could not reach any conclusion regarding the severity of genotype A as only one patient was genotyped as A although he had severe liver disease. Kumar et al. [23] reported that genotype A was more severe than D, being significantly more associated with high alanine transaminase levels, HBeAg positivity, anti-HBe negativity, and cirrhosis.

36 (39.56%) genotype D patients in this study had necrotising inflammation. None of the genotype F patients developed necrotising inflammation pointing to the milder form of disease associated with genotype F.

However, 10 (9.34%) samples still did not yield a PCR product. Similarly in a study done by Vivekanandan et al. [18], 5.2% of the strains could not be genotyped. Zekri et al. [37] also mentioned that in multiplex PCR indeterminate samples range from 2.9% to 4.5%. This phenomenon may be the result of intrinsic differences in the sensitivities of these PCRs or the result of the difference in the region of the HBV genome targeted. This may be due to the presence of infection with multiple genotypes or the viral load in these samples was not sufficient enough for the detection limit of these amplification protocols.

DNA sequencing and phylogenetic analysis confirmed genotype D in 10 of our representative samples. A clustering of genotype D in our study was observed with genotype D from India and Syria, D4 of Italy, D2 of Japan, D5 of Poland, and D6 of Germany [17]. The diversity in the clustering in our study may be due to the fact that population groups of Northern, Western, and Eastern India ethnically are of Caucasoid origin, who speak Indo-European languages and show close genetic affinities with populations of Eurasia and Europe [40]. Our study is amongst the few in North India confirming genotypes by DNA sequencing and tracing the roots of our samples [23].

Emergence of genotype F in India merits further study regarding its clinical implications and treatment modalities. Our study suggests it causes milder disease. Information about HBV genotypes can direct a clinician towards more informed management of HBV patients. Long term follow-up studies are required to understand the clinical, therapeutic and epidemiological differences among HBV genotypes especially genotype F in the Indian subcontinent.

Authors' Contribution

Hiba Sami, Meher Rizvi, and Mohd Azam contributed equally to this work.

References

[1] E. Szabó, C. Páska, P. K. Novák, Z. Schaff, and A. Kiss, "Similarities and differences in hepatitis B and C virus induced hepatocarcinogenesis," *Pathology and Oncology Research*, vol. 10, no. 1, pp. 5–11, 2004.

[2] K. Okuda, "Hepatocellular carcinoma," *Journal of Hepatology*, vol. 32, no. 1, pp. 225–237, 2000.

[3] World Health Organization, *Introduction of Hepatitis B Vaccine into Childhood Immunization Services*, Unpublished Document WHO/V&B/01.31, Department of Vaccines and Biologicals, World Health Organization, Geneva, Switzerland, 2001.

[4] H. Norder, B. Hammas, S. Lofdahl, A. M. Courouce, and L. O. Magnius, "Comparison of the amino acid sequences of nine

different serotypes of hepatitis B surface antigen and genomic classification of the corresponding hepatitis B virus strains," *Journal of General Virology*, vol. 73, part 5, pp. 1201–1208, 1992.

[5] J. Y. N. Lau and T. L. Wright, "Molecular virology and pathogenesis of hepatitis B," *The Lancet*, vol. 342, no. 8883, pp. 1335–1340, 1993.

[6] C. J. Chu and A. S. F. Lok, "Clinical significance of hepatitis B virus genotypes," *Hepatology*, vol. 35, no. 5, pp. 1274–1276, 2002.

[7] P. Arauz-Ruiz, H. Norder, B. H. Robertson, and L. O. Magnius, "Genotype H: a new Amerindian genotype of hepatitis B virus revealed in Central America," *Journal of General Virology*, vol. 83, no. 8, pp. 2059–2073, 2002.

[8] L. Stuyver, S. de Gendt, C. van Geyt et al., "A new genotype of hepatitis B virus: complete genome and phylogenetic relatedness," *Journal of General Virology*, vol. 81, no. 1, pp. 67–74, 2000.

[9] M. Mizokami, T. Nakano, E. Orito et al., "Hepatitis B virus genotype assignment using restriction fragment length polymorphism patterns," *FEBS Letters*, vol. 450, no. 1-2, pp. 66–71, 1999.

[10] C. Grandjacques, P. Pradat, L. Stuyver et al., "Rapid detection of genotypes and mutations in the pre-core promoter and the pre-core region of hepatitis B virus genome: correlation with viral persistence and disease severity," *Journal of Hepatology*, vol. 33, no. 3, pp. 430–439, 2000.

[11] H. Naito, S. Hayashi, and K. Abe, "Rapid and specific genotyping system for hepatitis B virus corresponding to six major genotypes by PCR using type-specific primers," *Journal of Clinical Microbiology*, vol. 39, no. 1, pp. 362–364, 2001.

[12] A. L. Ballard and E. H. Boxall, "Colourimetric point mutation assay: for detection of precore mutants of hepatitis B," *Journal of Virological Methods*, vol. 67, no. 2, pp. 143–152, 1997.

[13] S. Minamitani, S. Nishiguchi, T. Kuroki, S. Otani, and T. Monna, "Detection by ligase chain reaction of precore mutant of hepatitis B virus," *Hepatology*, vol. 25, no. 1, pp. 216–222, 1997.

[14] S. Usuda, H. Okamoto, H. Iwanari et al., "Serological detection of hepatitis B virus genotypes by ELISA with monoclonal antibodies to type-specific epitopes in the preS2-region product," *Journal of Virological Methods*, vol. 80, no. 1, pp. 97–112, 1999.

[15] R. K. Saiki, D. H. Gelfand, S. Stoffel et al., "Primer-directed enzymatic amplification of DNA with a thermostable DNA polymerase," *Science*, vol. 239, no. 4839, pp. 487–491, 1988.

[16] O. Kirschberg, C. Schüttler, R. Repp, and S. Schaefer, "A multiplex-PCR to identify hepatitis B virus—genotypes A–F," *Journal of Clinical Virology*, vol. 29, no. 1, pp. 39–43, 2004.

[17] A. Kramvis, M. Kew, and G. François, "Hepatitis B virus genotypes," *Vaccine*, vol. 23, no. 19, pp. 2409–2423, 2005.

[18] P. Vivekanandan, P. Abraham, G. Sridharan et al., "Distribution of hepatitis B virus genotypes in blood donors and chronically infected patients in a tertiary care hospital in Southern India," *Clinical Infectious Diseases*, vol. 38, no. 9, pp. e81–e86, 2004.

[19] V. Thakur, R. C. Guptan, S. N. Kazim, V. Malhotra, and S. K. Sarin, "Profile, spectrum and significance of HBV genotypes in chronic liver disease patients in the Indian subcontinent," *Journal of Gastroenterology and Hepatology*, vol. 17, no. 2, pp. 165–170, 2002.

[20] S. Chattopadhyay, B. C. Das, and P. Kar, "Hepatitis B virus genotypes in chronic liver disease patients from New Delhi, India," *World Journal of Gastroenterology*, vol. 12, no. 41, pp. 6702–6706, 2006.

[21] S. S. Gandhe, M. S. Chadha, and V. A. Arankalle, "Hepatitis B virus genotypes and serotypes in Western India: lack of clinical

significance," *Journal of Medical Virology*, vol. 69, no. 3, pp. 324–330, 2003.

[22] A. Banerjee, F. Kurbanov, S. Datta et al., "Phylogenetic relatedness and genetic diversity of hepatitis B virus isolates in Eastern India," *Journal of Medical Virology*, vol. 78, no. 9, pp. 1164–1174, 2006.

[23] A. Kumar, S. I. Kumar, R. Pandey, S. Naik, and R. Aggarwal, "Hepatitis B virus genotype A is more often associated with severe liver disease in northern India than is genotype D," *Indian Journal of Gastroenterology*, vol. 24, no. 1, pp. 19–22, 2005.

[24] S. Baig, A. Siddiqui, R. Chakravarty, and T. Moatter, "Hepatitis B virus subgenotypes D1 and D3 are prevalent in Pakistan," *BMC Research Notes*, vol. 2, article 1, 2009.

[25] M. V. A. Mora, C. M. Romano, M. S. Gomes-Gouvêa et al., "Molecular characterization of the Hepatitis B virus genotypes in Colombia: a Bayesian inference on the genotype F," *Infection, Genetics and Evolution*, vol. 11, no. 1, pp. 103–108, 2011.

[26] M. Devesa, C. L. Loureiro, Y. Rivas et al., "Subgenotype diversity of hepatitis B virus American genotype F in Amerindians from Venezuela and the general population of Colombia," *Journal of Medical Virology*, vol. 80, no. 1, pp. 20–26, 2008.

[27] H. Kato, K. Fujiwara, R. G. Gish et al., "Classifying genotype F of hepatitis B virus into F1 and F2 subtypes," *World Journal of Gastroenterology*, vol. 11, no. 40, pp. 6295–6304, 2005.

[28] M. von Meltzer, S. Vásquez, J. Sun et al., "A new clade of hepatitis B virus subgenotype F1 from Peru with unusual properties," *Virus Genes*, vol. 37, no. 2, pp. 225–230, 2008.

[29] F. Kurbanov, Y. Tanaka, and M. Mizokami, "Geographical and genetic diversity of the human hepatitis B virus," *Hepatology Research*, vol. 40, no. 1, pp. 14–30, 2010.

[30] J. Singh, V. Pahal, and R. Kumar, "First report of genotype F and novel mutations in the core promoter region of HBV isolates from a Northern Indian population," *Cancer Prevention Research*, vol. 1, no. 7, supplement, 2008.

[31] R. E. F. Rezende, B. A. L. Fonseca, L. N. Z. Ramalho et al., "The precore mutation is associated with severity of liver damage in Brazilian patients with chronic hepatitis B," *Journal of Clinical Virology*, vol. 32, no. 1, pp. 53–59, 2005.

[32] A. A. Abdo, B. M. Al-Jarallah, F. M. Sanai et al., "Hepatitis B genotypes: relation to clinical outcome in patients with chronic hepatitis B in Saudi Arabia," *World Journal of Gastroenterology*, vol. 12, no. 43, pp. 7019–7024, 2006.

[33] T. Guettouche and H. J. Hnatyszyn, "Chronic hepatitis B and viral genotype: the clinical significance of determining HBV genotypes," *Antiviral Therapy*, vol. 10, no. 5, pp. 593–604, 2005.

[34] A. Tsubota, Y. Arase, F. Ren, H. Tanaka, K. Ikeda, and H. Kumada, "Genotype may correlate with liver carcinogenesis and tumor characteristics in cirrhotic patients infected with hepatitis B virus subtype adw," *Journal of Medical Virology*, vol. 65, no. 2, pp. 257–265, 2001.

[35] C. T. Wai, R. J. Fontana, J. Polson et al., "Clinical outcome and virological characteristics of hepatitis B-related acute liver failure in the United States," *Journal of Viral Hepatitis*, vol. 12, no. 2, pp. 192–198, 2005.

[36] F. Rodriguez-Frias, M. Buti, R. Jardi et al., "Hepatitis B virus infection: precore mutants and its relation to viral genotypes and core mutations," *Hepatology*, vol. 22, no. 6, pp. 1641–1647, 1995.

[37] A. R. N. Zekri, M. M. Hafez, N. I. Mohamed et al., "Hepatitis B virus (HBV) genotypes in Egyptian pediatric cancer patients with acute and chronic active HBV infection," *Virology Journal*, vol. 4, article 74, 2007.

[38] S. E. Livingston, J. P. Simonetti, B. J. McMahon et al., "Hepatitis B virus genotypes in Alaska Native people with hepatocellular carcinoma: preponderance of genotype F," *Journal of Infectious Diseases*, vol. 195, no. 1, pp. 5–11, 2007.

[39] J. M. Sánchez-Tapias, J. Costa, A. Mas, M. Bruguera, and J. Rodés, "Influence of hepatitis B virus genotype on the long-term outcome of chronic hepatitis B in western patients," *Gastroenterology*, vol. 123, no. 6, pp. 1848–1856, 2002.

[40] P. P. Majumder, "People of India: biological diversity and affinities," *Evolutionary Anthropology*, vol. 6, no. 3, pp. 100–110, 1998.

10

Hepatitis Delta Virus: A Peculiar Virus

Carolina Alves, Cristina Branco, and Celso Cunha

Medical Microbiology Unit, Center for Malaria and Tropical Diseases, Institute of Hygiene and Tropical Medicine, Nova University, Rua da Junqueira 100, 1349-008 Lisbon, Portugal

Correspondence should be addressed to Celso Cunha; ccunha@ihmt.unl.pt

Academic Editor: Michael Bukrinsky

The hepatitis delta virus (HDV) is distributed worldwide and related to the most severe form of viral hepatitis. HDV is a satellite RNA virus dependent on hepatitis B surface antigens to assemble its envelope and thus form new virions and propagate infection. HDV has a small 1.7 Kb genome making it the smallest known human virus. This deceivingly simple virus has unique biological features and many aspects of its life cycle remain elusive. The present review endeavors to gather the available information on HDV epidemiology and clinical features as well as HDV biology.

1. Introduction

In 1977, a novel antigen was found in the nucleus of hepatocytes from patients with a more severe form of hepatitis B. It was first thought to be a previously unknown marker of hepatitis B virus (HBV). Only later, it was found that the then called delta antigen was not part of HBV but of a separate defective virus that requires the presence of HBV for infection. The newfound virus was designated hepatitis delta virus (HDV) and, by 1986, its RNA genome was cloned and sequenced (reviewed by [1]). This peculiar virus has been classified as the only member of the genus *Deltavirus* due to its uniqueness [2]. The HDV virion is a hybrid particle, composed of the delta antigen and HDV RNA enclosed by the surface antigens of HBV (HBsAgs). HDV has the smallest RNA genome of all known animal viruses. However, it is comparable, although larger, to viroid RNAs, pathogenic agents of higher plants.

2. Epidemiology

HDV infection is distributed worldwide, although not uniformly, and it is estimated that 5% of HBsAgs carriers are also infected with HDV, which signifies that there might be between 15 and 20 million HDV-infected individuals [3]. This is a very rough number because it lacks data from areas where HBV is highly prevalent and HDV is poorly studied.

HDV is highly endemic in Mediterranean countries, the Middle East, northern parts of South America, and Central Africa [4]. HDV also has high prevalence in Turkey [5], Central Asia [6], and the Amazonian region of Western Brazil [7].

In Southern Europe, HDV infection has been highly prevalent, with studies from the 1980s and 1990s showing that the incidence of HDV in HBsAgs positive individuals was higher than 20% [8]. With the implementation of HBV vaccination programs in the 1980s, HDV prevalence considerably decreased to 5–10% by the late 1990s [9]. However, in the beginning of the XXI century, the number of HDV-infected HBsAgs carriers in Europe increased to 8–12% [9, 10]. This increase has been attributed to immigration of individuals from highly endemic regions [10]. Another report claims that the increase in HDV incidence is not only due to immigration but also due to other factors associated with HDV modes of transmission [9]. Drug addiction and other risk behaviors, such as multiple sexual partners, tattooing and piercing, or uncontrolled medical procedures, have been shown to contribute to the spread of hepatitis D in Italy [9]. In fact, in western countries, the virus is highly prevalent in intravenous drug addicts with chronic HBV infection [9, 10].

More recent and reliable data are needed, especially from poorly studied regions. As an example, only recently are data starting to emerge from the United States of America. A 2013 survey has shown that in Northern California, 8% of 499 chronic HBV patients tested positive for HDV infection [11].

Based on nucleotide sequence analysis, eight HDV genotypes have been defined, some of which are distributed by distinct geographic regions [4, 12]. The divergence in nucleotide sequence between isolates of the same genotype is less than 15% and between different genotypes it can be as high as 40% [4]. HDV genotype 1 is the most common and prevalent worldwide, present mainly in Europe, Middle East, North America, and Northern Africa. It is associated with both severe and mild forms of the disease [13]. Genotype 2 is more common in the Far East, present in Japan, Taiwan, and parts of Russia [14]. Genotype 2 is associated with a milder disease course [13]. HDV genotype 3 is exclusively found in the Amazon Basin [7] and has been associated with the most aggressive forms of HDV infection. The combined infection of HDV genotype 3 and HBV genotype F was associated with fulminant hepatitis in South America [15]. Genotype 4, present in Japan and Taiwan [16], has variable pathogenicity. A genotype 4 isolate from Okinawa, Japan, has been associated with greater progression to cirrhosis than the genotype 4 predominant in Taiwan [17]. Genotypes 5 to 8 were found in African patients who had migrated to Northern Europe [4, 12]. Phylogenetic reconstructions based on the delta antigen coding sequence have shown a probable ancient radiation of African lineages [4].

3. Clinical Expression

Hepatitis delta virus is usually associated with a severe form of hepatitis, but the range of clinical manifestations is very wide going from asymptomatic cases to fulminant hepatitis.

Regarding HDV transmission, like its helper virus HBV, it is parenterally transmitted through exposure to infected blood or body fluids. Intrafamilial spread is naturally common in highly endemic regions.

HDV requires the presence of HBsAgs to form new infectious virions and propagate HDV infection. Thus, hepatitis D only occurs in individuals infected with HBV. Consequently, there are two major patterns of infection: "coinfection" with HBV and HDV or "superinfection" of patients already infected with HBV. A rare third pattern has been reported; it can occur after liver transplantation for an HDV-infected individual and is designated as "helper-independent latent infection" [18]. In this scenario, an initial HDV infection of the new liver occurs without any apparent help from HBV. Such an infection remains asymptomatic unless reactivated by HBV appearance [18].

For an HBV and HDV acute coinfection, the most common outcome (95%) is viral clearance [14]. However, it can be more severe than an acute HBV monoinfection, resulting in some cases in acute liver failure [19]. Acute hepatitis strikes after an incubation period of 3–7 weeks, beginning with a period of nonspecific symptoms such as fatigue, lethargy, or nausea [20].

HDV superinfection of chronic HBV patients also causes severe acute hepatitis, but in this case, for up to 80% of patients, it progresses to chronicity [21]. The processes, which determine whether a patient clears HDV spontaneously or becomes chronically infected, remain unclear. When chronic HDV infection is established, the preexisting liver disease caused by HBV is usually aggravated [22]. It has been claimed that, during the acute phase of HDV infection, HBV replication is suppressed to very low levels and that this suppression can persist once a chronic HDV infection is established [23]. Patients with HDV superinfection suffer a more rapid progression to cirrhosis [24, 25], increased liver decompensation, and eventually death [26], when compared with patients with HBV monoinfection. Despite the higher rates of progression to cirrhosis, not all published studies refer to an increased rate of hepatocellular carcinoma [27]. One explanation of this may be the abovementioned suppression of HBV replication by HDV, since other studies assert that higher HBV DNA serum levels correlate with a greater risk of carcinoma [28].

4. Diagnosis

Since HDV is a satellite virus of HBV, every HBsAgs positive patient should be screened for coinfection with HDV; that is, patients should be tested, at least once, for anti-HDV antibodies. A negative result does not justify testing for HDV RNA as, so far, it seems that every individual infected with HDV develops anti-HDV antibodies [29]. In contrast, a positive result for anti-HDV antibodies requires confirmation of continued HDV infection, through detection of HDV RNA in serum. Anti-HDV antibodies may be present even after HDV RNA has disappeared during recovery from the infection [29].

Currently, there is no need for quantification of the HDV RNA levels in serum during the diagnosis step. There is no evidence that a correlation exists between the stage of liver disease and the levels of HDV RNA [30]. Thus, a liver biopsy is still the major tool for evaluating the stage of delta hepatitis in patients [29]. However, a quantitative assay of HDV RNA is useful during the therapy stage to monitor the treatment response of patients undergoing therapy. Unfortunately, very few data are available on the levels of HDV RNA during the different stages of the disease. Thus, there is no accepted threshold level at which one might recommend treatment.

For some time, quantification of HDV RNA levels in clinical samples has suffered from the lack of a standardized test. Quantification of HDV RNA was done in specialized laboratories using in-house protocols, which unfortunately become irrelevant outside the laboratory of origin. Such assays typically lacked an internal control and were limited to only one genotype. Furthermore, there is no international reference standard to make results from different laboratories comparable. As proposed elsewhere, an HDV RNA reference preparation should be defined by the World Health Organization to be used as an international standard [31].

In 2012, two standard protocols were proposed to detect and quantify HDV RNA from clinical samples [32, 33]. One method is described as able to be automated to accurately quantify the major HDV genotypes present in Europe (genotype 1 and the migrant African strains 5–8; [32]). The other standardized test is described as being able to detect and quantify all HDV genotypes [33]. Both protocols use a commercial kit to extract nucleic acids from samples and include an internal control to enable monitoring of the overall performances of the assay. The one-step RT-qPCR makes use of another commercial kit [32, 33].

Application of standardized procedures is crucial to improve our understanding of HDV RNA kinetics during the course of disease. It will improve patient management, as data can be gathered which will help in the decision to start treatment, as well as monitoring the response to therapy in chronic patients. Also it will contribute to the screening of HDV infections in the endemic areas, providing more reliable epidemiological data. Overall, acceptance of standardization will help clarify the pathophysiology of HDV infections.

5. Treatment

Ideally, a successful treatment of an HDV infection eradicates HDV and its helper virus HBV. Clearance of HDV is obtained when both HDV RNA and HDAg in the liver become persistently undetectable and a complete resolution is achieved when HBsAgs clearance is also obtained.

However, at this time, there is no efficient therapy. Prolonged treatment with recombinant interferons is the only therapy that has shown antiviral activity against HDV. Such therapies, which last up to 2 years, have been reported as only 20–40% efficient [34].

In general, when searching for a treatment for viral disorders, the first and preferred targets analyzed are the viral components, such as enzymes involved in the virus replication cycle. But HDV lacks any specific enzymatic function to target. Since the only known enzymatic activity the virus possesses is a ribozyme, the virus relies on the host cell to provide for all other enzymatic activities needed for its life cycle. This represents a serious challenge in finding an HDV-specific therapeutic target.

Puzzlingly, the nucleoside and nucleotide analogues used for treatment of HBV infection are inefficient against HDV. Although they block HBV DNA synthesis in chronic patients, they have little impact on HDV and do not even enhance interferon treatments [34]. Famciclovir, lamivudine, and adefovir, all used in HBV treatment, have been shown to lack any significant antiviral activity against HDV [35–37]. Ribavirin, a nucleotide analogue, which inhibits HDV replication in cell culture, when administered alone or in combination with interferon, also failed to increase rates of HDV RNA clearance [38].

Interferon-α (IFN-α) has been used for treatment of HDV infections since the mid-1980s [39]. Several trials were carried out exploring different doses and durations. Responses to treatment varied and clearance occurred at different times from the beginning of treatment, occurring even after discontinuation of treatment [35]. Researchers have yet to identify pretreatment characteristics that determine responders and nonresponders to IFN-α therapy. It seems that 2 years of treatment with IFN-α is superior to shorter treatment durations to obtain HDV RNA clearance [35]. It has been reported that in a 1-year treatment, there is only a 10 to 20% chance of HDV clearance, and in a 2-year treatment trial, 20% of patients were cleared [40]. The rate of response is proportional to the dose of IFN-α; patients treated with doses of 9 million units responded better than those treated with only 3 million units,

and relapse was common when the IFN-α dose was reduced [35]. Unfortunately, a prolonged treatment with high doses of IFN-α is tolerated by only a minority of patients [29]. IFN-α side effects include flu-like symptoms, fatigue, and weight loss as well as severe psychiatric disturbances. Patients have a tendency to become deeply depressed; suicides and attempted suicides have been reported [35]. The severity of reactions tends to be proportional to IFN-α dose, and intermittent use of IFN-α, observed in drug abusers, increased incidence and severity of side effects [35].

By 2006, IFN-α was largely replaced by longer-lasting pegylated IFN-α (PEG-IFN-α) [38, 41, 42]. Clearance of HDV RNA was obtained for 6 out of 14 patients in a 1-year treatment plan [41]. However, in a similar study, only 2 patients in 12 were cured [42]. In a third study, 8 patients out of 38 became HDV RNA negative after 72 weeks of treatment [38]. Ribavirin was also used in this trial but without any apparent beneficial effect [38].

The Hep-Net International hepatitis D intervention trial, which included 90 patients from Germany, Greece, and Turkey, tested PEG-IFN-α2a alone or with adefovir and adefovir alone [36]. HDV RNA clearance was only observed in patients who had received treatment including PEG-IFN-α2a, showing an antiviral efficacy in more than 40% of patients, and 25% became HDV RNA negative [36]. Adefovir showed little efficacy in reducing HDV RNA levels, but a PEG-IFN-α2a plus adefovir therapy was superior in reducing HBsAgs serum levels [36].

Currently, it is usually recommended to treat chronic hepatitis D with PEG-IFN-α for one year or longer, if the patient can tolerate the adverse effects of such therapy [14]. For patients with advanced liver disease, liver transplantation is the only therapy available [40].

An optimization of the available treatment strategies is clearly needed, either regarding doses or duration, and also possible combinations such as PEG-IFN-α2a with adefovir to also tackle HBsAgs, crucial for HDV propagation. Most importantly, alternative treatments need to be explored, as the efficacy of the current therapies is clearly unsatisfactory. One of the most promising alternatives is the prenylation inhibitors since, as will be discussed subsequently, prenylation of HDAg is essential for interaction with HBsAgs. Furthermore, prenylation inhibitors have already been developed to treat a number of malignancies and were shown to be safe [43].

6. HDV Biology

6.1. HDV Virions and Putative Host Cell Receptors. An infectious HDV virion is an enveloped, roughly spherical particle, of around 36 nm in diameter [44]. The outer coat of the virion containing host lipids and the HBsAgs surrounds an inner nucleocapsid consisting of viral ribonucleoproteins (RNPs) with the genomic RNA and about 200 molecules of HDAg per genome [45].

Since HDV and HBV share the same envelope proteins, it is often assumed that attachment and cell entry occur via similar mechanisms.

Several studies have attempted to identify the regions of the HBsAgs required for HDV and HBV entry. The preS1 region of L-HBsAg is myristoylated at the N-terminus. This posttranslational modification and about 48 adjacent amino acids are essential for HBV and HDV entry into hepatocytes. Synthetic peptides that mimic this region are potent inhibitors of virus entry [46].

Many studies have aimed to discover the host receptors for HBV (and maybe HDV). Many candidates have been proposed but not confirmed [47].

It has been suggested that functional purinergic receptors are required for HDV entry as compounds that block the activation of such receptors inhibited HDV and HBV infection of primary human hepatocytes [48]. However, a different study has since reported that such blocking compounds interfere in HDV and HBV infection due to their charge and not because the receptors are directly involved in the process [49]. In fact, one of the blocking compounds used, ivermectin, reduced HDV infection when added after virus inoculation just as well as when added before inoculation [49]. This suggests that ivermectin focuses on a step of the HDV replication cycle other than the receptor binding stage. In this study, it has also been shown that HDV cell entry depends on binding to the glycosaminoglycan side chains of the hepatocyte heparan sulfate proteoglycans, much like what has been observed for HBV infection [49].

In contrast to all previous studies, an important new report by Yan and colleagues demonstrates that a necessary and sufficient receptor for HBV and HDV is the sodium taurocholate cotransporting polypeptide [50]. This protein is a multiple transmembrane transporter expressed in the liver. Silencing expression of this protein in primary hepatocytes using small interfering RNAs inhibited HBV and HDV infection. Expression of this protein in human liver cell lines rendered them susceptible to infection by HBV and HDV. Therefore, it is now possible for the first time to study the infection processes for these viruses *in vitro*, using established human liver cell lines, which are much more convenient and reproducible than primary hepatocyte cultures.

6.2. HDV RNAs. HDV has a small circular RNA genome with only ~1700 nucleotides; this sequence length varies by no more than 30 nucleotides among HDV isolates [51]. In native conditions, the RNA folds into an unbranched rod-like structure due to intramolecular base pairing involving around 74% of its nucleotides [52].

HDV contains one functional open reading frame (ORF), encoding the delta antigen [53]. This ORF is not encoded by the genomic RNA but by another RNA species that arises during replication, the HDV antigenome, an exact complement of the genome.

The delta antigen is translated from a third RNA species, a linear 0.8 Kb messenger RNA (mRNA) of antigenomic polarity and a $5'$-cap and $3'$-polyadenylated tail [54]. The different HDV RNA species are represented in Figure 1. In an infected cell, the three HDV RNA species accumulate in very different amounts, although genomic RNA is the only species assembled into HDV virions. HDV genomic RNA is the most abundant; around 300,000 copies accumulate in an infected

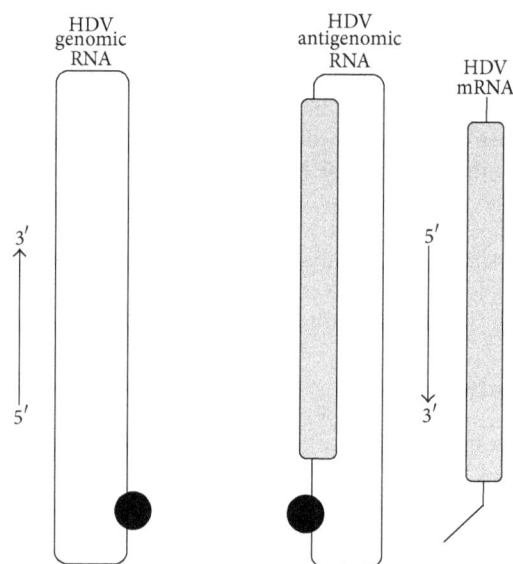

FIGURE 1: Schematic representation of three HDV RNA species. The HDV genomic RNA is a single-stranded circular RNA with ~1700 nucleotides. It forms an unbranched rod-like structure due to intramolecular base pairing. The HDV antigenomic RNA is the exact complement of the genomic RNA, and both RNAs have site-specific ribozymes indicated by the black circle. The HDV antigenomic RNA contains the open reading frame for the delta antigen, represented in grey, but the antigen is translated from another RNA species, the mRNA. The mRNA is ~800 nucleotides long with a $5'$-cap and a $3'$-polyadenylated tail.

cell whereas 100,000 copies of the antigenome are present [53]. The HDV mRNA is considerably less abundant with approximately 500 copies per cell [55].

Site-specific self-cleavage and ligation has been reported on antigenomic HDV RNA, showing that this RNA possesses ribozyme activity, just like plant viroids [56]. Both genomic and antigenomic RNAs display this ribozyme activity, which is comprised within a contiguous sequence of less than 100 nucleotides [57]. They enhance HDV RNA self-cleavage by a 10^6- to 10^7-fold when compared with uncatalyzed cleavage [58, 59]. Although ribozymes are characteristic of viroids, their structures are different from HDV ribozymes, which are actually more related to the cytoplasmic polyadenylation element-binding protein 3 ($CPEB_3$) ribozyme, a conserved mammalian sequence within an intron of the $CPEB_3$ gene [60]. In fact, numerous HDV-like ribozymes have since been found in several eukaryotic species [61].

6.3. HDV RNA Replication. HDV RNAs are transcribed in the nucleus of infected cells, but the details of this process remain poorly defined. The three RNA species that accumulate in infected cells are the product of posttranscriptional processing. The precursors, from which they arise, are thought to be transcribed by a double-rolling circle mechanism, exemplified in Figure 2. In this model, the circular genome RNA is used as a template to produce multimeric species of opposite polarity [62]. These greater than unit-length RNAs are subsequently self-cleaved by the HDV

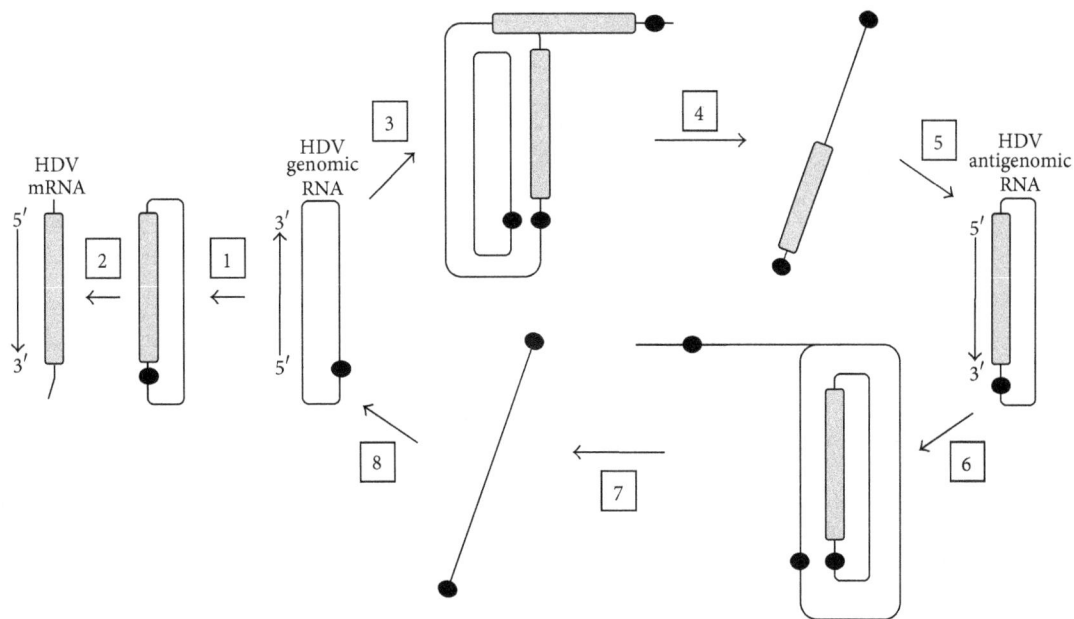

FIGURE 2: Model of HDV replication through a rolling-circle mechanism. The HDV genomic RNA is used as a template for the precursors of HDV mRNA (Steps 1-2) and also acts as a template for multimeric RNAs of antigenomic polarity (Step 3). These multimeric RNAs contain at least two copies of the HDV ribozyme and are thus self-cleaved to produce linear unit-length HDV antigenomes (Step 4), which are then ligated to produce circular antigenomic RNA (Step 5). In turn, the new antigenomic RNA is a template for multimeric RNAs of genomic polarity (Step 6) that are similarly self-cleaved and subsequently ligated to produce new circular genomic RNA (Steps 7-8).

ribozymes and religated, producing unit-length circular antigenomic RNAs. The religation step is thought to involve a host ligase [63] although it has been shown that the HDV ribozyme can self-ligate *in vitro* [64]. Through a similar mechanism, the unit-length circular antigenomic RNA acts as a template for the transcription of multimeric species, which are processed to produce genomic RNA. The genomic RNA also acts as a template for transcripts that are processed into mRNA.

Even though such a rolling-circle mechanism has been widely accepted as a model for HDV replication, critical details remain to be confirmed and/or clarified such as the host cell components involved (reviewed by [65]).

HDV has no known DNA intermediate, as observed for retroviruses [65], and the only HDV protein, the delta antigen, is too small to be a polymerase. This means that HDV RNA must somehow redirect host DNA-dependent RNA polymerases to use HDV RNAs as templates. How this is achieved and which host polymerase(s) is (are) involved have been extensively studied but the results remain somewhat controversial.

The host RNA polymerase II (pol II) seems to be required for genomic HDV RNA transcription. Nuclear run-on experiments on an endogenous HDV RNA template have shown that inhibition of pol II by low concentrations of the specific inhibitor α-amanitin blocks HDV RNA synthesis of both the genomic and antigenomic strands [66]. One possible explanation is that the rod-like conformation of HDV RNAs may trick pol II into accepting the RNA as a double-stranded DNA template. It has been shown, through immunoprecipitation assays, that pol II binds the terminal stem loop regions of HDV genome [67]. It has also been

reported that, after binding to the stem-loop, pol II is able to elongate multimeric RNA species, carrying out transcription [68]. Such elongation was observed on a partial antigenomic RNA stem loop and originated a chimeric molecule of newly synthesized transcript covalently bound to the $5'$-end of the template. Thus, it is not clear if such elongation is biologically relevant.

Despite being shown that pol II interacts with genomic HDV RNA, it has been suggested that a different host polymerase is responsible for the synthesis of antigenomic HDV RNA [69, 70]. The idea that at least two different host polymerases are involved in the HDV replication cycle is based on the observation that, in transfected cells, the synthesis of new HDV antigenomic RNA was not inhibited by concentrations of α-amanitin that would inhibit pol II activity [70]. This has led to the speculation that pol I copies genomic HDV RNA to produce new antigenomic RNA [70]. This is contrary to the aforementioned nuclear run-on assays, which have shown that both genomic and antigenomic RNA syntheses are sensitive to low doses of α-amanitin, consistent with pol II involvement [66]. Note that HDV RNA has been detected in the nucleoplasm of cultured cells with nucleolus exclusion [71]. This suggests that if another host polymerase, other than pol II, is involved in HDV RNA replication, it is pol III, rather than pol I, which is resistant to high concentrations of α-amanitin.

An additional complication arises from *in vitro* studies, which indicate that fragments of the HDV RNA genome interact not only with pol II but also with pol I and pol III [72]. However, such *in vitro* interactions may not have biological relevance, especially since they do not lead to RNA-directed transcription.

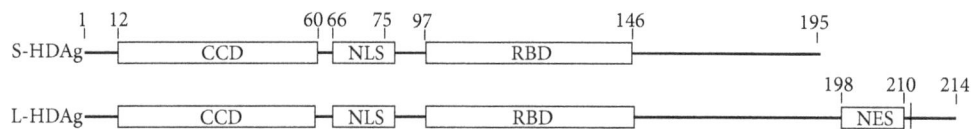

FIGURE 3: Functional domains of S-HDAg and L-HDAg. The delta antigens share most of their sequence differing only in the 19-amino-acid extension at the C-terminal of L-HDAg. They have, within the common sequence, as represented, a coiled-coil domain (CCD), a nuclear localization signal (NLS), and an RNA binding domain (RBD). Also indicated on L-HDAg are the nuclear export signal (NES) and the unique cysteine, residue 211, which is the target for farnesylation. The numbers indicate the position of the amino acid residues.

The HDV mRNA possesses characteristics of a pol II transcript that is processed to a mRNA, namely, a $5'$-cap structure and a $3'$-poly(A) tail. In fact, the role of pol II in HDV mRNA transcription has been generally accepted [66, 69, 70, 73, 74].

The controversy regarding the transcription process is thus limited to whether genomic RNA is transcribed by pol II or another polymerase, either pol I or pol II [75]. If different polymerases are involved, then distinct metabolic requirements, as well as accessory factors, are necessary to accomplish these processes (reviewed in [76]).

In addition to the posttranscriptional processing to make the abovementioned three HDV RNAs, there is an important RNA-editing event. During the virus replication cycle, some of the antigenomes are edited at a specific site by a host adenosine deaminase (ADAR1). This changes the adenosine in the amber codon to inosine. After subsequent RNA-directed RNA synthesis, it leads to the replacement of inosine with guanosine [77]. That is, the UAG stop codon is changed to a UGG tryptophan codon. In this way, the delta antigen ORF is extended by 19 amino acids, that is, to the next stop codon. The specificity of the editing site is in part directed by the specific folding of the HDV antigenomic RNA [78].

Therefore, although HDV has only one ORF, it encodes two proteins: the small delta antigen (S-HDAg) of 195 amino acids and the large delta antigen (L-HDAg) with 214 amino acids.

7. Delta Antigens

The two delta antigen isoforms share 195 amino acids and differ only in that the large form has 19 extra amino acids on the C-terminus. As such, S-HDAg and L-HDAg share several functional domains within the common amino acid sequence, as illustrated in Figure 3. The delta antigens contain a nuclear localization signal (NLS) comprised by amino acids 66 through to 75 [79]; a coiled-coil domain (CCD), also referred to as dimerization domain, within amino acids 12 to 60; and an RNA binding domain within amino acids 97 and 146 [80]. L-HDAg has, within its extra sequence, a nuclear export signal (NES) spanning amino acids 198 to 210 [81].

Both delta antigens undergo posttranslational modifications (PTMs) by several host enzymes. Several groups have investigated the impact these PTMs may have on the antigens' functions, but the precise significance of most of these modifications remains uncertain.

The exception is one PTM, characteristic only of L-HDAg, which has been shown to be essential. It occurs on cysteine residue 211 and is mediated by a host farnesyltransferase [82, 83]. This isoprenylation of L-HDAg is necessary, although not sufficient for viral packaging. It is somehow necessary for the interactions with HBsAgs, leading to the assembly of new viral particles [84, 85].

There are other PTM events, ones shared by both forms of the delta antigen. These involve phosphorylation, methylation, acetylation, and sumoylation (reviewed in [75]).

Phosphorylation has been observed at multiple sites, mostly at serine and threonine residues. Different phosphorylation patterns were observed for S-HDAg and L-HDAg, and, if relevant, the distinct patterns may in part account for their distinct biological functions [86]. Several host enzymes have been reported to phosphorylate delta antigens at different sites: casein kinase II on Ser2 and Ser213 [87]; double-stranded RNA-activated protein kinase R on residues Ser177, Ser180, and Thr182 [88]; extracellular signal-related kinases 1 and 2 (ERK1/2) on Ser177 [89]; and protein kinase C on residue Ser210 [87]. It has been alleged that S-HDAg phosphorylation increases replication of genomic HDV RNA from the antigenomic strand [89]. By enhancing the expression of ERK1/2 in cells transfected with plasmids expressing S-HDAg and dimeric HDV antigenomic RNA, an increase in the accumulation of HDV genomic RNA was observed but not for antigenomic RNA [89]. More recently, it has been suggested that phosphorylation of S-HDAg at Ser177 can work as a switch in HDV antigenomic RNA replication from the initiation to the elongation stage [90].

Acetylation of Lys72 on S-HDAg, by host p300 acetyltransferase, is thought to regulate nucleocytoplasmic shuttling of viral RNA [91, 92]. Note that this amino acid is within the NLS of the HDAgs [79]. Thus such a modification could be expected to have an impact on nuclear import. Acetylation of S-HDAg has also been suggested to function as a switch in the synthesis of the different viral RNA species as this PTM was reported to be essential for HDV genome and mRNA synthesis but dispensable for antigenomic RNA synthesis [93].

Methylation of Arg13 on S-HDAg, by protein arginine methyltransferase I, has been observed *in vitro* and has also been proposed to have a switching effect on HDV RNA replication [93, 94]. The studies were performed with S-HDAg with an R13A mutation, which failed to be methylated *in vitro*. In transfected cells, the mutant S-HDAg reduced genomic RNA synthesis and almost completely suppressed HDV mRNA synthesis [93, 94].

Finally, SUMOylation of multiple lysine sites, by small ubiquitin-related modifier isoform 1 (SUMO1), has been reported. Such PTM was detected on S-HDAg but not on L-HDAg [95]. And this PTM was proposed to enhance genomic RNA and mRNA synthesis based on experiments where SUMO1 was fused to S-HDAg, so as to mimic SUMOylated S-HDAg [95].

Although the two delta antigens share sequence and functional domains, they play very distinct roles in the HDV replication cycle. S-HDAg is essential for HDV RNA accumulation, whereas L-HDAg acts as a dominant negative inhibitor of HDV replication [96] and also is essential for the assembly, via HBsAgs, of HDV RNA into new virus particles. There is, however, a common function attributed to both antigens: it has been observed that both can downregulate HBV replication in cultured cells [97].

Regarding the appearance of L-HDAg, it is important to recall that because of the accumulated editing of the HDV antigenome, the proportion of this form, in relation to the total amount of accumulated HDAgs, increases during the replication cycle from 0% to around 30%, [98]. This is sufficient to suppress replication but allows the accumulation of viral genomes that can then be packaged into new infectious particles with the help of L-HDAg. The NES present in L-HDAg allows the viral RNP to be exported from the nucleus to the cytoplasm for packaging [81]. HDV RNPs then interact with HBsAgs at the endoplasmatic reticulum to form new infectious virions, which are then secreted to propagate further rounds of HDV infection [71]. Such assembly of new virions only occurs when HBsAgs are present; otherwise, the viral RNPs return to the nucleus [71].

S-HDAg has been more thoroughly studied than the large form, likely due to the fact that it is required for the accumulation of HDV RNA. Several roles have been attributed to this viral protein including putative and observed functions.

S-HDAg is present in the virions forming viral RNPs with the HDV genome. One of the first tasks it performs is the transport of the viral genome into the nucleus of infected cells, where RNA-directed RNA synthesis takes place. This transport is achieved by the presence of the previously described NLS and RBD. Nuclear import may be facilitated by karyopherin 2α, since this importin interacts with S-HDAg *in vitro* [99].

Another role attributed to S-HDAg is the regulation of HDV RNA editing, particularly the deamination by ADAR-1. This editing seems to occur at multiple locations on HDV RNAs, but it is focused on the antigenomic RNA at the stop codon adenosine [100]. S-HDAg has been found to suppress editing at this stop codon when expressed in transfected cells at levels close to those observed during HDV replication [100]. This observation suggests that the antigen plays a role in limiting HDV RNA editing, as excessive editing has been shown to inhibit HDV RNA accumulation [101].

It has been known for more than two decades that the small form of the delta antigen is essential for the accumulation of processed HDV RNAs [96]. Several theories have been proposed for the precise role(s) it may play, as will be discussed ahead.

S-HDAg has been shown to interact with host pol II. In a pull-down assay, both S-HDAg and L-HDAg fused with a glutathione S-transferase tag were able to bind pol II from HeLa nuclear extracts [102]. In the same study, S-HDAg was observed to enhance pol II elongation, presumably by displacing the subunit A of the negative elongation factor (NELF-A). S-HDAg was thus reported as an elongation enhancer of RNA-templated pol II transcription *in vitro* [102]. However, the observed enhancement appears to be limited to $3'$-OH end additions, rather than transcription. In a subsequent study, Yamaguchi et al. reported that S-HDAg functionally interacts with pol II suggesting that S-HDAg may be involved in facilitating the uncommon RNA-directed synthesis by an RNA polymerase that is normally DNA-directed [103]. They proposed that the interaction between pol II and S-HDAg loosens what, from molecular structure studies, is considered to be a pol II clamp, thereby reducing transcriptional fidelity and allowing the recognition of the atypical RNA template.

Amidst all the reports that S-HDAg actively participates in HDV RNA transcription, there is a contradictory result showing that the presence of S-HDAg is not required for the accumulation of processed short HDV transcripts, although full-length transcripts, genomic or antigenomic, do require S-HDAg, or even L-HDAg [104]. As an explanation, it was proposed that full-length HDV RNAs are susceptible to nucleolytic degradation in the absence of S-HDAg, and, due to their size, such RNAs are more prone to be degraded than smaller RNAs. In other words, S-HDAg interacts with HDV RNAs to protect them and thereby allow their accumulation in infected cells.

Another role attributed to S-HDAg is that of HDV RNA chaperone. *In vitro* studies have reported that S-HDAg can stimulate HDV RNA ribozyme activity [105]. From such studies, it is inferred that *in vivo* S-HDAg may be directly involved in posttranscriptional processing of nascent multimeric transcripts by enhancing cleavage into unit-length molecules. It should be noted, however, that the abovementioned studies of Lazinski and Taylor indicate that, *in vivo*, HDAg is not directly needed for ribozyme cleavage and subsequent ligation [104].

S-HDAg may also be involved in deviating/redirecting other host cell components to facilitate HDV RNA replication. S-HDAg is a rather promiscuous protein in that many cellular partners have been detected.

HDV has a very small RNA genome, as mentioned earlier, and encodes only one viral protein, HDAg. Albeit the fact that a second isoform of the HDAg appears later in the replication cycle, S-HDAg and L-HDAg are not sufficient for HDV to complete its replication cycle. HDV must rely extensively on host cell factors to complete its replication cycle.

A comprehensive study using immunopurification followed by mass spectrometry identified over 100 host proteins associated with a tagged S-HDAg [105]. This set included 9 of the 12 subunits of the pol II complex, further supporting the idea that pol II is involved in HDV RNA transcription [106]. In another study, a yeast two-hybrid approach identified 30 proteins encoded by a human liver cDNA library that interacted with S-HDAg [107]. Only three proteins from this study had also been identified by the previously mentioned immunopurification approach.

How can one small protein be involved in so many inter-actions and perform such different functions? The answer may be related to the S-HDAg's lack of structure. Based on S-HDAg's sequence, it has been predicted that the protein is extensively disordered [108]. The prediction that S-HDAg is an intrinsically disordered protein (IDP) has been exper-imentally confirmed by circular dichroism measurements, which have shown that the protein has little structure apart for its coiled-coil domain [108].

In fact, S-HDAg has several characteristics generally attributed to IDPs. These proteins are commonly nucleic acid binding proteins displaying chaperone activity, such as S-HDAg. Also, a high net charge is characteristic and S-HDAg has an estimated net charge of +12 [52]. Multimerization ability is yet another feature of IDPs that is present in S-HDAg.

The lack of a rigid 3-dimensional structure may account for the protein's promiscuity and ability to take part in several interactions with distinct partners. As such, a deceivingly simple virus, encoding only one protein, can hijack the host cell mechanisms required to complete its life cycle.

8. Conclusion

More than 30 years after its discovery, a lot of fundamental aspects of the HDV life cycle and interaction with the host still remain unknown. But its peculiar simplicity makes all its beauty.

Acknowledgments

The authors would like to thank John Taylor for constructive comments. Carolina Alves and Cristina Branco were sup-ported by Ph.D. grants by Fundação para a Ciência e Tecnolo-gia (Portugal).

References

[1] M. Rizzetto, "Hepatitis D: thirty years after," *Journal of Hepatol-ogy*, vol. 50, no. 5, pp. 1043–1050, 2009.

[2] F. A. Murphy, "Virus taxonomy," in *Fields Virology*, B. N. Fields, D. M. Knipe, and P. M. Howley, Eds., vol. 2, pp. 15–57, 3rd edition, 1996.

[3] M. Rizzetto and A. Ciancio, "Epidemiology of hepatitis D," *Seminars in Liver Disease*, vol. 32, no. 3, pp. 211–219, 2012.

[4] N. Radjef, E. Gordien, V. Ivaniushina et al., "Molecular phylo-genetic analyses indicate a wide and ancient radiation of african hepatitis delta virus, suggesting a *Deltavirus* genus of at least seven major clades," *Journal of Virology*, vol. 78, no. 5, pp. 2537–2544, 2004.

[5] I. H. Bahcecioglu, C. Aygun, N. Gozel, O. K. Poyrazoglu, Y. Bulut, and M. Yalniz, "Prevalence of hepatitis delta virus (HDV) infection in chronic hepatitis B patients in eastern Turkey: still

a serious problem to consider," *Journal of Viral Hepatitis*, vol. 18, no. 7, pp. 518–524, 2011.

[6] B. Tsatsralt-Od, M. Takahashi, T. Nishizawa, K. Endo, J. Inoue, and H. Okamoto, "High prevalence of dual or triple infection of hepatitis B, C, and delta viruses among patients with chronic liver disease in Mongolia," *Journal of Medical Virology*, vol. 77, no. 4, pp. 491–499, 2005.

[7] R. Paraná, A. Kay, F. Molinet et al., "HDV genotypes in the Western Brazilian Amazon region: a preliminary report," *American Journal of Tropical Medicine and Hygiene*, vol. 75, no. 3, pp. 475–479, 2006.

[8] P. Farci, "Delta hepatitis: an update," *Journal of Hepatology*, vol. 39, no. 1, pp. S212–S219, 2003.

[9] G. B. Gaeta, T. Stroffolini, A. Smedile, G. Niro, and A. Mele, "Hepatitis delta in Europe: vanishing or refreshing?" *Hepatol-ogy*, vol. 46, no. 4, pp. 1312–1313, 2007.

[10] H. Wedemeyer, B. Heidrich, and M. P. Manns, "Hepatitis D virus infection—not a vanishing disease in Europe," *Hepatology*, vol. 45, no. 5, pp. 1331–1332, 2007.

[11] R. G. Gish, D. H. Yi, S. Kane, M. Clask, M. Mangahas et al., "Coinfection with hepatitis B and D: epidemiology, prevalence and disease in patients in Northern California," *Journal of Gastroenterology and Hepatology*, vol. 28, no. 9, pp. 1521–1525, 2013.

[12] F. Le Gal, E. Gault, M.-P. Ripault et al., "Eighth major clade for hepatitis delta virus," *Emerging Infectious Diseases*, vol. 12, no. 9, pp. 1447–1450, 2006.

[13] C.-W. Su, Y.-H. Huang, T.-I. Huo et al., "Genotypes and viremia of hepatitis B and D viruses are associated with outcomes of chronic hepatitis D patients," *Gastroenterology*, vol. 130, no. 6, pp. 1625–1635, 2006.

[14] S. A. Hughes, H. Wedemeyer, and P. M. Harrison, "Hepatitis delta virus," *The Lancet*, vol. 378, no. 9785, pp. 73–85, 2011.

[15] M. S. Gomes-Gouvêa, M. C. P. Soares, G. Bensabath et al., "Hep-atitis B virus and hepatitis delta virus genotypes in outbreaks of fulminant hepatitis (Labrea black fever) in the western Brazilian Amazon region," *Journal of General Virology*, vol. 90, no. 11, pp. 2638–2643, 2009.

[16] J.-C. Wu, T.-Y. Chiang, and I.-J. Sheen, "Characterization and phylogenetic analysis of a novel hepatitis D virus strain dis-covered by restriction fragment length polymorphism analysis," *Journal of General Virology*, vol. 79, no. 5, pp. 1105–1113, 1998.

[17] H. Watanabe, K. Nagayama, N. Enomoto et al., "Chronic hepatitis delta virus infection with genotype IIb variant is correlated with progressive liver disease," *Journal of General Virology*, vol. 84, no. 12, pp. 3275–3289, 2003.

[18] A. Ottobrelli, A. Marzano, A. Smedile et al., "Patterns of hepati-tis delta virus reinfection and disease in liver transplantation," *Gastroenterology*, vol. 101, no. 6, pp. 1649–1655, 1991.

[19] S. Govindarajan, K. P. Chin, A. G. Redeker, and R. L. Peters, "Fulminant B viral hepatitis: role of delta agent," *Gastroenterol-ogy*, vol. 86, no. 6, pp. 1417–1420, 1984.

[20] P. Farci and G. Niro, "Clinical features of hepatitis D," *Seminars in Liver Disease*, vol. 32, pp. 228–236, 2012.

[21] A. Smedile, G. Verme, and A. Cargnel, "Influence of delta infection on severity of hepatitis B," *The Lancet*, vol. 2, no. 8305, pp. 945–947, 1982.

[22] A. Smedile, P. Dentico, and A. Zanetti, "Infection with the delta (δ) agent in chronic HBsAg carriers," *Gastroenterology*, vol. 81, no. 6, pp. 992–997, 1981.

[23] P. Farci, P. Karayiannis, M. E. Lai et al., "Acute and chronic hepatitis delta virus infection: direct or indirect effect on hepatitis B virus replication?" *Journal of Medical Virology*, vol. 26, no. 3, pp. 279–288, 1988.

[24] G. Fattovich, S. Boscaro, and F. Noventa, "Influence of hepatitis delta virus infection on progression to cirrhosis in chronic hepatitis type B," *Journal of Infectious Diseases*, vol. 155, no. 5, pp. 931–935, 1987.

[25] G. Saracco, F. Rosina, M. R. Brunetto et al., "Rapidly progressive HBsAg-positive hepatitis in Italy. The role of hepatitis delta virus infection," *Journal of Hepatology*, vol. 5, no. 3, pp. 274–281, 1987.

[26] G. Fattovich, G. Giustina, E. Christensen et al., "Influence of hepatitis delta virus infection on morbidity and mortality in compensated cirrhosis type B," *Gut*, vol. 46, no. 3, pp. 420–426, 2000.

[27] T. J. S. Cross, P. Rizzi, M. Horner et al., "The increasing prevalence of hepatitis delta virus (HDV) infection in South London," *Journal of Medical Virology*, vol. 80, no. 2, pp. 277–282, 2008.

[28] C.-J. Chen, H.-I. Yang, J. Su et al., "Risk of hepatocellular carcinoma across a biological gradient of serum hepatitis B virus DNA Level," *Journal of the American Medical Association*, vol. 295, no. 1, pp. 65–73, 2006.

[29] H. Wedemeyer and M. P. Manns, "Epidemiology, pathogenesis and management of hepatitis D: update and challenges ahead," *Nature Reviews Gastroenterology and Hepatology*, vol. 7, no. 1, pp. 31–40, 2010.

[30] K. Zachou, C. Yurdayin, H. Dienes, G. Dalekos, A. Erhardt et al., "Significance of HDV-RNA and HBsAg levels in delta hepatitis: first data of the Hep-Net/international HDV intervention trial," *Journal of Hepatology*, vol. 44, Supplement 2, p. S178, 2006.

[31] A. Olivero and A. Smedile, "Hepatitis delta virus diagnosis," *Seminars in Liver Disease*, vol. 32, pp. 220–227, 2012.

[32] C. Scholtes, V. Icard, M. Amiri, P. Chevallier-Queyron, M. A. Trabaud et al., "Standardized one-step real-time reverse transcription-PCR assay for universal detection and quantification of hepatitis delta virus from clinical samples in the presence of a heterologous internal-control RNA," *Journal of Clinical Microbiology*, vol. 50, pp. 2126–2128, 2012.

[33] R. B. Ferns, E. Nastouli, and J. A. Garson, "Quantitation of hepatitis delta virus using a single-step internally controlled real-time RT-qPCR and a full-length genomic RNA calibration standard," *Journal of Virological Methods*, vol. 179, no. 1, pp. 189–194, 2012.

[34] H. Wedemeyer, C. Yurdaydìn, G. N. Dalekos et al., "Peginterferon plus adefovir versus either drug alone for hepatitis delta," *The New England Journal of Medicine*, vol. 364, no. 4, pp. 322–331, 2011.

[35] G. A. Niro, F. Rosina, and M. Rizzetto, "Treatment of hepatitis D," *Journal of Viral Hepatitis*, vol. 12, no. 1, pp. 2–9, 2005.

[36] H. Wedemeyer, C. Yurdaydin, G. Dalekos, A. Erhardt, Y. Cakaloglu et al., "[4] 72 week data of the HIDIT-1 trial: a multicenter randomized study comparing peginterferon α-2a plus adefovir vs. peginterferon α-2a plus placebo vs. adefovir in chronic delta hepatitis," *Journal of Hepatology*, vol. 46, Supplement 1, p. S4, 2007.

[37] C. Yurdaydin, H. Bozkaya, S. Gürel et al., "Famciclovir treatment of chronic delta hepatitis," *Journal of Hepatology*, vol. 37, no. 2, pp. 266–271, 2002.

[38] G. A. Niro, A. Ciancio, G. B. Gaeta et al., "Pegylated interferon alpha-2b as monotherapy or in combination with ribavirin in chronic hepatitis delta," *Hepatology*, vol. 44, no. 3, pp. 713–720, 2006.

[39] M. Rizzetto, F. Rosina, and G. Saracco, "Treatment of chronic delta hepatitis with α-2 recombinant interferon," *Journal of Hepatology*, vol. 3, no. 2, pp. S229–S233, 1986.

[40] P. Farci, L. Chessa, C. Balestrieri, G. Serra, and M. E. Lai, "Treatment of chronic hepatitis D," *Journal of Viral Hepatitis*, vol. 14, no. 1, pp. 58–63, 2007.

[41] C. Castelnau, F. Le Gal, M.-P. Ripault et al., "Efficacy of peginterferon alpha-2b in chronic hepatitis delta: relevance of quantitative RT-PCR for follow-up," *Hepatology*, vol. 44, no. 3, pp. 728–735, 2006.

[42] A. Erhardt, W. Gerlich, C. Starke et al., "Treatment of chronic hepatitis delta with pegylated interferon-α2b," *Liver International*, vol. 26, no. 7, pp. 805–810, 2006.

[43] B. B. Bordier, J. Ohkanda, P. Liu et al., "In vivo antiviral efficacy of prenylation inhibitors against hepatitis delta virus," *Journal of Clinical Investigation*, vol. 112, no. 3, pp. 407–414, 2003.

[44] L.-F. He, E. Ford, P. H. Purcell, W. T. London, J. Phillips, and J. L. Gerin, "The size of the hepatitis delta agent," *Journal of Medical Virology*, vol. 27, no. 1, pp. 31–33, 1989.

[45] S. Gudima, J. Chang, G. Moraleda, A. Azvolinsky, and J. Taylor, "Parameters of human hepatitis delta virus genome replication: the quantity, quality, and intracellular distribution of viral proteins and RNA," *Journal of Virology*, vol. 76, no. 8, pp. 3709–3719, 2002.

[46] M. Engelke, K. Mills, S. Seitz et al., "Characterization of a hepatitis B and hepatitis delta virus receptor binding site," *Hepatology*, vol. 43, no. 4, pp. 750–760, 2006.

[47] D. Glebe and S. Urban, "Viral and cellular determinants involved in hepadnaviral entry," *World Journal of Gastroenterology*, vol. 13, no. 1, pp. 22–38, 2007.

[48] J. M. Taylor and Z. Han, "Purinergic receptor functionality is necessary for infection of human hepatocytes by hepatitis delta virus and hepatitis b virus," *PLoS ONE*, vol. 5, no. 12, Article ID e15784, 2010.

[49] O. L. Longarela, T. T. Schmidt, K. Schöneweis, R. Romeo, H. Wedemeyer et al., "Proteoglycans act as cellular hepatitis delta virus attachment receptors," *PLoS ONE*, vol. 8, Article ID e58340, 2013.

[50] H. Yan, G. Zhong, G. Xu, W. He, Z. Jing et al., "Sodium taurocholate cotransporting polypeptide is a functional receptor for human hepatitis B and D virus," *ELife*, vol. 1, Article ID e00049, 2012.

[51] P. Dény, "Hepatitis delta virus genetic variability: from genotypes I, II, III to eight major clades?" *Current Topics in Microbiology and Immunology*, vol. 307, pp. 151–171, 2006.

[52] M. Y. P. Kuo, J. Goldberg, L. Coates, W. Mason, J. Gerin, and J. Taylor, "Molecular cloning of hepatitis delta virus RNA from an infected woodchuck liver: sequence, structure, and applications," *Journal of Virology*, vol. 62, no. 6, pp. 1855–1861, 1988.

[53] P.-J. Chen, G. Kalpana, and J. Goldberg, "Structure and replication of the genome of the hepatitis δ virus," *Proceedings of the National Academy of Sciences of the United States of America*, vol. 83, no. 22, pp. 8774–8778, 1986.

[54] S.-Y. Hsieh and J. Taylor, "Regulation of polyadenylation of hepatitis delta virus antigenomic RNA," *Journal of Virology*, vol. 65, no. 12, pp. 6438–6446, 1991.

[55] S. Gudima, S.-Y. Wu, C.-M. Chiang, G. Moraleda, and J. Taylor, "Origin of hepatitis delta virus mRNA," *Journal of Virology*, vol. 74, no. 16, pp. 7204–7210, 2000.

[56] L. Sharmeen, M. Y. P. Kuo, G. Dinter-Gottlieb, and J. Taylor, "Antigenomic RNA of human hepatitis delta virus can undergo self-cleavage," *Journal of Virology*, vol. 62, no. 8, pp. 2674–2679, 1988.

[57] A. D. Branch, B. J. Benenfeld, B. M. Baroudy, F. V. Wells, J. L. Gerin, and H. D. Robertson, "An ultraviolet-sensitive RNA structural element in a viroid-like domain of the hepatitis delta virus," *Science*, vol. 243, no. 4891, pp. 649–652, 1989.

[58] M. D. Been, "HDV ribozymes," *Current Topics in Microbiology and Immunology*, vol. 307, pp. 47–65, 2006.

[59] A. R. Ferré-D'Amaré, K. Zhou, and J. A. Doudna, "Crystal structure of a hepatitis delta virus ribozyme," *Nature*, vol. 395, no. 6702, pp. 567–574, 1998.

[60] K. Salehi-Ashtiani, A. Lupták, A. Litovchick, and J. W. Szostak, "A genomewide search for ribozymes reveals an HDV-like sequence in the human CPEB3 gene," *Science*, vol. 313, no. 5794, pp. 1788–1792, 2006.

[61] C.-H. T. Webb, N. J. Riccitelli, D. J. Ruminski, and A. Lupták, "Widespread occurrence of self-cleaving ribozymes," *Science*, vol. 326, no. 5955, p. 953, 2009.

[62] J. M. Taylor, "Hepatitis delta virus: cis and trans functions required for replication," *Cell*, vol. 61, no. 3, pp. 371–373, 1990.

[63] C. E. Reid and D. W. Lazinski, "A host-specific function is required for ligation of a wide variety of ribozyme-processed RNAs," *Proceedings of the National Academy of Sciences of the United States of America*, vol. 97, no. 1, pp. 424–429, 2000.

[64] L. Sharmeen, M. Y.-P. Kuo, and J. Taylor, "Self-ligating RNA sequences on the antigenome of human hepatitis delta virus," *Journal of Virology*, vol. 63, no. 3, pp. 1428–1430, 1989.

[65] J. M. Taylor, "Chapter 3 replication of the hepatitis delta virus RNA genome," *Advances in Virus Research*, vol. 74, pp. 103–121, 2009.

[66] J. Chang, X. Nie, E. C. Ho, Z. Han, and J. Taylor, "Transcription of hepatitis delta virus RNA by RNA polymerase II," *Journal of Virology*, vol. 82, no. 3, pp. 1118–1127, 2008.

[67] V. S. Greco-Stewart, P. Miron, A. Abrahem, and M. Pelchat, "The human RNA polymerase II interacts with the terminal stem-loop regions of the hepatitis delta virus RNA genome," *Virology*, vol. 357, no. 1, pp. 68–78, 2007.

[68] J. Filipovska and M. M. Konarska, "Specific HDV RNA-templated transcription by pol II in vitro," *RNA*, vol. 6, no. 1, pp. 41–54, 2000.

[69] T. B. Macnaughton, S. T. Shi, L. E. Modahl, and M. M. C. Lai, "Rolling circle replication of hepatitis delta virus RNA is carried out by two different cellular RNA polymerases," *Journal of Virology*, vol. 76, no. 8, pp. 3920–3927, 2002.

[70] L. E. Modahl, T. B. Macnaughton, N. Zhu, D. L. Johnson, and M. M. C. Lai, "RNA-dependent replication and transcription of hepatitis delta virus RNA involve distinct cellular RNA polymerases," *Molecular and Cellular Biology*, vol. 20, no. 16, pp. 6030–6039, 2000.

[71] J. P. Tavanez, C. Cunha, M. C. A. Silva, E. David, J. Monjardino, and M. Carmo-Fonseca, "Hepatitis delta virus ribonucleoproteins shuttle between the nucleus and the cytoplasm," *RNA*, vol. 8, no. 5, pp. 637–646, 2002.

[72] V. S. Greco-Stewart, E. Schissel, and M. Pelchat, "The hepatitis delta virus RNA genome interacts with the human RNA polymerases I and III," *Virology*, vol. 386, no. 1, pp. 12–15, 2009.

[73] J. Chang and J. Taylor, "In vivo RNA-directed transcription, with template switching, by a mammalian RNA polymerase," *EMBO Journal*, vol. 21, no. 1-2, pp. 157–164, 2002.

[74] T.-B. Fu and J. Taylor, "The RNAs of hepatitis delta virus are copied by RNA polymerase II in nuclear homogenates," *Journal of Virology*, vol. 67, no. 12, pp. 6965–6972, 1993.

[75] J. Taylor, "Virology of hepatitis D virus," *Seminars in Liver Disease*, vol. 32, pp. 195–200, 2012.

[76] M. M. C. Lai, "RNA replication without RNA-dependent RNA polymerase: surprises from hepatitis delta virus," *Journal of Virology*, vol. 79, no. 13, pp. 7951–7958, 2005.

[77] A. G. Polson, B. L. Bass, and J. L. Casey, "Erratum: RNA editing of hepatitis delta virus antigenome by dsRNA-adenosine deaminase," *Nature*, vol. 381, no. 6580, p. 346, 1996.

[78] J. L. Casey, "Hepatitis delta virus RNA editing," in *Hepatitis Delta Virus*, H. Handa and Y. Yamaguchi, Eds., pp. 52–65, Georgetown Landes Bioscience, Austin, Tex, USA, 2006.

[79] C. Alves, N. Freitas, and C. Cunha, "Characterization of the nuclear localization signal of the hepatitis delta virus antigen," *Virology*, vol. 370, no. 1, pp. 12–21, 2008.

[80] C.-Z. Lee, J.-H. Lin, M. Chao, K. McKnight, and M. M. C. Lai, "RNA-binding activity of hepatitis delta antigen involves two arginine- rich motifs and is required for hepatitis delta virus RNA replication," *Journal of Virology*, vol. 67, no. 4, pp. 2221–2227, 1993.

[81] C.-H. Lee, S. C. Chang, C. H. H. Wu, and M.-F. Chang, "A novel chromosome region maintenance 1-independent nuclear export signal of the large form of hepatitis delta antigen that is required for the viral assembly," *The Journal of Biological Chemistry*, vol. 276, no. 11, pp. 8142–8148, 2001.

[82] J. S. Glenn, J. A. Watson, C. M. Havel, and J. M. White, "Identification of a prenylation site in delta virus large antigen," *Science*, vol. 256, no. 5061, pp. 1331–1333, 1992.

[83] J. C. Otto and P. J. Casey, "The hepatitis delta virus large antigen is farnesylated both in vitro and in animal cells," *The Journal of Biological Chemistry*, vol. 271, no. 9, pp. 4569–4572, 1996.

[84] S. B. Hwang and M. M. C. Lai, "Isoprenylation masks a conformational epitope and enhances trans-dominant inhibitory function of the large hepatitis delta antigen," *Journal of Virology*, vol. 68, no. 5, pp. 2958–2964, 1994.

[85] C.-Z. Lee, P.-J. Chen, M. M. C. Lai, and D.-S. Chen, "Isoprenylation of large hepatitis delta antigen is necessary but not sufficient for hepatitis delta virus assembly," *Virology*, vol. 199, no. 1, pp. 169–175, 1994.

[86] J.-J. Mu, H.-L. Wu, B.-L. Chiang, R.-P. Chang, D.-S. Chen, and P.-J. Chen, "Characterization of the phosphorylated forms and the phosphorylated residues of hepatitis delta virus delta antigens," *Journal of Virology*, vol. 73, no. 12, pp. 10540–10545, 1999.

[87] T.-S. Yeh, S. J. Lo, P.-J. Chen, and Y.-H. W. Lee, "Casein kinase II and protein kinase C modulate hepatitis delta virus RNA replication but not empty viral particle assembly," *Journal of Virology*, vol. 70, no. 9, pp. 6190–6198, 1996.

[88] C.-W. Chen, Y.-G. Tsay, H.-L. Wu, C.-H. Lee, D.-S. Chen, and P.-J. Chen, "The double-stranded RNA-activated kinase, PKR, can phosphorylate hepatitis D virus small delta antigen at functional serine and threonine residues," *The Journal of Biological Chemistry*, vol. 277, no. 36, pp. 33058–33067, 2002.

[89] Y.-S. Chen, W.-H. Huang, S.-Y. Hong, Y.-G. Tsay, and P.-J. Chen, "ERK1/2-mediated phosphorylation of small hepatitis delta antigen at serine 177 enhances hepatitis delta virus antigenomic RNA replication," *Journal of Virology*, vol. 82, no. 19, pp. 9345–9358, 2008.

[90] S.-Y. Hong and P.-J. Chen, "Phosphorylation of serine 177 of the small hepatitis delta antigen regulates viral antigenomic RNA replication by interacting with the processive RNA polymerase II," *Journal of Virology*, vol. 84, no. 6, pp. 1430–1438, 2010.

[91] W.-H. Huang, R.-T. Mai, and Y.-H. W. Lee, "Transcription factor YY1 and its associated acetyltransferases CBP and p300 interact with hepatitis delta antigens and modulate hepatitis delta virus RNA replication," *Journal of Virology*, vol. 82, no. 15, pp. 7313–7324, 2008.

[92] J.-J. Mu, Y.-G. Tsay, L.-J. Juan et al., "The small delta antigen of hepatitis delta virus is an acetylated protein and acetylation of lysine 72 may influence its cellular localization and viral RNA synthesis," *Virology*, vol. 319, no. 1, pp. 60–70, 2004.

[93] C.-H. Tseng, K.-S. Jeng, and M. M. C. Lai, "Transcription of subgenomic mRNA of hepatitis delta virus requires a modified hepatitis delta antigen that is distinct from antigenomic RNA synthesis," *Journal of Virology*, vol. 82, no. 19, pp. 9409–9416, 2008.

[94] Y.-J. Li, M. R. Stallcup, and M. M. C. Lai, "Hepatitis delta virus antigen is methylated at arginine residues, and methylation regulates subcellular localization and RNA replication," *Journal of Virology*, vol. 78, no. 23, pp. 13325–13334, 2004.

[95] C.-H. Tseng, T.-S. Cheng, C.-Y. Shu, K.-S. Jeng, and M. M. C. Lai, "Modification of small hepatitis delta virus antigen by SUMO protein," *Journal of Virology*, vol. 84, no. 2, pp. 918–927, 2010.

[96] M. Chao, S.-Y. Hsieh, and J. Taylor, "Role of two forms of hepatitis delta virus antigen: evidence for a mechanism of self-limiting genome replication," *Journal of Virology*, vol. 64, no. 10, pp. 5066–5069, 1990.

[97] V. Williams, S. Brichler, N. Radjef et al., "Hepatitis delta virus proteins repress hepatitis B virus enhancers and activate the alpha/beta interferon-inducible MxA gene," *Journal of General Virology*, vol. 90, no. 11, pp. 2759–2767, 2009.

[98] S. K. Wong and D. W. Lazinski, "Replicating hepatitis delta virus RNA is edited in the nucleus by the small form of ADAR1," *Proceedings of the National Academy of Sciences of the United States of America*, vol. 99, no. 23, pp. 15118–15123, 2002.

[99] H.-C. Chou, T.-Y. Hsieh, G.-T. Sheu, and M. M. C. Lai, "Hepatitis delta antigen mediates the nuclear import of hepatitis delta virus RNA," *Journal of Virology*, vol. 72, no. 5, pp. 3684–3690, 1998.

[100] A. G. Polson, H. L. Ley III, B. L. Bass, and J. L. Casey, "Hepatitis delta virus RNA editing is highly specific for the amber/W site and is suppressed by hepatitis delta antigen," *Molecular and Cellular Biology*, vol. 18, no. 4, pp. 1919–1926, 1998.

[101] G. C. Jayan and J. L. Casey, "Increased RNA editing and inhibition of hepatitis delta virus replication by high-level expression of ADAR1 and ADAR2," *Journal of Virology*, vol. 76, no. 8, pp. 3819–3827, 2002.

[102] Y. Yamaguchi, J. Filipovska, K. Yano et al., "Stimulation of RNA polymerase II elongation by hepatitis delta antigen," *Science*, vol. 293, no. 5527, pp. 124–127, 2001.

[103] Y. Yamaguchi, T. Mura, S. Chanarat, S. Okamoto, and H. Handa, "Hepatitis delta antigen binds to the clamp of RNA polymerase II and affects transcriptional fidelity," *Genes to Cells*, vol. 12, no. 7, pp. 863–875, 2007.

[104] D. W. Lazinski and J. M. Taylor, "Expression of hepatitis delta virus RNA deletions: cis And trans requirements for self-cleavage, ligation, and RNA packaging," *Journal of Virology*, vol. 68, no. 5, pp. 2879–2888, 1994.

[105] C.-C. Wang, T.-C. Chang, C.-W. Lin et al., "Nucleic acid binding properties of the nucleic acid chaperone domain of hepatitis delta antigen," *Nucleic Acids Research*, vol. 31, no. 22, pp. 6481–6492, 2003.

[106] C. Dan, D. Haussecker, H. Yong, and M. A. Kay, "Combined proteomic-RNAi screen for host factors involved in human hepatitis delta virus replication," *RNA*, vol. 15, no. 11, pp. 1971–1979, 2009.

[107] A. Casaca, M. Fardilha, E. Da Cruz E Silva, and C. Cunha, "The heterogeneous ribonuclear protein C interacts with the hepatitis delta virus small antigen," *Virology Journal*, vol. 8, article 358, 2011.

[108] C. Alves, H. Cheng, H. Roder, and J. Taylor, "Intrinsic disorder and oligomerization of the hepatitis delta virus antigen," *Virology*, vol. 407, no. 2, pp. 333–340, 2010.

11

Antiviral Activity of Resveratrol against Human and Animal Viruses

Yusuf Abba,[1,2] Hasliza Hassim,[3] Hazilawati Hamzah,[1] and Mohamed Mustapha Noordin[1]

[1]Department of Veterinary Pathology and Microbiology, Faculty of Veterinary Medicine, Universiti Putra Malaysia, 43400 Serdang, Selangor, Malaysia
[2]Department of Veterinary Pathology, Faculty of Veterinary Medicine, University of Maiduguri, PMB 1069, Maiduguri, Borno State, Nigeria
[3]Department of Veterinary Preclinical Sciences, Faculty of Veterinary Medicine, Universiti Putra Malaysia, 43400 Serdang, Selangor, Malaysia

Correspondence should be addressed to Mohamed Mustapha Noordin; noordinmm@upm.edu.my

Academic Editor: Robert C. Gallo

Resveratrol is a potent polyphenolic compound that is being extensively studied in the amelioration of viral infections both *in vitro* and *in vivo*. Its antioxidant effect is mainly elicited through inhibition of important gene pathways like the NF-$\kappa\beta$ pathway, while its antiviral effects are associated with inhibitions of viral replication, protein synthesis, gene expression, and nucleic acid synthesis. Although the beneficial roles of resveratrol in several viral diseases have been well documented, a few adverse effects have been reported as well. This review highlights the antiviral mechanisms of resveratrol in human and animal viral infections and how some of these effects are associated with the antioxidant properties of the compound.

1. Introduction

Resveratrol (RSV) is a naturally occurring polyphenol stilbene found mostly in fermented grapes, mulberry, red wine, and peanuts. It is available in the trans- and cis-isomer forms; however, the cis-resveratrol isomer is unstable and easily transformed into the trans-form when reacted to light. It is insoluble in water but soluble in polar solvents such as ethanol and dimethyl sulfoxide. Resveratrol scavenges for superoxide and hydroxyl *in vitro*, as well as lipid hydroperoxyl radicals [1]. Previous studies have shown resveratrol to enhance longevity, regulate lipid levels, and act as a prophylactic compound against cancers and related viral infections [2]. It also attenuates superoxide generation in the mitochondria and inhibits mitochondrial dysfunction induced by arachidonic acid [2, 3]. The antiviral mechanisms and effects of RSV have been widely studied in a number of viruses which include influenza virus, hepatitis C virus [4], respiratory syncytial virus [5–9], varicella zoster virus [10], Epstein-Barr virus [11, 12], herpes simplex virus [13–16], human immunodeficiency virus [17, 18], African swine fever virus, enterovirus,

human metapneumonia virus, and duck enteritis virus and in multiple sclerosis, whose animal models can be induced by viral infection. In almost all of these studies, RSV showed remarkable recession of the viral infection with the exception of multiple sclerosis and hepatitis C, where disease progression was worsened following administration of RSV [4, 19].

2. Structure, Bioavailability, and Function

Resveratrol (RV), or 3,5,4′-trihydroxy-trans-stilbene, is a natural bioflavonoid compound found in plants and fruits. Its chemical structure is made up of two phenolic rings which are bonded by a double styrene bond, thus forming the 3,5,4′-trihydroxystilbene with a molecular weight of 228.25 g/mol (Figure 1). Apart from its natural isomers, cis- and trans-forms, several synthetic and natural analogs of RSV exist, which exhibit similar or slightly varying pharmacological properties to RSV [20].

Resveratrol has poor water solubility and poor oral bioavailability and rapidly metabolized in the system. Its poor bioavailability is attributed to its rapid metabolism in

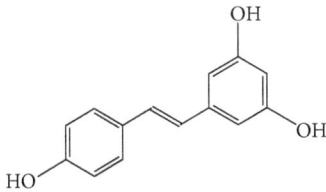

FIGURE 1: Chemical structure of resveratrol.

the liver into glucuronides and sulfates [21]. Even though the amount of oral dosing of RSV did not significantly affect its bioavailability in plasma, the type of food consumed and intraindividual differences in metabolism were shown to significantly affect its bioavailability [20, 22]. In a related study, the plasma bioavailability of RSV 30 minutes after oral consumption of red wine was only in trace amounts, while moments later RSV glucuronides were systemically abundant for a prolonged time [23]. Recently, research has been focused on developing structured nanoparticles that will enhance the bioavailability of RSV and prolong its release *in vivo*. Solid lipid nanoparticles (SLNs) and nanostructured lipid carriers (NLCs) loaded with RSV were shown to have an entrapment efficiency of 70% and stability lasting for over 2 months. *In vitro* simulation studies showed a slow sustained release of RSV at both stomach and intestinal pH levels [24]. Similarly, the use of zein-based nanoparticles was reported to enhance the *in vivo* delivery of RSV in mouse model of endotoxic shock [25].

Resveratrol has been shown to induce apoptosis by upregulating and downregulating numerous genes that are important in cellular function including but not limited to TRAIL-R2/DR5, TRAIL-R2/DR4, p53, Bim, Noxa, PUMA, Bak, Bax, Mcl-1, survivin, Bcl-XL, and Bcl-2. RSV has been shown to inhibit cellular growth at G1 and G1/S phases as well as be an anti-inflammatory mediator by inhibition of nuclear factor-kappa β (NF-$\kappa\beta$) activity, procyclooxygenase-2 activity, and prostaglandin production. Furthermore, its actions in delaying the onset of cardiovascular diseases and cancer progression as well as antiviral effects have been widely studied as well [2, 20].

3. Mechanism of Antioxidant Action

Resveratrol (RSV) scavenges for O_2^- and OH^- *in vitro* as well as lipid hydroperoxyl free radicals. One of its most important antioxidant effects is elicited through the inhibition of reactive oxygen species production and glutathione depletion. It also attenuates superoxide generation in the mitochondria and inhibits mitochondrial dysfunction induced by arachidonic acid. These actions have thus resulted in restoration of mitochondrial membrane through mediation of the AMPK-mediated inhibitory phosphorylation of GsK3β downstream of poly(ADP-ribose) polymerase-LKβ1 pathway [26, 27]. RSV also increased the levels of catalase, superoxide dismutase, and heme-oxygenase-1 in cultured epithelial cells exposed to H_2O_2. This shows its remarkable ability in mediating the major antioxidant enzymes

involved in the breakdown of H_2O_2 [28]. Similarly, RSV has been shown to have an 89.1% inhibition of lipid peroxidation of linolenic acid emulsion and has strong scavenging activities against 2,2-diphenyl-1-picrylhydrazyl, 2,2-azinobis-(3-ethylbenzothiazoline-6-sulfonic acid), N,N-dimethyl-p-phenylenediamine, O_2^-, and H_2O_2. It also has a high reducing power and Fe^{2+} chelating activity [3]. However, a recent study showed that the antioxidant effect of RSV is largely dependent on its dose as higher doses were shown to increase ROS production resulting in mitochondrial-dependent death in endothelial cells [29]. It could be thus said that RSV acts as both an antioxidant and a prooxidant depending on the dose administered.

4. Prophylactic and Therapeutic Uses in Virus Associated Conditions

The antiviral effects of RSV have been demonstrated in a number of pathogenic human and animal viruses. However, in most of these reports, inhibition of virus proliferation was not directly associated with its antioxidant activity but its ability to inhibit viral protein production and gene expression at various levels [41, 42].

4.1. Influenza Virus. In influenza virus infection, RSV was shown to actively block nuclear-cytoplasmic translocation of viral ribonucleoproteins in MDCK cells, thus decreasing the expression of late viral proteins related to inhibition of protein kinase C associated pathways. This activity was also found to be unassociated with glutathione-mediated antioxidant activity of the compound [30]. The various mechanisms of viral inhibition exerted by resveratrol on certain viruses are summarized in Table 1.

4.2. Epstein-Barr Virus. In Epstein-Barr virus (EBV) infection, RSV showed an enhanced inhibitory effect on EBV early antigen induction using Raji cells. It was also shown to reduce papilloma production in mouse by 60% after 20 weeks of inoculation [31]. In another study, RSV was shown to dose-dependently inhibit EBV lytic cycle by inhibition of transcription genes and proteins, Rta, Zta, and diffused early antigen (EA-D), as well as inhibiting the activity of EBV immediate-early protein: BRLF1 and BZLF1 promoters. This effect was seen to reduce virion production [12]. Similarly, another *in vitro* study confirmed the previous finding that RSV does inhibit lytic gene expression and viral particle production in a dose-dependent manner. Here its main antiviral mechanism was associated with inhibition of protein synthesis, reduction in ROS production, and inhibition of transcription factors NF-$\kappa\beta$ and AP1 [32]. Since EBV is one of the most renowned oncogenic viruses, it is pertinent to study the role of EBV in cellular transformation and cancer progression. RSV was thus shown to prevent transformation of EBV in human B-cells through downregulation of antiapoptotic proteins: Mcl and survivin. This was also linked to suppression of EBV induced signaling of NF-$\kappa\beta$ and STAT-3, as well as miR-155 and miR-34a in EBV infected cells [11].

TABLE 1: Mechanism of viral inhibition and exacerbation induced by resveratrol on different viruses.

Virus	Mode of propagation	Mechanism of resveratrol action	Effects on viral infection	References
Influenza virus	MDCK cells	Block nuclear-cytoplasmic translocation of viral ribonucleoproteins	Decrease in the expression of late viral proteins related to inhibition of protein kinase C associated pathways	[30]
Epstein-Barr virus	(i) Raji cells (ii) Mice (iii) P3HR1 cells (iv) Burkitt's lymphoma cell (v) Human B-cells	(i) Inhibition of early antigen induction (ii) Inhibition of early genes expression of lytic proteins (iii) Inhibition of lytic gene expression and viral particle production (iv) Inhibition of protein synthesis and reduction of ROS production and transcription of factors NF-κβ and AP1 (v) Downregulation of antiapoptotic proteins: Mcl and survivin (vi) Suppression of NF-κβ, STAT-3, miR-155, and miR-34a signaling	(i) Reduce papilloma production (ii) Inhibition of viral transcription (iii) Prevents transformation of EBV	[31] [12] [32] [11]
Herpes simplex virus	(i) Vero and MRC-5 cells (ii) Mice (iii) Mice (iv) HeLa, Vero, and H1299 cells (v) Vero cells	(i) Decreased production of early viral protein ICP4 (ii) Inhibition of interphase phase and prevention of virus reactivation (iii) Rapid and transient release of reactive oxygen species (iv) Reduction of mRNA of ICP0, ICP4, ICP8, and HSV-1 DNA polymerase (v) Reduction of mRNA of glycoprotein C and HSV late gene	(i) Reduction in viral yields (ii) Suppression of development of cutaneous lesions (iii) Prevented development of extravaginal lesions (iv) Inhibition of HSV replication through ROS generation (v) Inhibition of viral transcription and DNA synthesis	[15] [16] [14] [13] [33]
Respiratory syncytial virus	(i) Mice (ii) Lung epithelial cell (iii) Mice (iv) Mice (v) Mice	(i) Control of toll-like receptor 3 expression, inhibition of TRIF signaling, and induction of M2 receptor (ii) Inhibition of viral induced toll-like receptor domain and TANK binding kinase 1 protein expression (iii) Increased SARM and decreased TRIF expression	(i) Reduction in inflammation and levels of interferon-gamma (ii) Partial reduction in viral replication and decreased production of interleukin-6 (iii) Enhanced interferon-gamma expression and airway inflammatory response (iv) Decreased level of inflammatory cells and interferon-gamma (v) Increased TNF-α, IFN-γ, and IL-2 production	[9] [7] [6] [34] [8] [5]
Human immunodeficiency virus (HIV-1)	Primary peripheral blood lymphocytes	Inhibition of DNA synthesis in NL4-3 clone with mutant M184V RT	Inhibition of HIV-1 strain replication	[18]
Varicella zoster virus	MRC-5 cell	Reduction in synthesis of protein and mRNA levels of IE62	Decrease in viral production	[10]
Enterovirus (EV 71)	Rhabdosarcoma cell line	Inhibition of synthesis of viral protein 1 and phosphorylation of proinflammatory cytokines	Inhibition of IL-6 and IFN-γ in infected cells	[35]
Duck enteritis virus (DEV)	Duck embryo fibroblast	Suppression of nucleic acid replication, viral capsid formation, and viral early protein expression	Inhibition of DEV in host cells	[36]
Human metapneumonia virus (hMPV)	Alveolar type 2 cancerous cell line	Suppression of NF-κβ and interferon regulatory factor (IRF-3)	Inhibition of viral replication and reduction in cellular oxidative damage and proinflammatory mediators	[37]
African swine fever virus (ASFV)	Vero cell	Inhibition of early and late viral protein synthesis and virion formation	Reduced viral DNA replication resulting in 98–100% reduction in viral titers	[38]
Human rhinovirus (HRV-16)	HeLa cell and nasal epithelia (ex vivo)	Reversion of HRV-induced expression of ICAM-1	Exhibited high dose-dependent antiviral activity against HRV, leading to reduction in secretion of IL-6, IL-8, and RANTES	[39]

TABLE 1: Continued.

Virus	Mode of propagation	Mechanism of resveratrol action	Effects on viral infection	References
Cytomegalovirus	Human embryonic lung fibroblast (HEL 299)	(i) Prevention of production of immediate-early, early, and late viral proteins (ii) Reduced viral induced activation of epidermal growth factor receptor, phosphatidy-inositol-3-kinase signaling, and NF-$\kappa\beta$ and Sp1 transcription factor activation	Decreased viral replication	[40]
Hepatitis C virus	OR6 cells	(i) Dose-dependently enhanced HCV viral RNA replication (ii) Reversed antiviral effects of ribavirin and interferon	Increased viral RNA replication	[4]
Theiler's murine encephalomyelitis virus (TMEV)	Mice	Significantly exacerbated demyelination and inflammation without neuroprotection in the central nervous system	(i) Exacerbated clinical signs and histological findings in TMEV infected mice (ii) Resulted in a twofold increase in IL-17 and a twofold decrease in IFN-γ	[19]

4.3. Herpes Simplex Virus. RSV was shown to inhibit the replication of herpes simplex virus-1 and herpes simplex virus-2 (HSV-1 and HSV-2) in a dose-dependent and reversible way. In this study, the authors observed a reduction in virus yield as a result of inhibition of an early event in the replication cycle: decreased production of early viral protein ICP-4. RSV also delayed interphase stage of the cell cycle and prevented virus reactivation in neuron cells that were latently infected [15]. In another study by the author using nude mice, topical application of 12.5% and 15% resveratrol ointment suppressed the development of cutaneous lesions in abraded skin infected with HSV-1 [16]. Similarly, application of 19% RSV cream on the vagina of mouse infected with HSV-2 and HSV-1 completely prevented the development of vaginal lesions, while the mortality rate was 3% as compared to the placebo group where mortality rate was 37% [14]. These remarkable effects of RSV on HSV-1 and HSV-2 infections were reported to be due to the promotion of a rapid and sustained release of ROS, which resulted in the inhibition of NF-$\kappa\beta$ and extracellular signal-regulated kinases/mitogen-activated protein kinases (Erk/MAPK), as well as a blockade in the expression of immediate-early, early, and late HSV genes and viral DNA synthesis [13, 33].

4.4. Respiratory Syncytial Virus. Respiratory syncytial virus (RPSV) infection is one of the most important viral diseases of the respiratory system affecting humans and it has no specific treatment. Administration of RSV in mice infected with RPSV reduced the accompanying inflammation and levels of interferon-gamma (IFN-γ). The mechanism here was attributed to control of toll-like receptor 3 expression, inhibition of toll/IL-1R domain-containing adaptor inducing IFN (TRIF) signaling, and induction of muscarinic 2 receptor (M2R) [9]. In an *in vitro* study, RSV treatment in epithelial cells inoculated with RPSV resulted in decreased production of interleukin- (IL-) 6 and a partial reduction in viral replication. There was also an inhibition of viral induced toll-like receptor domain and TANK binding kinase 1 (TBK1) protein expression [7]. RSV treatment of mice infected with RPSV was shown to increase sterile-α- and armadillo motif-containing protein (SARM) expression and decrease matrix metalloproteinase 12 (MMP-12) and TIR-domain-containing adapter-inducing interferon-β (TRIF) expression; these in turn decreased IFN-γ expression and airway inflammation and hyperresponsiveness (AHR) [6, 34]. In a related study, RPSV infected mice treated with RSV also showed decreased levels of inflammatory cells and AHR. However, while RSV was able to drastically reduce the levels of nerve growth factor (NGF) after 21 days of infection, the level of brain derived neurotrophic factor (BDNF) was not significantly affected in both the treated and untreated groups [9]. Combination of RSV and baicalin (a flavonoid found in numerous species of *Scutellaria*) joint enema was shown to increase the levels of tumor necrosis factor-alpha (TNF-α), IFN-γ, and IL-2 in mice infected with RPSV, which is believed to be among its antiviral mechanisms [5].

4.5. Human Immunodeficiency Virus. In the treatment of HIV-1, a combination of RSV and decitabine (a nucleoside metabolic disorder used in the treatment of myelodysplastic syndromes) was found to be more potent than RSV alone as an anti-HIV-1 drug. However, the research also reported 15 other derivatives of RSV that were more potent as an anti-HIV-1 drug [17]. Inhibition of replication of the HIV molecular clone NL4-3 containing the mutant M184V reverse transcriptase (RT) by RSV (5 μM) was reported to be associated with inhibition DNA synthesis during the reverse transcription step of the HIV life cycle. This fact was proven when administration of RSV to NL4-3 clones without the mutant M184V RT failed to inhibit viral DNA synthesis [18].

4.6. Varicella Zoster and Enterovirus 71. Other viruses that were inhibited by RSV include varicella zoster, which was dose-dependently and reversibly inhibited in MRC-5 cells when added to culture within the first 30 hours of infection. Here, RSV was shown to decrease the synthesis of intermediate early protein (IE 62) [10]. Enterovirus 71 (EV 71) was also susceptible to RSV treatment as the compound effectively inhibited the synthesis of its viral protein 1 (VP1) and phosphorylation of proinflammatory cytokines (IKKα, IKKβ, IKKγ, IKBα, NF-$\kappa\beta$ p50, and NF-$\kappa\beta$ p65) in rhabdosarcoma cell line. Secretion of IL-6 and TNF-α was also inhibited in the infected cells by RSV [35].

4.7. Duck Enteritis Virus, Human Metapneumonia Virus, African Swine Fever Virus, Human Rhinovirus, and Cytomegalovirus. Duck viral enteritis (DVE) also known as duck herpes viral enteritis or duck plague is a highly fatal disease of ducks and ducklings caused by the duck enteritis virus (DEV), a herpesvirus [43, 44]. In DVE, viral replication was impaired by RSV via suppression of nucleic acid replication and suppression of viral capsid formation *in vitro*. Production of viral protein was also suppressed within the first 24 hours following infection [36]. RSV in combination with a bioflavonoid, quercetin, was shown to reduce cellular oxidative damage and secretion of proinflammatory mediators (IL-1α, IL-6, and TNF-α) and chemokines (CXCL10 and CCL4), through suppression of NF-$\kappa\beta$ and interferon regulating factor (IRF-3), as well as viral replication in human metapneumonia (hMPV) virus infection. However, RSV did not affect viral gene transcription and protein synthesis [37]. African swine fever virus (ASFV) causes an acute hemorrhagic disease in pigs that results in up to 100% mortality. Resveratrol and oxyresveratrol (a hydroxylated analog of resveratrol) were also found to have a dose-dependent effect on African swine fever virus *in vitro*. This was achieved through inhibition of early and late viral protein synthesis, reduced viral DNA replication, and virion formation. Hence a 98–100% reduction in viral titers was observed [38]. In human rhinovirus (HRV) infection of HeLa and nasal epithelial cells, RSV was found to exhibit a high dose-dependent antiviral activity against the virus, which was achieved through reversion of HRV-induced expression of ICAM-1. In addition, reduction in the secretion of IL-6, IL-8, and RANTES was also observed [39]. In an antiviral study of resveratrol on cytomegalovirus infection of human embryonic lung fibroblast (HEL 299), RSV prevented the synthesis of viral proteins and also inhibited virus induced activation of epidermal growth factor and phosphatidylinositol-3-kinase signal transduction. Furthermore, transcription factors of NF-$\kappa\beta$ and Sp1 were also inhibited. These mechanisms were observed to decrease the overall replication of the virus [40].

4.8. Hepatitis C Virus and Multiple Sclerosis. In a study conducted by Nakamura et al. [4], RSV was found to dose-dependently enhance viral RNA replication in hepatitis C virus infection *in vitro*. Interestingly, RSV was also reported to reverse the antiviral effects of ribavirin and interferon on HCV RNA replication and was considered nontherapeutic in the treatment of HCV infection [4]. Similarly, RSV was also found to exacerbate the clinical and histological signs of viral model of multiple sclerosis (MS), induced by Theiler's murine encephalomyelitis virus (TMEV), which belongs to the Picornaviridae [19]. Sato et al. also showed that RSV also exacerbated an autoimmune model for MS, experimental autoimmune encephalomyelitis (EAE). However, such studies are few and there are more studies highlighting the beneficial effects of RSV against viral infections, rather than its deleterious exacerbatory effects.

5. Conclusion

Resveratrol has shown a high antiviral potential that can be explored in both human and animal viral infections. Its main antiviral mechanisms were seen to be elicited through inhibition of viral protein synthesis, inhibition of various transcription and signaling pathways, and inhibition of viral related gene expressions. Even though there are still limitations on its bioavailability following intake, which is being widely studied, more studies should be focused on its direct use in the amelioration of viral infections in humans and companion animals.

Authors' Contribution

All authors contributed equally to this work and have read and approved the final paper.

Acknowledgments

The authors are grateful for funding from Universiti Putra Malaysia, Grant no. 9438740, and the Ministry of Higher Education Malaysia, Fundamental Research Grant 5524641.

References

[1] K. B. Pandey and S. I. Rizvi, "Plant polyphenols as dietary antioxidants in human health and disease," *Oxidative Medicine and Cellular Longevity*, vol. 2, no. 5, pp. 270–278, 2009.

[2] S. Shankar, G. Singh, and R. K. Srivastava, "Chemoprevention by resveratrol: molecular mechanisms and therapeutic potential," *Frontiers in Bioscience*, vol. 12, no. 13, pp. 4839–4854, 2007.

[3] I. Gülçin, "Antioxidant properties of resveratrol: a structure-activity insight," *Innovative Food Science and Emerging Technologies*, vol. 11, no. 1, pp. 210–218, 2010.

[4] M. Nakamura, H. Saito, M. Ikeda et al., "An antioxidant resveratrol significantly enhanced replication of hepatitis C virus," *World Journal of Gastroenterology*, vol. 16, no. 2, pp. 184–192, 2010.

[5] K. Cheng, Z. Wu, B. Gao, and J. Xu, "Analysis of influence of baicalin joint resveratrol retention enema on the TNF-α, SIgA, IL-2, IFN-γ of rats with respiratory syncytial virus infection," *Cell Biochemistry and Biophysics*, vol. 70, no. 2, pp. 1305–1309, 2014.

[6] T. Liu, N. Zang, N. Zhou et al., "Resveratrol inhibits the TRIF-dependent pathway by upregulating sterile alpha and armadillo motif protein, contributing to anti-inflammatory effects after respiratory syncytial virus infection," *Journal of Virology*, vol. 88, no. 8, pp. 4229–4236, 2014.

[7] X.-H. Xie, N. Zang, S.-M. Li et al., "Resveratrol inhibits respiratory syncytial virus-induced IL-6 production, decreases viral replication, and downregulates TRIF expression in airway epithelial cells," *Inflammation*, vol. 35, no. 4, pp. 1392–1401, 2012.

[8] N. Zang, S. Li, W. Li et al., "Resveratrol suppresses persistent airway inflammation and hyperresponsivess might partially via nerve growth factor in respiratory syncytial virus-infected mice," *International Immunopharmacology*, vol. 28, no. 1, pp. 121–128, 2015.

[9] N. Zang, X. Xie, Y. Deng et al., "Resveratrol-mediated gamma interferon reduction prevents airway inflammation and airway hyperresponsiveness in respiratory syncytial virus-infected immunocompromised mice," *Journal of Virology*, vol. 85, no. 24, pp. 13061–13068, 2011.

[10] J. J. Docherty, T. J. Sweet, E. Bailey, S. A. Faith, and T. Booth, "Resveratrol inhibition of varicella-zoster virus replication in vitro," *Antiviral Research*, vol. 72, no. 3, pp. 171–177, 2006.

[11] J. L. Espinoza, A. Takami, L. Q. Trung, S. Kato, and S. Nakao, "Resveratrol prevents EBV transformation and inhibits the outgrowth of EBV-immortalized human B cells," *PLoS ONE*, vol. 7, no. 12, Article ID e51306, 2012.

[12] C.-Y. Yiu, S.-Y. Chen, L.-K. Chang, Y.-F. Chiu, and T.-P. Lin, "Inhibitory effects of resveratrol on the Epstein-Barr virus lytic cycle," *Molecules*, vol. 15, no. 10, pp. 7115–7124, 2010.

[13] X. Chen, H. Qiao, T. Liu et al., "Inhibition of herpes simplex virus infection by oligomeric stilbenoids through ROS generation," *Antiviral Research*, vol. 95, no. 1, pp. 30–36, 2012.

[14] J. J. Docherty, M. M. Fu, J. M. Hah, T. J. Sweet, S. A. Faith, and T. Booth, "Effect of resveratrol on herpes simplex virus vaginal infection in the mouse," *Antiviral Research*, vol. 67, no. 3, pp. 155–162, 2005.

[15] J. J. Docherty, M. M. H. Fu, B. S. Stiffler, R. J. Limperos, C. M. Pokabla, and A. L. DeLucia, "Resveratrol inhibition of herpes simplex virus replication," *Antiviral Research*, vol. 43, no. 3, pp. 145–155, 1999.

[16] J. J. Docherty, J. S. Smith, M. M. Fu, T. Stoner, and T. Booth, "Effect of topically applied resveratrol on cutaneous herpes simplex virus infections in hairless mice," *Antiviral Research*, vol. 61, no. 1, pp. 19–26, 2004.

[17] C. L. Clouser, J. Chauhan, M. A. Bess et al., "Anti-HIV-1 activity of resveratrol derivatives and synergistic inhibition of HIV-1 by the combination of resveratrol and decitabine," *Bioorganic & Medicinal Chemistry Letters*, vol. 22, no. 21, pp. 6642–6646, 2012.

[18] A. Heredia, C. Davis, M. N. Amin et al., "Targeting host nucleotide biosynthesis with resveratrol inhibits emtricitabine-resistant HIV-1," *AIDS*, vol. 28, no. 3, pp. 317–323, 2014.

[19] F. Sato, N. E. Martinez, M. Shahid, J. W. Rose, N. G. Carlson, and I. Tsunoda, "Resveratrol exacerbates both autoimmune and viral models of multiple sclerosis," *The American Journal of Pathology*, vol. 183, no. 5, pp. 1390–1396, 2013.

[20] J. Gambini, M. Inglés, G. Olaso et al., "Properties of resveratrol: in vitro and in vivo studies about metabolism, bioavailability, and biological effects in animal models and humans," *Oxidative Medicine and Cellular Longevity*, vol. 2015, Article ID 837042, 13 pages, 2015.

[21] E. Wenzel and V. Somoza, "Metabolism and bioavailability of trans-resveratrol," *Molecular Nutrition & Food Research*, vol. 49, no. 5, pp. 472–481, 2005.

[22] T. Walle, "Bioavailability of resveratrol," *Annals of the New York Academy of Sciences*, vol. 1215, no. 1, pp. 9–15, 2011.

[23] P. Vitaglione, S. Sforza, G. Galaverna et al., "Bioavailability of trans-resveratrol from red wine in humans," *Molecular Nutrition & Food Research*, vol. 49, no. 5, pp. 495–504, 2005.

[24] A. R. Neves, M. Lúcio, S. Martins, J. L. C. Lima, and S. Reis, "Novel resveratrol nanodelivery systems based on lipid nanoparticles to enhance its oral bioavailability," *International Journal of Nanomedicine*, vol. 8, no. 1, pp. 177–187, 2013.

[25] R. Penalva, I. Esparza, E. Larraneta, C. J. González-Navarro, C. Gamazo, and J. M. Irache, "Zein-Based nanoparticles improve the oral bioavailability of resveratrol and its anti-inflammatory effects in a mouse model of endotoxic shock," *Journal of Agricultural and Food Chemistry*, vol. 63, no. 23, pp. 5603–5611, 2015.

[26] J.-T. Hwang, D. Y. Kwon, O. J. Park, and M. S. Kim, "Resveratrol protects ROS-induced cell death by activating AMPK in H9c2 cardiac muscle cells," *Genes & Nutrition*, vol. 2, no. 4, pp. 323–326, 2008.

[27] S. M. Shin, I. J. Cho, and S. G. Kim, "Resveratrol protects mitochondria against oxidative stress through AMP-activated protein kinase-mediated glycogen synthase kinase-3β inhibition downstream of poly (ADP-ribose) polymerase-LKB1 pathway," *Molecular Pharmacology*, vol. 76, no. 4, pp. 884–895, 2009.

[28] Y. Zheng, Y. Liu, J. Ge et al., "Resveratrol protects human lens epithelial cells against H_2O_2-induced oxidative stress by increasing catalase, SOD-1, and HO-1 expression," *Molecular Vision*, vol. 16, pp. 1467–1474, 2010.

[29] A. M. Posadino, A. Cossu, R. Giordo et al., "Resveratrol alters human endothelial cells redox state and causes mitochondrial-dependent cell death," *Food and Chemical Toxicology*, vol. 78, pp. 10–16, 2015.

[30] A. T. Palamara, L. Nencioni, K. Aquilano et al., "Inhibition of influenza A virus replication by resveratrol," *Journal of Infectious Diseases*, vol. 191, no. 10, pp. 1719–1729, 2005.

[31] G. J. Kapadia, M. A. Azuine, H. Tokuda et al., "Chemopreventive effect of resveratrol, sesamol, sesame oil and sunflower oil in the Epstein-Barr virus early antigen activation assay and the mouse skin two-stage carcinogenesis," *Pharmacological Research*, vol. 45, no. 6, pp. 499–505, 2002.

[32] A. De Leo, G. Arena, E. Lacanna, G. Oliviero, F. Colavita, and E. Mattia, "Resveratrol inhibits Epstein Barr Virus lytic cycle in Burkitt's lymphoma cells by affecting multiple molecular targets," *Antiviral Research*, vol. 96, no. 2, pp. 196–202, 2012.

[33] S. A. Faith, T. J. Sweet, E. Bailey, T. Booth, and J. J. Docherty, "Resveratrol suppresses nuclear factor-κB in herpes simplex virus infected cells," *Antiviral Research*, vol. 72, no. 3, pp. 242–251, 2006.

[34] X. Long, S. Li, J. Xie et al., "MMP-12-mediated by SARM-TRIF signaling pathway contributes to IFN-γ-independent airway inflammation and AHR post RSV infection in nude mice," *Respiratory Research*, vol. 16, no. 1, article 11, 2015.

[35] L. Zhang, Y. Li, Z. Gu et al., "Resveratrol inhibits enterovirus 71 replication and pro-inflammatory cytokine secretion in rhabdosarcoma cells through blocking IKKs/NF-κB signaling pathway," *PLoS ONE*, vol. 10, no. 2, Article ID e0116879, 2015.

[36] J. Xu, Z. Yin, L. Li et al., "Inhibitory effect of resveratrol against duck enteritis virus *in vitro*," *PLoS ONE*, vol. 8, no. 6, Article ID e65213, 2013.

[37] N. Komaravelli, J. P. Kelley, M. P. Garofalo, H. Wu, A. Casola, and D. Kolli, "Role of dietary antioxidants in human metapneumovirus infection," *Virus Research*, vol. 200, pp. 19–23, 2015.

[38] I. Galindo, B. Hernáez, J. Berná et al., "Comparative inhibitory activity of the stilbenes resveratrol and oxyresveratrol on African swine fever virus replication," *Antiviral Research*, vol. 91, no. 1, pp. 57–63, 2011.

[39] P. Mastromarino, D. Capobianco, F. Cannata et al., "Resveratrol inhibits rhinovirus replication and expression of inflammatory mediators in nasal epithelia," *Antiviral Research*, vol. 123, pp. 15–21, 2015.

[40] D. L. Evers, X. Wang, S.-M. Huong, D. Y. Huang, and E.-S. Huang, "3,4′,5-Trihydroxy-trans-stilbene (resveratrol) inhibits human cytomegalovirus replication and virus-induced cellular signaling," *Antiviral Research*, vol. 63, no. 2, pp. 85–95, 2004.

[41] M. Campagna and C. Rivas, "Antiviral activity of resveratrol," *Biochemical Society Transactions*, vol. 38, no. 1, pp. 50–53, 2010.

[42] T. Yang, S. Li, X. Zhang, X. Pang, Q. Lin, and J. Cao, "Resveratrol, sirtuins, and viruses," *Reviews in Medical Virology*, 2015.

[43] E. Kaleta, "Herpesviruses of birds—a review," *Avian Pathology*, vol. 19, no. 2, pp. 193–211, 2007.

[44] R. U. Rani and B. Muruganandan, "Outbreak of duck viral enteritis in a vaccinated duck flock," *Scholars Journal of Agriculture and Veterinary Sciences B*, vol. 2, no. 3, pp. 253–255, 2015.

12

Influenza Virus Aerosols in the Air and their Infectiousness

Nikolai Nikitin, Ekaterina Petrova, Ekaterina Trifonova, and Olga Karpova

Department of Virology, Lomonosov Moscow State University, 1/12 Leninskie Gory, Moscow 119234, Russia

Correspondence should be addressed to Nikolai Nikitin; nikitin@mail.bio.msu.ru

Academic Editor: Stefan Pöhlmann

Influenza is one of the most contagious and rapidly spreading infectious diseases and an important global cause of hospital admissions and mortality. There are some amounts of the virus in the air constantly. These amounts is generally not enough to cause disease in people, due to infection prevention by healthy immune systems. However, at a higher concentration of the airborne virus, the risk of human infection increases dramatically. Early detection of the threshold virus concentration is essential for prevention of the spread of influenza infection. This review discusses different approaches for measuring the amount of influenza A virus particles in the air and assessing their infectiousness. Here we also discuss the data describing the relationship between the influenza virus subtypes and virus air transmission, and distribution of viral particles in aerosol drops of different sizes.

1. Introduction

Influenza is one of the most contagious and rapidly spreading infectious diseases and an important global cause of hospital admissions and mortality [1]. Influenza virus concentration [2, 3], air circulation time, air temperature, and humidity [4] play an important role in overcoming the epidemic threshold.

Influenza virus particles are constantly circulating in the air (airborne) in different forms (within dust particles or aerosol droplets) [5, 6]. There are some amounts of the virus in the air constantly. These amounts are insufficient to cause disease in people (the immune system of healthy humans prevents infection). However, at a higher concentration of the airborne virus, the risk of human infection increases dramatically.

Early detection of the threshold virus concentration is essential for prevention of the spread of influenza infection. Furthermore, manufacturers are going to integrate detectors of virus particle numbers into hospital air control system equipment. This review discusses different approaches for measuring the amount of influenza A virus particles in the air and assessing their infectiousness.

One of the fundamental works focused on the definition of the harmful concentration of the influenza A virus in the air is a paper by Alford, with coworkers [7]. It is cited in many

recent reports [8–10]. A study was initiated to determine the minimum infectious aerosol dose and the resulting patterns of infection and illness. Observations made during experimental infections with human volunteers are particularly interesting and relevant. In studies conducted by Alford and colleagues [7], volunteers were exposed to carefully titrated aerosolized influenza virus suspensions by inhaling through a face mask. The demonstration of infection in participants of the study was achieved by recovery of infectious viruses from throat swabs, taken daily, or by seroconversion, that is, the development of neutralizing antibodies. The use of carefully titrated viral stocks enabled the determination of the minimal infectious dose by aerosol inoculation. The approximate 50% human infectious dose (HID_{50}-50% human infectious dose) of virus per volunteer was from 1 to 126 $TCID_{50}$ (the tissue culture 50% infectious dose). The dose for half of the volunteers was 5 $TCID_{50}$. The other half of the men, who had very low or nondetectable preinoculation antibody titers, were infected with 0.6 to 3 $TCID_{50}$. The study reliably shows that the human infectious dose of the influenza A virus, when administered by aerosol to subjects free of serum neutralizing antibodies, is approximately 3 $TCID_{50}$. The approaches used in this study allow the precise number of infectious particles in the total number of particles to be determined.

Ward, with coworkers [11], confirmed experimentally that three $\log_{10}$ copies/mL corresponded to 1 $TCID_{50}$/mL. That is, one $TCID_{50}$/mL contains 1000 copies of the viral genome.

According to other reports, the aerosol infection dose for humans was about 1.95×10^3 viral genome copies, for approximately 300–650 copies of human influenza viruses were contained in 1 $TCID_{50}$, according to previous studies [9, 12].

During the 2009-2010 influenza season (from December to April), Yang, with coworkers [10], collected samples from a health centre, a day-care centre, and airplanes. The concentrations of airborne influenza viruses (A/PR/8/34 (H1N1) and A/swine/Minnesota/1145/2007 (H3N2)) were measured. The influenza A virus RNA was quantified by RT-PCR. Fifty percent of the samples collected contained the influenza A virus, with concentrations ranging from 5.8×10^3 to 3.7×10^4 genome copies per m^3. The average concentration of the virus was $1.6 \pm 0.9 \times 10^4$ genome copies per m^3, corresponding to 35.4 ± 21.0 $TCID_{50}$ per m^3 air. According to Yang et al. [10], 1 $TCID_{50}$ of A/PR/8/34 (H1N1) stock was equivalent to 2.1×10^3 genome copies, and the ratio for the pandemic A/California/04/2009 (H1N1) strain was determined to be 452 ± 84 genome copies per $TCID_{50}$.

Using the measured airborne virus concentration and an adult breathing rate, Yang, with colleagues [10], estimated the inhalation doses during exposures of 1 h (e.g., the duration of a clinical visit), 8 h (a workday), and 24 h to be 1.35×10^4, 1.06×10^5, and 3.2×10^6 viral particles (or 30 ± 18, 236 ± 140, and 708 ± 419 $TCID_{50}$), respectively. Compared with the aerosol HID_{50} 0.6–3 $TCID_{50}$ [7], these doses are adequate to induce infection. In other words, over 1 h, the inhalation dose is estimated to be 30 ± 18 $TCID_{50}$ or about 16 000 particles of the influenza A virus, which is more than enough to induce infection.

2. RT-PCR is the Principal Method for Virus Particle Determination in the Air

To determine the concentration of virus particles in the air, a quantitative reverse transcription polymerase chain reaction (RT-PCR) method is often used [2, 10, 13–15]. Some detection limits for the influenza A virus matrix gene reported recently by PCR are 0.1 $TCID_{50}$/mL [16], 0.2 $TCID_{50}$/mL [17], and 0.006–0.02 $TCID_{50}$/mL [12] or 0.01–0.1 $TCID_{50}$/mL by Light-Cycler [18]. In some studies a difference in sensitivity of RT-PCR for different subtypes of the influenza A virus was observed. The RT-PCR showed sensitivity of 350 copies of H3N2 and 120 copies of H1N1 per reaction, representing the influenza A types in common circulation at the time of the study [19]. In another study [20], the influenza virus subtypes H1 and H3 have been successfully identified with equal efficiency.

However, this method does not always provide an adequate result. RT-PCR allows for the obtaining of information on the total number of viral particles, but not on the number of infectious particles. Simply testing aerosols by RT-PCR for detection of viral nucleic acid would not be sufficient to demonstrate that the viruses in fine particles remain infectious. Given the extensive debate in the literature [21, 22]

and the likelihood that a large percentage of viral copies detected by molecular methods are defective [23, 24], it would be important for new studies to quantify infectious viruses and not merely measure the total viral RNA copy numbers. Based on RT-PCR assay and the influenza virus stock used for calibration, Fabian, with colleagues [9], established a ratio of 300 copies per $TCID_{50}$, which is well within the previously published estimates of 100–350 or 650 [12, 25].

3. Distribution of Viral Particles in the Aerosol Depending on the Size of the Drops

Alford and colleagues [7] studied the aerosol particles of influenza virus suspensions with a diameter of 1–3 μm. Blachere, with colleagues [26], revealed that 46% of influenza virus particles were found in the first stage of the samplers, which collected particles with a diameter of >4 μm. However, 49% of the isolates were collected in the second stage, which collects particles with a diameter of 1–4 μm, and 4% were collected on the back-up filter, which collects particles with a diameter of <1 μm. These findings indicate that 99% of the total viral particles were found in the respiratory aerosol fraction. Coughing, sneezing, talking, and breathing generate a cloud of airborne particles with diameters that can range from a few millimeters to <1 μm [27–30]. Large droplets (>50 μm in diameter) settle on the ground almost immediately, and intermediate-sized droplets (10–50 μm) settle within several minutes. Small particles (<10 μm), including droplet nuclei from evaporated larger particles, can remain airborne for hours and are easily inhaled deep into the respiratory tract. Fabian, with coworkers [9], detected the influenza virus RNA in the exhaled breath of patients and found that >99% of exhaled particles were <5.0 μm in diameter. These findings regarding the influenza virus RNA suggest that the influenza virus may be contained in fine particles generated during tidal breathing and add to the body of literature suggesting that fine particle aerosols may play a role in influenza transmission. Calculation of Stokes' law on settling rate indicated that it took 67 min for a particle with an aerodynamic diameter of 5 μm to settle down from a 3 m height in the static environment; and particles of ≤5 μm could reach as far as pulmonary alveoli [6]. Lindsley et al. [13] found that a 4-μm particle takes 33 min to settle 1 m in still air, and a 1-μm particle takes 8 h; in addition, room air mixing and turbulence can keep these particles airborne even longer. Bischoff, with colleagues [31], later clarified that up to 89% of influenza virus-carrying particles were <4.7 μm in diameter. Other works confirming this data were carried out on different subtypes of the influenza A virus [2, 13, 14].

Infectious viruses and viral RNA can be detected in both larger particles of >5 μm and smaller particles of <5 μm [9, 14, 32]. Experimental studies have demonstrated that the influenza virus can remain infectious in small particle aerosols and can transit across rooms [24, 33]. Cowling et al. [33] found that aerosol transmission (particles <5 μm) accounted for approximately half of all transmission events. Infectious influenza was recovered in all aerosol fractions

(5.0% in >4 μm aerodynamic diameter, 75.5% in 1–4 μm, and 19.5% in <1 μm; $n = 5$) [24].

The aerosol fraction that is <4 μm (the "respirable fraction") is of particular concern because it can remain airborne for an extended time and disperse throughout a room occupied by a patient with influenza. Also, particles containing influenza RNA are small enough to be drawn down into the alveolar region of the lungs. The infectious dose required for inoculation by the aerosol route relative to contact or droplet transmission is unclear, but two reviews of previous studies concluded that the infectious dose by the aerosol route is likely to be considerably lower than the infectious dose by intranasal inoculation [21, 34] and that aerosol inoculation results in more severe symptoms [21], presumably because aerosol particles are able to deposit deeper in the respiratory tract. However, the viability of influenza viruses in particles of different sizes and the persistence of viable airborne viruses in the environment are not yet known.

4. Viability and Infectivity of Airborne Influenza Virus Particles Depend on Environmental Conditions

Numerous reports have shown that the viability of different airborne viruses is dependent on environmental conditions and on the methods of collection and handling of bioaerosol samples [35]. For example, the survival of airborne influenza was shown to greatly depend on the relative humidity (RH), as well as on ambient air temperature and ultraviolet radiation levels [34].

The infectivity of influenza virus particles is preserved depending on temperature, pH and salinity of the water, and UV irradiation. At 4°C, the half-life of infectivity is about 2-3 weeks in water. Due to the conformation of the lipid bilayer, survival under normal environmental conditions should be shorter. Infectivity of the influenza virus particle is easily inactivated by all alcoholic disinfectants, chlorine and aldehydes. As far as is known, temperatures above 70°C destroy infectivity in a few seconds [36].

Using the newly developed guinea pig model of the influenza virus transmission, Lowen and coauthors [37] tested the impact of ambient temperature and relative humidity (RH) on the efficiency of viral spread between hosts. When inoculated and exposed guinea pigs were housed in separate cages, transmission was found to be dependent on both temperature and RH [37–39]. Among the temperatures tested, transmission was highly efficient at 5°C but was blocked or inefficient at 30°C. Dry conditions (20% and 35% RH) were also found to be more favourable for spread than either intermediate (50% RH) or humid (80% RH) conditions. These identical results were obtained using a seasonal human strain, A/Panama/2007/1999 (H3N2) and A/Netherlands/602/2009 (H1N1). Yang and coauthors [40] propose that the effect of RH on virus viability is mediated by salt concentration within droplets: at high RH, physiological concentrations are maintained and viruses are relatively stable; at intermediate RH, evaporation leads to increased salt concentration, resulting in virus inactivation; and at low RH (<50%), salts crystallize out of solution, yielding low salt concentrations and high virion stability. Pica, with colleagues [41], tested two influenza B viruses transmission at low (5°C) versus intermediate (20°C) temperatures. The transmission was more efficient under colder conditions. Thus, transmission of human influenza viruses by a respiratory droplet or aerosol route in the guinea pig model proceeds most readily under cold, dry conditions. These findings suggested two means by which environmental factors could drive the wintertime seasonality of influenza.

Atkinson and Wein [8] created a mathematical model that describes aerosol (i.e. droplet-nuclei) and contact transmission of influenza A virus subtype H5N1 within a household containing one infected. It was demonstrated that in addition to the concentration of particles in the air that a person inhales, time plays a determining role in the influenza virus infection.

5. Relationship between the Influenza Virus Subtypes and Virus Air Transmission

Is there any difference in the influenza virus transmission depending on the virus subtype? In scientific publications, contradictory data obtained on laboratory animals only are presented. Studies using the guinea pig and the ferret models have demonstrated differences of the influenza virus transmission for different strains or genetic compositions by the aerosol route [37, 38, 42].

In the study by Chou, with coworkers [43], the aerosol transmission rate of an influenza virus A/California/04/09 (H1N1) and another H1N1 strain, A/Puerto Rico/8/34, was measured as the percentage of susceptible guinea pigs infected following exposure to inoculated animals. A/California/04/09 was found to spread more efficiently. Differences in the nucleotide sequence of the M segment of the virus genome were found to cause a difference in the aerosol transmission rate. Interestingly, the Eurasian avian-like swine viruses, which possess an M segment closely related to that of the A/California/04/09 (H1N1) virus, are not transmitted efficiently in humans [44].

Pearce, with colleagues [45], characterized four A(H3N2)v viruses isolated in 2009, 2010, and 2011, from patients with uncomplicated upper respiratory tract illnesses (A/Kansas/13/2009 (KS/09), A/Minnesota/11/10 (MN/10), A/Pennsylvania/14/, 10 (PA/10), and A/Indiana/08/11 (IN/11)) and demonstrated that the 2010-2011 A(H3N2)v viruses replicated efficiently in ferrets and readily transmitted in both the direct-contact (DC) and respiratory-droplet (RD) models, whereas the 2009 A(H3N2)v virus exhibited efficient DC transmission but less efficient RD transmission. Typically, the difference in efficiency of the infection of animals in all experiments was a delay in infection of 0.5–1 day. In this context, it is difficult to claim that a clear relationship between the influenza virus subtypes and air virus transmission can be revealed. Authors do not exclude that other genetic requirements must be met in order for the transfer to take place.

The study by Chan, with colleagues [46], has demonstrated that the sensitivity of the commercially available rapid influenza antigen detection tests did not depend on the subtype influenza virus. The analytical sensitivity of the detection tests for swine influenza virus (TCID$_{50}$ log$_{10}$ 3.3–4.7) was comparable with that of seasonal influenza A/HK/4039-46/09 (H1N1) virus (TCID$_{50}$ log$_{10}$ 4.0–4.5). Thus, differences between subtypes were not identified.

Clear correlation and dependence of the number of diseased subjects on the concentration of the influenza virus in the air were shown in various models. For example, a relationship between the number of infected pigs and the influenza detection in the air was identified in a study on a single H1N1 viral strain [47]. The chance of detection of an influenza positive air sample increased 2.2 times per each additional nasal secretion by a sick pig. This suggests that the risk of aerosolization and perhaps aerosol transmission increases as the number of infected pigs increase.

6. Exhaled Breath of Healthy Subjects also Contains Influenza Virus Particles

It is important to consider that the air exhaled by the healthy person also contains influenza virus particles. In studies of particles exhaled by healthy subjects during tidal breathing, researchers reported concentrations from 1 to over 1×10^4 particles per liter, with the majority of the particles being less than 0.3 μm in diameter [29, 48, 49]. One of these studies also reported that 55% of the population studied exhaled >98% of the particles in the air volume investigated and concluded that these subjects, classified as high producers, could, over time, exhale more particles during normal tidal breathing than during relatively infrequent coughing or sneezing events [49]. Concentrations in exhaled breath samples ranged from <48 to 300 influenza virus RNA copies per filter in the positive samples, corresponding to exhaled breath generation rates ranging from <3.2 to 20 influenza virus RNA copies per minute. Total particle concentrations ranged from 67 to 8.5×10^3 particles per liter of air.

7. Conclusions

The human infectious dose of the influenza A virus, when administered by aerosol to subjects free of serum neutralizing antibodies, ranges between 1.95×10^3 and 3.0×10^3 viral particles.

To determine the concentration of virus particles in the air, the RT-PCR method is often used. However, RT-PCR analysis provides information on the total number of viral particles, but not on the number of infectious particles. Influenza virus genomic segments are chosen and packaged at random, whereby only parts of the virions are infectious.

According to various scientific publications, data about the influence of the virus subtype on the effectiveness of influenza transmission are contradictory. The subtype-specific differences in influenza virus transmission were observed in animal models, and recipient animals did not exhibit a pre-existing influenza virus specific immune response. However,

the pathogenicity of a virus subtype depends on the immune status of the recipients (human). The second point is (when) how recently viruses of the same subtype circulated in the population previously.

Therefore, it is important to consider that the risk of acquiring influenza is determined by both the concentration of the influenza A virus infectious particles (not their total amount) in the air and the immune status of the exposed individuals.

Acknowledgment

This work was supported by LG Electronics Inc. (Republic of Korea).

References

[1] T. Vega, J. E. Lozano, T. Meerhoff et al., "Influenza surveillance in Europe: establishing epidemic thresholds by the Moving Epidemic Method," *Influenza and other Respiratory Viruses*, vol. 7, no. 4, pp. 546–558, 2013.

[2] G. Cao, F. M. Blachere, W. G. Lindsley, J. D. Noti, and D. H. Beezhold, "Development of a methodology to detect viable airborne virus using personal aerosol samplers," EPA/600/R-10/127, Environmental Protection Agency, Washington, DC, USA, 2010.

[3] I. Marois, A. Cloutier, É. Garneau, and M. V. Richter, "Initial infectious dose dictates the innate, adaptive, and memory responses to influenza in the respiratory tract," *Journal of Leukocyte Biology*, vol. 92, no. 1, pp. 107–121, 2012.

[4] J. McDevitt, S. Rudnick, M. First, and J. Spengler, "Role of absolute humidity in the inactivation of influenza viruses on stainless steel surfaces at elevated temperatures," *Applied and Environmental Microbiology*, vol. 76, no. 12, pp. 3943–3947, 2010.

[5] C. B. Hall, "The spread of influenza and other respiratory viruses: complexities and conjectures," *Clinical Infectious Diseases*, vol. 45, no. 3, pp. 353–359, 2007.

[6] R. Tellier, "Aerosol transmission of influenza A virus: a review of new studies," *Journal of the Royal Society Interface*, vol. 6, supplement 6, pp. S783–S790, 2009.

[7] R. H. Alford, J. A. Kasel, P. J. Gerone, and V. Knight, "Human influenza resulting from aerosol inhalation.," *Proceedings of the Society for Experimental Biology and Medicine*, vol. 122, no. 3, pp. 800–804, 1966.

[8] M. P. Atkinson and L. M. Wein, "Quantifying the routes of transmission for pandemic influenza," *Bulletin of Mathematical Biology*, vol. 70, no. 3, pp. 820–867, 2008.

[9] P. Fabian, J. J. McDevitt, W. H. DeHaan et al., "Influenza virus in human exhaled breath: an observational study," *PLoS ONE*, vol. 3, no. 7, Article ID e2691, 2008.

[10] W. Yang, S. Elankumaran, and L. C. Marr, "Concentrations and size distributions of airborne influenza A viruses measured indoors at a health centre, a day-care centre and on aeroplanes," *Journal of the Royal Society Interface*, vol. 8, no. 61, pp. 1176–1184, 2011.

[11] C. L. Ward, M. H. Dempsey, C. J. A. Ring et al., "Design and performance testing of quantitative real time PCR assays for

influenza A and B viral load measurement," *Journal of Clinical Virology*, vol. 29, no. 3, pp. 179–188, 2004.

[12] L. J. R. Van Elden, M. Nijhuis, P. Schipper, R. Schuurman, and A. M. van Loon, "Simultaneous detection of influenza viruses A and B using real-time quantitative PCR," *Journal of Clinical Microbiology*, vol. 39, no. 1, pp. 196–200, 2001.

[13] W. G. Lindsley, F. M. Blachere, K. A. Davis et al., "Distribution of airborne influenza virus and respiratory syncytial virus in an urgent care medical clinic," *Clinical Infectious Diseases*, vol. 50, no. 5, pp. 693–698, 2010.

[14] W. G. Lindsley, F. M. Blachere, R. E. Thewlis et al., "Measurements of airborne influenza virus in aerosol particles from human coughs," *PLoS ONE*, vol. 5, no. 11, Article ID e15100, 2010.

[15] K. W. Moon, E. H. Huh, and H. C. Jeong, "Seasonal evaluation of bioaerosols from indoor air of residential apartments within the metropolitan area in South Korea," *Environmental Monitoring and Assessment*, vol. 186, no. 4, pp. 2111–2120, 2014.

[16] B. Schweiger, I. Zadow, R. Heckler, H. Timm, and G. Pauli, "Application of a fluorogenic PCR assay for typing and subtyping of influenza viruses in respiratory samples," *Journal of Clinical Microbiology*, vol. 38, no. 4, pp. 1552–1558, 2000.

[17] R. A. M. Fouchier, T. M. Bestebroer, S. Herfst, L. Van der Kemp, G. F. Rimmelzwaan, and A. D. M. E. Osterhaus, "Detection of influenza a viruses from different species by PCR amplification of conserved sequences in the matrix gene," *Journal of Clinical Microbiology*, vol. 38, no. 11, pp. 4096–4101, 2000.

[18] S. K. Poddar, "Detection of type and subtypes of influenza virus by hybrid formation of FRET probe with amplified target DNA and melting temperature analysis," *Journal of Virological Methods*, vol. 108, no. 2, pp. 157–163, 2003.

[19] B. Stone, J. Burrows, S. Schepetiuk et al., "Rapid detection and simultaneous subtype differentiation of influenza A viruses by real time PCR," *Journal of Virological Methods*, vol. 117, no. 2, pp. 103–112, 2004.

[20] C. Tseng, L. Chang, and C. Li, "Detection of airborne viruses in a pediatrics department measured using real-time qPCR coupled to an air-sampling filter method," *Journal of Environmental Health*, vol. 73, no. 4, pp. 22–28, 2010.

[21] R. Tellier, "Review of aerosol transmission of influenza A virus," *Emerging Infectious Diseases*, vol. 12, no. 11, pp. 1657–1662, 2006.

[22] G. Brankston, L. Gitterman, Z. Hirji, C. Lemieux, and M. Gardam, "Transmission of influenza A in human beings," *The Lancet Infectious Diseases*, vol. 7, no. 4, pp. 257–265, 2007.

[23] P. Fabian, J. J. McDevitt, E. A. Houseman, and D. K. Milton, "Airborne influenza virus detection with four aerosol samplers using molecular and infectivity assays: considerations for a new infectious virus aerosol sampler," *Indoor Air*, vol. 19, no. 5, pp. 433–441, 2009.

[24] J. D. Noti, W. G. Lindsley, F. M. Blachere et al., "Detection of infectious influenza virus in cough aerosols generated in a simulated patient examination room," *Clinical Infectious Diseases*, vol. 54, no. 11, pp. 1569–1577, 2012.

[25] Z. Wei, M. Mcevoy, V. Razinkov et al., "Biophysical characterization of influenza virus subpopulations using field flow fractionation and multiangle light scattering: correlation of particle counts, size distribution and infectivity," *Journal of Virological Methods*, vol. 144, no. 1-2, pp. 122–132, 2007.

[26] F. M. Blachere, W. G. Lindsley, T. A. Pearce et al., "Measurement of airborne influenza virus in a hospital emergency depart-ment," *Clinical Infectious Diseases*, vol. 48, no. 4, pp. 438–440, 2009.

[27] R. G. Loudon and R. M. Roberts, "Droplet expulsion from the respiratory tract," *The American Review of Respiratory Disease*, vol. 95, no. 3, pp. 435–442, 1967.

[28] M. W. Jennison, "Atomizing of mouth and nose secretions into the air as revealed by high-speed photography," in *Aerobiology*, F. R. Moulton, Ed., pp. 106–128, American Association for the Advancement of Science, Washington, DC, USA, 1942.

[29] R. S. Papineni and F. S. Rosenthal, "The size distribution of droplets in the exhaled breath of healthy human subjects," *Journal of Aerosol Medicine: Deposition, Clearance, and Effects in the Lung*, vol. 10, no. 2, pp. 105–116, 1997.

[30] C. Y. H. Chao, M. P. Wan, L. Morawska et al., "Characterization of expiration air jets and droplet size distributions immediately at the mouth opening," *Journal of Aerosol Science*, vol. 40, no. 2, pp. 122–133, 2009.

[31] W. E. Bischoff, K. Swett, I. Leng, and T. R. Peters, "Exposure to influenza virus aerosols during routine patient care," *The Journal of Infectious Diseases*, vol. 207, no. 7, pp. 1037–1046, 2013.

[32] D. K. Milton, M. P. Fabian, B. J. Cowling, M. L. Grantham, and J. J. McDevitt, "Influenza virus aerosols in human exhaled breath: particle size, culturability, and effect of surgical masks," *PLoS Pathogens*, vol. 9, no. 3, Article ID e1003205, 2013.

[33] B. J. Cowling, D. K. Ip, V. J. Fang et al., "Aerosol transmission is an important mode of influenza A virus spread," *Nature Communications*, vol. 4, article 1935, 2013.

[34] T. P. Weber and N. I. Stilianakis, "Inactivation of influenza A viruses in the environment and modes of transmission: a critical review," *Journal of Infection*, vol. 57, no. 5, pp. 361–373, 2008.

[35] S. A. Sattar and M. K. Ijaz, "Airborne viruses," in *Manual of Environmental Microbiology*, C. J. Hurst, R. L. Crawford, M. J. McInerney, G. R. Knudsen, and L. D. Stetzenbach, Eds., pp. 871–883, ASM Press, Washington, DC, USA, 2002.

[36] L. Guertler, "Virology of human influenza," in *Influenza Report*, B. S. Kamps, C. Hoffmann, and W. Preiser, Eds., pp. 87–91, Flying, Paris, France, 2006.

[37] A. C. Lowen, S. Mubareka, T. M. Tumpey, A. García-Sastre, and P. Palese, "The guinea pig as a transmission model for human influenza viruses," *Proceedings of the National Academy of Sciences of the United States of America*, vol. 103, no. 26, pp. 9988–9992, 2006.

[38] A. C. Lowen, S. Mubareka, J. Steel, and P. Palese, "Influenza virus transmission is dependent on relative humidity and temperature," *PLoS Pathogens*, vol. 3, no. 10, pp. 1470–1476, 2007.

[39] J. Steel, P. Palese, and A. C. Lowen, "Transmission of a 2009 pandemic influenza virus shows a sensitivity to temperature and humidity similar to that of an H3N2 seasonal strain," *Journal of Virology*, vol. 85, no. 3, pp. 1400–1402, 2011.

[40] W. Yang, S. Elankumaran, and L. C. Marr, "Relationship between humidity and influenza's seasonality," *PLoS ONE*, vol. 7, no. 10, Article ID e46789, 2012.

[41] N. Pica, Y. Chou, N. M. Bouvier, and P. Palese, "Transmission of influenza B viruses in the Guinea pig," *Journal of Virology*, vol. 86, no. 8, pp. 4279–4287, 2012.

[42] N. van Hoeven, C. Pappas, J. A. Belser et al., "Human HA and polymerase subunit PB2 proteins confer transmission of an avian influenza virus through the air," *Proceedings of the National Academy of Sciences of the United States of America*, vol. 106, no. 9, pp. 3366–3371, 2009.

[43] Y. Chou, R. A. Albrecht, N. Pica et al., "The m segment of the 2009 new pandemic H1N1 influenza virus is critical for its high

transmission efficiency in the Guinea pig model," *Journal of Virology*, vol. 85, no. 21, pp. 11235–11241, 2011.

[44] T. T. Lam, H. Zhu, J. Wang et al., "Reassortment events among swine influenza a viruses in China: implications for the origin of the 2009 influenza pandemic," *Journal of Virology*, vol. 85, no. 19, pp. 10279–10285, 2011.

[45] M. B. Pearce, A. Jayaraman, C. Pappas et al., "Pathogenesis and transmission of swine origin A(H3N2)v influenza viruses in ferrets," *Proceedings of the National Academy of Sciences of the United States of America*, vol. 109, no. 10, pp. 3944–3949, 2012.

[46] K. H. Chan, S. T. Lai, L. L. M. Poon, Y. Guan, K. Y. Yuen, and J. S. M. Peiris, "Analytical sensitivity of rapid influenza antigen detection tests for swine-origin influenza virus (H1N1)," *Journal of Clinical Virology*, vol. 45, no. 3, pp. 205–207, 2009.

[47] C. A. Corzo, A. Romagosa, S. A. Dee, M. R. Gramer, R. B. Morrison, and M. Torremorell, "Relationship between airborne detection of influenza A virus and the number of infected pigs," *Veterinary Journal*, vol. 196, no. 2, pp. 171–175, 2013.

[48] C. I. Fairchild and J. F. Stampfer, "Particle concentration in exhaled breath," *The American Industrial Hygiene Association Journal*, vol. 48, no. 11, pp. 948–949, 1987.

[49] D. A. Edwards, J. C. Man, P. Brand et al., "Inhaling to mitigate exhaled bioaerosols," *Proceedings of the National Academy of Sciences of the United States of America*, vol. 101, no. 50, pp. 17383–17388, 2004.

13

Elucidating the Interacting Domains of *Chandipura* Virus Nucleocapsid Protein

Kapila Kumar,[1] **Sreejith Rajasekharan,**[1] **Sahil Gulati,**[1] **Jyoti Rana,**[1] **Reema Gabrani,**[1] **Chakresh K. Jain,**[1] **Amita Gupta,**[2] **Vijay K. Chaudhary,**[3] **and Sanjay Gupta**[1]

[1] *Center for Emerging Diseases, Department of Biotechnology, Jaypee Institute of Information Technology, A-10, Sector 62, Noida, Uttar Pradesh 201 307, India*
[2] *Department of Microbiology, University of Delhi, Benito Juarez Marg, New Delhi 110021, India*
[3] *Department of Biochemistry, University of Delhi, Benito Juarez Marg, New Delhi 110021, India*

Correspondence should be addressed to Sanjay Gupta; sanjay.gupta@jiit.ac.in

Academic Editor: Trudy Morrison

The nucleocapsid (N) protein of *Chandipura* virus (CHPV) plays a crucial role in viral life cycle, besides being an important structural component of the virion through proper organization of its interactions with other viral proteins. In a recent study, the authors had mapped the associations among CHPV proteins and shown that N protein interacts with four of the viral proteins: N, phosphoprotein (P), matrix protein (M), and glycoprotein (G). The present study aimed to distinguish the regions of CHPV N protein responsible for its interactions with other viral proteins. In this direction, we have generated the structure of CHPV N protein by homology modeling using SWISS-MODEL workspace and Accelrys Discovery Studio client 2.55 and mapped the domains of N protein using PiSQRD. The interactions of N protein fragments with other proteins were determined by ZDOCK rigid-body docking method and validated by yeast two-hybrid and ELISA. The study revealed a unique binding site, comprising of amino acids 1–30 at the N terminus of the nucleocapsid protein (N1) that is instrumental in its interactions with N, P, M, and G proteins. It was also observed that N2 associates with N and G proteins while N3 interacts with N, P, and M proteins.

1. Introduction

Chandipura virus (CHPV) is a recently recognized emerging human pathogen [1–3] of the genus *Vesiculovirus* and family Rhabdoviridae [4]. The ~11 kb genome of CHPV [5] is encapsidated by nucleocapsid (N) protein and serves as a template for both replication and transcription. The transcription of the genome by viral encoded RNA-dependent RNA polymerase (RdRp; L protein) produces five capped and polyadenylated mRNAs which code for five proteins nucleocapsid protein (N), phosphoprotein (P), matrix protein (M), glycoprotein (G), and large protein (L) in sequential order and in decreasing amounts [6]. Interactions among these proteins are essential for functioning of key processes during virus replication and pathogenesis. However, only few details of the molecular functions of these viral proteins that orchestrate the virus life cycle are known.

The N protein plays a pivotal role in virus biology by virtue of its interactions with other viral proteins. The interaction of monomeric N protein with P maintains it in the encapsidation competent soluble (active) form [7, 8]. In its active form, N protein tightly wraps the RNA genome and maintains the structural integrity along with the template function of the negative strand genome RNA. Within the virion, this encapsidated RNA (N-RNA) template is associated with the transcription factor (P protein) of the RNA polymerase complex to form the transcribing ribonucleoprotein (RNP) particle [9]. During transcription, the RNA polymerase complex (L and P) interacts with the N protein of the N-RNA template to transiently displace N and gain access to the genomic RNA [10]. The association of N (as part of the RNP particle) with M enables the condensation of the enwrapped genome thus giving the virion its characteristic bullet shape [11]. Earlier research had mapped certain interactions of CHPV such as NN and NP at the domain level [12, 13]. Deletion studies had shown that the N-terminal 47 amino acids (aa) together with residues 180–264 are indispensable for N protein oligomerization [12]. The N-terminal 180 aa and the C-terminal 102 aa of N protein have been mapped to be required for binding with P protein in its monomeric and RNA encapsidated state, respectively [13]. With the elucidation of interactions of N protein with M and G proteins [14], there arises a need to further map the specific regions of CHPV N protein involved in mediating its interactions.

In the current study, the structures of CHPV N and M proteins were generated by homology modeling using SWISS-MODEL workspace, G protein using I-TASSER, and P protein using a modified *abinitio* method. The nucleocapsid protein was divided into three fragments using PiSQRD—an N-terminal region N1 (Nt arm), a central region N2 (Nt lobe), and a C-terminal region N3 (Ct lobe). The involvement of N protein fragments in interactions of N with N, P, M, and G proteins was predicted using ZDOCK rigid-body docking method and further checked and validated by yeast two-hybrid system and ELISA-based assay.

2. Materials and Methods

2.1. Structure Elucidation of Chandipura Virus Proteins by Homology Modeling. The structural models of N and M proteins of CHPV were generated by SWISS-MODEL workspace. Vesicular stomatitis virus (VSV), Indiana strain N protein complex (PDB ID: 3HHW), and M protein (PDB ID: 1LG7) were used as templates for homology modeling of CHPV proteins as both these viruses belong to the same genus (genus *Vesiculovirus*) and their N and M proteins shares 71.3% and 28% sequence similarity, respectively. The structure of CHPV G protein in its perfusion state (530 aa) was generated using I-TASSER [15, 16] and the atomic clashes along with bond length errors within the structure were removed by using ModRefiner [17]. Threading-based modeling was used for G protein since its full-length template (VSV G) was not available in PDB, unlike N and M proteins, whose structures were available for VSV. Modeling of CHPV P protein structure using homology modeling and threading was not feasible due to the unavailability of full-length structural homologs. VSV P protein oligomerization domain (PDB ID: 2FQM) and C-terminal domain (PDB ID: 2 K47) were the only available templates that exhibited homology to the corresponding domains of CHPV P protein. The structure of CHPV P protein was hence modeled using *abinitio* method in combination with protein folding constraints. CHPV P protein was divided into four segments: N-terminal domain (residues 1–103), oligomerization domain (residues 104–168), interconnecting domain (residues 169–215), and C-terminal domain (residues 216–289). Oligomerization and C-terminal domains were modeled using SWISS-MODEL workspace while N-terminal and interconnecting domain models were obtained using I-TASSER and QUARK [18], respectively. The disulfide connectivity was predicted using DiANNA server [19]. The complete model was assembled manually by reproducing the torsion angles from the individual modeled segments to the extended polypeptide chain (289 amino acid) using Swiss-PDB Viewer. The torsion angles were changed wherever necessary based on disulfide connectivity and secondary structure prediction. CHPV protein structures thus generated were subjected to prepare protein protocol of Accelrys Discovery Studio Client 2.55 for the energy minimization, optimizing short/medium sized loop regions, and protonating the protein structures. The stereochemical quality was estimated using Ramachandran plot, Verify3D, and ERRAT.

2.2. Decomposing N Protein into Quasi-Rigid Dynamic Fragments. PiSQRD web resource was used to subdivide CHPV N protein structure in quasi-rigid fragments. This server uses an algorithm introduced by Potestio and coworkers [20] and subdivides proteins into regions that behave approximately as rigid units in the course of protein structural fluctuations. Default values were used for all the parameters with captured mobility threshold of 80% and 10 lowest energy essential modes.

2.3. Docking of CHPV N Protein. Docking of CHPV N protein with N, P, M, and G proteins was performed using the ZDOCK (Accelrys) [21, 22] rigid-body docking method. ZDOCK provides rigid body docking of two protein structures using the ZDOCK algorithm as well as clustering the poses according to the ligand position. RDOCK (Accelrys) [23, 24] refinement was performed on the top 100 poses of the filtered ZDOCK output of each interacting pair. A distance, dependent dielectric constant $4r$ (r being the distance) was used during refinement.

2.4. Construction of Yeast Two-Hybrid Plasmids. The putative nucleocapsid (N) fragments (N1, N2, and N3) were amplified using specific primers (as listed in Table 1) designed to incorporate *Nde*I and *Bam*HI restriction enzyme sequences at their 5′ ends to facilitate cloning in yeast expression vectors pGBKT7 (BD, bait) and pGADT7 (AD, prey), Clontech, USA. These primer pairs span nucleotides 1–90 bp (N1), 61–894 bp (N2), and 693–1260 bp (N3) of the complete nucleocapsid ORF cloned in pET33b vector [7]. The clone of CHPV N gene was a kind gift from Dr. Dhrubajyoti Chattopadhyay of Dr. B. C. Guha Centre for Genetic Engineering, Kolkata, India.

The N protein fragments were amplified by standard polymerase chain reaction (PCR) as described previously [14], purified using PCR clean up kit (Sigma Aldrich, USA), and digested with *Nde*I and *Bam*HI restriction enzymes(Fermentas, USA). The vectors, pGBKT7 and pGADT7, were linearized by the same enzyme combination and subsequently ligated with the digested N protein fragments using T4 DNA Ligase (5 U/μL, Fermentas, USA) to generate the respective bait and prey constructs. The resulting pGBKT7 constructs, BD-N1, BD-N2, and BD-N3, encoded the fragments N1 (30 amino acids), N2 (278 amino acids), and N3 (191 amino acids) fused in frame at the C terminus of BD domain and the pGADT7 constructs, AD-N1, AD-N2, and AD-N3, encoded the corresponding fragments fused

TABLE 1: Primers used for cloning N gene domains of CHPV.

S. no.	Construct	Oligo name	Primer sequence
1	BD-N1 and AD-N1	N1 F *Nde*I (BD/AD)	GGAAGTGA**CATATG**AGTTCTCAAGTATTCTGC
		N1 R *Bam*HI (BD/AD)	GCTAACA**GGATCC**GAAGAATGCCCCTGGAAAC
2	BD-N2 and AD-N2	N2 F *Nde*I (BD/AD)	GGAAGTGA**CATATG**GAAGACCCAGTGGAGTTTC
		N2 R *Bam*HI (BD/AD)	GCTAACA**GGATCC**ATGGAAACTGGGATTTTTTGTTG
3	BD-N3 and AD-N3	N3 F *Nde*I (BD/AD)	GGAAGTGA**CATATG**ACTCTGTCACACCTCCAG
		N3 R *Bam*HI (BD/AD)	GCTAACA**GGATCC**TCATGCAAAGAGTTTCCTGGC
4	GST-N1	N1 F *Bam*HI (pGEX-4T3)	GCTAACA**GGATCC**ATGAGTTCTCAAGTATTCTGC
		N1 R *Xho*I (pGEX-4T3)	GCTAACA**CTCGAG**GAAGAATGCCCCTGGAAAC
5	GST-N2	N2 F *Bam*HI (pGEX-4T3)	GCTAACA**GGATCC**ATGGAAGACCCAGTGGAGTTTC
		N2 R *Xho*I (pGEX-4T3)	GCTAACA**CTCGAG**ATGGAAACTGGGATTTTTTGTTG
6	GST-N3	N3 F *Bam*HI (pGEX-4T3)	GCTAACA**GGATCC**ATGACTCTGTCACACCTCCAG
		N3 R *Xho*I (pGEX-4T3)	GCTAACA**CTCGAG**TCATGCAAAGAGTTTCCTGGC

Primers used for PCR amplification of Chandipura Virus N1, N2, and N3 domains of nucleocapsid gene (F: forward primer and R: reverse primer). The names of the restriction enzymes are in italics and their recognition sequences in bold.

in frame downstream of AD domain. The complete ORFs encoding N, P, M, and G proteins of CHPV as both BD and AD fusions (BD-N, BD-P, BD-M, BD-G, AD-N, AD-P, AD-M, and AD-G) used in this study have been described earlier by the authors [14].

2.5. Construction of Bacterial Expression Plasmids. Deletion constructs for ELISA were generated by PCR amplification using primers corresponding to the appropriate end sequences with added *Bam*HI (5′) and *Xho*I (3′) sites (Table 1) and N-pET33b as template [7]. The amplified products were digested with *Bam*HI and *Xho*I, purified, and subcloned into pGEX-4T3 vector (GST tag) digested with the same enzyme combination. The recombinants were confirmed by restriction enzyme digestion. The pGEX-4T3 constructs called GST-N1, GST-N2, and GST-N3 contain the three fragments fused in frame with GST tag at the N terminus.

2.6. Yeast Transformation. Saccharomyces cerevisiae strains Y187 and AH109 were used for protein interaction analysis. These strains were transformed individually with BD and AD constructs, respectively, following lithium acetate yeast transformation protocol as explained by manufacturer (Matchmaker GAL4 two-hybrid system 3 and libraries user manual, protocol number PT3247-1). Successful transformants were screened on Synthetically Defined (SD, Clontech, USA) media lacking amino acids tryptophan and leucine (selection marker for BD and AD plasmids, resp.). The constructs were also screened for autologous activation of the reporter gene *HIS3* on SD media lacking tryptophan and histidine (SD/-Trp/-His) for bait constructs and on SD media lacking leucine and histidine (SD/-Leu/-His) for prey constructs.

2.7. Yeast Two-Hybrid Screening. Each of the bait construct in Y187 yeast strain was allowed to mate with each prey construct in AH109 yeast strain. All three fragments in BD/AD (N1, N2, and N3) vector were mated with four complete ORFs (N, P, M, and G) in AD/BD vector accounting

for a total of 24 mated combinations. Successfully mated diploids containing both bait and prey vectors were selected on SD media lacking tryptophan and leucine (SD/-Trp/-Leu) and tested for positive protein interaction by plating on SD/-Trp/-Leu/-His media. Simultaneously, the mated clones were screened on SD/-Trp/-Leu/-His/α-gal plate for α-galactosidase assay. The development of blue color in the presence of X-α-gal was indicative of positive interaction in this assay.

2.8. Enzyme Linked Immunosorbent Assay (ELISA) for Interaction Validation. ELISA was performed as a second independent method to check the interactions of N proteins fragments with N, P, M, and G proteins. Streptactin-coated microtiter plate (IBA-GmBH, Germany) was used to check the interactions between full-length CHPV proteins as Strep tag fusions (Strep-N, Strep-P, Strep-M, and Strep-G), generated by the authors in previous study [14], and nucleocapsid protein fragments as GST fusions (GST-N1, GST-N2, and GST-N3). The protocol involved has been described earlier by the authors [14, 25].

3. Results

3.1. Model Building and Validation. CHPV P protein structure was transformed locally to introduce disulfide links between Cys37–Cys172 and Cys57–Cys286 in concordance with the predicted disulfide linkages. Ramachandran plot analysis displayed approximately 98% of the residues in allowed region (favored + allowed regions) and the rest were outliers. Approximately 99.3% residues of CHPV N protein lie in the allowed regions. Verify 3D and ERRAT analysis indicated good quality of all the protein models with minimal interatomic clashes.

3.2. Rigid-Body Docking for Domain-Protein Assembly. CHPV full-length protein structures (N (Figure 1(a)), P (Figure 1(b)), M (Figure 1(c)), and G (Figure 1(d))) were docked on CHPV N protein model as rigid-bodies using

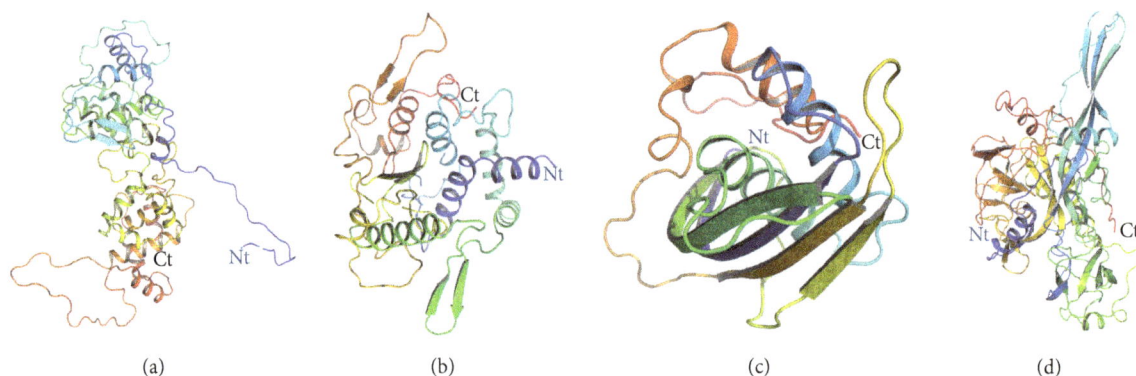

FIGURE 1: Structures of N, P, M, and G proteins of CHPV predicted using SWISS-MODEL workspace. (a) Nucleocapsid protein, (b) phosphoprotein, (c) matrix protein, and (d) glycoprotein of Chandipura virus rendered in cartoon (rainbow color).

ZDOCK at a 15° rotational sampling density. Top 2000 poses were further reranked (ZRank) using detailed electrostatics, van der Waals, and desolvation energy terms. The success of the resulting predictions was evaluated based on their RMSD values. An acceptable docking pose has been defined as one where the RMSD of one of the proteins is ≤10 Å from the cluster center (the choice of 10 Å RMSD to define an acceptable pose is in concordance with the ligand RMSD used by CAPRI to define an acceptable solution in protein-protein docking). The best pose with minimum E RDOCK score, generated after RDOCK refinement, was chosen as near-native structure for each interaction. It was observed that N terminal region of N protein (N1, Nt arm) interacts with N, P, and M proteins while the central N2 (Nt lobe) interacts with N, P, and G proteins and C terminal region N3 (Ct lobe) interacts with N, P, M, and G proteins (Figures 2 and 3). These putative interactions were experimentally checked by Y2H and ELISA methods.

3.3. Screening for Potential Positive Interactions by Y2H Analysis. The Y2H bait (pGBKT7) and prey (pGADT7) recombinants containing N1, N2, and N3 were transformed in haploid *S. cerevisiae* strains Y187 and AH109, respectively. The positive transformants were selected on SD/-Trp media for bait fusion and SD/-Leu media for prey fusion vectors. Prior to the Y2H interaction analysis, generated bait fusion constructs were checked for their ability to activate the expression of the *HIS3* reporter gene (autoactivation) on histidine-deficient SD media (SD/-Trp/-His for bait constructs and SD/-Leu/-His for prey constructs). None of the hybrid bait or prey vectors tested in the study showed background transcriptional activity (data not shown) and thus were found to be suitable for Y2H studies. However, N1 and N2 fragments as AD fusions were observed to interact with empty BD vector (control vector) to activate the reporter genes *HIS3* and *MEL1* (Figure 5, sectors 28 and 29). Thus, the interactions involving N1 and N2 fragments were analysed using BD-N1 and BD-N2. Full-length CHPV viral genes had been previously transformed in corresponding yeast strains and checked for autoactivation in earlier studies by the authors [14]. BD-P was found to activate the histidine reporter gene

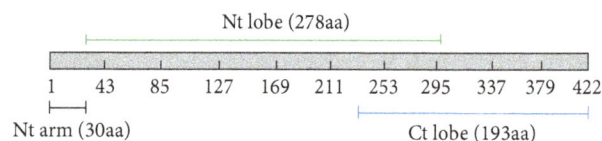

FIGURE 2: Schematic representation of the CHPV nucleocapsid (N) protein fragments. N terminal fragment 1 (N1) is of 30 amino acids (aa). Fragment 2 (N2) is of 278 aa. The N terminal 10 aa residues and the C terminal 68 aa residues of N2 overlaps with N1 and N3, respectively. Fragment 3 (N3) constitutes the C terminal 193 aa of the CHPV N protein.

and thus combinations involving P protein were studied in context of reverse direction taking P protein as AD fusion.

Yeast strains carrying bait and prey constructs were mated and the resulting diploids were screened under selective conditions on SD/-Trp/-Leu media. Each combination of proteins was considered in both directions, that is, as both BD as well as AD fusions to test for the interactions of N1, N2, and N3 with full-length N, P, M, and G proteins (Table 2). Mated diploids were checked for potential positive interaction by analyzing expression of reporter genes *HIS3* and *MEL1*. The expression of *HIS3* was analysed by growth on SD media lacking tryptophan, leucine, and histidine (SD/-Trp/-Leu/-His). The colonies which grew on histidine-deficient media were considered to be positive (Figure 4). Activation of another reporter gene, that is, *MEL1*, was also checked by plating on SD/-Trp/-Leu/-His/α-gal media and blue colored colonies indicated positive interactions (Figure 5). Tumor suppressor protein p53 and Simian virus large T-antigen encoded by pGBKT7-53 and pGADT7-T vectors as BD and AD fusions are known to be interacting proteins and thus were taken as positive control (Figure 5, sector 31), whereas pGBKT7-Lam and pGADT7-T encoding noninteracting Lamin protein and Simian virus large T-antigen served as negative interaction control (Figure 5, sector 32). It was observed that 8 (interaction of N1 with N, P, and M proteins; N2 with N and G proteins and N3 with N, P, and M proteins) out of the 10 putative positive interactions identified through computational approach were positive in Y2H screening. In addition to the 8 associations, interaction

(a)

(b)

(c)

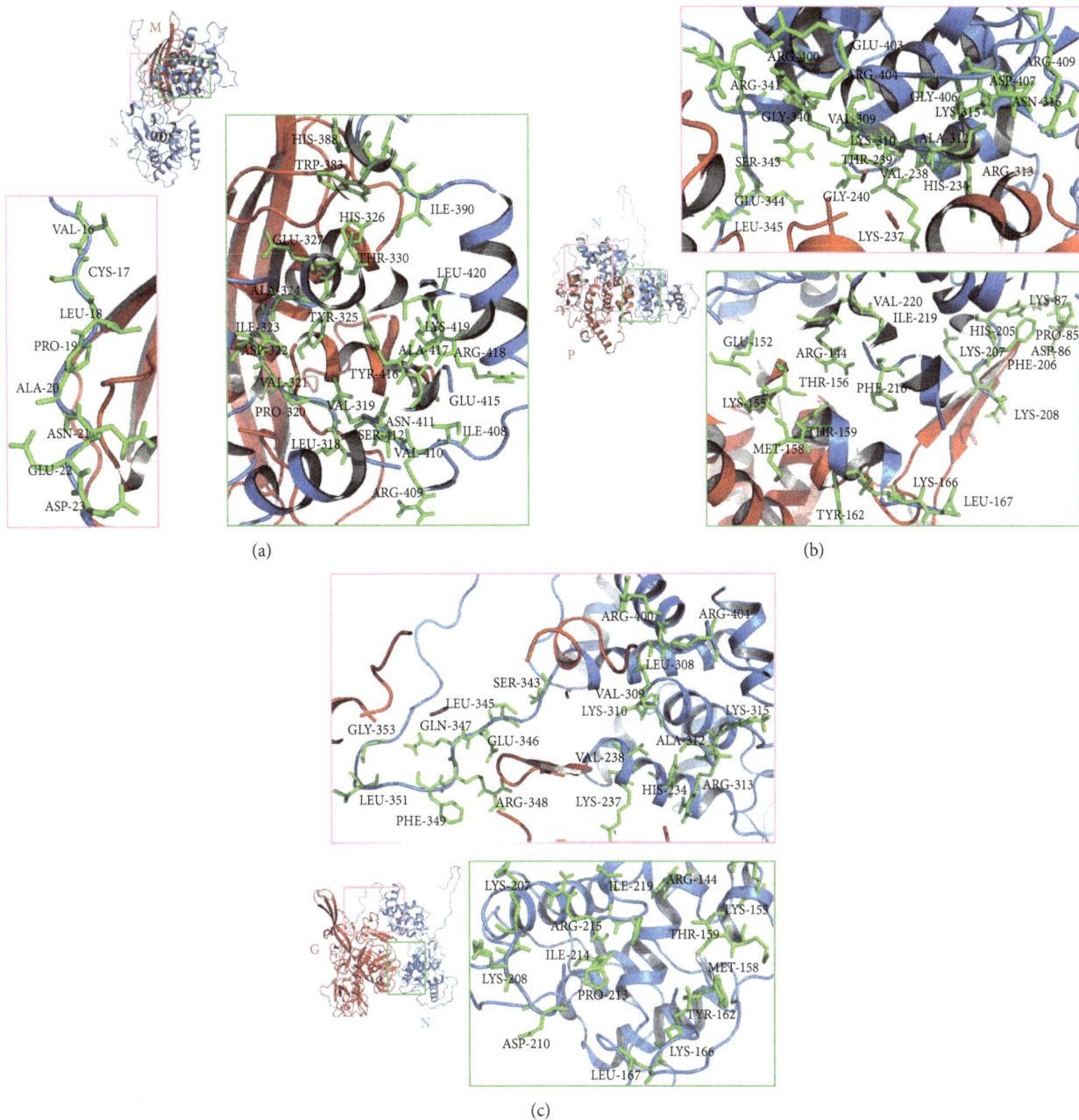

FIGURE 3: Docking results for NM, NP, and NG interactions using ZDOCK. (a) Top rank NM complex ZDOCK pose 684 (E_RDOCK score—31.356). (b) Top rank NP complex ZDOCK pose 220 (E_RDOCK score—27.932). (c) Top rank NG complex ZDOCK pose 8 (E_RDOCK score—40.982). CHPV proteins are represented as cartoons models. CHPV N protein is shown in blue and M, P, and G proteins are shown in red. (a) The residues of CHPV N Protein in the NM interface correspond to N1 and N2. (b) The residues of CHPV N Protein in the NP interface correspond to N1 and N3. (c) The residues of CHPV N protein in the NG interface correspond to N2 and N3 fragments. The NN interaction has been discussed previously by the authors [26].

of N1 fragment and G protein was also observed to be positive in Y2H.

3.4. Validation of Protein Interactions by ELISA Assay.

Each pairwise combination tested by Y2H analysis was checked independently by ELISA assay to add to the reliability of the data obtained. The authors have previously used ELISA as a method of choice for identification of protein interactions

[14, 25]. The truncated fragments of N protein cloned in pGEX-4T3 vector allowed for the expression of each fragment with a GST tag at the N terminus. Full-length CHPV N, P, M, and G proteins as Strep tag fusions were expressed and the cell lysates were prepared as described previously [14]. GST-N1, GST-N2, and GST-N3 were also expressed and the cell lysates were analysed for the solubility of these fragments. All three fragments were observed to be soluble (data not shown)

TABLE 2: Cumulative results of CHPV N protein fragments interaction analysis.

Protein pair used for interaction analysis	Interaction analysis by ZDOCK and RDOCK	Y2H assay		ELISA	Known from the Literature
		HIS3 reporter (nutritional selection)	MEL1 reporter (α-galactosidase assay)		
N1-N	√	√	√	√	√ [12]
N1-P	√	√	√	√	√ [13]
N1-M	√	√	√	√	—
N1-G	X	√	√	√	—
N2-N	√	√	√	√	√ [12]
N2-P	√	X	X	X	—
N2-M	X	X	X	X	—
N2-G	√	√	√	√	—
N3-N	√	√	√	√	√ [12]
N3-P	√	√	√	√	√ [13]
N3-M	√	√	√	√	—
N3-G	√	X	X	X	—

√ represents positive interaction.
X represents negative interaction.

and were used for the binding assay. Known interactions among full-length CHPV proteins interaction between P and N proteins involving GST-P+Strep-N and PP involving Strep-P+GST-P as positive controls, while non-interacting pairs PM as GST-P+Strep-M and PG involving GST-P+Strep-G as negative controls; 14 were taken as controls. The binding of N1 region with a nonrelevant protein-nonstructural protein 1 (nsP1) of Chikungunya virus (CHIKV) as Strep tag fusion [25] and the binding of Strep (Strep-N, Strep-P, Strep-M and Strep-G) and GST (GST-N1, GST-N2, and GST-N3) fusions directly with the Streptactin plate were also taken as experimental controls. Anti-GST antibody (primary antibody) followed by HRP-conjugated secondary antibody was used to check the protein interactions. The absorbance was measured at 450 nm after stopping the reaction of HRP with the substrate TMB using 2N HNO$_3$. The experiments were performed in triplicates and their mean ± SD have been graphically represented in Figure 6.

Analysis of 12 pairs of putative protein interactions among N protein fragments and CHPV proteins revealed a total of 9 positive interactions. N1 (Nt arm) has been shown to interact with all four viral proteins, that is, N, P, M, and G. N2 (Nt lobe) was observed to associate with N and G protein while N3 (Ct lobe) bound to N, P, and M proteins (Table 2). The interactions observed in ELISA were in concordance with both Y2H analysis and the data available from the literature.

4. Discussion

During the life cycle of viruses, the encoded proteins extensively interact with one another to perform their functions. These protein-protein interactions (PPIs) are achieved through specific regions that are responsible for the physical interactions. These regions which mediate different interactions are considered as building blocks of interaction networks. Domain mapping has been carried out for several viruses such as Herpes Simplex Virus type I (HSV), Epstein Barr Virus (EBV), Kaposi's Sarcoma associated Human Virus (KSHV) [27], Murine Coronavirus [28], Rabies Virus (RV) [29], VSV [8, 30], and Sendai virus [31] and for certain proteins of CHPV as well [12, 13]. These studies have highlighted the importance of identifying the regions which can be targeted for therapeutic strategies.

In a recent study on intraviral protein interactions among Chandipura viral proteins, the authors had reported the interaction of N protein with N, P, M, and G proteins of CHPV [14]. In order to investigate the interacting regions of N protein, a series of interaction studies employing bioinformatics-based docking studies, yeast two-hybrid system, and ELISA have been performed. The division of N protein into putative fragments was guided in large part by our prediction of the protein structure. These predictions provide insights into the less known structures of CHPV proteins.

Each combination of nucleocapsid protein fragments and CHPV full-length proteins was tested using ZDOCK, Y2H, and ELISA, allowing us to assess the reliability of the data for protein interactions obtained in this study. The study identified the interacting residues involved in NN association which are present in all the three regions of nucleocapsid protein considered in this study (N1, N2, and N3). However, the central 278 aa region (N2) essential for interaction with G protein is shown to be dispensable for interactions with M and P proteins. The interaction dataset generated by Y2H and ELISA correlates with 75% of the ZDOCK based predictions and 100% with the data known from literature (Table 2).

Earlier mapping studies of CHPV N protein involved the generation of N protein fragments by enzymatic digestion

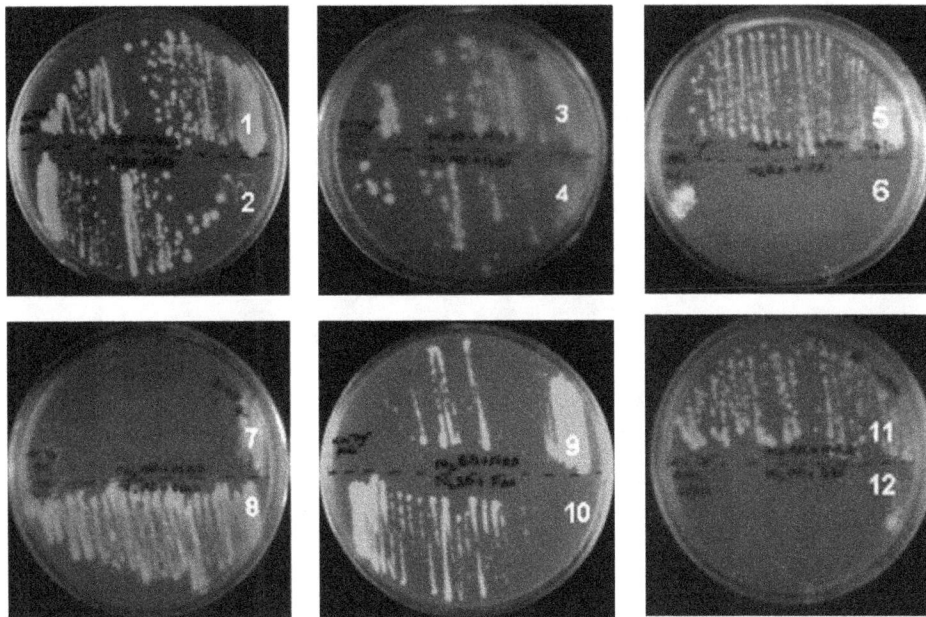

Sector	Bait	Prey	Sector	Bait	Prey
1	N1	N	7	N2	M
2	N1	P	8	N2	G
3	N1	M	9	N3	N
4	N1	G	10	N3	P
5	N2	N	11	N3	M
6	N2	P	12	N3	G

FIGURE 4: Screening of CHPV N1, N2, and N3 interactions with N, P, M, and G proteins by Y2H. All four viral genes cloned individually as DNA-binding domain fusion (bait) and transformed in Y187 yeast strain were systematically mated with N1, N2, and N3 cloned individually as DNA-activation domain fusions (prey), transformed in AH109 yeast strain. The diploid cells formed were selected on SD medium lacking amino acids tryptophan and leucine. Putative positive interactions were tested on SD media lacking tryptophan, leucine, and histidine All combinations were tested in at least 2 independent experiments. Presence of growth on medium (sectors 1, 2, 3, 4, 5, 8, 9, 10, and 11) indicated interaction between the proteins while absence (sectors 6, 7, and 12) suggested no interaction.

using chymotrypsin [12, 13]. Nevertheless, our choice of boundaries for putative nucleocapsid protein fragments was based on precise structural and biophysical criteria. The interacting residues involved in NN association (previously reported by the authors; 31) lie within the N terminal and central regions of the N monomer as shown by Mondal and coworkers [12]. However, our bioinformatics predictions have narrowed down these regions to smaller peptides including residues 8 to 22 at the N terminus and 245 to 256 in the central region. Moreover, with evidence from the oligomerisation studies of CHPV N [26] and VSV N protein [30], we suggest the involvement of intermittent residues from 321 to 395 at the C terminus in the oligomerisation of N protein. Deletion studies in CHPV had shown the involvement of N terminal aa 1 to 180 and C terminal aa 320 to 390 of N protein in NP interaction. The present work after considering the N protein oligomerisation as well as RNA binding constraints suggests smaller peptides within these regions—residues 2 to 30, 140 to 165, 205 to 240, and 320 to 343, to be indispensable

for NP association. In addition to identifying the interacting residues of N involved in NN and NP associations, we have also predicted the regions of N protein responsible for NM and NG interactions. The NM interaction involves the aa residues 16 to 20 and 318 to 420, while NG binding requires aa 144 to 240 in the central region of the N protein. Our data corroborates well with the previously identified interacting regions involved in NN and NP interactions for both CHPV [12, 13] and VSV [30], thus validating our approach of interaction analysis.

Although important data has been generated by mapping studies, the biological significance of these interactions is the scope of further experimentation. Nevertheless, these associations can prove to be valuable starting points for understanding CHPV biology and designing antiviral strategies. Components blocking the N protein interacting regions may represent a novel class of molecules suitable for a therapeutic intervention in Chandipura-mediated disease.

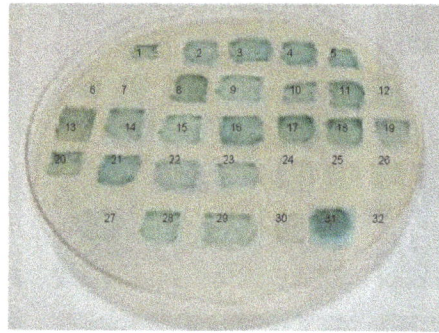

Sector	Bait	Prey	Sector	Bait	Prey	Sector	Bait	Prey	Sector	Bait	Prey
1	N1	N	10	N3	P	19	M	N2	28	Empty	N1
2	N1	P	11	N3	M	20	G	N2	29	Empty	N2
3	N1	M	12	N3	G	21	N	N3	30	Empty	N3
4	N1	G	13	N	N1	22	P	N3	31	p53	SV40 T antigen
5	N2	N	14	P	N1	23	M	N3	32	Lamin	SV40 T antigen
6	N2	P	15	M	N1	24	G	N3			
7	N2	M	16	G	N1	25	N1	Empty			
8	N2	G	17	N	N2	26	N2	Empty			
9	N3	N	18	P	N2	27	N3	Empty			

FIGURE 5: X-Alpha galactosidase assay for interaction confirmation. All possible interacting pairs of CHPV nucleocapsid protein with other viral proteins with controls were plated in an array format on plate containing X-α-gal (SD/-Trp/-Leu/-His/α-gal). Interactions were considered in both directions, that is, as BD and AD fusions. The results obtained were consistent in both directions except for BD-P which was autoactivating (sectors; 14, 18, and 22). Production of blue color reconfirmed the positive interactions observed in Y2H analysis while absence of color and growth is an indication of a noninteracting protein pair.

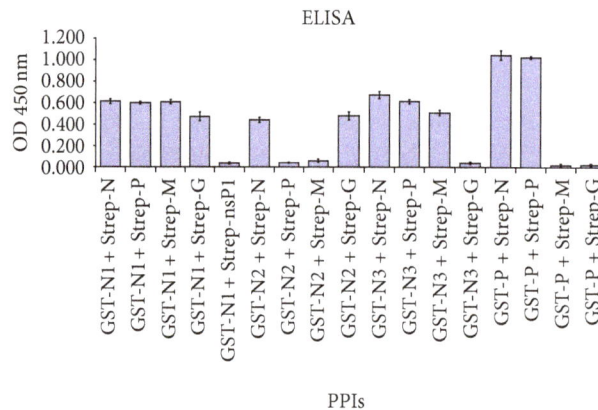

FIGURE 6: Validation of N protein fragment interactions with CHPV full-length proteins by ELISA. ELISA revealing the interaction between N1, N2, and N3 and full-length CHPV N, P, M, and G proteins. Only Strep tag fusion proteins (Strep-N, Strep-P, Strep-M, Strep-G, and Strep-nsP1) and only GST tag fusion proteins (GST-N1, GST-N2, GST-N3, and GST-P) were considered as background controls. A nonrelevant Strep tag fusion protein (Strep-nsP1) of Chikungunya virus was taken to test the specificity of N1 interactions. The data shown is mean of three independent experiments (mean ± SD). The known interaction pairs (PN and PP) and noninteractors (PM and PG) of CHPV were included as experimental controls. The absorbance at 450 nm is plotted on the y axis (10 division = 0.2 OD) and the test protein pairs are considered on the x axis.

Acknowledgments

The authors thank Professor Dhrubajyoti Chattopadhyay, Dr. B. C. Guha Centre for Genetic Engineering, Kolkata, India, for the kind gift of CHPV gene clones and his critical suggestions in this paper. This work was funded by research grant from Department of Science and Technology, Government of India (no. SR/SO/HS/0046/2008).

References

[1] B. L. Rao, A. Basu, N. S. Wairagkar et al., "A large outbreak of acute encephalitis with high fatality rate in children in Andhra Pradesh, India, in 2003, associated with Chandipura virus," *The Lancet*, vol. 364, no. 9437, pp. 869–874, 2004.

[2] M. S. Chadha, V. A. Arankalle, R. S. Jadi et al., "An outbreak of Chandipura virus encephalitis in the eastern districts of

Gujarat State, India," *American Journal of Tropical Medicine and Hygiene*, vol. 73, no. 3, pp. 566–570, 2005.

[3] B. V. Tandale, S. S. Tikute, V. A. Arankalle et al., "Chandipura virus: a major cause of acute encephalitis in children in North Telangana, Andhra Pradesh, India," *Journal of Medical Virology*, vol. 80, no. 1, pp. 118–124, 2008.

[4] A. K. Banerjee, "Transcription and replication of rhabdoviruses," *Microbiological Reviews*, vol. 51, no. 1, pp. 66–87, 1987.

[5] S. Basak, A. Mondal, S. Polley, S. Mukhopadhyay, and D. Chattopadhyay, "Reviewing chandipura: a vesiculovirus in human epidemics," *Bioscience Reports*, vol. 27, no. 4-5, pp. 275–298, 2007.

[6] L. E. Iverson and J. K. Rose, "Localized attenuation and discontinuous synthesis during vesicular stomatitis virus transcription," *Cell*, vol. 23, no. 2, pp. 477–484, 1981.

[7] A. Majumder, S. Basak, T. Raha, S. P. Chowdhury, D. Chattopadhyay, and S. Roy, "Effect of osmolytes and chaperone-like action of P-protein on folding of nucleocapsid protein of chandipura virus," *The Journal of Biological Chemistry*, vol. 276, no. 33, pp. 30948–30955, 2001.

[8] M. Chen, T. Ogino, and A. K. Banerjee, "Interaction of vesicular stomatitis virus P and N proteins: Identification of two overlapping domains at the N terminus of P that are involved in N 0-P complex formation and encapsidation of viral genome RNA," *Journal of Virology*, vol. 81, no. 24, pp. 13478–13485, 2007.

[9] A. K. Gupta and A. K. Banerjee, "Expression and purification of vesicular stomatitis virus N-P complex from Escherichia coli: role in genome RNA transcription and replication in vitro," *Journal of Virology*, vol. 71, no. 6, pp. 4264–4271, 1997.

[10] T. J. Green and M. Luo, "Structure of the vesicular stomatitis virus nucleocapsid in complex with the nucleocapsid-binding domain of the small polymerase cofactor, P," *Proceedings of the National Academy of Sciences of the United States of America*, vol. 106, no. 28, pp. 11713–11718, 2009.

[11] C. E. Mire, D. Dube, S. E. Delos, J. M. White, and M. A. Whitt, "Glycoprotein-dependent acidification of vesicular stomatitis virus enhances release of matrix protein," *Journal of Virology*, vol. 83, no. 23, pp. 12139–12150, 2009.

[12] A. Mondal, R. Bhattacharya, T. Ganguly et al., "Elucidation of functional domains of Chandipura virus Nucleocapsid protein involved in oligomerization and RNA binding: implication in viral genome encapsidation," *Virology*, vol. 407, no. 1, pp. 33–42, 2010.

[13] A. Mondal, A. Roy, S. Sarkar, J. Mukherjee, T. Ganguly, and D. Chattopadhyay, "Interaction of chandipura virus n and p proteins: Identification of two mutually exclusive domains of n involved in interaction with p," *PLoS ONE*, vol. 7, no. 4, Article ID e34623, 2012.

[14] K. Kumar, J. Rana, R. Sreejith et al., "Intraviral protein interactions of Chandipura Virus," *Archives of Virology*, vol. 157, no. 10, pp. 1949–1957, 2012.

[15] Y. Zhang, "I-TASSER server for protein 3D structure prediction," *BMC Bioinformatics*, vol. 9, article 40, 2008.

[16] A. Roy, A. Kucukural, and Y. Zhang, "I-TASSER: a unified platform for automated protein structure and function prediction," *Nature protocols*, vol. 5, no. 4, pp. 725–738, 2010.

[17] D. Xu and Y. Zhang, "Improving the physical realism and structural accuracy of protein models by a two-step atomic-level energy minimization," *Biophysical Journal*, vol. 101, no. 10, pp. 2525–2534, 2011.

[18] D. Xu and Y. Zhang, "Ab initio protein structure assembly using continuous structure fragments and optimized knowledge-based force field," *Proteins*, vol. 80, pp. 1715–1735, 2012.

[19] F. Ferrè and P. Clote, "DiANNA: a web server for disulfide connectivity prediction," *Nucleic Acids Research*, vol. 33, no. 2, pp. W230–W232, 2005.

[20] R. Potestio, F. Pontiggia, and C. Micheletti, "Coarse-grained description of protein internal dynamics: an optimal strategy for decomposing proteins in rigid subunits," *Biophysical Journal*, vol. 96, no. 12, pp. 4993–5002, 2009.

[21] R. Chen, L. Li, and Z. Weng, "ZDOCK: an initial-stage protein-docking algorithm," *Proteins*, vol. 52, no. 1, pp. 80–87, 2003.

[22] R. Chen and Z. Weng, "A novel shape complementarity scoring function for protein-protein docking," *Proteins*, vol. 51, no. 3, pp. 397–408, 2003.

[23] L. Li, R. Chen, and Z. Weng, "RDOCK: refinement of Rigid-body Protein Docking Predictions," *Proteins*, vol. 53, no. 3, pp. 693–707, 2003.

[24] C. Zhang, G. Vasmatzis, J. L. Cornette, and C. DeLisi, "Determination of atomic desolvation energies from the structures of crystallized proteins," *Journal of Molecular Biology*, vol. 267, no. 3, pp. 707–726, 1997.

[25] R. Sreejith, J. Rana, N. Dudha et al., "Mapping of interactions among Chikungunya virus nonstructural proteins," *Virus Research*, vol. 169, no. 1, pp. 231–236, 2012.

[26] R. Sreejith, S. Gulati, and S. Gupta, "Interfacial Interactions involved in the biological assembly of Chandipura virus nucleocapsid protein," *Virus Genes*, vol. 46, no. 3, pp. 535–537, 2013.

[27] Z. Itzhaki, "Domain-domain interactions underlying herpesvirus-human protein-protein interaction networks," *PLoS ONE*, vol. 6, no. 7, Article ID e21724, 2011.

[28] K. R. Hurst, C. A. Koetzner, and P. S. Masters, "Identification of in vivo-interacting domains of the murine coronavirus nucleocapsid protein," *Journal of Virology*, vol. 83, no. 14, pp. 7221–7234, 2009.

[29] M. Mavrakis, S. Méhouas, E. Réal et al., "Rabies virus chaperone: identification of the phosphoprotein peptide that keeps nucleoprotein soluble and free from non-specific RNA," *Virology*, vol. 349, no. 2, pp. 422–429, 2006.

[30] X. Zhang, T. J. Green, J. Tsao, S. Qiu, and M. Luo, "Role of intermolecular interactions of vesicular stomatitis virus nucleoprotein in RNA encapsidation," *Journal of Virology*, vol. 82, no. 2, pp. 674–682, 2008.

[31] J. Curran, J.-B. Marq, and D. Kolakofsky, "An N-terminal domain of the Sendai paramyxovirus P protein acts as a chaperone for the NP protein during the nascent chain assembly step of genome replication," *Journal of Virology*, vol. 69, no. 2, pp. 849–855, 1995.

14

Structural Differences Observed in Arboviruses of the Alphavirus and Flavivirus Genera

Raquel Hernandez,[1] Dennis T. Brown,[1] and Angel Paredes[2]

[1] Department of Molecular and Structural Biochemistry, North Carolina State University, Raleigh, NC 27695, USA
[2] U.S. FDA/National Center for Toxicological Research, Department of Health and Human Services, Jefferson, AR 72079, USA

Correspondence should be addressed to Raquel Hernandez; raquel_hernandez@ncsu.edu

Academic Editor: Trudy Morrison

Arthropod borne viruses have developed a complex life cycle adapted to alternate between insect and vertebrate hosts. These arthropod-borne viruses belong mainly to the families Togaviridae, Flaviviridae, and Bunyaviridae. This group of viruses contains many pathogens that cause febrile, hemorrhagic, and encephalitic disease or arthritic symptoms which can be persistent. It has been appreciated for many years that these viruses were evolutionarily adapted to function in the highly divergent cellular environments of both insect and mammalian phyla. These viruses are hybrid in nature, containing viral-encoded RNA and proteins which are glycosylated by the host and encapsulate viral nucleocapsids in the context of a host-derived membrane. From a structural perspective, these virus particles are macromolecular machines adapted in design to assemble into a packaging and delivery system for the virus genome and, only when associated with the conditions appropriate for a productive infection, to disassemble and deliver the RNA cargo. It was initially assumed that the structures of the virus from both hosts were equivalent. New evidence that alphaviruses and flaviviruses can exist in more than one conformation postenvelopment will be discussed in this review. The data are limited but should refocus the field of structural biology on the metastable nature of these viruses.

1. Background

1.1. Arbovirus Evolution. The arboviruses are not a taxonomic classification, but rather a grouping based on viral transmission through an insect vector to infection of a vertebrate host. The arboviruses contain members of the Togaviridae, Flaviviridae, Bunyaviridae, Rhabdoviridae, Reoviridae, and Orthomyxoviridae and are also represented by a single DNA virus, African swine fever virus family Asfarviridae of genus Asfivirus http://ictvonline.org/virusTaxonomy.asp. Evidence exists that arboviruses from the alphavirus lineage evolved from plant viruses [1, 2] which adapted to growth in insects [3]. Hematophagous insect viruses then acquired the ability to infect vertebrates, thus adapting from separate kingdoms (plant to insect) as well as phyla (insect to vertebrate) [4]. Members of the Bunyaviridae still maintain the plant to insect cycle [5–7] as well as the insect only cycle [8–10]. Arbovirus members of the flaviviruses are believed to have emerged about 1000 years ago in a nonhuman primate to mosquito cycle [11, 12] from predecessors that date at least 85,000 years [8]. It has been suggested that each of the 4 dengue serotypes (DEN1-4) adapted to humans independently only a few hundred years ago [13]. It is believed that this capability to diversify so broadly must have arisen from the inherent error-prone nature of the RNA-dependent RNA polymerases [14] while also limiting the evolution of arboviruses to certain families within the RNA virus class that are highly error-prone [14–16]. It is thought that the ability of viruses from each of these families to use or infect vertebrate hosts arose independently [17]. For these viruses to be able to cycle between insect and vertebrate hosts, their genomes must be compatible to hosts of two divergent phyla. This has been achieved by the evolutionary selection of virus that represents a consensus sequence able to function in both hosts. Thus, the arboviruses represent genomes selected by multiple mechanisms of adaptation and are exposed to repeated selection.

For the families Togaviridae, Flaviviridae, and Bunyaviridae, which comprise the bulk of arboviruses, the structure of the glycoprotein E1, E, and possibly Gc, respectively, appear to have arisen from an ancient predecessor [3]. While the sequences of the E1 cognate glycoproteins have diverged in the Togaviridae, Flaviviridae, and Bunyaviridae, the function and structures of these viruses have been retained [18]. Evolution of the protein structure has been constrained by adapting to both arthropod and vertebrate hosts. This difference in the rates of genomic divergence has been seen in a nonarbovirus member of the Togavirus family, Rubivirus [19] in which the known structure of E1 appears to have diverged relative to the arbovirus members of this family [20]. This observation suggests that the consensus sequence of the arboviral genome is maintained by eliminating genetic drift, which impacts fitness in each host. In other words, the virus sequence evolves more slowly when divergent hosts are continuously selecting for virus fitness. Collectively, the available information suggests that the mosquito-borne viruses acquired the form we now see from the arthropod vectors and did so concurrent with becoming hematophagous, presumably to optimize egg maturation [21].

1.2. Arbovirus Structure. Of the seven families of arboviruses, three (Togaviridae, Flaviviridae, and Bunyaviridae) are icosahedral, membrane-containing plus-stranded RNA viruses. It is interesting that though the respective glycoproteins E1, E, and Gc of the alpha, flavi, and bunyaviruses encode the same basic protein fold, these proteins assemble into different icosahedral structures. Alphaviruses are $T = 4$, (Figure 1) [22], flaviviruses are $T = 3$ (Figure 3) [23], and bunyaviruses (*Phlebovirus*) are $T = 12$ [24, 25]. This capability of viruses to assemble an inherently similar global fold into their different structures is dependent on the specific functions of their structural proteins. Rhabdoviruses are enveloped but assume a rod or bullet shaped structure that demonstrates some helical symmetry [26]. Reoviruses ($T = 13$ l, laevorotatory, turning toward the left) are icosahedral but do not contain a membrane [27]. Orthomyxoviruses are enveloped viruses that display pleomorphic structures ranging from spherical such as influenza A to filamentous as in influenza C [28]. In those influenza structures assuming a spherical shape, that shape is not due to icosahedral symmetry. In addition, orthomyxoviruses have more lipid-membrane relative to their envelopes than do enveloped icosahedral viruses resulting in the pleomorphism displayed [29]. The Asfarviridae viruses are large double-stranded DNA viruses, spherical to pleomorphic, and 175–215 nm in diameter and exhibit icosahedral symmetry ($T = 189$–217). These viruses are composed of two icosahedral protein shells that contain an intervening lipid structure [30]. These structural differences illustrate that there is no common virus structure adopted by the arboviruses which by itself would explain how these viruses infect both arthropods and vertebrates. There are two systems of virus classification currently in use. While morphology is a useful basis for virus identification and classification, at present, two classification systems exist. The hierarchical virus classification system, the International committee on taxonomy of viruses (ICTV) [31], and the Baltimore classification system. The Baltimore classification system is based on nucleic acid type which places viruses

into seven groups in hierarchical classification [32]. The ICTV uses evolutionary relationships as the classification scheme [33]. ICTV classification places viruses by phenotype; morphology, protein composition, and genotype; nucleic acid type, sequence, mode of replication, host organisms infected, and the type of disease they cause. Capsid structure has also been utilized to organize virus groups based on the hypothesis that only a limited number of protein folds are available to self-assemble a nucleocapsid [24, 34]. This review will be limited to the group arboviral members of the alphaviruses and flaviviruses for which considerable structural information exists. Bunyaviruses are less well studied and are proposed to be functionally analogous to flaviviruses by computational, proteomic, and indirect biochemical analysis [35]. Studies on whole virus particles will be the focus of this discussion.

1.3. Alphaviruses and Flavivirus Virion and Genome Structure. The alphaviruses are a genus in the family Togaviridae. They are icosahedral viruses of $T = 4$ geometry and contain a +polarity single-stranded RNA genome. The genome is organized with the 4 nonstructural genes found at the $5'$ end of the capped RNA genome followed by the structural proteins capsid (C), PE2, 6K, and E1 (Figure 2 [Virus DB: VBRC genome browser accession VG0000908]). The virus is composed of the structural proteins C, E2, and E1 that are synthesized from subgenomic RNA made from an internal promoter [36]. This RNA is then translated as a polyprotein and processed by viral and host enzymes during maturation and comprises the virion. The glycoproteins are assembled within the ER. The $5'$ capsid protein is autoproteolytically processed from the proprotein and organizes the genomic RNA into a nucleocapsid. During virus maturation, PE2 is converted to E3 and E2 by furin [37]. In most cases E3 does not remain associated with the virus [38]. E1 and E2 form trimers of heterodimers that envelope the nucleocapsid which assembles independently [39]. By regulating RNA and protein synthesis in a temporal manner, the Sindbis virus is able to quickly replicate to as many as 10^6 particles/cell [40]. Alphaviruses bud from the cell surface in mammalian cells but are assembled in vacuoles in mosquito cells and mature via the exocytic pathway [41]. These viruses are hybrid in nature acquiring lipids, carbohydrates, and other modifications from the host cell while their proteins are virus encoded. The final assembled structure is that of two nested spheres separated by an intervening membrane bilayer held together by protein associations between the E2 protein endodomain and the capsid protein.

Flaviviruses are in the family Flaviviridae and are also membrane-containing viruses, but they assemble into pseudo $T = 3$ icosahedra. The genome is organized with the structural proteins at the $5'$ end of the +strand RNA molecule and is translated into a single polypeptide which is processed during maturation by viral and host-encoded enzymes into multifunctional proteins. The three structural and seven nonstructural proteins are cleaved by a series of virus and host-encoded proteases. The structural proteins starting at the $5'$ end of the monocistronic genome include C, preM, and E (Figure 4 [Virus DB: VBRC genome browser]). As with the alphaviruses, the flavivirus polyproteins are inserted into the

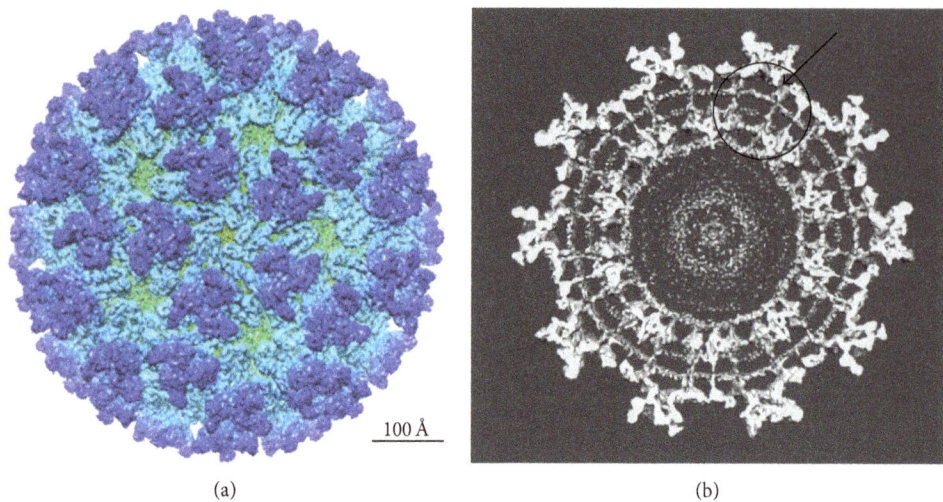

(a)   (b)

FIGURE 1: Shown in (a) is the cryo-EM reconstruction of the TC-83 strain of Venezuelan equine encephalitis virus (VEE) viewed down the strict 5-fold axis. Resolution is 4.4 Å. The trimers consist primarily of E2 with the smoother "skirt" comprised of E1. In (b) is shown a slice down the 2-fold axis. Note the transmembrane domains (arrow) and the extensive organization of the nucleocapsid (circle) compared to that seen in the flaviviruses (Figure 3(b)). With permission from Zhang et al. [22]. For full resolution images see EM DATA BANK (EMDB)/5275.

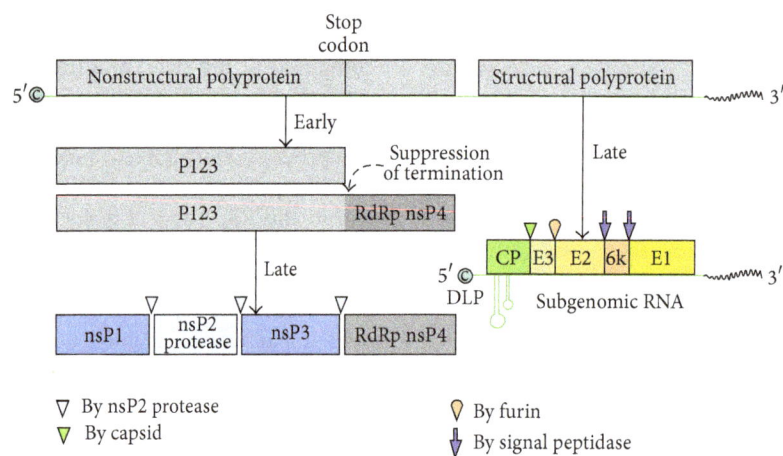

FIGURE 2: Alphavirus genome organization. Alphaviruses are organized with the (nonstructural) ns proteins at the 5′ end of the genome The ns proteins are translated from the genomic RNA while the structural proteins are translated from a subgenomic RNA that includes the 3′ end of the genome. The RNA polymerase nsP4 is a read-through protein present early during infection. After replication is established, protein production switches to the structural proteins C, E3, E2, 6K, and E1. The C protein autoproteolytically cleaves itself from the remaining polyprotein. Only E1, E2, and C are found in the mature virion. In some cases, E3 may also be associated with the virus particles [Virus DB: VBRC genome browser accession VG0000908].

ER of the host cell. However, the flaviviruses are also assembled in the (endoplasmic reticulum) ER with concurrent assembly of the nucleocapsid RNA/C structure [42]. preM and E form dimers during protein maturation resulting in further processing of preM to M as the glycoproteins mature [43]. Little is known about the process of encapsidation into the mature virion, but incorporation of the nucleocapsid of these viruses is thought to involve nonstructural proteins [42]. Flavivirus particles differ from alphavirus particles in that the nucleocapsid association is more peripheral to the intervening membrane, and no organized structure is observed [42, 44]. The particle structure is also nested with

an intervening membrane, but no strong contacts with the M and E membrane domains are formed [45, 46].

2. Review

2.1. The Plasticity of Alphavirus Variants as Seen by Ultrastructure. The alphavirus virion assembles into a stable structure that shields the genome from the adverse effects of the surrounding environment. Virus particles are assembled into a high energy state in which infectious particles are poised to deliver their genomic cargo after the appropriate stimuli are encountered through specific interaction(s) with the host

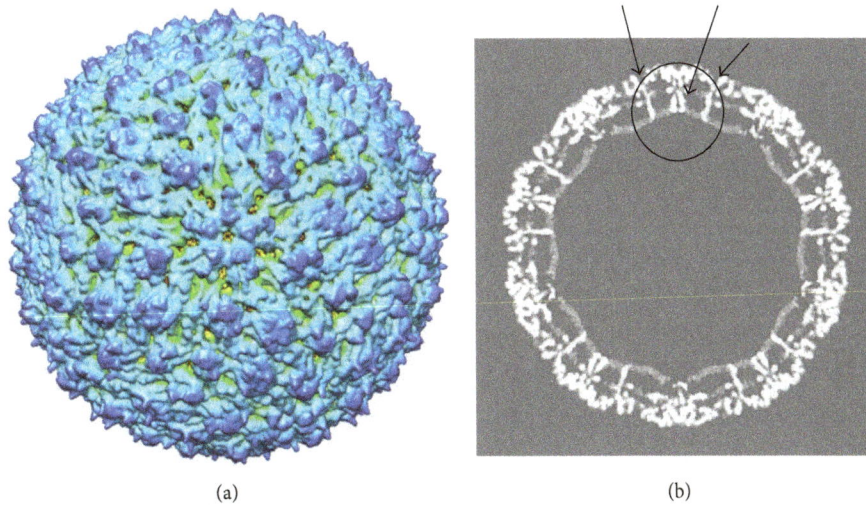

(a)   (b)

FIGURE 3: (a) Dengue 2 New Guinea C cryoEM reconstruction rendered at 3.5 Å, EMD 5520. The view is down the icosahedral 5-fold axis. Although this is termed a mature virus because of the smooth appearance, preM may still be associated with the virus. The surface of the structure has few distinctive features; however, it has been proposed that, upon maturation, the virus E protein undergoes major conformational changes as infection is initiated. (b) This image is a slice of the virus down the 2-fold axis. M (black arrow) and E (red arrow), transmembrane domains are seen below the outer surface. Distortion of the lipid membrane is seen where the transmembrane domains penetrate and no organized capsid structure is detected (circled). Compare to the alphavirus slice in Figure 1 where the nucleocapsid structure is clearly delineated. Reprinted with permission from Zhang et al. [23]. See EM DATA BANK (EMDB)/5520 for full resolution images.

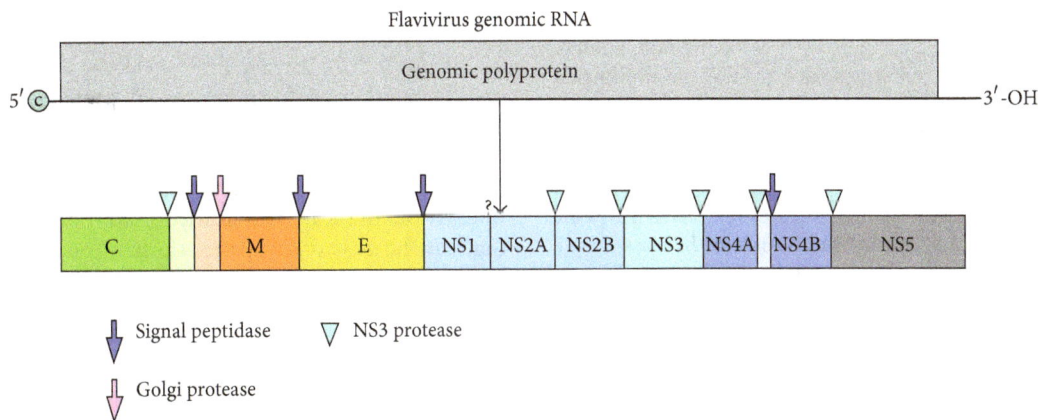

FIGURE 4: The $5'$ terminus of the genome is capped and the polyprotein is expressed as a single polypeptide. The $3'$ terminus is not polyadenylated but rather forms a loop structure. The locations of the nsP and structural proteins are switched with respect to the $5'$ end of the RNA compared with the alphaviruses. It has been proposed that this switch occurred via recombination in an ancient precursor, thus retaining the structural glycoprotein folds. Capsid protein, preM, and E proteins are proteolytically processed as indicated by the enzymes shown by the arrows and arrowheads [Virus DB: VBRC genome browser].

cell. Thus, the infectious virion is a metastable intermediate that assumes sequentially different conformations depending on the pH, temperature, and host environments it encounters (discussed below). The plasticity of the virion structure underscores the flexibility of the viral proteins. This flexibility is most likely not due to global changes in the structural proteins but rather local changes in the metastable domains of the proteins [48]. For structural stability, the virion as a whole must contain protein domains with structural stability imparted by the protein sequence itself or via stabilizing protein-protein interactions. We describe two examples of the alphavirus Sindbis laboratory constructed mutants

producing structural variants that illustrate the ability of point mutations in the structural proteins to acquire novel architecture.

The first example is that of the Sindbis virus capsid protein mutant Y180S/E183G. Sindbis virus is a macromolecular structure composed of RNA, protein, and lipid. The inner protein shell, the nucleocapsid, is bound to the outer protein shell via interactions to the E2 endodomain. Genetic and structural evidence suggest that the nucleocapsid interacts with the E2 endodomain by aromatic amino acid interactions between Y420 in the E2 protein and Y180 and W 247 in the capsid protein [49, 50]. Mutations in the capsid protein,

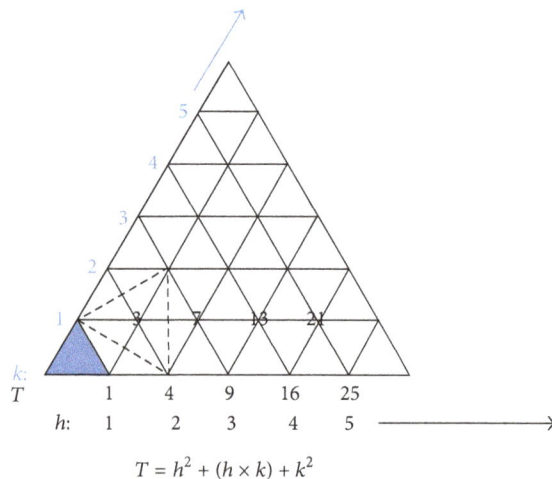

$$T = h^2 + (h \times k) + k^2$$

FIGURE 5: A family of icosahedra. Shown at the bottom of the large triangle, in blue, are the members of the $P = 1$ icosahedra. The triangulation number represents the number of non- or quasiequivalent structural morphological units per asymmetric unit of the icosahedral face. $T = 1$ is the only structure with a single equivalent morphological unit per icosahedral face, shown in blue. Neighboring 5-fold vertices of the 3-fold triangular face are connected to one another by three 2-fold axes giving the structure an overall 5-fold, 3-fold, and 2-fold symmetry. The T number, also determined by the formula $T = h^2 + (h \times k) + k^2$, indicates the relative position of each 5 fold vertex as a function of the number of quasiequivalent morphological units per icosahedral face. Sindbis, with four quasiequivalent conformations of the E1E2 heterodimer, is a $T = 4$ structure. In the capsid mutant Y180S/E183G, the flexibility of the capsid protein is altered allowing the mutant protein to adopt a wider range of quasiequivalent conformations that increases the triangular number of the structure. Superimposed on the $P = 1$ lattice is a class 3 triangular face outlined in red. Flaviviruses are arranged in this $T = 3$ symmetry and belong to the class $P = 3$ icosahedra shown in black on the $k = 1$ axis.

Y180S/E183G, result in the assembly of the virus structural proteins into icosahedra of increasing triangulation numbers. The triangulation numbers calculated for these morphological variants, follow the sequence $T = 4, 9, 16, 25$, and 36 [51–53]. All fall into the class $P = 1$ of icosadeltahedra. It has been suggested that the $T = 4$ structure of the nucleocapsid organizes the outer glycoprotein layer [47, 54]. These observations support models suggesting that the geometry of the preformed Sindbis nucleocapsid organizes the assembly of the virus membrane proteins into a structure of identical conformation. It has been proposed that these two mutations in the capsid protein endow the protein with the flexibility to increase the number of capsid proteins incorporated into the nucleocapsid before formation of the next five-fold vertex. This is seen more readily in the smaller icosahedra, examples of which are shown in Figure 5. The structures of increasing size and triangulation number also form in mosquito C7-10 cells suggesting that the capsid protein mutation and not the host cell or the temperature of assembly is involved in the expression of this phenotype (unpublished data). This mutant demonstrates the capability of a capsid protein fold to "evolve" a new triangulation number and structure. These

larger capsid structures can package large RNA molecules [55]. While the structures formed by this mutant are not stable, they are infectious and it is possible that subsequent mutations could stabilize any of the resulting new icosahedra (Figure 6).

A second example of a Sindbis virus structural protein mutant able to organize into nonnative structures is that of the furin mutants in the glycoproteins E1 and E2 shown in Figure 7 [53]. Furin protease recognition sites, Arg-X-Arg/Lys-Arg, were installed into the E1 sequence at amino acid positions 392 and E2-341. The numbering scheme refers to the position of the first mutated Arg to create the furin site. The furin double mutant was found to produce virus particles of normal infectivity and structure when grown in mammalian cells (particles composed of the requisite number of wild type proteins, see [37]) (Figure 7(A)) as well as virions that developed long tubular appendages of varying lengths (virions incorporating aberrant proteins) [53]. Virus production from insect cells was of insufficient titer to allow microscopic examination. The tubular structures are 73 nm in diameter, consistent with the size of the wild type virion. Scanning electron microscopy of double mutant infected cells revealed that maturing particles were initiated with spherical structures that probably contained a capsid (Figure 7(B) arrow). The large amorphous structures as seen in Figure 7(K) may contain multiple capsids with associated membrane. It was concluded that the tubular structures initiated by organizing around the icosahedral nucleocapsid and depending on the number of furin processed proteins incorporated, terminated the envelopment process by normal membrane fission (budding) or by repeated incorporation of hexagonal protein arrays. In the absence of the quasi-equivalent protein interactions of the icosahedral five-fold vertex, the only lattice that can form is a hexagonal sheet in one dimension or a hexagonal tube in 3 dimensions. The geometry of the tubular helices was determined by calculating the optical contrast reinforcement of the EM images of particles such as those seen in Figure 8. Thus, when densities are in phase, they will reinforce one another and a repeating distance can be calculated. This was seen at 16 and 20 nm, distances consistent with the spacing of the wild type virus 6-fold symmetry, shown in Figure 9, indicating that the geometry of the mutant particles is relevant biologically. For the furin mutant E1-392/E2-341, we see that a loss of function mutation in the structural glycoproteins E1 and E2 can result in the ability of the virion to maintain the protein-protein interactions that enable the formation of a helical array. Again, the flexibility of these viral proteins to produce variant structures underscores the plasticity of these proteins and illustrates the ability of viruses to explore possible new structures to enable efficient survival and facilitate evolutionary divergence.

These mutant virus structures help us to understand the metastable structure of alphaviruses by illustrating that assembly of the virion proceeds through structural intermediates. In the case of the C mutant, this protein can fold into native or the mutant conformations and assemble multiple forms. For this change to occur, the mutant protein must be able to assume an energetically stable intermediate. This is thought to happen during the folding of the protein and

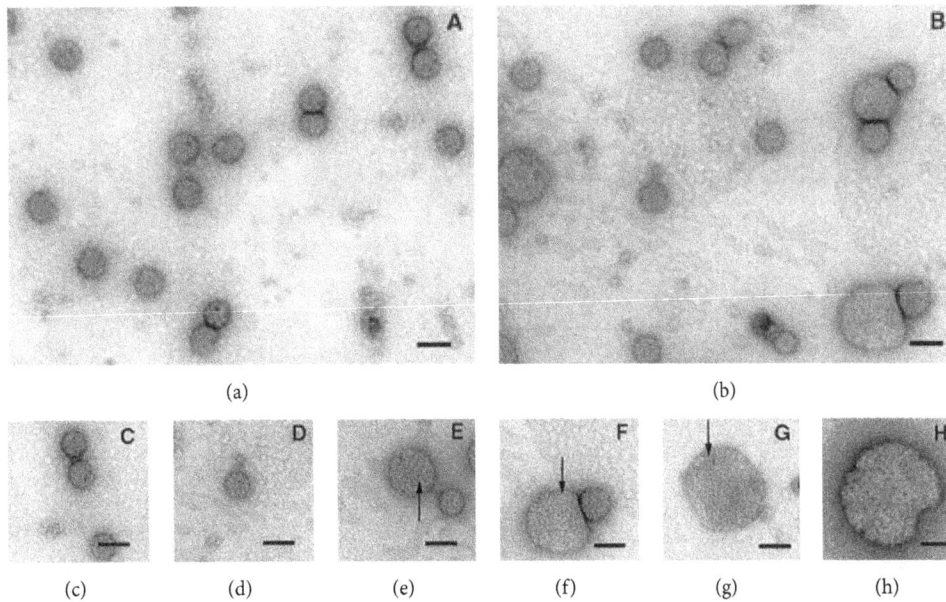

FIGURE 6: Electron micrographs of negatively stained virus particles recovered from the supernatant of BHK cells transfected with either Sindbis virus S420Y ((a) and (c)) or Y180S/E183G ((b), (d), (e), (f), (g), and (h)). (a) and (b): a wider view of the fields. (c)–(h): individual particles arranged to show examples of the various sizes found for the Y180S/E183G mutant ((d)–(h)) as compared to the control S420Y particle (c). Arrows point to morphological units. Bars, 100 nm. Ferreira et al. 2003 [47].

assembly, in this case, of the nucleocapsid. The E1-392/E2-341 mutant demonstrates that the metastability of the virus extends to both shells and can trap the glycoproteins into a conformation that only allows the formation of the hexagonal lattice.

2.2. Protein Structure.

Protein conformations have been reported for many of these virus structural proteins as single proteins and in protein aggregates (http://www.ncbi.nlm.nih.gov/). The metastable state of the virion, however, suggests caution in interpreting structures of the isolated structural proteins. Virus particles are macromolecular machines poised to deliver their genomic cargo once the appropriate stimuli are encountered. These particles must also shield the genome from the adverse effects of the surrounding environment. These requirements are accomplished by the virion assembling into intermediates which refold into a highly energy rich infectious conformation, which is a sufficiently stable structure [56]. Thus, the infectious virion is a metastable intermediate that can exist in more than one conformational state depending on pH, temperature, and the complex host biochemical environments (discussed below). With the correct stimulus, such as interaction with a receptor, the virus undergoes a conformational change that is propagated through the virion to initiate infection. Not all virions are infectious because the structural integrity of the particle is compromised during assembly or virus preparation. Thus, in a field of virus particles the number of noninfectious particles can outnumber infectious particles by as many as 1000 to 1. This large particle/pfu ratio can only complicate study of the native, infectious structure. Only stable or transiently stable conformations of these "metastable" proteins can be

crystalized or imaged by cryoEM as was done with flu virus [57–59]. The isolated flu structural proteins comprise limited conformations of very flexible proteins with multiple functions. While crystal structures for Sindbis and Chikungunya virus structural proteins have been reported, these may not be native because of the lack of the transmembrane domain and many more conformations of infectious intermediates may exist for these viruses. This is evident from both quasiequivalence and from the different conformations of E1 that have been identified in Sindbis virus [56, 60–66]. It has been reported that, during Sindbis virus infection, disulfide exchange may occur to enable the formation of the fusogenic conformation [67] and is important for the proper assembly of virions [56, 68]. The conformational changes induced by disulfide exchange during infection are not understood but are known to be required for infectivity of the virus [66]. Although there is no independent evidence to verify that the crystal structure of E1 is a native structure, there is independent evidence that different forms of E1 exist in the alphavirus structure when whole infectious virus is analyzed [63]. The argument that the crystal structures "fit well" into the cryoEM electron density maps gives much leeway to the interpretation of the structure. Because of quasiequivalence, a single high resolution X-ray crystal structure frequently cannot be fitted into a cryoEM electron density map of a high T-number virus without distorting the X-ray structure in order to make the fit. For Sindbis virus, a single alphavirus E1 structure is commonly used in virion structural reconstruction analysis, at least four other conformations of the E1 protein are known to exist as the protein is folding and unfolding [56, 60, 69]. Structural intermediates of disulfide bridged proteins that are detected

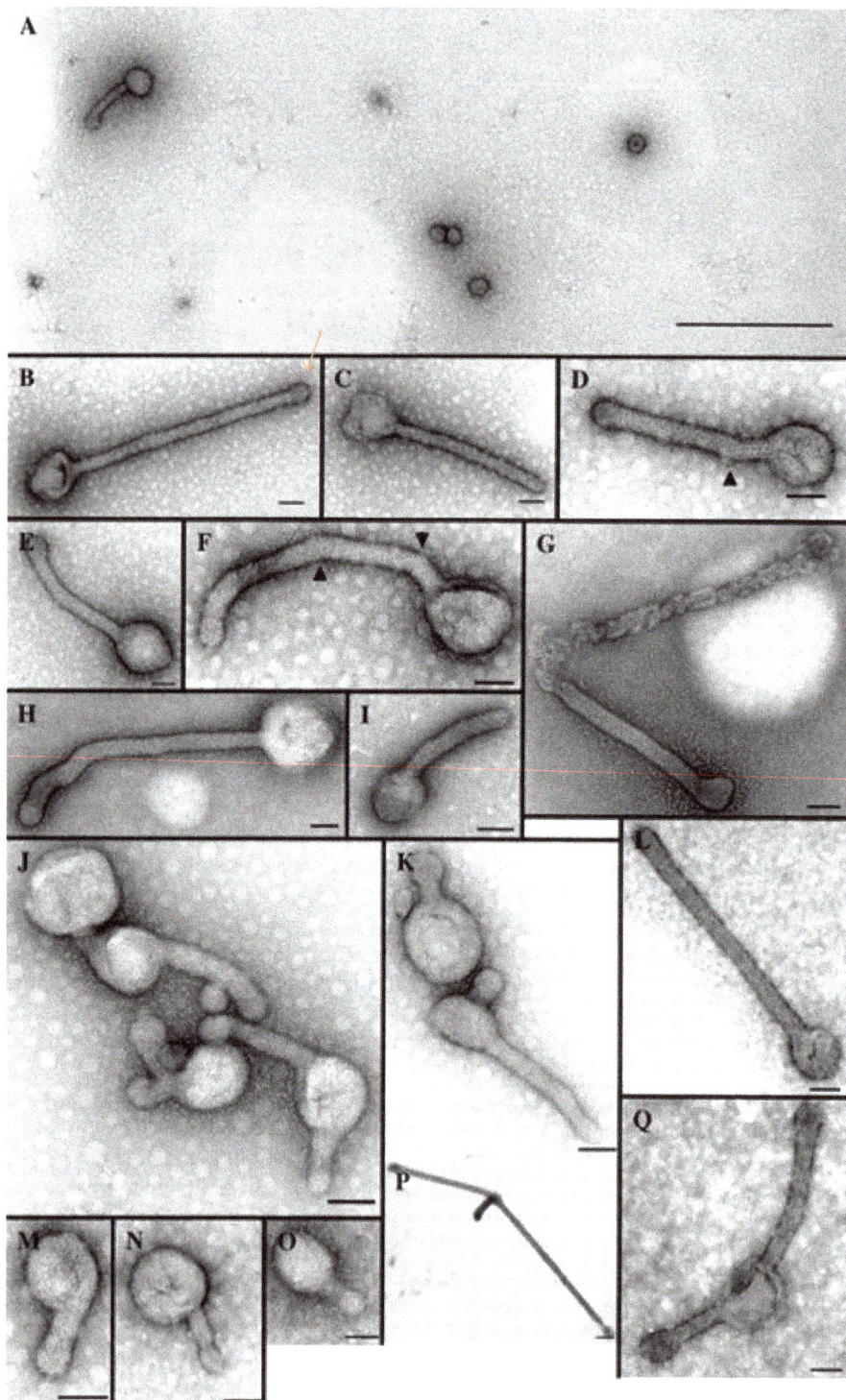

FIGURE 7: Electron micrographs of negatively stained furin sensitive double mutant E1-392/E2-341 containing virus recovered from the supernatant of transfected BHK-21 cells. (A) Low magnification of a typical sample of double mutant E1-392/E2-341. Bar, 1 μm. ((B)–(Q)). Selected particles demonstrating the variations seen in one preparation of E1-392/E2-341. Arrowheads show points of helical disruption and reinitiation. Bars, 100 nm. Kononchik et al. 2009 [53].

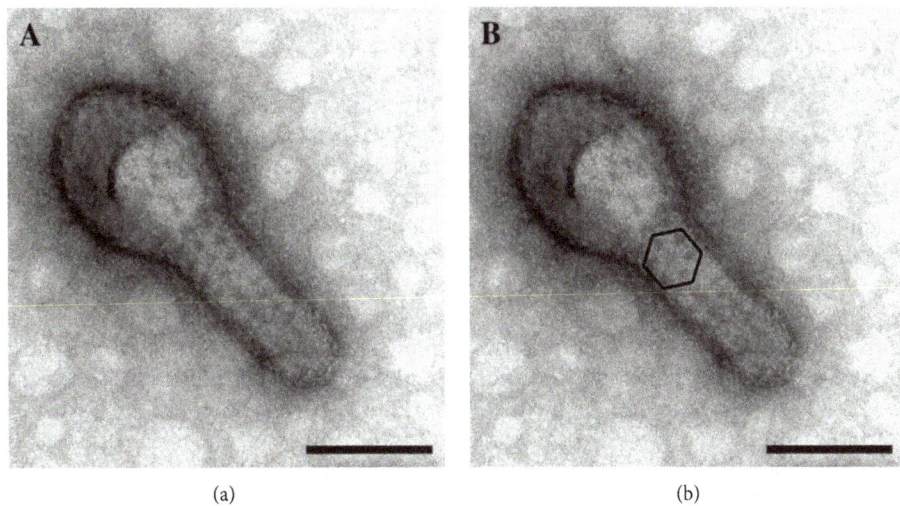

FIGURE 8: Electron micrograph of a negatively stained furin double mutant E1-392/E2-341. (a) A normal E1-392/E2-341 particle that shows surface detail. (b) The same particle (a) highlighting a hexagonal array clearly visible on the tubular structure. Bars, 100 nm. Kononchik et al. 2009 [53].

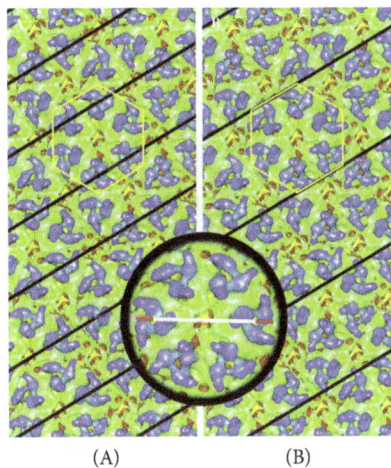

FIGURE 9: Helical reconstruction of the tubular section of the furin sensitive E1-392/E2-341 created by 6-fold rotational arrays (highlighted). Black lines illustrate the 16 nm repeat (A) and the 20 nm repeat (B) registers seen on the tubular section of the mutants. Illustrations are drawn to scale with the average width of the tubular structures. The insert illustrates the distances taken from a cryoelectron microscopy reconstruction of wild-type Sindbis virus measuring across a strict 2-fold axis of a hexagonal array: white is 16 nm and white with red is 20 nm. Kononchik et al. 2009 [53].

on nonreducing PAGE after extraction of the E1 protein from infectious virus also form multiple intermediates [56]. During alphavirus glycoprotein maturation, E1 and PE2 form integrated trimers of heterodimers. E2 is a molecular scaffold as well as the escort protein that delivers multimers to the cell surface [69]. Mass spectroscopy analysis of cysteines found in infectious virus was not the same as those reported for the alphavirus E1 crystal [66]. For these reasons it is critical to use infectious virus to assess structure because alphavirus E1 infectious conformations only exist in the intact, metastable

virus [66]. Stated succinctly, native intermediates of E1 cannot exist in solution because cysteines will reassort as the high energy form becomes stable. While whole virus analysis has proven difficult because the resolution of most whole virus reconstructions is not at atomic levels even when the resolution is increased by "fitting" a crystal new technology can address these problems.

The cryomicroscopes of today include new advancements such as a phase plate used to negate the contrast transfer function (CTF) of the microscope so that images do not require CTF correction. The phase plate technology is still under development because those that work well contaminate quickly when used [70–73]. Additionally, Charged-couple device (CCD) cameras are being replaced with direct (electron) detectors. These detectors capture hundreds of images per second and once captured, the images are averaged together to reduce noise and to correct for specimen drift to boost both resolution in the image and signal. As a consequence, data are now being collected that are used to determine structures of biological samples below 4 angstroms resolution. This is close enough to atomic resolution that the carbon backbone of amino acids, including the R-side chains, can be followed in the structure. Depending on size, this technology will eliminate the need for pseudocrystal structures for biological samples [74, 75].

As seen in Figure 3(b) [23] flaviviruses are more fragile and therefore display a larger particle to pfu ratio than alphaviruses [76, 77]. This is the result, in part, from the weaker association of the nucleocapsid with the structural proteins. Organization of the RNA within the C protein is unclear. How the nucleocapsid interacts with the glycoproteins preM and E during assembly is also vague [78, 79]. No nonstructural proteins are incorporated into the virion although nonstructural proteins have been implicated in encapsidation and budding of the virus from the ER [80, 81]. The preM protein functions as a chaperone during E folding [42, 82, 83]. During transport of the virion to the cell exterior, preM is cleaved by a furin-like protein resulting in virus particles with M protein in its mature form [43, 84]. The processing of preM to M by furin is not as efficient as that seen for the alphavirus PE2 protein processed to E2 and E3; thus, particles with preM are detected [85, 86]. It has been shown that the lack of preM results in poor protein folding and poor immunogenicity [87]. Because E is the primary a immunogen, this implies that the native conformation of the E protein can be compromised when expressed *in vitro* or outside the context of the virus. Soluble E protein X-ray structures have been solved for many of the flaviviruses [88–91]. The structure is divided into 3 domains, I, II, and III [92]. Domain I is the N terminal portion of the protein and is centrally located within the crystal structure. Domain II contains the fusion peptide and the dimerization domain, while domain III is an immunoglobulin like domain and is thought to contain the receptor binding site [93]. This 3D structure is similar to that of the alphaviruses with analogous functional domains [94]. Unlike the alphavirus E1 protein, the flavivirus TBE and dengue E protein crystalized into protein dimers [95]. E protein from WNV is crystalized as a monomer but is fit into cryoEM as a dimer [96]. It is widely held that flavivirus E protein is induced by low pH to reorganize into a fusogenic trimer which initiates infection from an acidified endosome. While much indirect evidence has been reported to support this model of infection for alpha and flaviviruses, by comparison to flu, [92, 97–99], a second model of direct penetration by the virus at the host cell surface, determined by direct observation, is largely dismissed in favor of the fusion model. Ideally, a working model of Togavirus penetration should address all experimental evidence. Two issues will be discussed, first the use of indirect evidence from structural models which describe a fusion pathway and second, direct observation by ultrastructural and biochemical analysis that provide empiric evidence that infection by these agents is direct penetration at the plasma membrane.

2.3. Electron Cryomicroscopy. As electron optics and electron cryomicroscopy in general have improved, it has become possible to take images of frozen hydrated viruses from electron microscopes and use them to reconstruct the three-dimensional structures of infectious virus to ever higher resolutions by cryoEM (reviewed in [100]). Electron cryomicroscopy or cryoEM has allowed the placement of the different structural components of the glycoprotein shells of arboviruses for which structural data is available. The alphavirus Semliki Forest virus was first imaged by Vogel et al. [101, 102]. In 1987, Fuller reported the first cryoEM structure of Sindbis virus to be a $T = 4$ icosahedron surrounding a $T = 3$ core [103]; however, the core was later shown to be $T = 4$ [104]. The first cryoEM image of Sindbis virus with sufficient resolution to image the trimeric spikes was produced in 1993 by Paredes et al. [104]. This reconstruction confirmed a structure that had previously been postulated genetically and biochemically, [105–107]. The most outwardly protruding structure in the cryoEM image was a 3-fold trimer with laterally associated proteins. It was not possible at that time to identify which proteins corresponded to what

part of the structure, although it was known that the virus particles were held together by an E1-E1 protein lattice [106, 108]. Several studies have since used cryoEM to study the structure of E2 in alphaviruses. In 1995, Cheng et al. reported their cryoEM reconstruction of Ross River virus, in which the authors concluded that the capsid protein bound as a monomer to the E1-E2 dimer in a 1:1 stoichiometry [109]. The alphavirus E2 density was also probed using Ross River virus and anti-Ross River E2 neutralizing FAbs that blocked attachment. In these cryoEM structures, the E2 FAb labeled E2 on the outermost density of the tip of the bilobed spike protein [110]. The position of E2 was later confirmed in a 1998 study by Paredes et al., when a mutant of Sindbis virus defective in processing the N-terminal E3 protein from the PE2 precursor was reconstructed using cryoEM. The reconstruction identified the additional E3 density at the tips of the trimeric spikes and identified the general locations of E2 in the spike region with E1 involved in the protein lattice at the base of the spike [111]. The general location of both proteins was reported in 2001 by deleting the carbohydrate modification sites from E1 and E2 singly and together in several nonglycosylated mutants. By comparing the cryoEM densities of the nonglycosylated mutants to wild type virus, it was possible to determine the relative positions of E1 and E2 on the virus surface [112].

In a 1990 study by Flynn et al., conformational changes in Sindbis virus E1 and E2 were observed as virus engaged the plasma membrane at neutral pH in cells that were not acidified [113]. These rearrangements were detected by (monoclonal antibody) MAb and corresponded to transitional epitopes. These epitopes could also be detected in a time and temperature dependent manner. Subsequent studies [114] showed that structural rearrangements seen in the previous study were closely mimicked by three artificial treatments of purified virions. Structural rearrangements of virus exposed to brief incubation at 51°C, treatment with 1–5 mM dithiothreitol, or incubation at pH 5.8 to 6.0 were probed using a panel of MAbs specific for Sindbis virus E1 and E2 glycoproteins. Infectivity was retained after all three treatments. These observations were interpreted to suggest that Sindbis virions are metastable and can exist in at least two infectious conformations. The authors concluded that these intermediate structures may represent different conformations of a complex pathway that leads to productive infection and was an early indication that infection proceeded through protein structural intermediates induced by virus-cell interactions in the absence of low pH at the cell surface. None of these structural intermediates can be inferred from a single rigid X-ray structure.

A high resolution alphavirus structure was recently reported by Zhang et al. and is shown in Figure 1 [22]. At 4.4 Å resolution, this structure of VEEV was determined by a combination of homology and *de novo* modeling. The final VEEV model was compared to a pseudomodel in which CHIKV E1 and E2 X-ray structures were fit separately into the VEEV model. The results indicate that E2 (residues 1–341) had a higher RMSD value (4.2) than E1 (residues 1–391) with an RMSD of 1.8 Å, representing the difference from identical molecules with an RMSD of 0.0 [115]. The low pH SINV

E1-E2 crystal structure was also fit into the VEEV cryo-EM structure (PDB ID: 3MUU, chain A) showing an RMSD of 2.4 for E2 and 2.9 for E1 ectodomain. The transmembrane domain and the E2 endodomain were modeled *de novo*. In addition, the structure of the capsid protein as determined by cryoEM reveals a predicted α helix at residues 115–124 that is missing in the capsid structure as revealed by X-ray crystallography. Thus, this cryoEM reconstruction has served to refine the structures of viral components by independently using improved methods to define the protein structures. Advances in single-particle cryo-EM have pushed the limit to near atomic resolution of ~3.3 Å [116–119]. Methods are improving, and soon it should be possible to build a virus model in the absence of structural artifacts without the need to dissociate virions. Subtomogram averaging from *in situ* cryo-EM has the potential of looking at virus proteins without the need for crystallization. Unlike X-ray crystallography, this method ensures that the viral proteins imaged in this manner are in their native conformations. While this method yields lower resolution images than single particle reconstructions, advancements in this technology will undoubtedly improve the resolution.

2.4. pH Studies and Viral Penetration. A low pH study of Sindbis virus was undertaken by Paredes et al. in 2004 which hypothesized that low pH triggers the same or similar conformational rearrangements as does contact of the virus with the cell receptor. Sindbis virus treated at low pH was investigated by cryoEM. SVHR is a laboratory strain selected for heat resistance (Sindbis virus heat resistant) [120] which also confers the ability of the virus to produce virus which is ~100% infectious. Infectious, BHK-grown SVHR virus of a particle to pfu ~1 at pH 7.4 was exposed to pH 5.3, returned to neutral pH, and prepared for microscopy. This study revealed that low pH treatment triggered a substantial rearrangement of both E1 and E2 spike proteins and there was a significant formation of nobs of E1 density protruding from each of the 5-fold axes (see arrow in Figure 10(b)). This is the only structure of an alphavirus at a pH which establishes the conditions required for membrane fusion after return to neutral pH. Returning to neutral pH did not restore the native structure and resulted in noninfectious virus. The observation that low pH inactivated alphaviruses had been made early on [121] and can be explained by the inability of the low pH form of the virus to reorganize to the native conformation.

That low pH inactivates Sindbis virus in solution is consistent with the observation that treated virus does not return to its infectious conformation [121–124]. For this reorganization of E1 to occur, the virus must be taken to the pH 5.3 threshold since pH changes above this do not affect infectivity of SVHR or establish conditions required for membrane fusion [122, 123]. In another whole virus study using SFV clone pSFV4, low pH structures by Haag et al. only exposed the virus to pH 5.9, not taking the virus through the requisite 5.3 pH for producing the low pH structures and subsequently resulted in very little change in the virus conformation [125]. This is because at the pH required for fusion, concentrated virus samples precipitate. The Paredes et al. study in 2004 proposed a new model

for membrane penetration of infectious alphaviruses. Given that the resolution of the low pH structure was 28 Å, it was deemed possible that the knob of density appearing at pH 5.3 may house a proteinaceous pore (Figure 10(b)). This protruding density is ~52 Å in width and ~60 Å in length and is generated from the surrounding "skirt" region at the five-fold axis. This is of sufficient size to house a pore from which RNA could be extruded, roughly 10 Å internal diameter [126]. This hypothesis was substantiated by electron micrographs of infectious Sindbis particles interacting with the host cell surface, creating a pore-like structure through which RNA was seen to extrude (Figure 11). Evidence was also shown that the virion may be interacting via the five-fold axis as suggested by the cryoEM structure. Taken in conjunction with emergence of the new structure of E1 at the five-fold axis and the direct visualization of pore formation at the host cell plasma membrane using EM, it was concluded that virus entry may proceed by an ancient pathway proposed for bacteriophage, direct cell penetration.

A recent study by Cao and Zhang reported an early-stage fusion intermediate of Sindbis virus using cryoEM to reconstruct the low pH intermediate [127]. The strain TE12, which fuses at pH 6.5, was used to conduct this research although this strain has a particle to pfu ratio of ~100. Sindbis virus was then treated to pH 6.4 and the virus was found to retain its $T = 4$ structure. This is surprising since this pH should have induced a global conformational change in both E1 and E2 as was shown previously with SVHR displaying ~100% infectivity [122]. This lack of reorganization of the TE12 strain could be a reflection of a large particle to pfu number or too high a pH to induce the conformational change. Neither virus titer nor particle to pfu ratios were reported for this study. TE12 was then mixed with liposomes at neutral or pH 6.4. At the lower pH, virus was seen to interact with liposomes or vice versa via bridge-like densities spanning the distance between the liposome and the virus. These structures are ~160 Å in length which is a greater span by 10 Å than the soluble low pH structure [127]. When the SVHR strain of Sindbis virus was treated at a pH of 5.2, E1 density at all 12, 5-fold vertices formed knobs of density ~60 Å in length, not the trimeric spikes seen with soluble E1. However, comparisons to the low pH SVHR structure were not made. The authors also report that no specific orientation is required for the virus to bind a target membrane; however, this lack of orientation is determined with liposomes in the absence of the virus receptor and without intact particles. This was not the case with the SVHR study which showed EM evidence of SVHR particles binding to cells at a 5 fold vertex and penetrating the cell at the surface in the absence of low pH [122]. Fusogenic properties of membrane containing virus is often studied using liposomes. Lipid fusion of alpha- and flaviviruses with liposomes is well documented and occurs with very specific admixtures of lipid [124, 128–130]. However, these liposomes are not representative of the composition or structure of the host cell membranes (discussed below). There is direct evidence that Sindbis virus can become noninfectious but retain its fusogenic ability [66, 131], thus separating the E1 fusion function from its infectivity. It is also well documented that alpha and flaviviruses can undergo low pH mediated

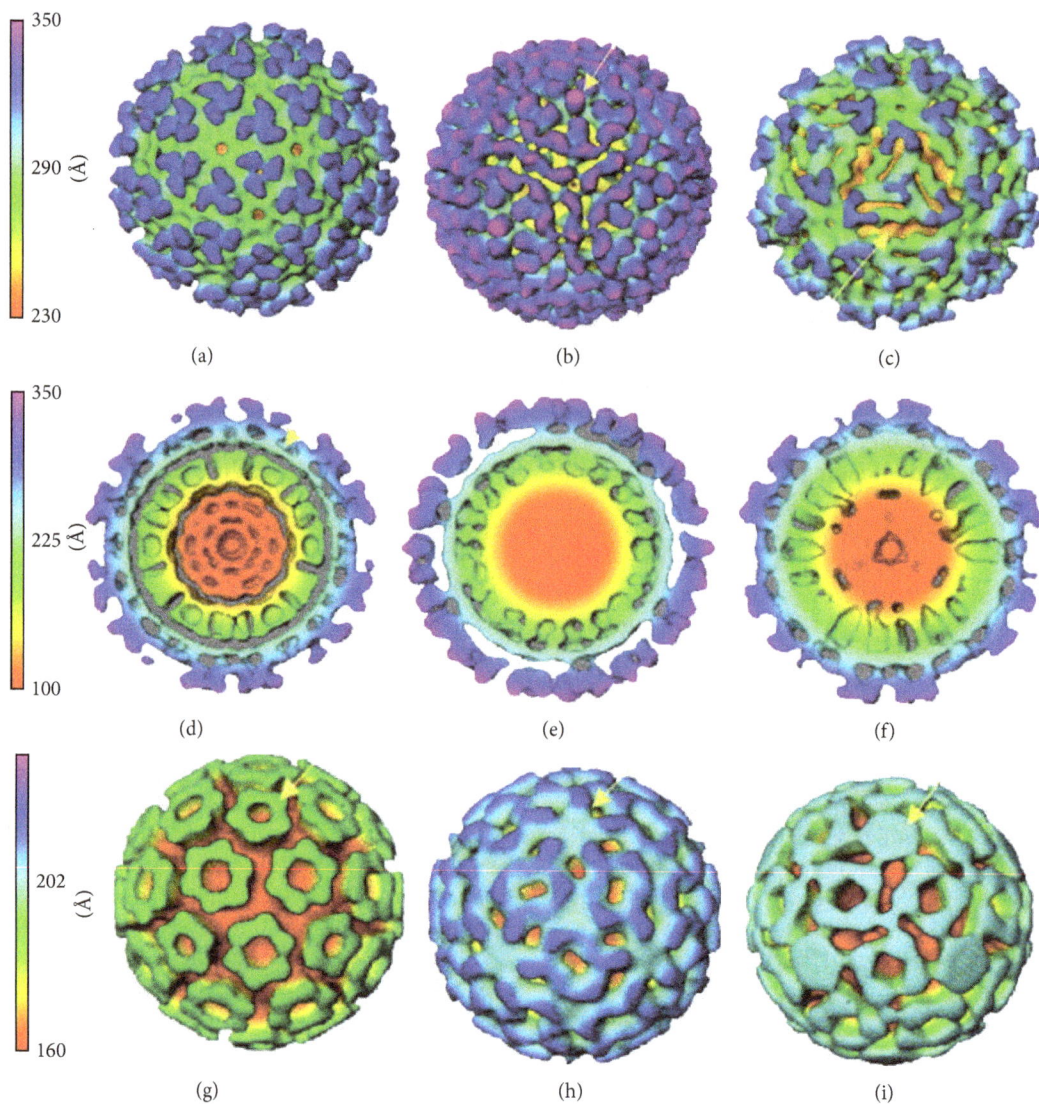

FIGURE 10: Conformational changes in Sindbis virus after exposure to various pH conditions. The three-dimensional structures of Sindbis virus surface at 28 Å resolution viewed along icosahedral threefold axes ((a)–(c)). Central cross section of Sindbis virus ((d)–(f)). Three-dimensional structure of the Sindbis virus capsid (inside the membrane bilayer) viewed along the icosahedral threefold axes ((g)–(i)). The reconstructions are colored according to a range of radii (key displayed) at different pH conditions ((a), (d), and (g)). Sindbis virus at pH 7.2. ((b), (e), and (h)) Sindbis virus at pH 5.3. ((c), (f), and (i)) Sindbis virus exposed to pH 5.3 (5 min) and returned to pH 7.2. Arrows: (b) protrusion at the fivefold axis; (c) fissure at twofold axis; ((d), (e), and (f)) fivefold axis; (g) region of cross-section occupied by the membrane bilayer. Paredes et al. 2004 [122].

fusion from within and from without; however, it is possible that the Togaviruses infect cells by a mechanism more similar to that used by polio virus [132] rather than that of influenza [122, 131, 133].

Cell-mediated endosomal uptake of alpha and flaviviruses followed by acidification and membrane fusion with the virus membrane is currently the favored model of alphavirus penetration and entry [134–136]. This mechanism is supported by indirect biochemical evidence and structures of E1 trimers extracted from virus-infected cells [137, 138] or expressed as soluble protein [139–141]. Current evidence of low pH structures for E1 includes studies done using liposomes to extract the proteins from virus during cofloatation [142–144]. While liposomes are useful for studying fusion in other systems in which the virus proteins can be expressed independently such as with flu, these artificial membranes may not be a suitably stringent reagent for the study of virus penetration in the case of alpha- and flaviviruses. This is because (1) no virus receptors are present, (2) high concentrations of cholesterol are required [145–147], and (3) the trimer of E1 has not been seen by any other method other than extraction or expression of E1 followed by liposome interaction. Insects are cholesterol auxotrophs [148] and do not contain the amounts of cholesterol required for liposome fusion [149]. Finally, the fusion of virus with liposomes is a non-leaky process [150] which is not the case with virus

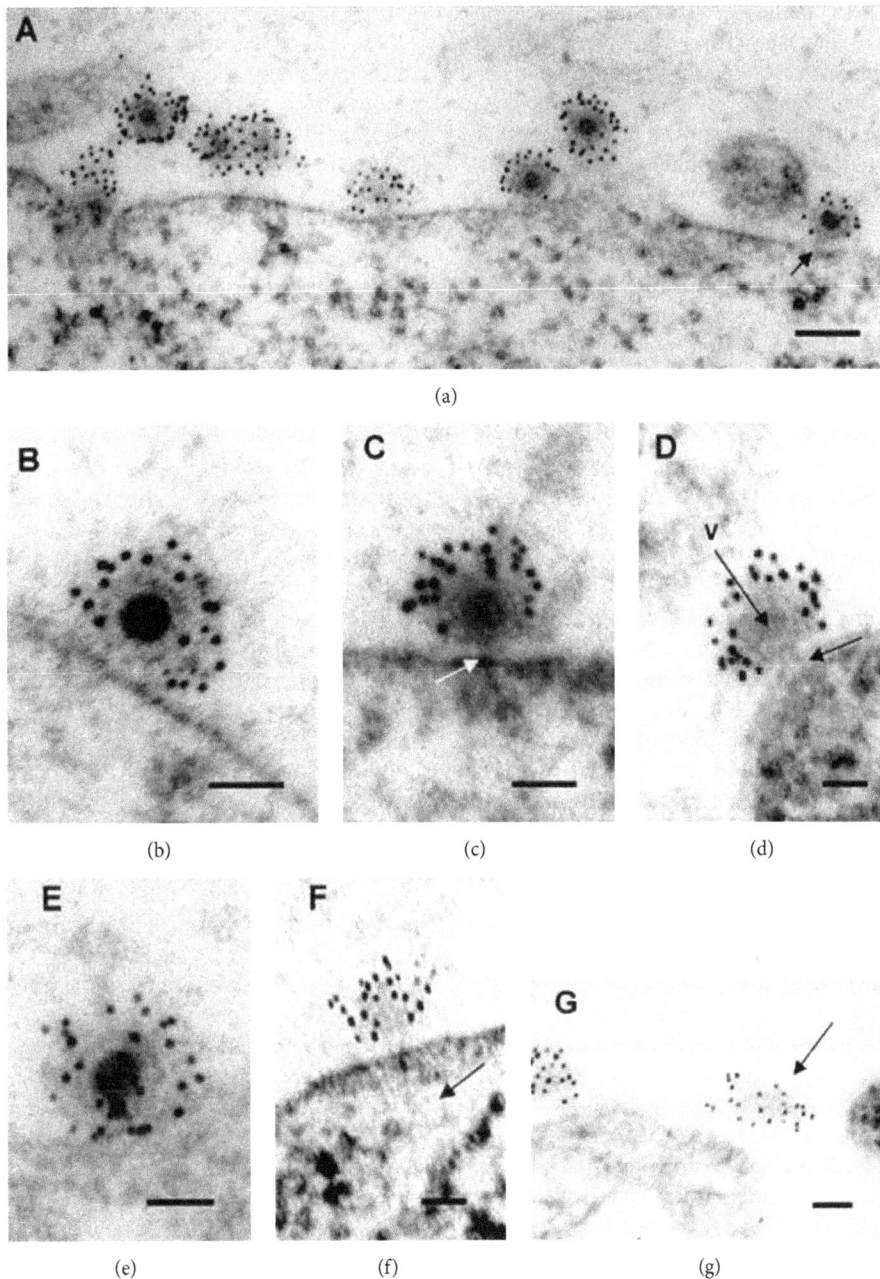

(a)

(b)　　　　　(c)　　　　　(d)

(e)　　　　　(f)　　　　　(g)

FIGURE 11: Electron micrographs of thin sections of Sindbis virus-cell complexes at pH 7.2. (a) Low magnification showing "full" and "empty" particles and a particle attached by a pore to the cell surface (arrow). (b) A virion attached to the cell surface before pore formation. (c) A virion with an electron dense core attached to the cell surface by a pore structure (arrow). (d) The pore at the vertex (V) of the protein shell penetrates the cell membrane (arrow). The virion has reduced electron density in the core region. (e) Reorganization of virus RNA into the developing pore. (f) An empty particle with a possible RNA molecule entering the cell (arrow). (g) An empty virion that has lost structure. Magnification scale bar (a) = 1000 Å, all others = 500 Å. Paredes et at. 2004 [122].

infections of host cells [151]. Interestingly, we have also shown that certain clones of mosquito cells derived from Singh's original isolate U4.4 are not susceptible to fusion from without Sindbis virus but are readily infected [123, 152]. Because the details of the fusion model of alpha and flavivirus penetration are predominant in the literature, evidence for direct virus penetration will be further discussed.

2.5. Ultrastructural Evidence of Direct Virus Penetration. As early as 1978, Fan and Sefton proposed that virus entry for Sindbis and VSV involved a mechanism which did not require fusion [153]. This evidence gave way to the more popular model of membrane fusion [129, 136]. The *a priori* belief that enveloped virus structures must encode a fusion loop and penetrate cells by fusion, now dominates the field to the exclusion of alternate modes of virus entry. E1 of the Alphaviruses and E of the flaviviruses are referred to as group II fusion proteins [139, 154, 155]. Indirect evidence has been used to develop a model proposing that these types of proteins insert a small fusion loop into the host endosomal membrane after endocytosis and a shift to low pH. The fusion loop is seen in Chikungunya virus [Alphavirus E1 (2ALA)] and flavivirus [Dengue E (1TG8)] crystal structures. Notably, pestiviruses and hepaciviruses, which belong the Family *Flaviviridae*, do not encode a fusion loop [156, 157] and investigators are in search of alternate fusion domains or fusion mechanisms [158]. The Bunyaviridae have been predicted to encode a fusion loop and by homology have been predicted to be class II fusion proteins [159].

A large volume of work has focused on the ability of alpha and flaviviruses to fuse artificial membranes and to elucidate the mechanism of low pH-mediated fusion of virus. The model of infection for these two families posits that membrane-containing viruses infect cells via low pH-mediated fusion within cell endosomes. The membrane-fusion mechanism of virus infection has been studied extensively for the influenza (flu) hemagglutinin and has been shown by direct evidence to form structural intermediates involved in virus penetration [97, 160–162]. Influenza, however, differs significantly from the alpha and flaviviruses in that the structure of the virus is amorphous with the structural proteins associated with large areas of exposed lipid. Additionally, HA and N do not form heterodimeric associations. Unlike influenza, there is no direct biochemical or structural evidence for membrane fusion by arboviruses, and no thermodynamics of the fusion process or the induction of the fusion intermediates. *The ability of alpha and flaviviruses to fuse membranes is not disputed; however, this may not result in virus penetration and infection.*

Our work on virus penetration has shown that Sindbis, West Nile virus (WNV), and dengue virus can penetrate cells at temperatures that do not allow membrane fusion [131, 133, 163, 164]. Using direct observation and biochemical methods, we have demonstrated that Sindbis and dengue virus infect cells in a time and temperature dependent manner (Figure 12).

Recent data show that Sindbis virus can penetrate mosquito C7-10 cells even more quickly than what has been seen in BHK cells. By 60 min. postinfection at 4°C, 90% of the virus was empty as compared to the 75% seen in BHK cells (see Figure 12(b)). Cultured insect cells contain less cholesterol than mammalian cells and are less viscous at 4°C, which may facilitate the process. The temperature kinetics of this reaction can be fit to Arrhenius plots, suggesting that the process of entry of the RNA into the cell is not force driven and that the energy to form the pore structure likely resides in the virus proteins. The energy of activation is calculated to be 27 kcal/mole. The entry process only requires a membrane potential and is affected by the chemistry of the host cell membrane (Vancini, personal communication). The data show that 70% of Sindbis virions are empty after one hour at 4°C (Figure 13). Sindbis virus carrying a green fluorescent protein will infect BHK cells at 15°C in the absence of fusion, or endocytosis producing fluorescent cells without return

through higher temperatures [133]. The obvious implication of these data is that studies in which virus has been allowed to attach on ice for one hour during the infection phase may not have synchronized the infection, as proposed in these studies, but rather allowed infectious particles to be internalized [122, 131, 133, 165], reviewed in [165, 166] (Figure 14). However, the effect of 15°C on formation of the replication complex has not been reported for Sindbis virus but it is possible that synchronization occurs at the level of RNA synthesis.

Strong biochemical support for the model of direct penetration at the cell surface comes from studies showing that cells become permeable to ions and small molecules as they are infected with alphaviruses [126, 151]. Virus infections leave pores on cell membranes [167] that allow the penetration of the cell membrane by small proteins such as the toxin α-sarcin (17 kDa) [168]. These results show that pores created in the plasma membrane as entry takes place affect membrane permeability. Fusion, by contrast, is a nonleaky process, does not compromise membranes, and does not leave pores in the membranes [150, 169].

Evidence of direct penetration has been presented for both mammalian and insect cells [165, 166]. This work has led to the alternate model that proposes that alpha and flaviviruses penetrate the host membrane bilayer through host cell triggered rearrangement of E1 or E proteins at a 5 fold axis resulting in the formation of a proteinaceous pore. This pore allows the release of the RNA into the cell cytoplasm, thus initiating the infection process. In the 1994 study by Guinea and Carrasco [170, 171], it was concluded that an ion gradient was required for viral entry of vesicular stomatitis virus, Semliki Forest virus, and influenza virus. This observation has been confirmed for Sindbis virus (Vancini, unpublished). We propose that membrane-containing icosahedral viruses incorporated lipid into their structure as an assembly scaffold allowing the cell exocytic mechanism to process and present maturing structural proteins to the nucleocapsid for envelopment. *This point is crucial for the development of prophylactics for pathogenic arboviruses.*

2.6. Native Virus Structure Analyzed Using Small Angle Neutron Scattering. In the paper by He et al. [40], we used small angle neutron scattering to explore the nature of Sindbis virus (alphavirus) particles produced by mammalian and insect cells. This method has the advantage over other methods of observing structure in that lipid and RNA densities are easily detected through a technique called contrast variation [172]. The findings were significantly distinct from what was expected because virus particles from these two hosts have important structural differences. Using virus particles purified in deuterium, a highly concentrated solution of virus suspension was made from virus grown in mosquito C7-10 or mammalian BHK cells. The particles were then analyzed using small angle neutron scattering (SANS), a nondestructive technique [173, 174]. The R_g (radius of gyration) indicated that the BHK-grown virus is less compact than that grown from mosquito cells. The diameter of the BHK grown virus was found to be 689 Å, compared to that of the insect grown virus which was 670 Å. The mass at the center of the BHK particles was less centrally distributed than that seen in the

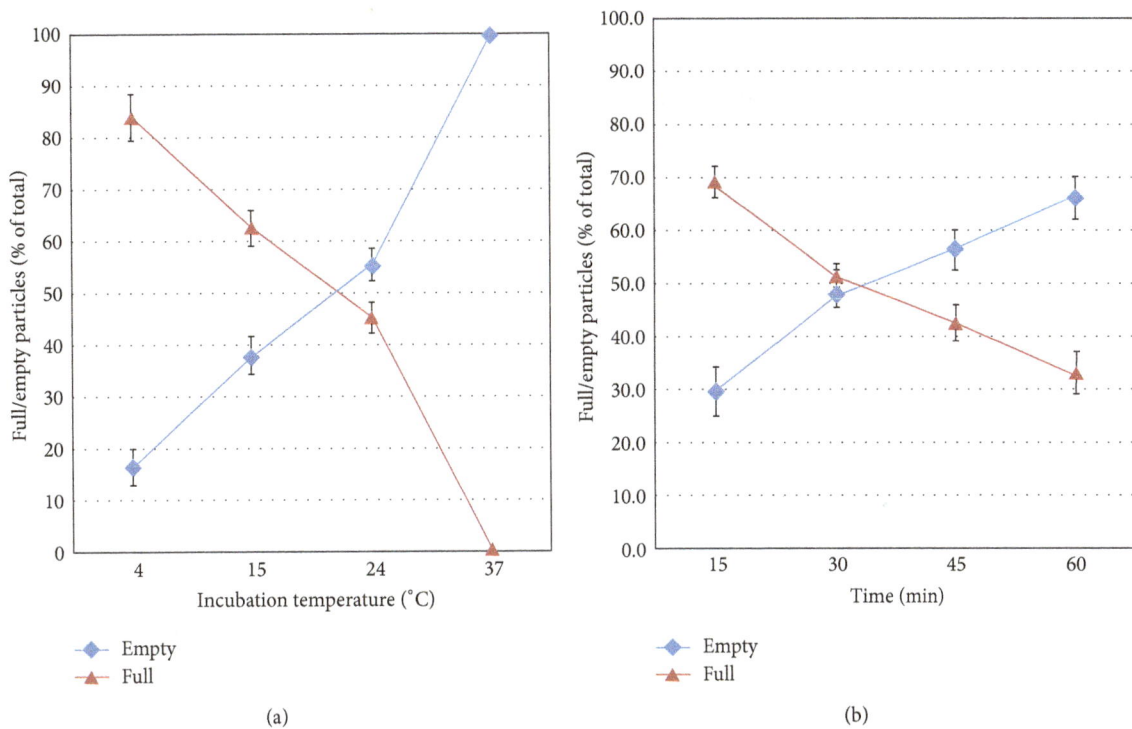

FIGURE 12: Temperature and time dependence of Sindbis virus genome delivery into BHK cells. (a) Sindbis virus-cell complexes incubated and fixed at 15 min postinfection at several temperatures were analyzed by electron microscopy. The graph shows that, as temperature rises from 4°C to 37°C, there is a progressive increase in the population of empty particles at the cell surface (◆), representing the total population at 37°C, and a decrease in the population of full particles (▲), reaching zero at 37°C. (b) Interaction of Sindbis virus with BHK cells at 4°C for 15 to 60 min. The percentage of empty particles increases from 30% at 15 min to 66.6% at 60 min (◆), and the population of full particles decreases from 70% to 33.4% over the same time period (▲). Data are shown as the means of triplicate samples of two independent experiments. Error bars are standard errors of the means (SEM). From Vancini et al. 2013 [133].

C7-10 virus. It was also determined that while the radial position of the lipid bilayer did not change significantly, the membrane had significantly more cholesterol when the virus was grown in mammalian cells than in insect cells. This property has been shown to affect the virus stability and, with it, infectivity [149]. Distribution of the densities of the particles was modeled using a four shell analysis representing the distribution of the biochemical components of the virus: RNA, capsid protein, lipid with protein, and glycoprotein. Comparing the shell thickness for each virus showed that the outer protein shell was more extended in the mammalian Sindbis virus than that from the insect virus. The SANS data also demonstrated that the RNA and nucleocapsid protein share a closer interaction in the mammalian cell-grown virus than in the virus from the insect host. It is possible that the nucleocapsid structure from mammalian cells is organized more closely to the virus membrane and that the center of the virus is mainly solvent. The biological consequences of the structural differences uncovered by this new technique are not known. It may be that the temperature of assembly may contribute to these differences, given the different biochemical environments. It is of interest to note that Zhang et al. [23] reported a structure of dengue virus at 3.5 Å resolution when grown from C6/36 mosquito cells grown at 33°C (see Figure 3) but did not report a difference in the size of the virus. This may suggest that virus conformation

and size may be the result of the host cell biochemistry and not temperature. This will be discussed further.

2.7. Temperature Studies and Modeling Crystal Structures into Cryo-EM Reconstructions.

Dengue virus particles are structurally classified as (1) mature, (2) partially mature, or (3) immature, depending on the surface morphology seen in cryoEM preparations prior to averaging of the particles [176]. The classification of particles into mature, partially mature, or immature status does not correlate with infectivity; it is estimated that infectious virus contains 30–40% uncleaved preM [84, 177]. Mature dengue has been reported to have a particle diameter of 500 Å [94]. Two recent studies have reported that temperature also affects viral conformation for dengue virus 2 strains 16681 and New Guinea C [175, 178]. In both these DV2 studies, one by Zhang et al. [178] and one by Fibriansah et al. [175], it was determined by cryoEM reconstruction of virus grown in insect C6/36 cells at 28°C then heated to 37°C that virus expanded by 40–50 Å compared to the virus remaining at ambient temperature [175]. In each study, both heated and unheated virus was found to have equivalent infectivity, indicating that both virus structures represent infectious intermediates. In the Fibriansah study, several categories of 37°C treated virus particles were seen in the cryoEM images, resulting in sorting of the virus into several classes based on the surface morphology, shown in Figure 15. Class II particles

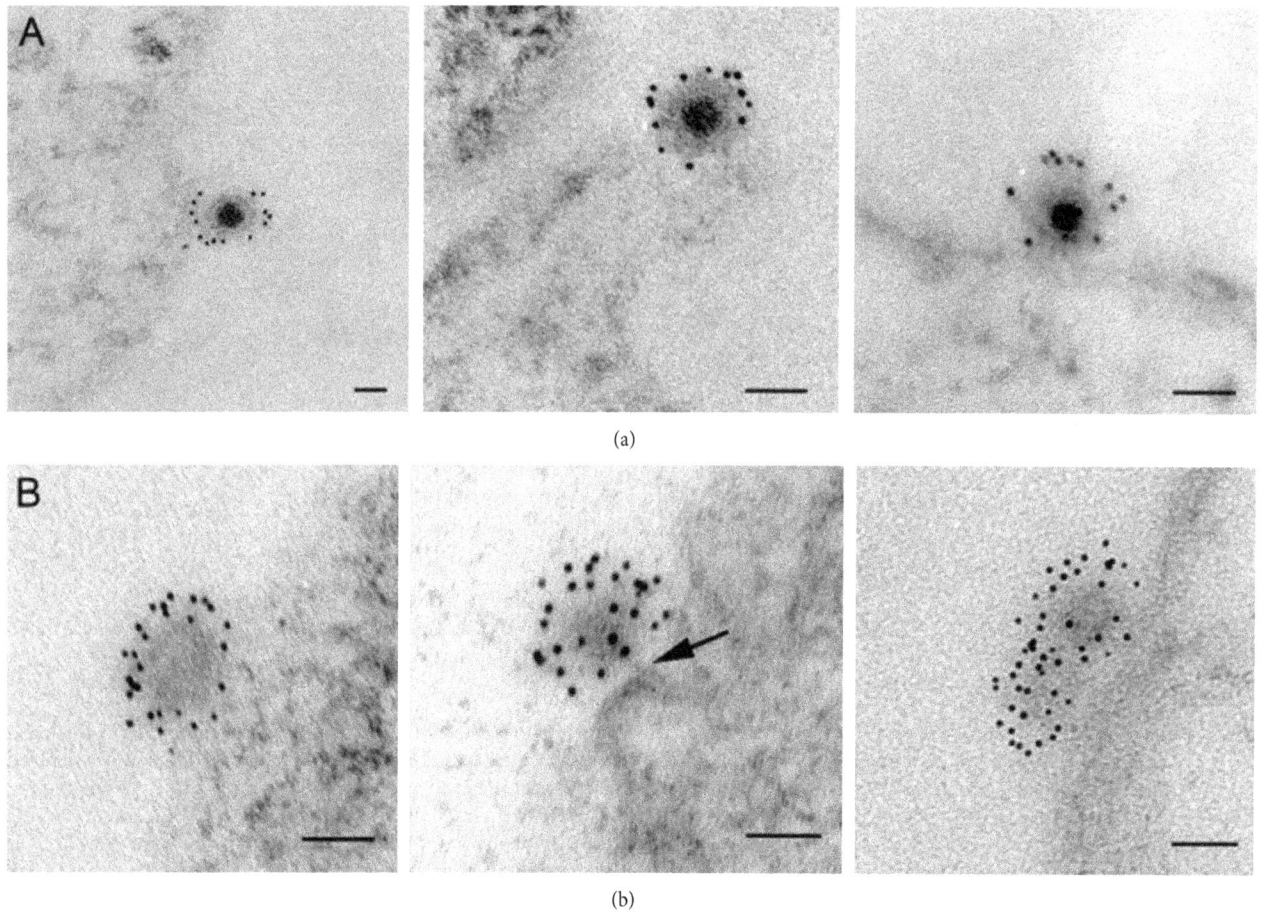

FIGURE 13: Thin-section electron microscopy of Sindbis virus-cell complexes at pH 7.2. A panel of images obtained by TEM illustrates two representative populations of particles at different incubation temperatures. High-magnification images show examples of electron-dense full particles at the cell surface at 4°C (row (a)) and empty particles with loss of RNA electron density predominantly at 37°C (row (b)). In row (b), middle panel, there is indication of a stalk connecting the virus and the cell (arrow). Bars, 50 nm. From Vancini et al. 2013 [133].

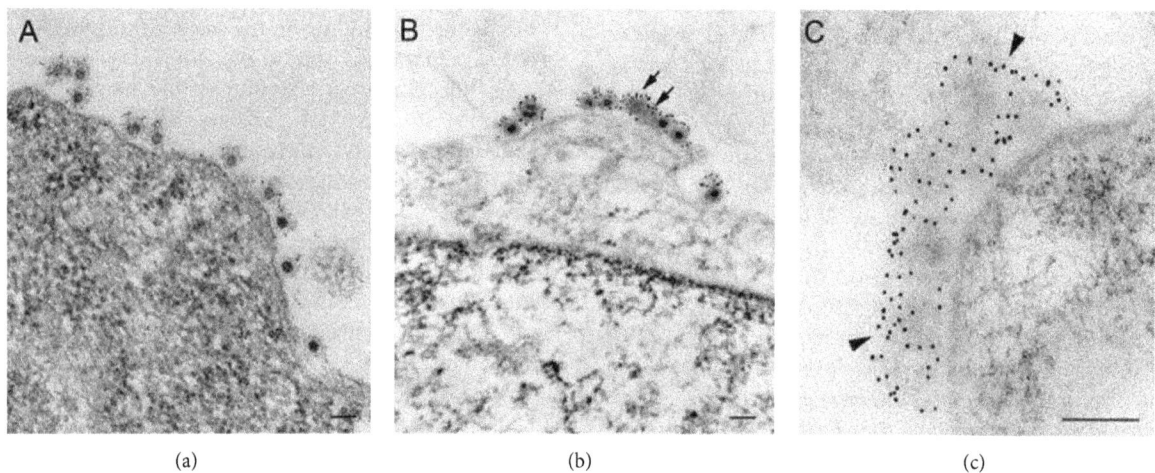

FIGURE 14: Overview of Sindbis virus-cell complexes. Low magnifications of Sindbis virus-cell complexes at different temperatures. (a) Interaction at 4°C; (b) interaction at 15°C; note the presence of both full and empty (arrows) particles; (c) interaction at 37°C, in which most particles lose their electron density and their well-defined structure (arrowheads). Bars, 100 nm. From Vancini et al. 2013 [133].

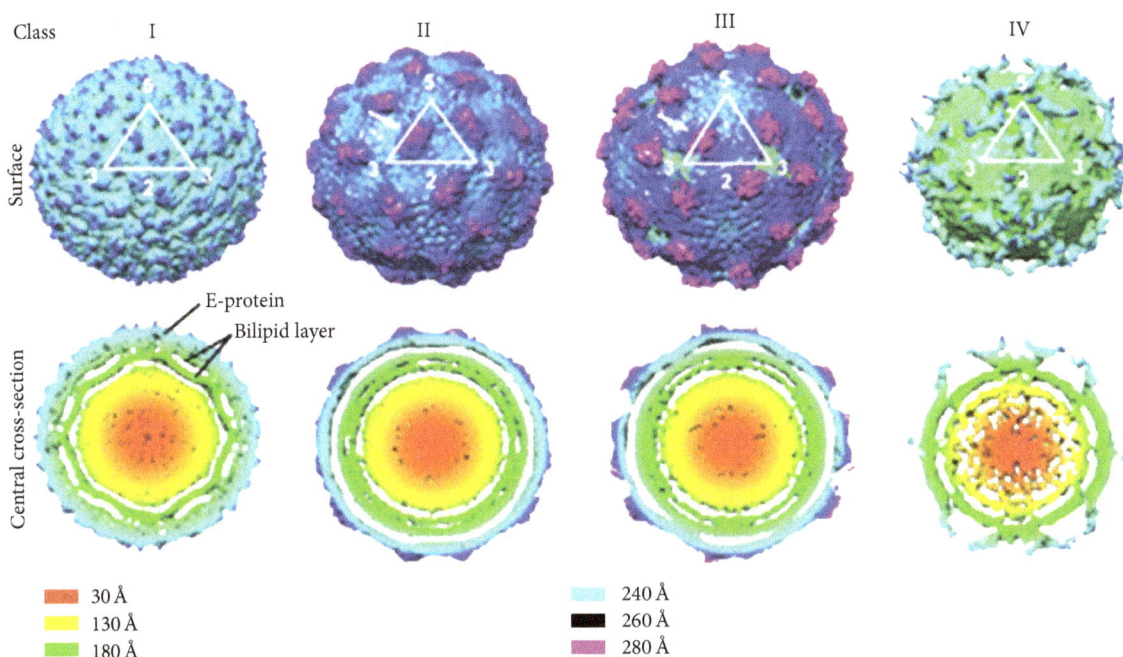

FIGURE 15: Cryo-EM maps reconstructed from class I to IV particles present in the DENV2 sample incubated at 37°C. The surfaces of the maps (above) and their central cross-sections (below) are colored according to their radii (shown in the lower panel). The white triangle represents an icosahedral asymmetric unit, and the corresponding 2-, 3-, and 5-fold symmetry vertices are indicated. Maps generated from the class I particles are similar to the previous published DENV-2 maps. Class II and III maps showed that the particles in these classes have bigger radii, indicating that the virus had expanded. There are protruding densities on the virus surface between the 5- and 3-fold vertices (white arrows). The class IV map showed very poor density on the E protein layer, indicating that the E protein had lost its icosahedral symmetry. With permission from Fibriansah et al. [175].

have a smooth surface while Class III display a surface with protrusions. Class IV structures are smaller than the other classes. Class I particles are those grown in mosquito cells at 28°C. In class II viruses, E and M, density was seen to reorganize to produce protruding structures between the 5- and 3-fold vertices. In the class III structures, the protrusions are more extended at the 2-fold axes, and "holes" appear at the strict 3-fold axes. While the structure of the class IV viruses was not further investigated by these authors, particles at 37°C continued to change conformation reorganizing density toward the 5-fold axes until a nonreversible endpoint was achieved. Thus, heating of the class I particles resulted in three different structures. In most icosahedral virus reconstructions from cryoEM, the signal is improved and the noise reduced by imposing icosahedral symmetry on the image data after refinement. For protein density to be visible after imposing this symmetry, the protein structure has to occupy more than 60% of the icosahedrally related positions [179, 180]. Conversely, structures that occupy very few of these positions, such as a portal complex, are averaged out of 3D reconstructions. Thus, while the amount of preM is averaged out during a cryoEM reconstruction, the presence of this protein may still have an important role to play in the structure, infectivity, and immunogenicity of the virus. It was not determined whether this structural difference was a result of a difference in the host environment or a function of the temperature of assembly. Insect host specific biochemistry, which functions optimally at 28°C, may be responsible for

assembly of the compact structure which can adopt a less compact form at 37°C. As was discussed above, virus size differences were seen with Sindbis virus from different hosts (mammalian or insect) [40].

The structure by Zhang et al. is similar to the one observed by Fibriansah et al, in which E and M density is seen to move away from the core of the virus after heating insect grown virus to 37°C. The former structure was solved to 35 Å, while the latter had a resolution of 14 Å. In the work by Zhang et al. the same experiment is performed; insect grown virus is heated to 37°C. Both virus structures were found to be equally infectious. These authors reconstructed the heated form to a fit with E protein into trimers, however, unlike Fibriansah who fit E into the traditional herringbone pattern. In the Zhang study, the authors propose that this larger 37°C structure represents the prefusogenic form of the virus that was previously just a predicted structure [94]. In Zhang et al., virus expansion was found to be reversible upon heating to 35°C, and that the end point stable intermediate does not occur until the temperature reaches 37°C, after which the expansion is not reversible. Both studies reported the introduction of holes in the virus surface at the 3' axes not seen in previous reconstructions.

Both these flavivirus studies speculated that the change in structure may affect the response of the mammalian host by revealing previously unexposed epitopes in virus from insect cells. This is important for mapping neutralizing epitopes *in vitro*, and it was previously not understood how known neutralizing Ab (NAb) anti-DV2 epitopes on the virus bound Ab to sequences that were occluded in the virus structure. This anomaly was explained by proposing that superficially hidden epitopes could be unmasked during thermal flux or "breathing" of the virus [181]. This quandary is now resolved by fitting the proteins into a larger expanded structure formed by heating insect virus to 37°C. However, the E fold in the crystal structure may not be native resulting in difficulty with a fit to the smaller structure. In addition to epitope exposure, what is the biological relevance of these observations? A more compact form of insect virus does not affect an *in vivo* mammalian infection because the more compact 28°C structure will only be exposed to the host for one round of replication, after which the second, 37°C structure will be adopted. Because both compact and expanded forms of the viruses from both hosts are infectious, these virions represent metastable structural intermediates primed for infection, and these two distinct infectious intermediates may represent the optimal form of the structure from its specific host biochemical environment.

One interpretation of these observations is that the mammalian virus is "born" in the prefusogenic-like structure, and the mosquito virus is not. While there is no direct evidence for any function of these predicted intermediate models, mosquito cells were speculated to be infected by virus adopting the fusogenic form via exposure to low pH or contact with the receptor, thus triggering the rearrangement. Again, as in the former study, the Zhang study demonstrated that virus incubated at both temperatures is infectious. This becomes an important point because arboviruses do not always cycle between insect and vertebrate hosts. For mosquito-borne viruses, there exists a "mosquito only" zoonotic cycle in the natural transmission cycle, during which mosquito and virus would only experience ambient temperature [182, 183]. Arboviruses carried by mosquitoes are transmitted from vector to vector horizontally by venereal transmission or vertically as infected eggs. This mosquito cycle is required especially in temperate climates for the virus to over winter. If fusion is required for infectivity, then the "insect only" virus cannot acquire the fusion intermediate in the absence of invoking other mechanisms of structural alteration. Whatever the interpretation, the present data argue that for dengue 2, (1) two infectious dengue virus intermediates are found in nature, one at 28°C and another at 37°C as a result of the temperature of assembly, or (2) virus architecture is determined by host components, which have a temperature component resulting in two structural intermediates, both of which are biologically relevant as was reported in He et al. [40]. Neither of these alternatives need be mutually exclusive. It is of note that the structure reported by Zhang et al. (shown in Figure 3) was made from images of virus grown in insect C6/36 cells at 33°C. At this temperature according to the Zhang et al. study, 50% of the particles should have expanded if temperature were the only contributing factor to particle expansion; however, no increase in particle size was reported [178].

In addition to interpreting the high temperature structures, these papers modeled crystal structures of isolated E into the cryoEM reconstructions. Both studies used E from Tick Borne Encephalitis virus PDB 1SVB [95]. Zhang et al. reconstructed the heated form of the structure to a fit with E protein in trimers of dimers as seen in [94] to 35 Å resolution. Zhang et al. propose that the larger 37°C structure represents the prefusogenic form of the virus that reorganizes the herringbone array into trimers after the virus is exposed to low pH. This structure containing the trimer pattern was previously only a predicted structure [94]. In a separate study, Fibriansah reconstructed the DV2 expanded structure into a 14 Å model by separating the images into three size classes. However, in the Fibriansah study that displayed higher resolution E was modeled into the traditional herringbone pattern in the larger class III structure. The question arises as to which of either of the two models represents a biologically relevant structure. Fab 1A1D-2 [175, 178] and MAb E 111 epitopes are suggested to become more exposed and accessible to add weight to each predicted model. An additional complication to the DV2 expansion story is that Kostyuchenko et al. have reported that dengue 1 and 4 do not undergo heat-related expansion of the virus when insect cell grown virus is exposed to 37°C [184, 185]. If virus expansion is required for infection and the structures are similar in structure and function, it is reasonable to assume that it would happen in all four serotypes. It is suggested, however, that stability of dengue 1 and 4 imparted to these serotypes by surface charges on the virus could explain the lack of conformational changes at higher temperatures. This implies that DV2 is more unstable to heat, but no biochemical evidence was provided. The possibility that these viruses may contain unprocessed preM was not discussed or assessed though preM is known to block infection. It is puzzling that these viruses would not respond to heat but supposedly respond to low pH induced conformations to become infectious.

3. Conclusions and Comments

As has been discussed, the structural proteins of the icosahedral arboviruses display a remarkable amount of plasticity. From the same conserved folds, $T = 3$ symmetry is assembled in the flaviviruses, and $T = 4$ for the alphaviruses, representing an approximate 200 Å increase in virion size while the genome lengths are not significantly larger. As discussed for the Sindbis virus C mutant, Y180S/E183G, even larger T numbers are possible. Not discussed, but pertinent to this point, is the $T = 12$ structure found in the bunyaviruses that deviate to pleomorphic structures in the other members of the Bunyaviridae. These viruses are also presumed to have arisen from an ancient common fold. The relationship of this adaptation is not understood because the $T = 12$ structure genome size ranges from 10.5 to 22.7 kbp and is composed of a tripartite, single-stranded, negative-sense RNA genome that could theoretically be packaged in a much smaller virion [34]. Plasticity has also been documented in virus assembled from the insect versus the vertebrate host [40]. This was an unexpected outcome but in retrospect not surprising because these host systems are so biochemically and genetically divergent. Alpha- and flaviviruses contain glycoproteins with many cysteines, and more than one disulfide bonded form of these glycoproteins could exist. This is difficult to analyze/detect in the intact virus and, for the most part, has been ignored and the configuration of the crystal structures accepted as native. Additionally, the cryoEM structures of the pH-treated viruses have been interpreted to assume the same structural reorganizations of the soluble E protein in the prefusion conformation which forms a trimer. However, these structures are supportive of indirect evidence for trimerization. Thus, in light of the conflicting evidence for fusion as the mechanism of virus penetration, these data should suggest that the structure of the isolated protein does not always predict the conformation in the intact metastable virus. Even at atomic resolution, reconstruction of an infectious virus cannot lead to definitive evidence of the mechanism of virus penetration without direct biochemical and genetic testing of the model.

It has been a long quest to solve the atomic structure of a membrane-containing virus. As technology improves, atomic details will certainly be resolved. However, there are certain impediments to the process that are not directly related to the technology. First, it has become a canon of some investigators that crystal structures of proteins and their multimeric states represent a native structure. This is a huge assumption for metastable virus proteins because in the absence of control structures or direct functional assays this assumption cannot be tested. Most virus samples also contain a large particle to pfu ratio. The most rigorous experiments would not require the dissolution of the virus particle [186, 187]. This is not to say that structures of proteins are not informative but that the process needs to include a caveat that crystals do not represent the entirety of the many conformations through which macromolecular metastable dynamic viruses proceed to reach the infectious intermediate and then deliver, upon infection, the genomic cargo. This is especially important if the crystal structure is of a single protein of a macromolecular complex that does not naturally exist in solution, such as one that contains membrane proteins. There is a precept that if a protein can be expressed by any expression system, the subsequent folds are representative of the native structure. This assumption may be too simplistic because it is well established biochemically and genetically that assembly of membrane-containing virus requires both host and virus chaperones, including lipids and post translational modifications in a complex and temporal manner to assemble an infectious virus. This exclusive process is not available in any heterologous expression system. An example of this is seen in a study of Sindbis virus E1 in which whole infectious virions were analyzed by mass spectrometry, and it was determined that E1 has cysteines not seen in the alphavirus crystal structure in which all Cys are found as disulfides [66]. The key to these analyses is not the technology employed but rather the starting infectious material. This observation suggests that other problems may exist with the current alphavirus E1 model as was also suggested above for flavivirus E. Very few protein structures are used to build a "fit" into a cryo-EM reconstruction because they are assumed to be interchangeable. If the fit does not work, either the process is abandoned or the protein domains are reoriented. An example of this is found in Fibriansah et al. in which they state that E protein could not be fit into the class II particles due to

the lack of structural features required for fitting, although it is possible that the cryo-EM contains data on a structural intermediate; that is, that the structure is expanding in a way that does not fit the current methods, model, or crystal. This implies that there may exist an intermediate structure which adopts an unpredicted structure. These structures are models of a very complicated assembly system. Cryo-EM of heated and low pH-treated viruses is revealing an extraordinary reorganization of the contour of alphaviruses and flaviviruses at ~20 Å resolutions. It seems plausible that, during these conformational reorganizations, structural proteins may proceed through several energetic states until the end point state is achieved. Much more work is required to fully appreciate these intricate structures and their mechanics of assembly and infection.

Authors' Contribution

Raquel Hernandez, Dennis T. Brown, and Angel Paredes wrote this review.

Acknowledgments

Authors Raquel Hernandez and Dennis T. Brown are supported by the Foundation for Research, Carson City, Nevada, and the North Carolina Agricultural Research Service.

References

[1] E. G. Strauss, R. Levinson, C. M. Rice, J. Dalrymple, and J. H. Strauss, "Nonstructural proteins nsP3 and nsP4 of Ross River and O'Nyong-nyong viruses: sequence and comparison with those of other alphaviruses," *Virology*, vol. 164, no. 1, pp. 265–274, 1988.

[2] J. H. Strauss and E. G. Strauss, "Virus evolution: how does an enveloped virus make a regular structure?" *Cell*, vol. 105, no. 1, pp. 5–8, 2001.

[3] R. Goldbach and J. Wellink, "Evolution of plus-strand RNA viruses," *Intervirology*, vol. 29, no. 5, pp. 260–267, 1988.

[4] R. W. Schlesinger, "New opportunities in biological research offered by arthropod cell cultures. I. Some speculations on the possible role of arthropods in the evolution of arboviruses," *Current Topics in Microbiology and Immunology*, vol. 55, pp. 241–245, 1971.

[5] C. A. Stafford, G. P. Walker, and D. E. Ullman, "Infection with a plant virus modifies vector feeding behavior," *Proceedings of the National Academy of Sciences of the United States of America*, vol. 108, no. 23, pp. 9350–9355, 2011.

[6] I. C. Bezerra, R. D. O. Resende, L. Pozzer, T. Nagata, R. Kormelink, and A. C. De Ávila, "Increase of tospoviral diversity in Brazil with the identification of two new tospovirus species, one from chrysanthemum and one from zucchini," *Phytopathology*, vol. 89, no. 9, pp. 823–830, 1999.

[7] S. Adkins, R. Quadt, T.-J. Choi, P. Ahlquist, and T. German, "An RNA-dependent RNA polymerase activity associated with virions of tomato spotted wilt virus, a plant- and insect-infecting bunyavirus," *Virology*, vol. 207, no. 1, pp. 308–311, 1995.

[8] J. H.-O. Pettersson, I. Golovljova, S. Vene, and T. Jaenson, "Prevalence of tick-borne encephalitis virus in ixodes ricinus ticks in northern europe with particular reference to southern Sweden," *Parasites & Vectors*, vol. 7, article 102, 2014.

[9] J. J. L. Lutomiah, C. Mwandawiro, J. Magambo, and R. C. Sang, "Infection and vertical transmission of Kamiti river virus in laboratory bred Aedes aegypti mosquitoes," *Journal of Insect Science*, vol. 7, article 55, 2007.

[10] G. Kuno, G.-J. J. Chang, K. R. Tsuchiya, N. Karabatsos, and C. B. Cropp, "Phylogeny of the genus flavivirus," *Journal of Virology*, vol. 72, no. 1, pp. 73–83, 1998.

[11] E. C. Holmes and S. S. Twiddy, "The origin, emergence and evolutionary genetics of dengue virus," *Infection, Genetics and Evolution*, vol. 3, no. 1, pp. 19–28, 2003.

[12] S. Cook, G. Moureau, A. Kitchen et al., "Molecular evolution of the insect-specific flaviviruses," *Journal of General Virology*, vol. 93, no. 2, pp. 223–234, 2012.

[13] J. P. Messina, O. J. Brady, T. W. Scot et al., "Global spread of dengue virus types: mapping the 70 year history," *Trends in Microbiology*, vol. 22, pp. 138–146, 2014.

[14] A. T. Ciota and L. D. Kramer, "Insights into arbovirus evolution and adaptation from experimental studies," *Viruses*, vol. 2, no. 12, pp. 2594–2617, 2010.

[15] L. L. Coffey, N. Forrester, K. Tsetsarkin, N. Vasilakis, and S. C. Weaver, "Factors shaping the adaptive landscape for arboviruses: implications for the emergence of disease," *Future Microbiology*, vol. 8, no. 2, pp. 155–176, 2013.

[16] G. D. Ebel, "Toward an activist agenda for monitoring virus emergence," *Cell Host & Microbe*, vol. 15, no. 6, pp. 655–656, 2014.

[17] S. C. Weaver, R. Winegar, I. D. Manger, and N. L. Forrester, "Alphaviruses: population genetics and determinants of emergence," *Antiviral Research*, vol. 94, no. 3, pp. 242–257, 2012.

[18] M. G. Rossmann and R. R. Reuckert, "What does the molecular structure of viruses tell us about viral functions?" *Microbiological Sciences*, vol. 4, no. 7, pp. 206–214, 1987.

[19] E. G. Westaway, M. A. Brinton, S. Y. Gaidamovich et al., "Togaviridae," *Intervirology*, vol. 24, no. 3, pp. 125–139, 1985.

[20] R. M. Dubois, M.-C. Vaney, M. A. Tortorici et al., "Functional and evolutionary insight from the crystal structure of rubella virus protein E1," *Nature*, vol. 493, no. 7433, pp. 552–556, 2013.

[21] P. F. Mattingly, "Genetical aspects of the aedes aegypti problem I. Taxonomy and bionomics," *Annals of Tropical Medicine and Parasitology*, vol. 51, no. 4, pp. 392–408, 1957.

[22] R. Zhang, C. F. Hryc, Y. Cong et al., "4.4 Å cryo-EM structure of an enveloped alphavirus venezuelan equine encephalitis virus," *EMBO Journal*, vol. 30, no. 18, pp. 3854–3863, 2011.

[23] X. Zhang, P. Ge, X. Yu et al., "Cryo-EM structure of the mature dengue virus at 3.5-Å resolution," *Nature Structural and Molecular Biology*, vol. 20, no. 1, pp. 105–110, 2013.

[24] D. H. Bamford, "Do viruses form lineages across different domains of life?" *Research in Microbiology*, vol. 154, no. 4, pp. 231–236, 2003.

[25] M. B. Sherman, A. N. Freiberg, M. R. Holbrook, and S. J. Watowich, "Single-particle cryo-electron microscopy of Rift Valley fever virus," *Journal of Virology*, vol. 387, no. 1, pp. 11–15, 2009.

[26] J. C. Brown, W. W. Newcomb, and G. W. Wertz, "Helical virus structure: the case of the rhabdovirus bullet," *Viruses*, vol. 2, no. 4, pp. 995–1001, 2010.

[27] P. Metcalf, "The symmetry of the reovirus outer shell," *Journal of Ultrastructure Research*, vol. 78, no. 3, pp. 292–301, 1982.

[28] R. Couch, *Orthomyxoviruses*, University of Texas Medical Branch at Galveston, Galveston, Tex, USA, 1996.

[29] R. G. Webster, W. J. Bean, O. T. Gorman, T. M. Chambers, and Y. Kawaoka, "Evolution and ecology of influenza A viruses," *Microbiological Reviews*, vol. 56, no. 1, pp. 152–179, 1992.

[30] I. Rouiller, S. M. Brookes, A. D. Hyatt, M. Windsor, and T. Wileman, "African swine fever virus is wrapped by the endoplasmic reticulum," *Journal of Virology*, vol. 72, no. 3, pp. 2373–2387, 1998.

[31] A. Lwoff, R. Horne, and P. Tournier, "A system of viruses," *Cold Spring Harbor Symposia on Quantitative Biology*, vol. 27, pp. 51–55, 1962.

[32] D. Baltimore, "Expression of animal virus genomes," *Bacteriological Reviews*, vol. 35, no. 3, pp. 235–241, 1971.

[33] A. King, E. Lefkowitz, M. Adams, and E. Carstens, *Virus Taxonomy, Ninth Report of the International Committee on Taxonomy of Viruses*, Elsevier, Waltham, Mass, USA, 1st edition, 2012.

[34] R. V. Mannige and C. L. Brooks III, "Periodic table of virus capsids: implications for natural selection and design," *PLoS ONE*, vol. 5, no. 3, Article ID e9423, 2010.

[35] M. Dessau and Y. Modis, "Crystal structure of glycoprotein c from rift valley fever virus," *Proceedings of the National Academy of Sciences of the United States of America*, vol. 110, no. 5, pp. 1696–1701, 2013.

[36] J. H. Strauss and E. G. Strauss, "The alphaviruses: gene expression, replication, and evolution," *Microbiological Reviews*, vol. 58, no. 3, pp. 491–562, 1994.

[37] S. Nelson, R. Hernandez, D. Ferreira, and D. T. Brown, "In vivo processing and isolation of furin protease-sensitive alphavirus glycoproteins: a new technique for producing mutations in virus assembly," *Virology*, vol. 332, no. 2, pp. 629–639, 2005.

[38] H. Garoff, K. Simons, and O. Renkonen, "Isolation and characterization of the membrane proteins of semliki forest virus," *Journal of Virology*, vol. 61, no. 2, pp. 493–504, 1974.

[39] M. Carleton, H. Lee, M. Mulvey, and D. T. Brown, "Role of glycoprotein PE2 in formation and maturation of the sindbis virus spike," *Journal of Virology*, vol. 71, no. 2, pp. 1558–1566, 1997.

[40] L. He, A. Piper, F. Meilleur et al., "The structure of sindbis virus produced from vertebrate and invertebrate hosts as determined by small-angle neutron scattering," *Journal of Virology*, vol. 84, no. 10, pp. 5270–5276, 2010.

[41] J. B. Gliedman, J. F. Smith, and D. T. Brown, "Morphogenesis of sindbis virus in cultured aedes albopictus cells," *Journal of Virology*, vol. 16, no. 4, pp. 913–926, 1975.

[42] C. L. Murray, C. T. Jones, and C. M. Rice, "Architects of assembly: roles of Flaviviridae non-structural proteins in virion morphogenesis," *Nature Reviews Microbiology*, vol. 6, no. 9, pp. 699–708, 2008.

[43] S. Elshuber, S. L. Allison, F. X. Heinz, and C. W. Mandl, "Cleavage of protein prM is necessary for infection of BHK-21 cells by tick-borne encephalitis virus," *Journal of General Virology*, vol. 84, no. 1, pp. 183–191, 2003.

[44] B. D. Lindenbach and C. M. Rice, "Flaviviridae: the viruses and their replication," in *Fields Virology*, D. M. Knipe and P. M. Howley, Eds., pp. 1043–1125, Lippincott Williams & Wilkins, Philadelphia, Pa, USA, 2001.

[45] M. M. Samsa, J. A. Mondotte, N. G. Iglesias et al., "Dengue virus capsid protein usurps lipid droplets for viral particle formation," *PLoS Pathogens*, vol. 5, no. 10, Article ID e1000632, 2009.

[46] W. Zhang, P. R. Chipman, J. Corver et al., "Visualization of membrane protein domains by cryo-electron microscopy of dengue virus," *Nature Structural Biology*, vol. 10, no. 11, pp. 907–912, 2003.

[47] D. Ferreira, R. Hernandez, M. Horton, and D. T. Brown, "Morphological variants of sindbis virus produced by a mutation in the capsid protein," *Virology*, vol. 307, no. 1, pp. 54–66, 2003.

[48] B. V. V. Prasad and M. F. Schmid, "Principles of virus structural organization," *Advances in Experimental Medicine and Biology*, vol. 726, pp. 17–47, 2012.

[49] S. Lee, K. E. Owen, H.-K. Choi et al., "Identification of a protein binding site on the surface of the alphavirus nucleocapsid and its implication in virus assembly," *Structure*, vol. 4, no. 5, pp. 531–541, 1996.

[50] R. Hernandez, H. Lee, C. Nelson, and D. T. Brown, "A single deletion in the membrane-proximal region of the Sindbis virus glycoprotein E2 endodomain blocks virus assembly," *Journal of Virology*, vol. 74, no. 9, pp. 4220–4228, 2000.

[51] C. H. VonBonsdorff and S. C. Harrison, "Sindbis virus glycoproteins form a regular icosahedral surface lattice," *Journal of Virology*, vol. 28, article 578, 1978.

[52] D. L. Caspar, R. Dulbecco, A. Klug et al., "Proposals," *Cold Spring Harbor Symposia on Quantitative Biology*, vol. 27, pp. 49–50, 1962.

[53] J. P. Kononchik Jr., S. Nelson, R. Hernandez, and D. T. Brown, "Helical virus particles formed from morphological subunits of a membrane containing icosahedral virus," *Journal of Virology*, vol. 385, no. 2, pp. 285–293, 2009.

[54] D. T. Brown, "The assembly of alphaviruses," in *The Togaviruses*, R. W. Schlesinger's, Ed., pp. 473–501, Academic Press, New York, NY, USA, 1980.

[55] K. Nanda, R. Vancini, M. Ribeiro, D. T. Brown, and R. Hernandez, "A high capacity alphavirus heterologous gene delivery system," *Journal of Virology*, vol. 390, no. 2, pp. 368–373, 2009.

[56] M. Mulvey and D. T. Brown, "Formation and rearrangement of disulfide bonds during maturation of the Sindbis virus E1 glycoprotein," *Journal of Virology*, vol. 68, no. 2, pp. 805–812, 1994.

[57] R. Xu and I. A. Wilson, "Structural characterization of an early fusion intermediate of influenza virus hemagglutinin," *Journal of Virology*, vol. 85, no. 10, pp. 5172–5182, 2011.

[58] F. Ni, X. Chen, J. Shen, and Q. Wang, "Structural insights into the membrane fusion mechanism mediated by influenza virus hemagglutinin," *Biochemistry*, vol. 53, no. 5, pp. 846–854, 2014.

[59] J. Chen, J. J. Skehel, and D. C. Wiley, "N- and C-terminal residues combine in the fusion-pH influenza hemagglutinin HA2 subunit to form an N cap that terminates the triple-stranded coiled coil," *Proceedings of the National Academy of Sciences of the United States of America*, vol. 96, no. 16, pp. 8967–8972, 1999.

[60] M. Carleton and D. T. Brown, "Disulfide bridge-mediated folding of Sindbis virus glycoproteins," *Journal of Virology*, vol. 70, no. 8, pp. 5541–5547, 1996.

[61] M. Carleton and D. T. Brown, "The formation of intramolecular disulfide bridges is required for induction of the sindbis virus mutant ts23 phenotype," *Journal of Virology*, vol. 71, no. 10, pp. 7696–7703, 1997.

[62] M. Mulvey and D. T. Brown, "Involvement of the molecular chaperone BiP in maturation of Sindbis virus envelope glycoproteins," *Journal of Virology*, vol. 69, no. 3, pp. 1621–1627, 1995.

[63] M. Mulvey and D. T. Brown, "Assembly of the Sindbis virus spike protein complex," *Virology*, vol. 219, no. 1, pp. 125–132, 1996.

[64] B. S. Phinney, K. Blackburn, and D. T. Brown, "The surface conformation of sindbis virus glycoproteins E1 and E2 at neutral and low pH, as determined by mass spectrometry-based mapping," *Journal of Virology*, vol. 74, no. 12, pp. 5667–5678, 2000.

[65] B. S. Phinney and D. T. Brown, "Sindbis virus glycoprotein E1 is divided into two discrete domains at amino acid 129 by disulfide bridge connections," *Journal of Virology*, vol. 74, no. 19, pp. 9313–9316, 2000.

[66] C. B. Whitehurst, E. J. Soderblom, M. L. West, R. Hernandez, M. B. Goshe, and D. T. Brown, "Location and role of free cysteinyl residues in the Sindbis virus E1 and E2 glycoproteins," *Journal of Virology*, vol. 81, no. 12, pp. 6231–6240, 2007.

[67] B. A. Abell and D. T. Brown, "Sindbis virus membrane fusion is mediated by reduction of glycoprotein disulfide bridges at the cell surface," *Journal of Virology*, vol. 67, no. 9, pp. 5496–5501, 1993.

[68] A. J. Snyder, K. J. Sokoloski, and S. Mukhopadhyay, "Mutating conserved cysteines in the alphavirus E2 glycoprotein causes virus-specific assembly defects," *Journal of Virology*, vol. 86, no. 6, pp. 3100–3111, 2012.

[69] M. Carleton and D. T. Brown, "Events in the endoplasmic reticulum abrogate the temperature sensitivity of sindbis virus mutant ts23," *Journal of Virology*, vol. 70, no. 2, pp. 952–959, 1996.

[70] R. Danev and K. Nagayama, "Transmission electron microscopy with Zernike phase plate," *Ultramicroscopy*, vol. 88, no. 4, pp. 243–252, 2001.

[71] R. Danev, R. M. Glaeser, and K. Nagayama, "Practical factors affecting the performance of a thin-film phase plate for transmission electron microscopy," *Ultramicroscopy*, vol. 109, no. 4, pp. 312–325, 2009.

[72] R. Danev and K. Nagayama, "Optimizing the phase shift and the cut-on periodicity of phase plates for TEM," *Ultramicroscopy*, vol. 111, no. 8, pp. 1305–1315, 2011.

[73] K. Nagayama, "Another 60 years in electron microscopy: development of phase-plate electron microscopy and biological applications," *Journal of Electron Microscopy*, vol. 60, no. 1, pp. S43–S62, 2011.

[74] A.-C. Milazzo, A. Cheng, A. Moeller et al., "Initial evaluation of a direct detection device detector for single particle cryo-electron microscopy," *Journal of Structural Biology*, vol. 176, no. 3, pp. 404–408, 2011.

[75] X. Li, P. Mooney, S. Zheng et al., "Electron counting and beam-induced motion correction enable near-atomic-resolution single-particle cryo-em," *Nature Methods*, vol. 10, no. 6, pp. 584–590, 2013.

[76] H. M. Van Der Schaar, M. J. Rust, B.-L. Waarts et al., "Characterization of the early events in dengue virus cell entry by biochemical assays and single-virus tracking," *Journal of Virology*, vol. 81, no. 21, pp. 12019–12028, 2007.

[77] K. M. Smith, K. Nanda, C. J. Spears et al., "Structural mutants of dengue virus 2 transmembrane domains exhibit host-range phenotype," *Virology Journal*, vol. 8, article no. 289, 2011.

[78] Y. Zhang, J. Corver, P. R. Chipman et al., "Structures of immature flavivirus particles," *EMBO Journal*, vol. 22, no. 11, pp. 2604–2613, 2003.

[79] M. Lobigs, "Inefficient signalase cleavage promotes efficient nucleocapsid incorporation into budding flavivirus membranes," *Journal of Virology*, vol. 78, no. 1, pp. 178–186, 2004.

[80] J. Y. Leung, G. P. Pijlman, N. Kondratieva, J. Hyde, J. M. Mackenzie, and A. A. Khromykh, "Role of nonstructural protein NS2A in flavivirus assembly," *Journal of Virology*, vol. 82, no. 10, pp. 4731–4741, 2008.

[81] J. M. Mackenzie, M. K. Jones, and P. R. Young, "Immunolocalization of the Dengue virus nonstructural glycoprotein NS1 suggests a role in viral RNA replication," *Virology*, vol. 220, no. 1, pp. 232–240, 1996.

[82] M. Lobigs, "Flavivirus premembrane protein cleavage and spike heterodimer secretion require the function of the viral proteinase NS3," *Proceedings of the National Academy of Sciences of the United States of America*, vol. 90, no. 13, pp. 6218–6222, 1993.

[83] I. C. Lorenz, S. L. Allison, F. X. Heinz, and A. Helenius, "Folding and dimerization of tick-borne encephalitis virus envelope proteins prM and E in the endoplasmic reticulum," *Journal of Virology*, vol. 76, no. 11, pp. 5480–5491, 2002.

[84] I. A. Zybert, H. van der Ende-Metselaar, J. Wilschut, and J. M. Smit, "Functional importance of dengue virus maturation: infectious properties of immature virions," *Journal of General Virology*, vol. 89, no. 12, pp. 3047–3051, 2008.

[85] P. Plevka, A. J. Battisti, J. Junjhon et al., "Maturation of flaviviruses starts from one or more icosahedrally independent nucleation centres," *EMBO Reports*, vol. 12, pp. 602–606, 2011.

[86] I. A. Rodenhuis-Zybert, J. Wilschut, and J. M. Smit, "Partial maturation: an immune-evasion strategy of dengue virus?" *Trends in Microbiology*, vol. 19, no. 5, pp. 248–254, 2011.

[87] A. T. Lehrer and M. R. Holbrook, "Tick-borne encephalitis," in *Vaccines for Biodefense and Emerging and Neglected Diseases*, A. D. T. Barrett and L. R. Stanberry, Eds., vol. 1, Academic Press, Burlington, Mass, USA, 1st edition, 2009.

[88] F. X. Heinz, C. W. Mandl, H. Holzmann et al., "The flavivirus envelope protein E: isolation of a soluble form from tick-borne encephalitis virus and its crystallization," *Journal of Virology*, vol. 65, no. 10, pp. 5579–5583, 1991.

[89] F. A. Rey, "Dengue virus envelope glycoprotein structure: new insight into its interactions during viral entry," *Proceedings of the National Academy of Sciences of the United States of America*, vol. 100, no. 12, pp. 6899–6901, 2003.

[90] L. Li, S.-M. Lok, I.-M. Yu et al., "The flavivirus precursor membrane-envelope protein complex: structure and maturation," *Science*, vol. 319, no. 5871, pp. 1830–1834, 2008.

[91] V. C. Luca, J. AbiMansour, C. A. Nelson, and D. H. Fremont, "Crystal structure of the Japanese encephalitis virus envelope protein," *Journal of Virology*, vol. 86, no. 4, pp. 2337–2346, 2012.

[92] Y. Modis, S. Ogata, D. Clements, and S. C. Harrison, "Structure of the dengue virus envelope protein after membrane fusion," *Nature*, vol. 427, no. 6972, pp. 313–319, 2004.

[93] Y. Modis, S. Ogata, D. Clements, and S. C. Harrison, "A ligand-binding pocket in the dengue virus envelope glycoprotein," *Proceedings of the National Academy of Sciences of the United States of America*, vol. 100, no. 12, pp. 6986–6991, 2003.

[94] R. J. Kuhn, W. Zhang, M. G. Rossmann et al., "Structure of dengue virus: implications for flavivirus organization, maturation, and fusion," *Cell*, vol. 108, no. 5, pp. 717–725, 2002.

[95] F. A. Rey, F. X. Heinz, C. Mandl, C. Kunz, and S. C. Harrison, "The envelope glycoprotein from tick-borne encephalitis virus at 2 Å resolution," *Nature*, vol. 375, no. 6529, pp. 291–298, 1995.

[96] G. E. Nybakken, C. A. Nelson, B. R. Chen, M. S. Diamond, and D. H. Fremont, "Crystal structure of the West Nile virus envelope glycoprotein," *Journal of Virology*, vol. 80, no. 23, pp. 11467–11474, 2006.

[97] P. A. Bullough, F. M. Hughson, J. J. Skehel, and D. C. Wiley, "Structure of influenza haemagglutinin at the pH of membrane fusion," *Nature*, vol. 371, no. 6492, pp. 37–43, 1994.

[98] M. Kielian, "Structural surprises from the flaviviruses and alphaviruses," *Molecular Cell*, vol. 9, no. 3, pp. 454–456, 2002.

[99] M. Kielian and S. Jungerwirth, "Mechanisms of enveloped virus entry into cells," *Molecular Biology and Medicine*, vol. 7, no. 1, pp. 17–31, 1990.

[100] J. Dubochet, "Cryo-EM—the first thirty years," *Journal of Microscopy*, vol. 245, no. 3, pp. 221–224, 2012.

[101] R. H. Vogel and S. W. Provencher, "Three-dimensional reconstruction from electron micrographs of disordered specimens II. Implementation and results," *Ultramicroscopy*, vol. 25, no. 3, pp. 223–239, 1988.

[102] R. H. Vogel, S. W. Provencher, C.-H. von Bonsdorff, M. Adrian, and J. Dubochet, "Envelope structure of Semliki Forest virus reconstructed from cryo-electron micrographs," *Nature*, vol. 320, no. 6062, pp. 533–535, 1986.

[103] S. D. Fuller, "The T=4 envelope of sindbis virus is organized by interactions with a complementary T=3 capsid," *Cell*, vol. 48, no. 6, pp. 923–934, 1987.

[104] A. M. Paredes, D. T. Brown, R. Rothnagel et al., "Three-dimensional structure of a membrane-containing virus," *Proceedings of the National Academy of Sciences of the United States of America*, vol. 90, no. 19, pp. 9095–9099, 1993.

[105] M. J. Schlesinger, S. Schlesinger, and B. W. Burge, "Identification of a second glycoprotein in Sindbis virus," *Virology*, vol. 47, no. 2, pp. 539–541, 1972.

[106] R. P. Anthony and D. T. Brown, "Protein-protein interactions in an alphavirus membrane," *Journal of Virology*, vol. 65, no. 3, pp. 1187–1194, 1991.

[107] A. M. Paredes, M. N. Simon, and D. T. Brown, "The mass of the Sindbis virus nucleocapsid suggests it has $T = 4$ icosahedral symmetry," *Virology*, vol. 187, no. 1, pp. 329–332, 1992.

[108] R. P. Anthony, A. M. Paredes, and D. T. Brown, "Disulfide bonds are essential for the stability of the Sindbis virus envelope," *Journal of Virology*, vol. 190, no. 1, pp. 330–336, 1992.

[109] R. H. Cheng, R. J. Kuhn, N. H. Olson et al., "Nucleocapsid and glycoprotein organization in an enveloped virus," *Cell*, vol. 80, no. 4, pp. 621–630, 1995.

[110] T. J. Smith, R. H. Cheng, N. H. Olson et al., "Putative receptor binding sites on alphaviruses as visualized by cryoelectron microscopy," *Proceedings of the National Academy of Sciences of the United States of America*, vol. 92, no. 23, pp. 10648–10652, 1995.

[111] A. M. Paredes, H. Heidner, P. Thuman-Commike, B. V. V. Prasad, R. E. Johnston, and W. Chiu, "Structural localization of the E3 glycoprotein in attenuated Sindbis virus mutants," *Journal of Virology*, vol. 72, no. 2, pp. 1534–1541, 1998.

[112] S. V. Pletnev, W. Zhang, S. Mukhopadhyay et al., "Locations of carbohydrate sites on alphavirus glycoproteins show that e1 forms an icosahedral scaffold," *Cell*, vol. 105, no. 1, pp. 127–136, 2001.

[113] D. C. Flynn, W. J. Meyer, J. M. Mackenzie Jr., and R. E. Johnston, "A conformational change in Sindbis virus glycoproteins E1 and E2 is detected at the plasma membrane as a consequence of early virus-cell interaction," *Journal of Virology*, vol. 64, no. 8, pp. 3643–3653, 1990.

[114] W. J. Meyer, S. Gidwitz, V. K. Ayers, R. J. Schoepp, and R. E. Johnston, "Conformational alteration of sindbis virion glycoproteins induced by heat, reducing agents, or low pH," *Journal of Virology*, vol. 66, no. 6, pp. 3504–3513, 1992.

[115] O. Carugo and S. Pongor, "Recent progress in protein 3D structure comparison," *Current Protein and Peptide Science*, vol. 3, no. 4, pp. 441–449, 2002.

[116] Y. Zhu, B. Carragher, R. M. Glaeser et al., "Automatic particle selection: results of a comparative study," *Journal of Structural Biology*, vol. 145, no. 1-2, pp. 3–14, 2004.

[117] M. C. Scott, C.-C. Chen, M. Mecklenburg et al., "Electron tomography at 2.4-ångström resolution," *Nature*, vol. 483, no. 7390, pp. 444–447, 2012.

[118] E. V. Orlova and H. R. Saibil, "Structural analysis of macromolecular assemblies by electron microscopy," *Chemical Reviews*, vol. 111, no. 12, pp. 7710–7748, 2011.

[119] E. V. Orlova and H. R. Saibil, "Structure determination of macromolecular assemblies by single-particle analysis of cryo-electron micrographs," *Current Opinion in Structural Biology*, vol. 14, no. 5, pp. 584–590, 2004.

[120] B. W. Burge and E. R. Pfefferkorn, "Isolation and characterization of conditional-lethal mutants of Sindbis virus," *Journal of Virology*, vol. 30, no. 2, pp. 204–213, 1966.

[121] J. Edwards, E. Mann, and D. T. Brown, "Conformational changes in Sindbis virus envelope proteins accompanying exposure to low pH," *Journal of Virology*, vol. 45, no. 3, pp. 1090–1097, 1983.

[122] A. M. Paredes, D. Ferreira, M. Horton et al., "Conformational changes in Sindbis virions resulting from exposure to low pH and interactions with cells suggest that cell penetration may occur at the cell surface in the absence of membrane fusion," *Journal of Virology*, vol. 324, no. 2, pp. 373–386, 2004.

[123] J. Edwards and D. T. Brown, "Sindbis virus-mediated cell fusion from without is a two-step event," *Journal of General Virology*, vol. 67, no. 2, pp. 377–380, 1986.

[124] L. Wessels, M. W. Elting, D. Scimeca, and K. Weninger, "Rapid membrane fusion of individual virus particles with supported lipid bilayers," *Biophysical Journal*, vol. 93, no. 2, pp. 526–538, 2007.

[125] L. Haag, H. Garoff, L. Xing, L. Hammar, S.-T. Kan, and R. H. Cheng, "Acid-induced movements in the glycoprotein shell of an alphavirus turn the spikes into membrane fusion mode," *EMBO Journal*, vol. 21, no. 17, pp. 4402–4410, 2002.

[126] A. Koschinski, G. Wengler, and H. Repp, "The membrane proteins of flaviviruses form ion-permeable pores in the target membrane after fusion: identification of the pores and analysis of their possible role in virus infection," *Journal of General Virology*, vol. 84, no. 7, pp. 1711–1721, 2003.

[127] S. Cao and W. Zhang, "Characterization of an early-stage fusion intermediate of Sindbis virus using cryoelectron microscopy," *Proceedings of the National Academy of Sciences of the United States of America*, vol. 110, no. 33, pp. 13362–13367, 2013.

[128] A. Salminen, J. M. Wahlberg, M. Lobigs, P. Liljestrom, and H. Garoff, "Membrane fusion process of Semliki Forest virus II: cleavage-dependent reorganization of the spike protein complex controls virus entry," *The Journal of Cell Biology*, vol. 116, no. 2, pp. 349–357, 1992.

[129] J. White, J. Kartenbeck, and A. Helenius, "Fusion of Semliki forest virus with the plasma membrane can be induced by low pH," *The Journal of Cell Biology*, vol. 87, no. 1, pp. 264–272, 1980.

[130] S. W. Gollins and J. S. Porterfield, "pH-dependent fusion between the flavivirus West Nile and liposomal model membranes," *Journal of General Virology*, vol. 67, part 1, pp. 157–166, 1986.

[131] G. Wang, R. Hernandez, K. Weninger, and D. T. Brown, "Infection of cells by Sindbis virus at low temperature," *Virology*, vol. 362, no. 2, pp. 461–467, 2007.

[132] B. Brandenburg, L. Y. Lee, M. Lakadamyali, M. J. Rust, X. Zhuang, and J. M. Hogle, "Imaging poliovirus entry in live cells," *PLoS Biology*, vol. 5, article e183, 2007.

[133] R. Vancini, G. Wang, D. Ferreira, R. Hernandez, and D. T. Brown, "Alphavirus genome delivery occurs directly at the plasma membrane in a time- and temperature-dependent process," *The Journal of Virology*, vol. 87, no. 8, pp. 4352–4359, 2013.

[134] J. M. Smit, R. Bittman, and J. Wilschut, "Low-pH-dependent fusion of Sindbis virus with receptor-free cholesterol- and sphingolipid-containing liposomes," *The Journal of Virology*, vol. 73, no. 10, pp. 8476–8484, 1999.

[135] M. Kielian and F. A. Rey, "Virus membrane-fusion proteins: more than one way to make a hairpin," *Nature Reviews Microbiology*, vol. 4, no. 1, pp. 67–76, 2006.

[136] A. Helenius, J. Kartenbeck, K. Simons, and E. Fries, "On the entry of Semliki Forest virus into BHK-21 cells," *The Journal of Cell Biology*, vol. 84, no. 2, pp. 404–420, 1980.

[137] J. M. Wahlberg, R. Bron, J. Wilschut, and H. Garoff, "Membrane fusion of Semliki Forest virus involves homotrimers of the fusion protein," *Journal of Virology*, vol. 66, no. 12, pp. 7309–7318, 1992.

[138] J. M. Wahlberg and H. Garoff, "Membrane fusion process of Semliki Forest virus I: low pH-induced rearrangement in spike protein quaternary structure precedes virus penetration into cells," *The Journal of Cell Biology*, vol. 116, no. 2, pp. 339–348, 1992.

[139] J. Lescar, A. Roussel, M. W. Wien et al., "The fusion glycoprotein shell of Semliki Forest virus: an icosahedral assembly primed for fusogenic activation at endosomal pH," *Cell*, vol. 105, no. 1, pp. 137–148, 2001.

[140] S. W. Metz, C. Geertsema, B. E. Martina et al., "Functional processing and secretion of Chikungunya virus E1 and E2 glycoproteins in insect cells," *Virology Journal*, vol. 8, article 353, 2011.

[141] G. Roman-Sosa and M. Kielian, "The interaction of alphavirus E1 protein with exogenous domain III defines stages in virus-membrane fusion," *Journal of Virology*, vol. 85, no. 23, pp. 12271–12279, 2011.

[142] K. Stiasny, S. L. Allison, J. Schalich, and F. X. Heinz, "Membrane interactions of the tick-borne encephalitis virus fusion protein E at low pH," *Journal of Virology*, vol. 76, no. 8, pp. 3784–3790, 2002.

[143] A. Roussel, J. Lescar, M.-C. Vaney, G. Wengler, and F. A. Rey, "Structure and interactions at the viral surface of the envelope protein E1 of semliki forest virus," *Structure*, vol. 14, no. 1, pp. 75–86, 2006.

[144] C. Sánchez-San Martín, H. Sosa, and M. Kielian, "A stable prefusion intermediate of the alphavirus fusion protein reveals critical features of class II membrane fusion," *Cell Host & Microbe*, vol. 4, no. 6, pp. 600–608, 2008.

[145] D. L. Gibbons, M.-C. Vaney, A. Roussel et al., "Conformational change and protein-protein interactions of the fusion protein of semliki forest virus," *Nature*, vol. 427, no. 6972, pp. 320–325, 2004.

[146] J. White and A. Helenius, "pH-Dependent fusion between the Semliki Forest virus membrane and liposomes," *Proceedings of the National Academy of Sciences of the United States of America*, vol. 77, no. 6, pp. 3273–3277, 1980.

[147] R. Bron, J. M. Wahlberg, H. Garoff, and J. Wilschut, "Membrane fusion of semliki forest virus in a model system: correlation between fusion kinetics and structural changes in the envelope glycoprotein," *EMBO Journal*, vol. 12, no. 2, pp. 693–701, 1993.

[148] A. Luukkonen, M. Brummer-Korvenkontio, and O. Renkonen, "Lipids of cultured mosquito cells (*Aedes albopictus*): comparison with cultured mammalian fibroblasts (BHK 21 cells)," *Biochimica et Biophysica Acta*, vol. 326, no. 2, pp. 256–261, 1973.

[149] A. Hafer, R. Whittlesey, D. T. Brown, and R. Hernandez, "Differential incorporation of cholesterol by sindbis virus grown in mammalian or insect cells," *Journal of Virology*, vol. 83, no. 18, pp. 9113–9121, 2009.

[150] J. M. Smit, G. Li, P. Schoen et al., "Fusion of alphaviruses with liposomes is a non-leaky process," *FEBS Letters*, vol. 521, no. 1–3, pp. 62–66, 2002.

[151] V. Madan, M. A. Sanz, and L. Carrasco, "Requirement of the vesicular system for membrane permeabilization by Sindbis virus," *Virology*, vol. 332, no. 1, pp. 307–315, 2005.

[152] J. Edwards and D. T. Brown, "Sindbis virus induced fusion of tissue cultured Aedes albopictus (mosquito) cells," *Virus Research*, vol. 1, no. 8, pp. 703–711, 1984.

[153] D. P. Fan and B. M. Sefton, "The entry into host cells of Sindbis virus, vesicular stomatitis virus and Sendai virus," *Cell*, vol. 15, no. 3, pp. 985–992, 1978.

[154] D. J. Schibli and W. Weissenhorn, "Class I and class II viral fusion protein structures reveal similar principles in membrane fusion," *Molecular Membrane Biology*, vol. 21, no. 6, pp. 361–371, 2004.

[155] M. Kielian, "Class II virus membrane fusion proteins," *Journal of Virology*, vol. 344, no. 1, pp. 38–47, 2006.

[156] D. Moradpour and F. Penin, "Hepatitis C virus proteins: from structure to function," *Current Topics in Microbiology and Immunology*, vol. 369, pp. 113–142, 2013.

[157] K. El Omari, O. Iourin, K. Harlos, J. M. Grimes, and D. I. Stuart, "Structure of a pestivirus envelope glycoprotein E2 clarifies its role in cell entry," *Cell Reports*, vol. 3, no. 1, pp. 30–35, 2013.

[158] T. Krey, H.-J. Thiel, and T. Rümenapf, "Acid-resistant bovine pestivirus requires activation for pH-triggered fusion during entry," *Journal of Virology*, vol. 79, no. 7, pp. 4191–4200, 2005.

[159] M. Rusu, R. Bonneau, M. R. Holbrook et al., "An assembly model of Rift Valley fever virus," *Frontiers in Microbiology*, vol. 3, article 254, 2012.

[160] P. A. Bullough, F. M. Hughson, A. C. Treharne, R. W. H. Ruigrok, J. J. Skehel, and D. C. Wiley, "Crystals of a fragment of influenza haemagglutinin in the low pH induced conformation," *Journal of Molecular Biology*, vol. 236, no. 4, pp. 1262–1265, 1994.

[161] X. Han, J. H. Bushweller, D. S. Cafiso, and L. K. Tamm, "Membrane structure and fusion-triggering conformational change of the fusion domain from influenza hemagglutinin," *Nature Structural Biology*, vol. 8, no. 8, pp. 715–720, 2001.

[162] J. J. Skehel and D. C. Wiley, "Receptor binding and membrane fusion in virus entry: the influenza hemagglutinin," *Annual Review of Biochemistry*, vol. 69, pp. 531–569, 2000.

[163] J. Lippincott-Schwartz, T. H. Roberts, and K. Hirschberg, "Secretory protein trafficking and organelle dynamics in living cells," *Annual Review of Cell and Developmental Biology*, vol. 16, pp. 557–589, 2000.

[164] R. Hernandez, C. Sinodis, M. Horton, D. Ferreira, C. Yang, and D. T. Brown, "Deletions in the transmembrane domain of a sindbis virus glycoprotein alter virus infectivity, stability, and host range," *The Journal of Virology*, vol. 77, no. 23, pp. 12710–12719, 2003.

[165] J. P. Kononchik, R. Hernandez, and D. T. Brown, "An alternative pathway for alphavirus entry," *Virology Journal*, vol. 8, article 304, 2011.

[166] D. T. Brown and R. Hernandez, "Infection of cells by alphaviruses," *Advances in Experimental Medicine and Biology*, vol. 726, pp. 181–199, 2012.

[167] M. A. Sanz, L. Pérez, and L. Carrasco, "Semliki forest virus 6K protein modifies membrane permeability after inducible expression in *Escherichia coli* cells," *The Journal of Biological Chemistry*, vol. 269, no. 16, pp. 12106–12110, 1994.

[168] A. Muñoz, J. L. Castrillo, and L. Carrasco, "Modification of membrane permeability during Semliki Forest virus infection," *Virology*, vol. 146, no. 2, pp. 203–212, 1985.

[169] H. J. Risselada, G. Bubnis, and H. Grubmüller, "Expansion of the fusion stalk and its implication for biological membrane fusion," *Proceedings of the National Academy of Sciences of the United States of America*, 2014.

[170] R. Guinea and L. Carrasco, "Concanamycin A: a powerful inhibitor of enveloped animal-virus entry into cells," *Biochemical and Biophysical Research Communications*, vol. 201, no. 3, pp. 1270–1278, 1994.

[171] R. Guinea and L. Carrasco, "Requirement for vacuolar proton-ATPase activity during entry of influenza virus into cells," *The Journal of Virology*, vol. 69, no. 4, pp. 2306–2312, 1995.

[172] W. T. Heller and K. C. Littrell, "Small-angle neutron scattering for molecular biology: basics and instrumentation," *Methods in Molecular Biology*, vol. 544, pp. 293–305, 2009.

[173] G. Zaccaï and B. Jacrot, "Small angle neutron scattering," *Annual Review of Biophysics and Bioengineering*, vol. 12, pp. 139–157, 1983.

[174] W. T. Heller, "Small-angle neutron scattering and contrast variation: a powerful combination for studying biological structures," *Acta Crystallographica D: Biological Crystallography*, vol. 66, no. 11, pp. 1213–1217, 2010.

[175] G. Fibriansah, T.-S. Ng, V. A. Kostyuchenko et al., "Structural changes in dengue virus when exposed to a temperature of 37°C," *Journal of Virology*, vol. 87, no. 13, pp. 7585–7592, 2013.

[176] Y. Zhang, W. Zhang, S. Ogata et al., "Conformational changes of the flavivirus E glycoprotein," *Structure*, vol. 12, no. 9, pp. 1607–1618, 2004.

[177] H. M. van der Schaar, M. J. Rust, B.-L. Waarts et al., "Characterization of the early events in dengue virus cell entry by biochemical assays and single-virus tracking," *Journal of Virology*, vol. 81, no. 21, pp. 12019–12028, 2007.

[178] X. Zhang, J. Sheng, P. Plevka, R. J. Kuhn, M. S. Diamond, and M. G. Rossmann, "Dengue structure differs at the temperatures of its human and mosquito hosts," *Proceedings of the National Academy of Sciences of the United States of America*, vol. 110, no. 17, pp. 6795–6799, 2013.

[179] R. Hernandez, A. Paredes, and D. T. Brown, "Sindbis virus conformational changes induced by a neutralizing anti-E1 monoclonal antibody," *Journal of Virology*, vol. 82, no. 12, pp. 5750–5760, 2008.

[180] R. Hernandez and A. Paredes, "Sindbis virus as a model for studies of conformational changes in a metastable virus and the role of conformational changes in in vitro antibody

neutralisation," *Reviews in Medical Virology*, vol. 19, no. 5, pp. 257–272, 2009.

[181] S.-M. Lok, V. Kostyuchenko, G. E. Nybakken et al., "Binding of a neutralizing antibody to dengue virus alters the arrangement of surface glycoproteins," *Nature Structural and Molecular Biology*, vol. 15, no. 3, pp. 312–317, 2008.

[182] S. Cook, G. Moureau, R. E. Harbach et al., "Isolation of a novel species of flavivirus and a new strain of Culex flavivirus (Flaviviridae) from a natural mosquito population in Uganda," *Journal of General Virology*, vol. 90, no. 11, pp. 2669–2678, 2009.

[183] M. Calzolari, L. Zé-Zé, D. Růžek et al., "Detection of mosquito-only flaviviruses in Europe," *Journal of General Virology*, vol. 93, pp. 1215–1225, 2012.

[184] V. A. Kostyuchenko, Q. Zhang, J. L. Tan, T.-S. Ng, and S.-M. Lok, "Immature and mature dengue serotype 1 virus structures provide insight into the maturation process," *Journal of Virology*, vol. 87, no. 13, pp. 7700–7707, 2013.

[185] V. A. Kostyuchenko, P. L. Chew, T. S. Ng, and S. M. Lok, "Near-atomic resolution cryo-electron microscopic structure of dengue serotype 4 virus," *Journal of Virology*, vol. 88, no. 1, pp. 477–482, 2014.

[186] J. A. G. Briggs, "Structural biology in situ-the potential of subtomogram averaging," *Current Opinion in Structural Biology*, vol. 23, no. 2, pp. 261–267, 2013.

[187] T. A. M. Bharat, J. D. Riches, L. Kolesnikova et al., "Cryo-electron tomography of marburg virus particles and their morphogenesis within infected cells," *PLoS Biology*, vol. 9, no. 11, Article ID e1001196, 2011.

15

Screening of Viral Pathogens from Pediatric Ileal Tissue Samples after Vaccination

Laura Hewitson,[1,2] **James B. Thissen,**[3] **Shea N. Gardner,**[4] **Kevin S. McLoughlin,**[4] **Margaret K. Glausser,**[1] **and Crystal J. Jaing**[3]

[1] *The Johnson Center for Child Health and Development, 1700 Rio Grande Street, Austin, TX 78701, USA*
[2] *Department of Psychiatry, University of Texas Southwestern, Dallas, TX 75390, USA*
[3] *Physical & Life Sciences Directorate, Lawrence Livermore National Laboratory, Livermore, CA 94550, USA*
[4] *Computations Directorate, Lawrence Livermore National Laboratory, Livermore, CA 94550, USA*

Correspondence should be addressed to Laura Hewitson; lhewitson@johnson-center.org

Academic Editor: Subhash Verma

In 2010, researchers reported that the two US-licensed rotavirus vaccines contained DNA or DNA fragments from porcine circovirus (PCV). Although PCV, a common virus among pigs, is not thought to cause illness in humans, these findings raised several safety concerns. In this study, we sought to determine whether viruses, including PCV, could be detected in ileal tissue samples of children vaccinated with one of the two rotavirus vaccines. A broad spectrum, novel DNA detection technology, the Lawrence Livermore Microbial Detection Array (LLMDA), was utilized, and confirmation of viral pathogens using the polymerase chain reaction (PCR) was conducted. The LLMDA technology was recently used to identify PCV from one rotavirus vaccine. Ileal tissue samples were analyzed from 21 subjects, aged 15–62 months. PCV was not detected in any ileal tissue samples by the LLMDA or PCR. LLMDA identified a human rotavirus A from one of the vaccinated subjects, which is likely due to a recent infection from a wild type rotavirus. LLMDA also identified human parechovirus, a common gastroenteritis viral infection, from two subjects. Additionally, LLMDA detected common gastrointestinal bacterial organisms from the *Enterobacteriaceae*, *Bacteroidaceae*, and *Streptococcaceae* families from several subjects. This study provides a survey of viral and bacterial pathogens from pediatric ileal samples, and may shed light on future studies to identify pathogen associations with pediatric vaccinations.

1. Introduction

Rotavirus is the most common cause of severe diarrhea among infants and young children [1]. Prior to the introduction of rotavirus vaccines, rotavirus infection was estimated to cause approximately 2.7 million cases of severe gastroenteritis in children, almost 60,000 hospitalizations, and around 37 deaths each year in the USA alone [2]. Three vaccines against rotavirus have been developed: Rotashield (Wyeth-Lederle Vaccines and Pediatrics, [3]), RotaTeq (Merck, [4]), and Rotarix (GlaxoSmithKline, [5]). Rotashield, a rhesus-based tetravalent rotavirus vaccine, was recommended for routine vaccination of US infants in 1999 [6] but was withdrawn from the US market within 1 year of its introduction because of its association with intussusception [7]. RotaTeq, a human-bovine reassortant rotavirus vaccine [8], was

recommended for vaccination of US infants in 2006 [9] with 3 doses administered orally at ages 2, 4, and 6 months [10]. In 2008, Rotarix, a monovalent vaccine based on an attenuated human rotavirus [11], was licensed in the USA for pediatric use as a 2-dose series and recommended for vaccination of US infants in June 2008 [12]. Since the introduction of rotavirus vaccines, there has been a dramatic reduction in the incidence and severity of rotavirus infections both in the US and globally [13–16].

During the course of developing novel virus detection techniques, researchers at the San Francisco Blood Research Systems Institute and Lawrence Livermore National Laboratory (LLNL) unexpectedly identified nucleic acids from an adventitious virus in Rotarix [17]. The detected virus shared 98% homology with porcine circovirus-1 (PCV-1) and covered the complete circular genome [17]. PCV infection is common in pigs and the virus is often detected in human stool samples [18] but is not believed to cause illness among humans [19–21]. Contamination of Rotarix with PCV-1 was subsequently confirmed by the vaccine manufacturer. In March 2010, in light of these findings, the US Food and Drug Administration (FDA) recommended temporarily suspending the use of Rotarix [22]. On May 6, 2010, the FDA reported preliminary findings that the RotaTeq vaccine also contained detectable PCV material [23].

On May 7, 2010, the FDA Vaccines and Related Biological Products Advisory Committee (VRBPAC) met to review whether the contaminated rotavirus vaccines could pose risks to human health. The committee concluded that based on the available evidence, the hypothetical risk of PCV infection among humans does not outweigh the observed benefits of rotavirus vaccines in preventing severe acute gastroenteritis among infants. The committee expressed reassurance that the detection of DNA and DNA fragments from PCV in rotavirus vaccines was not likely to cause harm to humans and recommended that information on this topic be provided to parents prior to vaccination. The committee did, however, recommend that the vaccine manufacturers work to develop rotavirus vaccines free of PCV1 and PCV2 contaminants. On May 14, 2010, the FDA issued a recommendation for pediatricians to resume use of Rotarix and to continue use of RotaTeq [24]. Subsequent testing by the vaccine manufacturers identified that the PCV material was introduced into both rotavirus vaccines through porcine-derived trypsin, a reagent used in the cell-culture growth process of vaccine production, commencing very early in the development process [17, 25]. The use of cells or biological products from other species in the production of vaccines can lead to leakage of cellular DNA and the introduction of noninfectious proviral DNA [17].

While the recent publicity about potential safety concerns over rotavirus vaccines does not appear to have had a negative impact on vaccine administration practices of most physicians, it has raised safety concerns among some parents [26]. The goal of this study was to use a novel DNA detection technology, the Lawrence Livermore Microbial Detection Array (LLMDA), [27, 28] to determine whether viral or bacterial DNA or DNA fragments, including PCV, could be detected in ileal tissue samples from children following vaccination with rotavirus vaccines.

The LLMDA is a nucleic acid detection technology that contains a total of 388,000 probes, designed to detect 2,200 viral species (38,000 target sequences) and 900 bacterial species (3,500 target sequences). This microarray uses long (50–65 base-pair) oligonucleotide probes to enable sensitive detection of known viral and bacterial species and the detection of divergent species with homology to sequenced organisms. The array data is analyzed using a novel composite likelihood maximization method to predict the strains and species that are most likely present in a sample. Each target detected has a log likelihood score, which is estimated from the BLAST similarity scores of the probes to each of the possible target sequences, together with the probe sequence complexity and other covariates derived from the BLAST results. Targets are color-coded and grouped by taxonomic

family. This array has been used to detect a wide array of viral infections from various clinical samples [27].

Though various nucleic acid detection technologies including TaqMan PCR and Luminex bead based systems are able to identify selected pathogens at the species or strain level rapidly, they do not have the capability to provide broad spectrum detection about known or novel organisms. While sequencing provides the most in-depth information to characterize a microbial genome, the costs, labor, and time associated with library preparation, sequencing, bioinformatic analysis, and data storage may be prohibitive when analyzing many isolates to screen for pathogens. Microarrays provide a means for broad surveillance of sequenced pathogens with assay time and cost close to PCR assays. In this study, we first used the LLMDA to screen for viral and bacterial pathogens in human ileal samples, and then used PCR to confirm the microarray findings.

2. Materials and Methods

2.1. Ethics Statement. This study was approved by the Austin Multi-Institutional Review Board (AMIRB). Subjects were scheduled to undergo a diagnostic ileocolonoscopy for chronic GI symptoms during the period from January 2008 to December 2010. They were recruited from a single pediatric gastroenterology clinic and written informed consent was received from the parent or guardian of all subjects prior to enrollment. Twenty-one subjects aged 16 to 52 months were included in this study and represented 15 males and 6 females. Subjects were (i) vaccinated against rotavirus ($n = 9$) using one of two rotavirus vaccines (Rotarix or RotaTeq); (ii) vaccinated but not against rotavirus ($n = 8$); or (iii) unvaccinated ($n = 4$). Subject demographics and vaccine status are shown in Table 1.

2.2. Sample Collection and Processing. A pinch biopsy from the terminal ileum was collected using a standard disposable forceps biopsy device, in accordance with routine diagnostic biopsy protocol. Each biopsy retrieved was immediately dissected so that at least half of the biopsy was fixed for subsequent histological examination for clinical pathology. The remaining sample was placed directly into RNAlater (Qiagen Inc., Valencia, CA) for between 24 and 48 hours and subsequently stored at −80°C until processing. All samples were coded and were blinded in regard to vaccination status. Samples were shipped to LLNL on dry ice. One sample of the reportedly PCV-contaminated RotaTeq live, oral, pentavalent vaccine (lot 0147Z) was also included in the analyses. The PCV-contaminated Rotarix was not available for analysis.

2.3. Nucleic Acid Extraction from Human Ileum and Vaccine Samples

2.3.1. Extraction from Human Ileum. One ileum sample was extracted per patient. Each ileum was roughly 20 mg and cut into approximately four smaller pieces prior to being placed in a 2 mL bead beating tube containing 0.5 g of 1.0 mm zirconia beads and 500 μL of chilled Hank's buffered salt solution.

TABLE 1: Subject demographics. Demographics of study subjects include gender, year of birth, vaccination status, and age in months at sample collection. Rota$^-$, vaccinated but not against rotavirus; Rota$^+$, fully vaccinated including against rotavirus.

Subject ID	Sex	Year of birth	Vaccination status	Age (months) at collection
1	Male	2007	Vaccinated/Rota$^-$	16.3
2	Female	2006	Unvaccinated	21.8
3	Female	2007	Vaccinated/Rota$^+$	17.3
4	Female	2006	Vaccinated/Rota$^-$	23.9
5	Male	2006	Vaccinated/Rota$^+$	27.9
6	Female	2006	Vaccinated/Rota$^-$	33.2
7	Male	2006	Unvaccinated	31.7
8	Female	2006	Vaccinated/Rota$^+$	29.4
9	Female	2006	Vaccinated/Rota$^-$	30
10	Male	2006	Vaccinated/Rota$^+$	34.2
11	Male	2006	Vaccinated/Rota$^-$	41.3
12	Male	2006	Unvaccinated	47.5
13	Male	2007	Vaccinated/Rota$^-$	33
14	Male	2006	Vaccinated/Rota$^-$	41.1
15	Male	2007	Vaccinated/Rota$^+$	36.1
16	Male	2006	Unvaccinated	47.3
17	Male	2006	Vaccinated/Rota$^+$	52.3
18	Male	2006	Vaccinated/Rota$^-$	46.5
19	Male	2006	Vaccinated/Rota$^+$	50.9
20	Male	2008	Vaccinated/Rota$^+$	32.2
21	Male	2006	Vaccinated/Rota$^+$	47.8

The tubes were bead beat for 30 sec at 25 speed. Following bead beating the samples were clarified by centrifuging for 5 min at 15,000 ×g. The supernatant was transferred to a new 1.5 mL tube to continue nucleic acid extraction. Due to the small amount of ileal tissue available for this study, no DNase treatment or filtration to remove bacterial or host cells was performed. Nucleic acids were extracted using the Qiagen QIAamp UltraSens Virus Kit (Qiagen) according to the manufacturer's instructions. Following extraction the nucleic acid concentration was determined using the Invitrogen Qubit fluorometer (Invitrogen, Grand Island, NY). Approximately 400 ng of DNA and 1.4 μg of RNA were obtained from each ileum sample after extraction.

2.3.2. Extraction from RotaTeq Vaccine. A RotaTeq vaccine sample was extracted for analysis. One dose contained 2 mL of the vaccine; therefore, two 1 mL extractions were performed. Each 1 mL extraction was performed using the QIAamp UltraSens Virus Kit following the manufacturer's protocol. Following extraction each vaccine sample was combined and nucleic acid concentration was determined using a Qubit fluorometer.

2.4. Microarray Processing

2.4.1. Whole Genome Amplification and Purification. The extracted ileum and vaccine samples were whole genome amplified using a random amplification protocol as described previously [26]. Briefly, 50 ng of DNA from each terminal ileal

sample and 10 ng of DNA from the RotaTeq vaccine sample were used in the amplification procedure. The amplification procedure was performed by incubating each sample with 1 μL of random primer 5′-GATGAGGGAAGATGGGGN-NNNNNNNN-3′ (100 pmole/μL) for 2 min at 85°C and immediately placed on ice for 2 min. To each reaction, 4 μL 5x Superscript III buffer, 1 μL dNTP (12.5 mM), 2 μL DTT (0.1 M), 1 μL Invitrogen Superscript III reverse transcriptase, and 1 μL Ultrapure DEPC water (Invitrogen) were added. The samples were placed in a Tetrad PTC-225 thermal cycler (MJ Research, Quebec, Canada) with the following conditions: 25°C for 10 min, 42°C for 2 hours, and 70°C for 5 min. Following first strand synthesis, each 20 μL sample was mixed with 2.4 μL 10x Klenow buffer (New England Biolabs, Ipswich, MA) and 0.5 μL 12.5 mM dNTP (New England Biolabs). Next, the samples were incubated for 2 min at 85°C and immediately placed on ice for 2 min. Lastly, 1 μL Klenow buffer was added to the samples and allowed to incubate at 37°C for 60 min followed by 70°C for 20 min.

Samples were amplified by combining 5 μL of the double-stranded cDNA product with 5 μL 10x Sigma Taq buffer, 1 μL dNTP (12.5 mM), 1 μL primer 5′-GATGAGGGAAGATGG-GG-3′ (100 pmole/μL), 1 μL Sigma KlenTaq LA polymerase, and 37 μL water. Reactions were placed in a thermocycler (Tetrad Thermal Cycler, MJ Research, Quebec, Canada) with the following conditions: 94°C for 2 min; 40 cycles of 94°C for 30 sec, 50°C for 1 min, and 68°C for 1 min; and 72°C for 10 min. Amplified samples were purified using the Qiaquick PCR Purification Columns (Qiagen) according to manufacturer's instructions. Samples were eluted in 40 μL of Buffer EB from the Qiagen kit and nucleic acid concentration determined by Qubit fluorometer.

2.4.2. Microarray Hybridization. We used the LLMDA v2 for analysis of viral or bacterial content from the tissue or vaccine samples. This array contains 388,000 probes to detect all sequenced viruses and bacteria that we sequenced before April of 2007 [28]. Additionally, we analyzed a subset of the samples using a multiplexed format of the LLMDA v2 printed on the Roche NimbleGen (Roche NimbleGen, Madison, WI) 4 × 72 K array format. Samples 1–8 and 10 were run on the 4 × 72 K format of the LLMDA. The other samples and RotaTeq were run on the 388 K format of the LLMDA.

For each sample, 1 μg of amplified product was fluorescently labeled using the One-Color DNA Labeling Kit (Roche NimbleGen) according to the recommended protocols. The DNA was purified after labeling and hybridized using the NimbleGen Hybridization Kit to the LLMDA according to manufacturers' instructions. The microarrays were allowed to hybridize for 17 hours and washed using the NimbleGen Wash Buffer Kit according to manufacturer's instructions. Microarrays were scanned on an Axon GenePix 4000B 5 μm scanner (Molecular Devices, Sunnyvale, CA). The scanned tiff image files were aligned using the NimbleScan Version 2.4 software and pair text files were exported for data analysis.

2.5. Microarray Data Analysis. Data was analyzed using the automated LLMDA analysis algorithm—composite likelihood maximization algorithm [28]. A threshold of 99% was used in the data analysis to analyze only the probes with signal intensity above 99% of random controls. Random controls are probes that do not hybridize to any known target sequences and were designed to match the overall GC content and thermodynamics of the target probes.

2.6. PCR Primer Design for Confirmation of Viral Pathogens from the LLMDA. Taqman signatures were designed using the run Primux triplet script that is part of the PriMux software [29] for the viruses detected in ileum samples 5, 7, 9, 11, and 14. Target sets were comprised of the available complete sequences for the following viruses: Torque teno virus (TTV)-like minivirus (6 genomes, validation for sample 7), human parechoviruses (44 genomes, validation for sample 9), small anelloviruses and Torque teno midi viruses (TTMV) (20 genomes, samples 11 and 14), echovirus 9 (7 genomes, validation for sample 5), and rotavirus A (7077 sequences, all segments, for sample 5). Predicted targets were identified using simulate_PCR.pl (submitted, https://sourceforge.net/projects/simulatepcr) based on comparison to the LLNL large (48 GB) internal database of all available finished and assembled microbial genomes from NCBI, multiple public and university sequencing centers (e.g., Broad, JCVI, IMG, Sanger, Singapore, etc.), and from collaborators, currently over 48 GB of sequence data. From the multiple signatures designed for each target set, one was selected that was predicted to detect the virus and its near neighbors that were reported by LLMDA results for that

sample. Primer sequences and expected amplicon sizes can be seen in Table 2. For PCR to detect PCV-2, an 84 bp PCR assay from previous studies of the RotaTeq vaccine [25] was used.

2.7. PCR Analysis. PCR primers were ordered through Integrated DNA Technologies. Each real-time PCR reaction consisted of 2.5 μL 10x PCR buffer (Invitrogen), 1 μL forward/reverse primer mix (10 μM), 1.75 μL MgCl$_2$ (50 mM), 1 μL BSA (2 μg/μL), 0.5 μL dNTPs (10 mM), 0.25 μL Invitrogen Platinum Taq polymerase (5 U/μL), and 13 μL water. All reactions were carried out on a Tetrad PTC-225 thermal cycler with the following conditions: 95°C for 3 min; 40 cycles of 95°C for 20 sec, 60°C for 30 sec, and 72°C for 30 sec; and 72°C for 2 min. Reactions with the RotaTeq vaccine and other controls were run with 1 ng of total DNA or cDNA, while reactions with ileum samples were run with 10 ng of total DNA or cDNA. Each sample was run in duplicate. The PCR products from the duplicate reactions were then combined and run on a 4% agarose gel.

3. Results

3.1. Microbial Detection Array Results. The viruses detected from the samples by the LLMDA array are shown in Figure 1. Microarray data analysis parameters were set to give both bacterial and viral results with probe signal intensity above 99% of random control probes. In sample 5, LLMDA detected probes that hybridized to several segments of the human rotavirus A and an echovirus 9. The detected rotavirus segments all appear to be from human origin. LLMDA also identified a human parechovirus 1 from sample 9. Small anellovirus 2 was detected in samples 11 and 14. TTV-like minivirus was detected in sample 7. Human endogenous retroviruses (HERVs) were detected in most ileal samples (data not shown). This is likely due to the residual human genomic DNA present in the samples. No other viral targets were identified from other ileal samples. A summary of the viral results is shown in Table 3. LLMDA identified several segments of human rotavirus A (segments 7, 9, and 3) from the RotaTeq vaccine (Figure 1). Several bovine rotavirus sequence segments including segment 1, 2, and 6 were also detected. Additionally, a baboon endogenous virus strain M7 was detected, likely due to the monkey cell line in which RotaTeq was produced from.

Additionally, LLMDA also detected bacteria from some of the ileum samples, summarized in Table 4. *Bacteroides* species were identified in samples 1, 5, 6, 15, 18, and 20. Plasmids from the *Shigella* species were detected in samples 5 and 15. *Streptococcus agalactiae* was detected in sample 15, and *Streptomyces coelicolor* was detected in sample 20.

3.2. PCR Assay to Confirm Microarray Results. The viruses detected by microarrays and the negative PCV results from microarrays were all confirmed by PCR assays (Figure 2). The PCV-2 PCR results are shown in Figure 2(a). None of the human ileum samples showed any band at 84 bp, the expected size of the PCR amplicon. Both the RotaTeq sample

TABLE 2: PCR assay primer sequences and expected amplicon sizes. The primers for echovirus, TTV like minivirus, small anellovirus, parechovirus, and rotavirus A were designed by LLNL as part of this study. The primers for PCV-2 were obtained from McClenahan et al. [25].

Oligo name	Sequence	Amplicon Size
Echovirus9_F	GCC CCT GAA TGC GGC TAA	112 bp
Echovirus9_R	AAA CAC GGA CAC CCA AAG TAG T	
TTV-like mini_F	CGA ATG GCT GAG TTT ATG CC	146 bp
TTV-like mini_R	GTT TCT TGC CCG TTC CGC	
Small anello_F	CTG AGT TTA CCC CGC TAG AC	118 bp
Small anello_R	CCG AAT TGC CCC TAG ACC	
Parechovirus_F	CCC AYG AAG GAT GCC CAG	113 bp
Parechovirus_R	TTG GCC CAC TAG ACG TTT T	
Rotavirus_seg10_F	CCA ADW GAA GTG ACY GCA	130 bp
Rotavirus_seg10_R	GCG ATA TGR YTG ACT DTG GCT	
PCV2_F	AGCAATCAGAYCCCGTTG	84 bp
PCV2_R	CCAAGGAVGTAATCCTCCGATA	

TABLE 3: Summary of vaccination status, viruses detected by microarray, and PCR confirmation results from human ileal samples. (+, vaccinated; −, unvaccinated).

Sample ID	Vaccination status	LLMDA viral results	PCR confirmation
1	+	Not detected	
2	−	Not detected	
3	+	Not detected	
4	+	Not detected	
5	+	Human rotavirus A human echovirus 9	Yes
6	+	Not detected	
7	−	TTV-like minivirus	Yes
8	+	Not detected	
9	+	Human parechovirus	Yes
10	+	Not detected	
11	+	Small anellovirus 2	Yes
12	−	Not detected	
13	+	Not detected	
14	+	Small anellovirus 2	Yes
15	+	Not detected	
16	−	Not detected	
17	+	Not detected	
18	+	Not detected	
19	+	Not detected	
20	+	Not detected	
21	+	Not detected	

and a positive control sample from an ATCC cell line (PK-15) showed bands around 84 bp on the gel.

PCR of rotavirus A from sample 5 and the RotaTeq vaccine gave the expected band at 140 bp (Figure 2(b)). PCR of TTV-like minivirus from sample 7 and a positive control from a previous study detected the expected band at 112 bp. Additionally, echovirus 9 PCR from sample 5 detected an expected product at 112 bp. PCR of human parechoviruses 1 from sample 9 detected the expected product at 113 bp. Small anelloviruses PCR from samples 11 and 14 produced expected band size at 118 bp (Figure 2(c)).

4. Discussion

We analyzed 21 human terminal ileum samples, obtained from children undergoing routine colonoscopy, for the presence of any viral and bacterial pathogens, and evaluated any association of specific pathogens with vaccination. The samples were analyzed on the LLMDA v2, which contains DNA probes to detect all sequenced viruses and bacteria [28]. The viruses detected by the LLMDA were subsequently confirmed by PCR assays. The intestinal mucosa is an ideal tissue for the study of virus—host interactions, as it is the site of ileal Peyer's patches composed of lymphoid cells, which

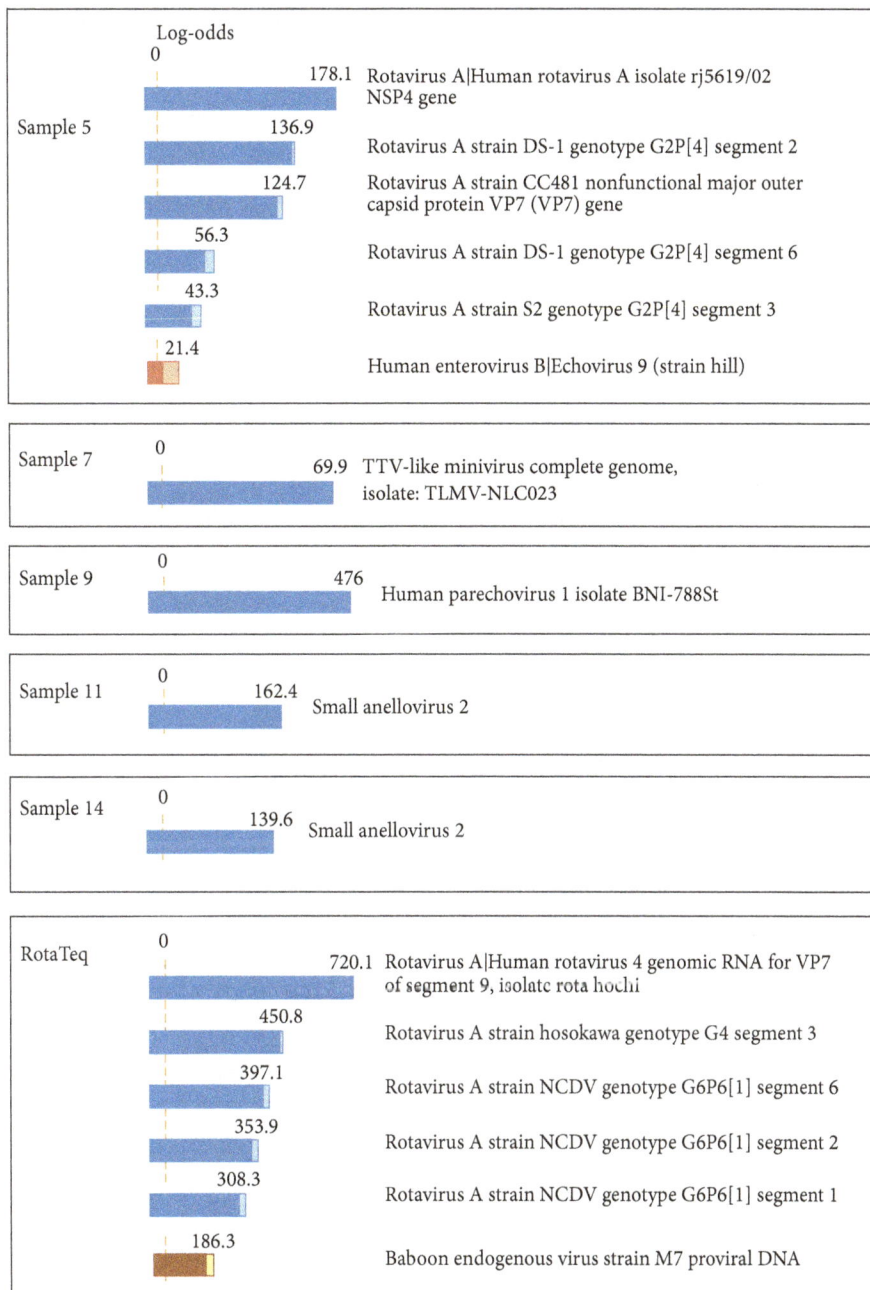

FIGURE 1: LLMDA viral results from human ileum samples and RotaTeq. Microarray data were analyzed using the composite likelihood maximization method developed at Lawrence Livermore National Laboratory [28]. The log likelihood for each of the possible targets is estimated from the BLAST similarity scores of the probe and target sequences, together with the probe sequence complexity and other covariates derived from the BLAST results [28].

are important in immune surveillance of the intestinal lumen. Ileal samples were collected from children that had been (i) fully vaccinated including against rotavirus; (ii) previously vaccinated but not against rotavirus; or (iii) completely unvaccinated.

Overall, no correlation between specific pathogens and vaccination status was identified from this study, nor was a correlation identified between pathogens and vaccination of rotavirus vaccines. PCV was not detected in any ileal samples either by microarray or PCR analyses.

A sample of the RotaTeq vaccine that had been previously shown to contain PCV DNA was included for analysis but the PCV-contaminated Rotarix vaccine was not available for analysis. The sample of RotaTeq vaccine tested positive for rotavirus A and baboon endogenous virus, as previously reported by Victoria and colleagues [17]. The origin of

TABLE 4: Bacterial sequences detected from the human ileum samples by the LLMDA array. Microarray data was analyzed using the CliMax software as described [28]. (+, vaccinated; −, unvaccinated).

Sample ID	Vaccination status	Bacterial results
1	+	*Bacteroides thetaiotaomicron*
		Bacteroides coprocola
2	−	Not detected
3	+	Not detected
4	+	Not detected
5	+	*Shigella sonnei*
		Klebsiella pneumoniae
		Shigella dysenteriae
		Bacteroides intestinalis
6	+	*Bacteroides fragilis*
		Bacteroides vulgatus
		Bacteroides plebeius
7	−	Not detected
8	+	Not detected
9	+	Not detected
10	+	Not detected
11	+	Not detected
12	−	Not detected
13	+	Not detected
14	+	Not detected
15	+	*Shigella sonnei*
		Bacteroides thetaiotaomicron
16	−	*Streptococcus agalactiae*
17	+	Not detected
18	+	*Bacteroides thetaiotaomicron*
19	+	Not detected
20	+	*Bacteroides thetaiotaomicron*
		Streptomyces coelicolor
21	+	Not detected

the baboon endogenous virus is assumed to be related to the African green monkey-derived Vero cell line used in its manufacture and cross-hybridization of its endogenous retroviruses to the baboon endogenous retrovirus probes [17].

Microarray analysis did not detect PCV from the RotaTeq vaccine, which confirmed the previous results from Victoria et al. that LLMDA detected PCV from Rotarix but did not detect PCV from the RotaTeq vaccine [17]. However, PCV2 in RotaTeq vaccine was detected by PCR assays. RotaTeq contained small PCV-1 and PCV-2 genome fragments but did not contain detectable larger portions of PCV genomes [30]. Studies have shown that the amount of PCV in RotaTeq was about 4000 times lower than the PCV in Rotarix, with the PCV in RotaTeq being barely detectable [25, 31, 32]. A case study by Ranucci et al. has reported that the concentration of PCV-2 DNA fragment in clinical consistency lots was in the range of below limit of detection to 6.4×10^3 copies/mL when measured by QPCR, and that PCV1 was below limit of detection (0.1–0.8×10^3 copies/mL) [30]. The current study showed that the PCV-2 signal was close to or above

the limit of detection of PCR, but below detection limit of LLMDA. PCV was not detected in any ileum samples either by microarray or PCR analyses. It is also likely that the PCV fragments from the RotaTeq vaccine have already been eliminated from the body, thus no PCV remains in the ileal samples.

Human rotavirus A was detected in one ileum sample (sample 5) by microarray and confirmed by PCR assay. This sample came from a fully vaccinated child and the infection likely from a recent rotavirus infection. It is unlikely that this is the remaining rotavirus since the child was vaccinated with RotaTeq about two years ago. RotaTeq (Merck) is a pentavalent vaccine that contains five live-attenuated strains with genotypes G1P[5], G2P[5], G3P[5], G4P[5], and G6P[8], derived through laboratory reassortment of human rotavirus strains with a bovine G6P[5] rotavirus strain (WC3) [33]. The LLMDA detected several segments of the virus, all from human origin. The genotype detected in this sample, G2P[4], is not a vaccine genotype and it has been previously identified in G1P[8] vaccinated patients [34].

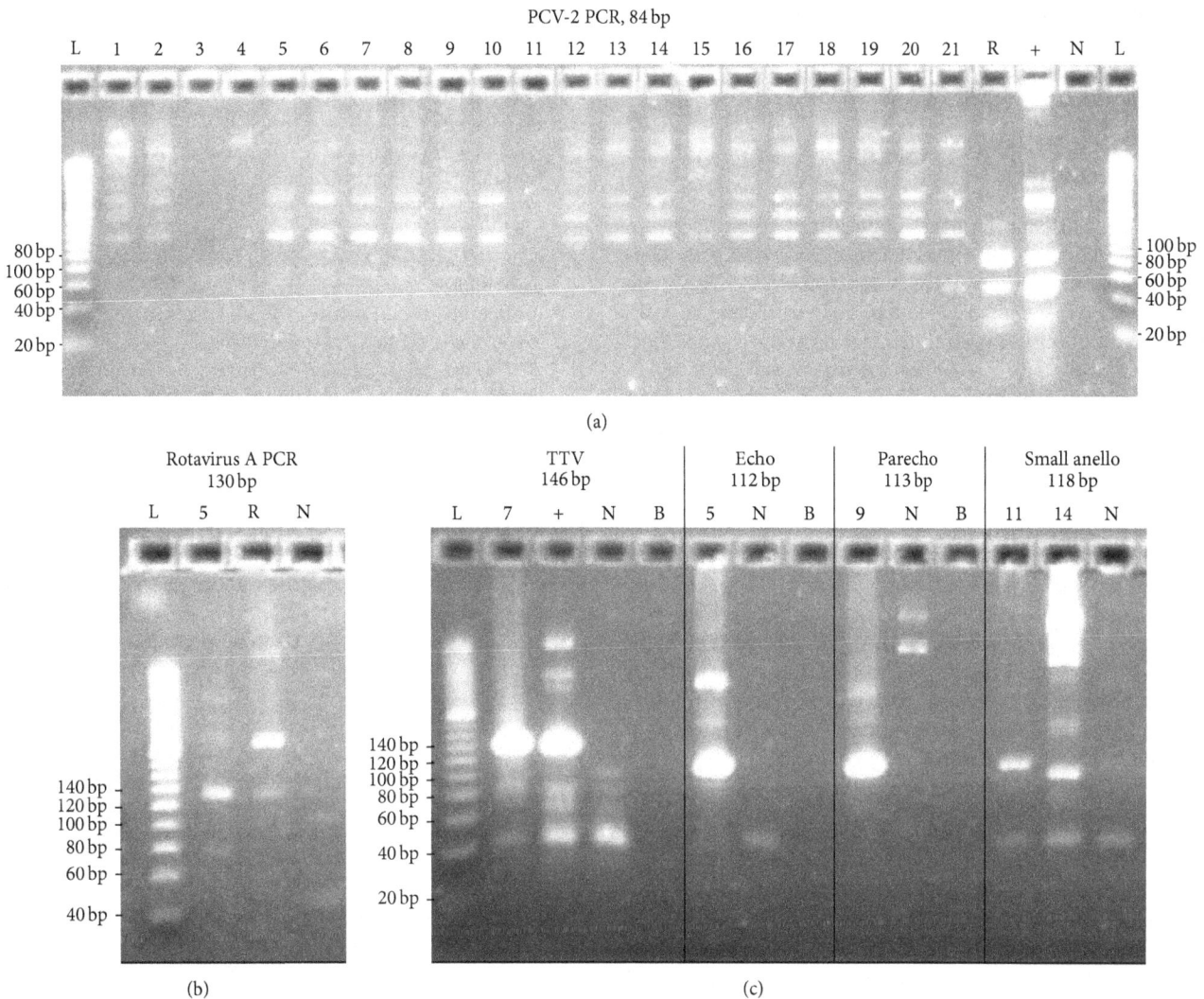

FIGURE 2: Results of PCR confirmation of PCV-2, rotavirus A, parechovirus, echovirus, TTV, and small anellovirus from selected samples. (a) PCV-2 PCR. Lanes 1–21 are human ileum samples. R: RotaTeq vaccine; +: PK-15 cell control; N: NTC; L: 20 bp ladder. Both RotaTeq and positive control gave expected size at 84 bp. (b) Rotavirus A PCR. Sample 5 and RotaTeq gave expected band at 130 bp. No product was detected in NTC. (c) PCR results of TTV, Echovirus, parechovirus, and small anellovirus. B: blank. Sample 7 and a positive control from a previous study gave expected band size at 146 bp. Echovirus PCR from sample 5 detected an expected band size at 112 bp. Parechovirus PCR from sample 9 detected an expected band size at 113 bp. Small anellovirus PCR from samples 11 and 14 produced expected band size at 118 bp.

In the same sample 5, echovirus 9 was also identified. Additionally, a closely related human parechovirus 1 was identified in sample 9. Both the echovirus 9 and human parechovirus 1 detection by microarray were confirmed by echovirus 9 and human parechovirus 1 specific PCR assays. Echovirus is a subspecies of the human enterovirus B found in the gastrointestinal tract. Human enteroviruses cause mild, gastrointestinal, or respiratory illness [35] and are commonly spread such that more than 95% of children are infected within two to five years of age [35, 36]. Nyström et al. have recently found human enterovirus species B in ileocecal Crohns' disease [37], suggesting that this viral species could play a role in Crohn's disease onset or progression.

Small anellovirus 2 was detected in two patient samples. Small anellovirus 2 is also referred to as Torque teno midi

virus (TTMV). TTMV and TTVs are ubiquitous in >90% of adults worldwide but no human pathogenicity of TTV has been fully established [38, 39]. No significant viruses were identified in any other samples. Analysis of a larger number of ileum samples will help further identify any additional coinfecting pathogens, as well as the frequency of occurrence of these pathogens.

The application of the LLMDA technology provides an effective means to survey vaccines for the presence of adventitious agents. In this study, LLMDA did not detect any PCV DNA sequences from the pediatric ileal samples; however, LLMDA detected wild type rotavirus, human enterovirus B, small anellovirus, TTMV, and common gastrointestinal bacteria including *Bacteroides, Shigella,* and *Streptococcus* from some samples, suggesting that LLMDA could be used as a tool to monitor the effectiveness of rotavirus vaccines and to detect reinfections and coinfections with other gastrointestinal viruses or bacteria that could cause pediatric gastrointestinal problems.

5. Conclusions

We analyzed 21 children ileal samples from colonoscopy on the LLMDA array to screen for bacterial and viral pathogens and possible adventitious agents that could be associated with vaccination. We detected a wild type rotavirus, parechovirus, and several common gastrointestinal bacterial agents, *Bacteroides*, *Shigella*, and *Streptococcus* from several ileal samples. This study shows that the broad spectrum technology, such as the LLMDA, could be used as a surveillance tool for vaccine safety and effectiveness.

Acknowledgments

The authors are very grateful to the families whose children participated in this study. They also thank Dr. A. Krigsman for providing the ileal research samples used in this study. The views expressed in this paper are those of the authors and do not reflect the official policy or position of the U.S. government, the Department of Energy, the Department of Defense, or the U.S. Air Force.

References

[1] P. H. Dennehy, "Transmission of rotavirus and other enteric pathogens in the home," *The Pediatric Infectious Disease Journal*, vol. 19, no. 10, pp. S103–S105, 2000.

[2] T. K. Fischer, C. Viboud, U. Parashar et al., "Hospitalizations and deaths from diarrhea and rotavirus among children <5 years of age in the United States, 1993–2003," *The Journal of Infectious Diseases*, vol. 195, no. 8, pp. 1117–1125, 2007.

[3] A. Z. Kapikian, Y. Hoshino, R. M. Chanock, and I. Perez-Schael, "Efficacy of a quadrivalent rhesus rotavirus-based human rotavirus vaccine aimed at preventing severe rotavirus diarrhea in infants and young children," *The Journal of Infectious Diseases*, vol. 174, supplement 1, pp. S65–S72, 1996.

[4] D. O. Matson, "The pentavalent rotavirus vaccine, RotaTeq," *Seminars in Pediatric Infectious Diseases*, vol. 17, no. 4, pp. 195–199, 2006.

[5] D. I. Bernstein and R. L. Ward, "Rotarix: development of a live attenuated monovalent human rotavirus vaccine," *Pediatric Annals*, vol. 35, no. 1, pp. 38–43, 2006.

[6] CDC, "Rotavirus vaccine for the prevention of rotavirus gastroenteritis among children," *Morbidity and Mortality Weekly Report*, vol. 48, pp. 1–20, 1999.

[7] CDC, "Withdrawal of rotavirus vaccine recommendation," *Morbidity and Mortality Weekly Report*, vol. 48, p. 1007, 1999.

[8] P. M. Heaton, M. G. Goveia, J. M. Miller, P. Offit, and H. F. Clark, "Development of a pentavalent rotavirus vaccine against prevalent serotypes of rotavirus gastroenteritis," *The Journal of Infectious Diseases*, vol. 192, supplement 1, pp. S17–S21, 2005.

[9] CDC, "Prevention of rotavirus gastroenteritis among infants and children. Recommendations of the Advisory Committee on Immunization Practices (ACIP)," *Morbidity and Mortality Weekly Report*, vol. 55, pp. 1–13, 2006.

[10] CDC, "Recommended immunization schedules for persons aged 0–18 years—United States," *Morbidity and Mortality Weekly Report*, vol. 55, pp. Q1–Q4, 2007.

[11] B. de Vos, T. Vesikari, A. C. Linhares et al., "A rotavirus vaccine for prophylaxis of infants against rotavirus gastroenteritis," *The Pediatric Infectious Disease Journal*, vol. 23, no. 10, pp. S179–S182, 2004.

[12] CDC, "Prevention of rotavirus gastroenteritis among infants and children. Recommendations of the Advisory Committee on Immunization Practices (ACIP)," *Morbidity and Mortality Weekly Report*, vol. 55, pp. 1–13, 2009.

[13] A. T. Curns, C. A. Steiner, M. Barrett, K. Hunter, E. Wilson, and U. D. Parashar, "Reduction in acute gastroenteritis hospitalizations among US children after introduction of rotavirus vaccine: analysis of hospital discharge data from 18 US States," *The Journal of Infectious Diseases*, vol. 201, no. 11, pp. 1617–1624, 2010.

[14] M. O'Ryan, J. Diaz, N. Mamani, M. Navarrete, and C. Vallebuono, "Impact of rotavirus infections on outpatient clinic visits in Chile," *The Pediatric Infectious Disease Journal*, vol. 26, no. 1, pp. 41–45, 2007.

[15] J. E. Tate, M. M. Patel, A. D. Steele et al., "Global impact of rotavirus vaccines," *Expert Review of Vaccines*, vol. 9, no. 7, pp. 395–407, 2010.

[16] M. M. Patel, D. Steele, J. R. Gentsch, J. Wecker, R. I. Glass, and U. D. Parashar, "Real-world impact of rotavirus vaccination," *The Pediatric Infectious Disease Journal*, vol. 30, no. 1, pp. S1–S5, 2011.

[17] J. G. Victoria, C. Wang, M. S. Jones et al., "Viral nucleic acids in live-attenuated vaccines: detection of minority variants and an adventitious virus," *Journal of Virology*, vol. 84, no. 12, pp. 6033–6040, 2010.

[18] L. Li, A. Kapoor, B. Slikas et al., "Multiple diverse circoviruses infect farm animals and are commonly found in human and chimpanzee feces," *Journal of Virology*, vol. 84, no. 4, pp. 1674–1682, 2010.

[19] J. A. Ellis, B. M. Wiseman, G. Allan et al., "Analysis of seroconversion to *Porcine circovirus* 2 among veterinarians from the United States and Canada," *Journal of the American Veterinary Medical Association*, vol. 217, no. 11, pp. 1645–1646, 2000.

[20] G. M. Allan, F. Mcneilly, I. Mcnair et al., "Absence of evidence for *Porcine circovirus* type 2 in cattle and humans, and lack of seroconversion or lesions in experimentally infected sheep," *Archives of Virology*, vol. 145, no. 4, pp. 853–857, 2000.

[21] K. Hattermann, C. Roedner, C. Schmitt, T. Finsterbusch, T. Steinfeldt, and A. Mankertz, "Infection studies on human cell lines with *Porcine circovirus* type 1 and *Porcine circovirus* type 2," *Xenotransplantation*, vol. 11, no. 3, pp. 284–294, 2004.

[22] FDA, Components of Extraneous Virus Detected in Rotarix Vaccine, No Known Safety Risk, 2010, http://www.fda.gov/News-Events/Newsroom/PressAnnouncements/ucm205625.htm.

[23] FDA, "Vaccines and Related Biological Products Advisory Committee Meeting Background Material," 2010, http://www.fda.gov/BiologicsBloodVaccines/Vaccines/ApprovedProducts/ucm211101.htm.

[24] FDA, "Update on Recommendations for the Use of Rotavirus Vaccines," 2010, http://www.fda.gov/BiologicsBloodVaccines/Vaccines/ApprovedProducts/ucm212140.htm.

[25] S. D. McClenahan, P. R. Krause, and C. Uhlenhaut, "Molecular and infectivity studies of *Porcine circovirus* in vaccines," *Vaccine*, vol. 29, no. 29-30, pp. 4745–4753, 2011.

[26] D. C. Payne, S. Humiston, D. Opel et al., "A multi-center, qualitative assessment of pediatrician and maternal perspectives on rotavirus vaccines and the detection of *Porcine circovirus*," *BMC Pediatrics*, vol. 11, article 83, 2011.

[27] L. Erlandsson, M. W. Rosenstierne, K. McLoughlin, C. Jaing, and A. Fomsgaard, "The microbial detection array combined with random Phi29-amplification used as a diagnostic tool for virus detection in clinical samples," *PLoS ONE*, vol. 6, no. 8, Article ID e22631, 2011.

[28] S. N. Gardner, C. J. Jaing, K. S. McLoughlin, and T. R. Slezak, "A microbial detection array (MDA) for viral and bacterial detection," *BMC Genomics*, vol. 11, article 668, 2010.

[29] D. Hysom, P. Naraghi-Arani, M. Elsheikh, A. C. Carrillo, P. L. Williams, and S. N. Gardner, "Skip the alignment: degenerate, multiplex primer and probe design using k-mer matching instead of alignments," *PLoS ONE*, vol. 7, no. 4, Article ID e34560, 2012.

[30] C. S. Ranucci, T. Tagmyer, and P. Duncan, "Adventitious agent risk assessment case study: evaluation of RotaTeq for the presence of *Porcine circovirus*," *PDA Journal of Pharmaceutical Science and Technology*, vol. 65, no. 6, pp. 589–598, 2011.

[31] S. M. Gilliland, L. Forrest, H. Carre et al., "Investigation of *Porcine circovirus* contamination in human vaccines," *Biologicals*, vol. 40, no. 4, pp. 270–277, 2012.

[32] S. A. Baylis, T. Finsterbusch, N. Bannert, J. Blumel, and A. Mankertz, "Analysis of *Porcine circovirus* type 1 detected in Rotarix vaccine," *Vaccine*, vol. 29, no. 4, pp. 690–697, 2011.

[33] D. C. Payne, M. Wikswo, and U. D. Parashar, "Manual for the surveillance of vaccine-preventable diseases," in *Rotavirus*, S. W. Roush, L. McIntyre, and L. M. Baldy, Eds., chapter 13, Centers for Disease Control and Prevention, Atlanta, Ga, USA, 5th edition, 2012.

[34] H. Antunes, A. Afonso, M. Iturriza et al., "G2P[4] the most prevalent rotavirus genotype in 2007 winter season in an European non-vaccinated population," *Journal of Clinical Virology*, vol. 45, no. 1, pp. 76–78, 2009.

[35] G. Stanway, P. Joki-Korpela, and T. Hyypia, "Human parechoviruses—biology and clinical significance," *Reviews in Medical Virology*, vol. 10, pp. 57–69, 2000.

[36] P. Joki-Korpela and T. Hyypia, "Parechoviruses, a novel group of human picornaviruses," *Annals of Medicine*, vol. 33, no. 7, pp. 466–471, 2001.

[37] N. Nyström, T. Berg, E. Lundin et al., "Human enterovirus species B in ileocecal Crohn's disease," *Clinical and Translational Gastroenterology*, vol. 4, article e38, 2013.

[38] S. Hino and H. Miyata, "Torque teno virus (TTV): current status," *Reviews in Medical Virology*, vol. 17, no. 1, pp. 45–57, 2007.

[39] Z. Burián, H. Szabó, G. Székely et al., "Detection and follow-up of torque teno midi virus ("small anelloviruses") in nasopharyngeal aspirates and three other human body fluids in children," *Archives of Virology*, vol. 156, no. 9, pp. 1537–1541, 2011.

16

Characterization of the Protective HIV-1 CTL Epitopes and the Corresponding HLA Class I Alleles: A Step towards Designing CTL based HIV-1 Vaccine

Sajib Chakraborty,[1] Taibur Rahman,[1] and Rajib Chakravorty[2]

[1] *Department of Biochemistry and Molecular Biology, Faculty of Biological Sciences, University of Dhaka, Dhaka 1000, Bangladesh*
[2] *Department of EEE, University of Melbourne, National ICT Australia, Melbourne, VIC 3010, Australia*

Correspondence should be addressed to Taibur Rahman; taibur@du.ac.bd

Academic Editor: Syed Hani Abidi

Human immunodeficiency virus (HIV) possesses a major threat to the human life largely due to the unavailability of an efficacious vaccine and poor access to the antiretroviral drugs against this deadly virus. High mutation rate in the viral genome underlying the antigenic variability of the viral proteome is the major hindrance as far as the antibody based vaccine development is concerned. Although the exact mechanism by which CTL epitopes and the restricting HLA alleles mediate their action towards slow disease progression is still not clear, the important CTL restricted epitopes for controlling viral infections can be utilized in future vaccine design. This study was designed for the characterization the HIV-1 optimal CTL epitopes and their corresponding HLA alleles. CTL epitope cluster distribution analysis revealed only two HIV-1 proteins, namely, Nef and Gag, which have significant cluster forming capacity. We have found the role of specific HLA supertypes such as HLA B*07, HLA B*58, and HLA A*03 in selecting the hydrophobic and conserved amino acid positions within Nef and Gag proteins, to be presented as epitopes. The analyses revealed that the clusters of optimal epitopes for Nef and p24 proteins of HIV-1 could potentially serve as a source of vaccine.

1. Introduction

Human immunodeficiency virus (HIV), a retrovirus that belongs to the Lentiviridae family, is the causative agent of acquired immunodeficiency syndrome (AIDS). HIV genome is composed of 9.8 Kb positive-sense, single-stranded RNA which is reverse transcribed by the enzyme reverse transcriptase to viral DNA upon its entry into the host cell [1]. Between the two types of HIV (HIV-1 and HIV-2), HIV-1 is more virulent and responsible for most of the HIV infections globally. Human immunodeficiency virus-1 (HIV-1) has infected more than 60 million people and caused nearly 30 million deaths worldwide [2]. In Asia, an estimated 4.9 million people were living with HIV in 2009, about the same as 5 years earlier. Most national HIV epidemics appear to have stabilized. Incidence fell by more than 25% in India, Nepal, and Thailand between 2001 and 2009. The epidemic remained stable in Malaysia and Sri Lanka during this time period.

Incidence increased by 25% in Bangladesh and Philippines between 2001 and 2009 even as the countries continue to have relatively low epidemic levels [3]. Although the antiretroviral therapy has proven to be effective in controlling the infection in the developed world, only one-fourth populations in the developing world can afford these medications due to less accessibility. Consequently vast majority of people are living with a constant threat of HIV infection and death by AIDS. In this devastating situation of world AIDS epidemic, there is an urgent need of developing effective HIV vaccine as no vaccine is proved to be efficacious to control HIV infection. To combat this deadly virus, its genome, proteome, pathogenesis, and mechanisms of evasion of immune response should be studied in great detail.

HIV possesses complex RNA genome and contains nine genes which can be classified into 3 functional groups. Among these genes, Gag, Pol, and Env are structural genes, Tat and Rev are regulatory genes, whereas the rest of the genes (Vpu, Vpr, Vif, and Nef) fall into the accessory category of genes [4]. Early HIV replication cycle begins with the recognition of the target cells (mainly CD4$^+$ T cells) by the mature virion and continues as virion core particles enters and facilitates its integration to the genomic DNA of the chromosome of the host cell. The late phase begins with the regulated expression of the integrated proviral genes and ends up with virus budding and maturation.

Gag gene encodes for 3 proteins: matrix (p17), capsid (p24), and nucleocapsid (p7) which are translated as polyproteins and later undergo a cleavage at specific site to give rise to three individual proteins. Pol gene also encodes for a polyprotein which has similar fate like Gag-poly-protein as it is also cleaved by viral protease into three different proteins: reverse transcriptase, protease, and integrase, whereas the Env gene encodes for a single glycoprotein (gp160) which later is cleaved into two proteins: surface glycoprotein gp120 and transmembrane protein gp41. Besides these some other regulatory proteins are also included in the HIV proteome such as Nef, Vpr, Tat, and Ref. In Table 1, we have collected functions of HIV proteins from Uniprot database of HIV, http://www.uniprot.org/uniprot/P04585.

High antigenic variability that results from the high mutation rate can be considered as the characteristic features of retrovirus such as HIV. This vast genetic heterogeneity of HIV not only helps the virus to evade selective pressures exerted by immune response and drug but also facilitates the viral evolution in a faster speed [5]. Phylogenetic studies showed the presence of three distinct groups: M (Major), O (Outliers), and N (non-M and non-O). M is the most predominant group of HIV-1 around the world [6]. Within the M group there are nine subtypes: A–D, F–H, J, and K. Among these, subtypes B are prevalent in most regions of the world such as USA, Europe, South East Asia, Australia, and South Africa [6].

Endogenous pathway of antigen processing and presentation is used to present endogenously synthesized cellular peptides as well as viral protein fragments via the MHC class I molecule to the cytotoxic-T-lymphocytes (CTLs). In this pathway, the proteins that are destined for the presentation are marked by the ubiquitinylation and subjected to proteolytic cleavage by the immunoproteasome. The fragments of peptides are transported to lumen of ER by the help of TAP (TAP1 and TAP2). These TAP proteins also help the loading of the short peptides with appropriate length (9 amino acids) into the groove of MHC class I molecule [7]. Although proteasome is the main player in generating the bulk of the CTL epitopes, cytosolic endopeptidases may also be involved in the production of certain CTL epitopes [8].

Peptides are typically tightly associated along their entire length in MHC class I groove. The N and C termini of the peptide are firmly H-bonded to the conserved residues of the MHC groove. The analysis of the naturally occurring peptides extracted from the MHC-peptide complex revealed that these are mostly 8-9 residues long and in certain key positions amino acids tend to be conserved within the peptide. These are called anchor positions which are proved to be essential for the binding of peptides to MHC class I molecules in allele specific manner. There are typically two (sometimes three) major anchor positions for the class I binding peptides. One is located at the C terminal end and the other one usually lies in the position 2 (P2) but also occurs in P3, 5, or 7 [7].

Cytotoxic-T-lymphocytes (CTLs) are one of the vital components of cellular immunity and play crucial role in eliminating viral infection. CTLs recognize viral antigen on the surface of virally infected cells in combination with appropriate major histocompatibility complex (MHC) molecule and exert their effect by killing the infected cells either by lysis or inducing apoptosis. Previous studies suggested that the HIV infection process can be divided into 3 stages. These are (1) acute viremia, (2) a latency period of variable time period, and (3) clinical AIDS. At the later stages of HIV infection CD4$^+$ T cells counts drops down below 200 cells/mm^3 which causes the complete collapse of immune response and consequently the opportunistic infectious agents such as *Pneumocystis carinii* come into the play [9]. There is now increasing body of evidence that CTLs play an important role in controlling the HIV infection. Analysis of the immune responses of the HIV infected patients revealed that antiviral CTL activity is correlated with clearance of virus particles during the acute phase of infection and a decline in the CTL activity is associated with the disease progression [9, 10]. Two types of antiviral CTL responses have been documented so far: one is classical viral epitope dependent-MHC restricted killing of virally infected cells and the other one is the noncytolytic response in which CTLs control the viral infection by inhibiting the viral replication [11]. Furthermore, studies of the SIV-macaque model, in which the administration of anti-CD8 monoclonal antibodies hinders the decline in viremia, provided strong evidence for the crucial role of CTLs in controlling HIV infection during acute phase [12]. Recently Goulder and Watkins have suggested three additional lines of evidence which signifies the potential role of CTLs in suppressing HIV infection: first they argued that specific HLA class I molecules are consistently associated with particular HIV disease outcomes. Secondly, they highlighted the fact that more rapid disease progression is observed in individuals with HLA class I homozygosity, and lastly they provided evidence that the loss of immune control over HIV infection arises when viral mutants escape CD8$^+$ T-cell recognition [13]. All the above mentioned evidence signifies the important antagonizing role of CTL immune response in HIV disease progression.

2. Analysis of the HLA Class I Restricted CTL Epitopes in HIV Proteome

Design and development of HIV vaccine largely depend on our understanding of complex dynamics between host immune response and viral adaptation to selective pressure exerted by the host. Understanding how the CTL epitopes interact with particular HLA alleles can give an insight into the mechanisms of success or failure of immune control of a pathogen, such as HIV-1, for which clearance of virus particles depends on CTL activity. So the vaccine development strategies for HIV should be focused on identifying the epitopes presented by HLA alleles prevalent

TABLE 1: Function of different HIV proteins.

Protein	Precursor	Functions
P17	Gag	Matrix protein p17 has two main functions. Firstly it targets Gag-pol polyproteins to the plasma membrane by the help of a membrane-binding signal which contains myristoylated N-terminus. Secondly it plays an essential role in the nuclear localization of the viral genome.
P24	Gag	Protein p24 forms the nucleocapsid that encapsulates the viral genomic RNA in the virion. The core is disassembled immediately after the entry of virion into host cell.
P7	Gag	Nucleocapsid protein p7 encapsulates viral genomic RNA and hence provides protection to viral genome. It binds these RNAs through its zinc finger motifs. It also acts as a nucleic acid chaperone as it tends to facilitate the rearrangement of nucleic acid secondary structure during reverse transcription of genomic RNA.
RT	Gag-pol	Reverse transcriptase/ribonuclease H (RT) is a multifunctional enzyme that facilitates the reverse transcription of viral RNA genome into dsDNA in the cytoplasm, shortly after virus entry into the cell. This enzyme also displays a DNA polymerase activity that can copy either DNA or RNA templates, and a ribonuclease H (RNase H) activity that cleaves the RNA strand of RNA-DNA heteroduplexes.
Integrase	Gag-pol	Integrase catalyzes integration of viral DNA into the host chromosome, by a multistep process involving DNA cutting and joining reactions.
Protease	Gag-pol	Cleavage of viral precursor polyproteins into mature proteins
Gp120	GP160	The surface protein gp120 (SU) facilitates the anchoring of the virus to the host target cell (CD4+) by binding to the primary receptor CD4. This interaction induces a change in the conformation exposing a high affinity binding site for a chemokine coreceptor (CXCR4 and/or CCR5) and promotes subsequent interaction between the envelope protein and CXCR4 and/or CCR5.
Tat		Tat acts as a nuclear transcriptional activator of viral gene expression that is essential for viral transcription from the LTR promoter. It also directs the components of the cellular transcription machinery into the viral RNA to promote transcription by the RNA polymerase complex.
Vif		It ensures the downregulation of APOBEC3G by recruiting the ubiquitin-proteasome machinery that targets APOBEC3G for degradation. It also binds to viral RNA and affects the stability of viral nucleoprotein core.
Vpr		It is largely involved in the transport of the viral preintegration (PIC) complex to the nucleus during the early phase of the infection. It probably interacts with karyopherin alpha/KPNA1 and KPNA2 thereby increasing their affinity for basic-type nuclear localization signal harboring proteins such as viral matrix protein, thus facilitating the translocation of the viral proteins into the nucleus.
Vpu		It promotes virion budding, by targeting human CD4 and CD317 to proteasomal degradation. CD4 degradation hinders any possible interactions between viral Env and human CD4 in the endoplasmic reticulum. It helps the proper Env assembly into virions.
Nef		(1) Downregulation of surface MHC-I molecules. (2) Downregulation of cell surface CD4 antigen. It interacts with the Src family kinase LCK and induces LCK-CD4 dissociation. Subsequently it causes clathrin-dependent endocytosis of CD4 antigen. Ultimately, the CD4 are decreased and infected cells. (3) It decreases the number of viral receptors and hence prevents reinfection by more HIV particles. (4) It prevents the apoptosis of the infected cell by inhibiting the Fas and TNFR-mediated death signals. It also interacts with p53 and protects the infected cell against p53-mediated apoptosis. Furthermore, it regulates the Bcl-2 family proteins through the formation of a Nef/PI3-kinase/PAK2 complex that induces phosphorylation of Bad.

in populations severely affected by the global HIV epidemic. In recent years, development of new technologies such as measuring interferon-gamma (IFNγ) release by the enzyme linked immunospot (ELISPOT) assay and flow cytometry ensured the efficient evaluation of CTL responses against HIV epitopes [14]. Moreover, development of overlapping pooled peptide technology (OLP) now provides the opportunity for the detailed and precise analyses of HIV-1-specific cellular immune responses by elucidation of the T-cell epitopes and the identification of immunodominant regions of HIV-1 gene products. Identification and characterization of the CTL epitopes as well as the corresponding HLA alleles can play a major role in elucidating the nature of protective CTL response and mechanism of the immune evasion of HIV. A large number of HIV CTL epitopes have been

identified and deposited into various databases. Apart from the experimental methods, various computational tools are now available which can predict CTL epitopes within viral proteome by using different sets of algorithms, for example, artificial neural network (ANN), average relative binding (ARB), stabilized matrix method (SMM), and so forth. The first CTL epitope was identified in 1988 by using synthetic peptide technology [15]. Since then, over 1200 HLA class I restricted HIV-1 epitopes were identified in HIV proteome (http://www.hiv.lanl.gov/content/immunology/index.html). In Table 2, a list of HLA class I allele restricted optimal CTL epitopes for HIV along with their corresponding HLA alleles and clades is given. For the identification of optimal epitopes, two criteria were imposed as described by Llanoa et al. [16]. These criteria include the unequivocal experimental

TABLE 2: List of optimal CTL epitopes for HIV-1 (taken and modified from HIV molecular immunology database (http://www.hiv.lanl.gov/content/immunology/tables/optimal_ctl_summary.html).

HIV protein	AA position	HLA	Sequence	Clade
gp160	2–10	B*0801 (B8)	RVKEKYQHL	—
gp160	31–39	B*1801 (B18)	AENLWVTVY	B
gp160	31–39	B44	AENLWVTVY	B
gp160	31–40	B*4402 (B44)	AENLWVTVYY	—
gp160	37–46	A*0301 (A3)	TVYYGVPVWK	A, B, C, D
gp160	42–51	B*5501 (B55)	VPVWKEATTT	—
gp160	42–52	B*3501 (B35)	VPVWKEATTTL	B
gp160	52–61	A*2402 (A24)	LFCASDAKAY	—
gp160	59–69	B58	KAYETEVHNVW	C
gp160	61–69	B*1801 (B18)	YETEVHNVW	B
gp160	78–86	B*3501 (B35)	DPNPQEVVL	B
gp160	104–112	B*3801 (B38)	MHEDIISLW	B
gp160	199–207	A*1101 (A11)	SVITQACPK	B
gp160	209–217	A*2902 (A29)	SFEPIPIHY	B, D
gp160	298–307	B*0702 (B7)	RPNNNTRKSI	B, C
gp160	310–318	A*3002 (A30)	HIGPGRAFY	B
gp160	311–320	A*0201 (A2)	RGPGRAFVTI	A, B, C
gp160	375–383	B*1516 (B63)	SFNCGGEFF	A, B, C
gp160	375–383	Cw*0401 (Cw4)	SFNCGGEFF	A, B, C
gp160	416–424	B*5101 (B51)	LPCRIKQII	B
gp160	419–427	A*3201 (A32)	RIKQIINMW	B, C
gp160	511–519	Cw18	YRLGVGALI	C
gp160	557–565	Cw*0304 (Cw10)	RAIEAQQHL	A, B, C, D
gp160	557–565	Cw8	RAIEAQQHM	A, B, C, D
gp160	557–565	Cw15	RAIEAQQHL	C
gp160	584–592	B*1402 (B14)	ERYLKDQQL	A, B, C, D
gp160	585–593	A23	RYLKDQQLL	B, C
gp160	585–593	A*2402 (A24)	RYLKDQQLL	B, C
gp160	586–593	B*0801 (B8)	YLKDQQLL	A, B
gp160	606–614	B*3501 (B35)	TAVPWNASW	B
gp160	698–707	A*3303 (A33)	VFAVLSIVNR	B
gp160	703–712	A*2501 (A25)	EIIFDIRQAY	—
gp160	704–712	A*3002 (A30)	IVNRNRQGY	B
gp160	770–780	A*0301 (A3)	RLRDLLLIVTR	B, C
gp160	770–780	A*3101 (A31)	RLRDLLLIVTR	B, C
gp160	777–785	A*6802 (A68)	IVTRIVELL	B
gp160	786–795	B*2705 (B27)	GRRGWEALKY	B
gp160	787–795	A*0101 (A1)	RRGWEVLKY	B
gp160	794–802	A*3002 (A30)	KYCWNLLQY	B
gp160	805–814	B*4001 (B60)	QELKNSAVSL	B
gp160	813–822	A*0201 (A2)	SLLNATDIAV	B
gp160	831–838	A*3303 (A33)	EVAQRAYR	B
gp160	843–851	B*0702 (B7)	IPRRIRQGL	A, B, C, D
gp160	846–854	A*0205 (A2)	RIRQGLERA	B
gp160	848–856	B8	RQGLERALL	—
Integrase	28–36	B42	LPPIVAKEI	B, C
Integrase	66–74	B*1510 (B71)	THLEGKIIL	B, C

TABLE 2: Continued.

HIV protein	AA position	HLA	Sequence	Clade
Integrase	123–132	B57	STTVKAACWW	B
Integrase	135–143	B*1503 (B72)	IQQEFGIPY	B, C
Integrase	165–172	Cw18	VRDQAEHL	C
Integrase	173–181	B*5701 (B57)	KTAVQMAVF	B
Integrase	179–188	A*0301 (A3)	AVFIHNFKRK	B, multiple
Integrase	179–188	A*1101 (A11)	AVFIHNFKRK	B, multiple
Integrase	185–194	B*1503 (B72)	FKRKGGIGGY	B, C
Integrase	203–211	A*1101 (A11)	IIATDIQTK	B
Integrase	219–227	A*3002 (A30)	KIQNFRVYY	AE, B, C, D
Integrase	260–268	B42	VPRRKAKII	—
Integrase	263–271	B*1503 (B72)	RKAKIIRDY	B, C
Nef	13–20	B*0801 (B8)	WPTVRERM	B
Nef	19–27	B62	RMRRAEPAA	B
Nef	37–45	B*4001 (B60)	LEKHGAITS	B
Nef	37–45	B50	LEKHGAITS	B
Nef	68–76	B*0702 (B7)	FPVTPQVPL	B
Nef	68–77	B*0702 (B7)	FPVTPQVPLR	B
Nef	71–79	B*0702 (B7)	TPQVPLRPM	B
Nef	71–79	B*4201 (B42)	RPQVPLRPM	B, C
Nef	73–82	A*0301 (A3)	QVPLRPMTYK	A, B, C, D
Nef	73–82	A*1101 (A11)	QVPLRPMTYK	A, B, C, D
Nef	74–81	B*3501 (B35)	VPLRPMTY	A, B, C, D
Nef	75–82	A*1101 (A11)	PLRPMTYK	B
Nef	77–85	B*0702 (B7)	RPMTYKAAL	B
Nef	82–91	Cw8	KAAVDLSHFL	B
Nef	83–91	A*0205 (A2)	GAFDLSFFL	A
Nef	83–91	Cw3	AALDLSHFL	B
Nef	83–91	Cw*0802 (Cw8)	AAVDLSHFL	B, C
Nef	84–92	A*0301 (A3)	AVDLSHFLK	A, B, D, F
Nef	84–92	A*1101 (A11)	AVDLSHFLK	A, B, D, F
Nef	90–97	B*0801 (B8)	FLKEKGGL	A, B, C, D
Nef	92–100	B*4001 (B60)	KEKGGLEGL	B, C
Nef	92–100	B*4002 (B61)	KEKGGLEGL	B, C
Nef	105–114	B*2705 (B27)	RRQDILDLWI	B
Nef	105–115	B18	RRQDILDLWVY	B
Nef	105–115	Cw7	KRQEILDLWVY	B, C
Nef	106–114	B13	RQDILDLWI	B
Nef	106–115	B*0702 (B7)	RQDILDLWIY	—
Nef	116–124	B57	HTQGYFPDW	B, C
Nef	116–125	B*5701 (B57)	HTQGYFPDWQ	B, C
Nef	117–127	B*1501 (B62)	TQGYFPDWQNY	B, C
Nef	120–128	A29	YFPDWQNYT	B, C
Nef	120–128	B*3701 (B37)	YFPDWQNYT	B, C
Nef	120–128	B*5701 (B57)	YFPDWQNYT	B, C
Nef	120–128	Cw6	YFPDWQNYT	B, C
Nef	127–135	B57	YTPGPGIRY	B, C
Nef	127–135	B63	YTPGPGIRY	B, C
Nef	128–137	B*0702 (B7)	TPGPGVRYPL	B, C

TABLE 2: Continued.

HIV protein	AA position	HLA	Sequence	Clade
Nef	128–137	B*4201 (B42)	TPGPGVRYPL	B, C
Nef	133–141	A33	TRYPLTFGW	B
Nef	134–141	A*2402 (A24)	RYPLTFGW	B, C
Nef	135–143	B*1801 (B18)	YPLTFGWCY	B, C, D
Nef	135–143	B53	YPLTFGWCF	B
Nef	135–143	B*5301 (B53)	YPLTFGWCY	B
Nef	136–145	A*0201 (A2)	PLTFGWCYKL	B
Nef	137–145	B57	LTFGWCFKL	A, B, C
Nef	137–145	B63	LTFGWCFKL	A, B, C
Nef	180–189	A*0201 (A2)	VLEWRFDSRL	B
Nef	183–191	B*1503 (B72)	WRFDSRLAF	B
p17	11–19	B*4002 (B61)	GELDRWEKI	B
p17	18–26	A*0301 (A3)	KIRLRPGGK	A, B
p17	19–27	B*2705 (B27)	IRLRPGGKK	B
p17	20–28	A*0301 (A3)	RLRPGGKKK	A, B
p17	20–29	A*0301 (A3)	RLRPGGKKKY	B
p17	24–32	B*0801 (B8)	GGKKKYKLK	B, F
p17	28–36	A*2402 (A24)	KYKLKHIVW	B, C, F
p17	33–41	Cw*0804 (Cw8)	HLVWASREL	C
p17	34–44	A30	LVWASRELERF	B, C
p17	36–44	B*3501 (B35)	WASRELERF	B, C
p17	74–82	B*0801 (B8)	ELRSLYNTV	F
p17	76–86	A*3002 (A30)	RSLYNTVATLY	B, C, F
p17	76–86	B58	RSLYNTVATLY	B, C, F
p17	76–86	B63	RSLYNTVATLY	B, C, F
p17	77–85	A*0201 (A2)	SLYNTVATL	A, B, C, D, F, G, K
p17	77–85	A*0202 (A2)	SLYNTVATL	A, B, C, D, F, G, K
p17	77–85	A*0205 (A2)	SLYNTVATL	A, B, C, D, F, G, K
p17	78–85	Cw14	LYNTVATL	B, D
p17	78–86	A*2902 (A29)	LYNTVATLY	B, C
p17	78–86	B*4403 (B44)	LYNTVATLY	B, C
p17	84–91	A*1101 (A11)	TLYCVHQK	—
p17	92–101	B*4001 (B60)	IEIKDTKEAL	B, F
p17	124–132	B*3501 (B35)	NSSKVSQNY	B
p24	3–11	B13	VQNLQGQMV	B, C
p24	12–20	B*1510 (B71)	HQAISPRTL	B
p24	13–23	A*2501 (A25)	QAISPRTLNAW	B
p24	15–23	B*5701 (B57)	ISPRTLNAW	A, C
p24	15–23	B63	ISPRTLNAW	A, B, C, D
p24	16–24	B*0702 (B7)	SPRTLNAWV	B
p24	24–32	B*1503 (B72)	VKVIEEKAF	B, C
p24	28–36	B*4415 (B12)	EEKAFSPEV	A, B, C, D
p24	30–37	B*5703 (B57)	KAFSPEVI	B
p24	30–40	B*5701 (B57)	KAFSPEVIPMF	A, B, C, G
p24	30–40	B*5703 (B57)	KAFSPEVIPMF	A, B, C, G
p24	30–40	B63	KAFSPEVIPMF	A, B, C, G
p24	32–40	B57	FSPEVIPMF	B, C
p24	35–43	A*2601 (A26)	EVIPMFSAL	A, B, C, D

TABLE 2: Continued.

HIV protein	AA position	HLA	Sequence	Clade
p24	36–43	Cw*0102 (Cw1)	VIPMFSAL	B, D
p24	44–52	B*4001 (B60)	SEGATPQDL	B
p24	48–56	B*0702 (B7)	TPQDLNTML	A, B, C, D
p24	48–56	B*3910 (B39)	TPQDLNTML	A, B, C, D
p24	48–56	B*4201 (B42)	TPQDLNTML	A, B, C, D
p24	48–56	B*5301 (B53)	TPYDINQML	A
p24	48–56	B*8101 (B81)	TPQDLNTML	A, B, C, D
p24	48–56	Cw*0802 (Cw8)	TPQDLNTML	A, B, C, D
p24	61–69	B*1510 (B71)	GHQAAMQML	B, C
p24	61–69	B*3901 (B39)	GHQAAMQML	B, C
p24	70–78	B*4002 (B61)	KETINEEAA	B
p24	71–80	A*2501 (A25)	ETINEEAAEW	A, B, D
p24	78–86	B*4002 (B61)	AEWDRVHPV	B
p24	84–92	B7	HPVHAGPIA	B, C, D, F
p24	94–104	B13	GQMREPRGSDI	B, C
p24	108–117	B*5701 (B57)	TSTLQEQIGW	B, C
p24	108–117	B*5801 (B58)	TSTLQEQIGW	B, C
p24	122–130	B*3501 (B35)	PPIPVGDIY	A, B, C
p24	128–135	B*0801 (B8)	EIYKRWII	B
p24	131–140	B*2703 (B27)	RRWIQLGLQK	—
p24	131–140	B*2705 (B27)	KRWIILGLNK	A, B, C, D
p24	137–145	B*1501 (B62)	GLNKIVRMY	A, B
p24	142–150	Cw18	VRMYSPVSI	B, C, F
p24	143–150	B*5201 (B52)	RMYSPTSI	B, F
p24	161–169	Cw18	FRDYVDRFF	C
p24	161–170	B*1801 (B18)	FRDYVDRFYK	B, D
p24	162–172	A*2402 (A24)	RDYVDRFFKTL	A
p24	162–172	B*4402 (B44)	RDYVDRFYKTL	B, D
p24	164–172	Cw*0303 (Cw9)	YVDRFFKTL	A, C, D
p24	164–172	A*0207 (A2)	YVDRFYKTL	B
p24	164–172	B*1503 (B72)	YVDRFFKTL	A, C, D
p24	164–172	Cw*0304 (Cw10)	YVDRFFKTL	A, C, D
p24	166–174	B*1402 (B14)	DRFYKTLRA	B, D
p24	174–184	B*4402 (B44)	AEQASQDVKNW	B, C, D
p24	174–185	Cw5	AEQASQEVKNWM	—
p24	176–184	B*5301 (B53)	QASQEVKNW	B, D
p24	176–184	B*5701 (B57)	QASQEVKNW	C
p24	197–205	B*0801 (B8)	DCKTILKAL	B
p24	217–227	A*1101 (A11)	ACQGVGGPGHK	B
p24	223–231	B*0702 (B7)	GPGHKARVL	B, C, D, F
Protease	3–11	A*6802 (A68)	ITLWQRPLV	A, B, C, D
Protease	3–11	A*7401 (A19)	ITLWQRPLV	A, B, C, D
Protease	30–38	A*6802 (A68)	DTVLEEWNL	D
Protease	34–42	B44	EEMNLPGRW	B
Protease	57–66	B13	RQYDQILIEI	B
Protease	68–76	B*1503 (B72)	GKKAIGTVL	BC
Protease	70–77	B57	KAIGTVLV	BC
Protease	76–84	A*0201 (A2)	LVGPTPVNI	B

TABLE 2: Continued.

HIV protein	AA position	HLA	Sequence	Clade
Protease	80–90	B81	TPVNIIGRNML	C
Rev	14–23	B*5701 (B57)	KAVRLIKFLY	B
Rev	14–23	B*5801 (B58)	KAVRLIKFLY	B
Rev	14–23	B63	KAVRLIKFLY	B
Rev	41–50	B7	RPAEPVPLQL	A, B, C, D, F
Rev	57–66	A*0301 (A3)	ERILSTYLGR	B
Rev	67–75	Cw*0501	SAEPVPLQL	B
RT	5–12	B*4001 (B60)	IETVPVKL	B
RT	18–26	B*0801 (B8)	GPKVKQWPL	A, B, C, D
RT	33–41	A*0201 (A2)	ALVEICTEM	B
RT	33–43	A*0301 (A3)	ALVEICTEMEK	B
RT	42–50	B*5101 (B51)	EKEGKISKI	B
RT	73–82	A*0301 (A3)	KLVDFRELNK	B
RT	93–101	A*0301 (A3)	GIPHPAGLK	B
RT	107–115	B*3501 (B35)	TVLDVGDAY	AG, B
RT	118–127	B*3501 (B35)	VPLDEDFRKY	B, C
RT	127–135	A2	YTAFTIPSV	—
RT	128–135	B*5101 (B51)	TAFTIPSI	B
RT	137–146	B18	NETPGIRYQY	B, C
RT	142–149	B*1401 (B14)	IRYQYNVL	C
RT	156–164	B7	SPAIFQSSM	A, B, C, D
RT	158–166	A*0301 (A3)	AIFQSSMTK	A, B, C, D
RT	158–166	A*1101 (A11)	AIFQSSMTK	—
RT	173–181	A*3002 (A30)	KQNPDIVIY	B
RT	175–183	B18	NPEIVIYQY	C
RT	175–183	B*3501 (B35)	HPDIVIYQY	A, B
RT	179–187	A*0201 (A2)	VIYQYMDDL	A, B, C, D
RT	202–210	B*4001 (B60)	IEELRQHLL	B
RT	244–252	B*5701 (B57)	IVLPEKDSW	B
RT	260–271	B*1501 (B62)	LVGKLNWASQIY	B
RT	263–271	A*3002 (A30)	KLNWASQIY	B, C
RT	269–277	A*0301 (A3)	QIYPGIKVR	B, C
RT	271–279	B*4201 (B42)	YPGIKVRQL	B, C
RT	309–317	A*0201 (A2)	ILKEPVHGV	A, B, C, D
RT	309–318	B*1501 (B62)	ILKEPVHGVY	A, B, D
RT	333–341	B13	GQGQWTYQI	B
RT	341–350	A*1101 (A11)	IYQEPFKNLK	B, C
RT	356–365	A*3002 (A30)	RMRGAHTNDV	B
RT	356–366	A*0301 (A3)	RMRGAHTNDVK	B
RT	375–383	B*5801 (B58)	IAMESIVIW	B, C
RT	392–401	A*3201 (A32)	PIQKETWETW	B
RT	436–445	A*6802 (A68)	GAETFYVDGA	B, C
RT	438–448	A66	ETFYVDGAANR	B, C
RT	449–457	A*2601 (A26)	ETKLGKAGY	B
RT	495–503	Cw*0802 (Cw8)	IVTDSQYAL	—
RT	496–505	B*1503 (B72)	VTDSQYALGI	—
RT	520–528	A*1101 (A11)	QIIEQLIKK	B
RT	560–568	B81	LFLDGIDKA	—

TABLE 2: Continued.

HIV protein	AA position	HLA	Sequence	Clade
Tat	2–11	B*5301 (B53)	EPVDPRLEPW	B
Tat	2–11	B58	EPVDPRLEPW	B
Tat	30–37	Cw12	CCFHCQVC	B
Tat	38–47	B*1503 (B72)	FQTKGLGISY	C
Tat	39–49	A*6801 (A68)	ITKGLGISYGR	B
Vif	17–26	A*0301 (A3)	RIRTWKSLVK	B
Vif	28–36	A*0301 (A3)	HMYISKKAK	B
Vif	31–39	B*5701 (B57)	ISKKAKGWF	B
Vif	48–57	B*0702 (B7)	HPRVSSEVHI	B
Vif	57–66	B51	IPLGDAKLII	B
Vif	79–87	B*1510 (B71)	WHLGHVSI	B
Vif	79–87	B*3801 (B38)	WHLGQGVSI	B
Vif	102–111	B*1801 (B18)	LADQLIHLHY	B
Vif	158–168	A*0301 (A3)	KTKPPLPSVKK	B
Vpr	29–37	B51	EAVRHFPRI	B
Vpr	30–38	B*5701 (B57)	AVRHFPRIW	B, C
Vpr	31–39	B27	VRHFPRIWL	B
Vpr	34–42	B*0702 (B7)	FPRIWLHGL	B
Vpr	34–42	B*8101 (B81)	FPRIWLHGL	B
Vpr	48–57	A*6802 (A68)	ETYGDTWTGV	C
Vpr	52–62	A*6801 (A68)	DTWAGVEAIIR	B
Vpr	59–67	A*0201 (A2)	AIIRILQQL	B
Vpu	29–37	A*3303 (A33)	EYRKILRQR	B
Gag-Pol	24–31	Cw*0102 (Cw1)	NSPTRREL	—
p2p7p1p6	1–10	B*4501 (B45)	AEAMSQVTNS	—
p2p7p1p6	42–50	B14	CRAPRKKGC	B
p2p7p1p6	64–71	B*4002 (B61)	TERQANFL	B
p2p7p1p6	66–74	B13	RQANFLGKI	B, C
p2p7p1p6	70–79	A*0201 (A2)	FLGKIWPSYK	B
p2p7p1p6	118–126	B*4001 (B60)	KELYPLTSL	B

validation of the epitope restriction by a specific HLA class I allele and the definition of the optimal epitope length (8 to 10 amino acid long). Analysis of the CTL epitopes listed in Table 2 reveals that epitopes from 5 HIV proteins (gp160, Nef, p24, p17, and RT) contributed 77% of the total epitopes listed in Table 2. The remaining percentage of the epitopes was derived from the eight other HIV proteins (Integrase, p2p7p1p6, Protease, Rev, Tat, Vpu, Vif and Vpr). The highest number of optimal epitopes was found for p24 (54) while the only one optimal epitope was identified for vpu (Figure 1). The epitope number for gp160, Nef, RT, and p17 were 45, 43, 41, and 23, respectively. The number of unique alleles recognized by these epitopes was also analyzed and found to be correlated with number of epitopes for each HIV protein (Figure 1). For instance, 54 p24 CTL epitopes were restricted cumulatively by 35 unique HLA class I alleles. Similarly, for the epitopes of gp160, Nef, p17, and RT, the numbers of

unique HLA class I alleles were found to be 31, 29, and 25, respectively.

3. Clustering of CTL Epitopes in HIV Proteome

Analysis of the HIV-1 proteins reveals that HLA class I restricted epitopes form overlapping clusters known as epitope rich/dense region whereas the regions deficient of any epitope clusters are called the epitope poor regions [17]. Yusim et al. have identified four short overlapping clusters in Nef protein of HIV-1 which was found to be multirestricted indicating that the clusters contain several epitopes recognized by different class I HLA molecules [18]. In another study, Currier et al. identified CTL epitope distribution patterns in the Gag and Nef proteins of HIV-1 from subtype-A

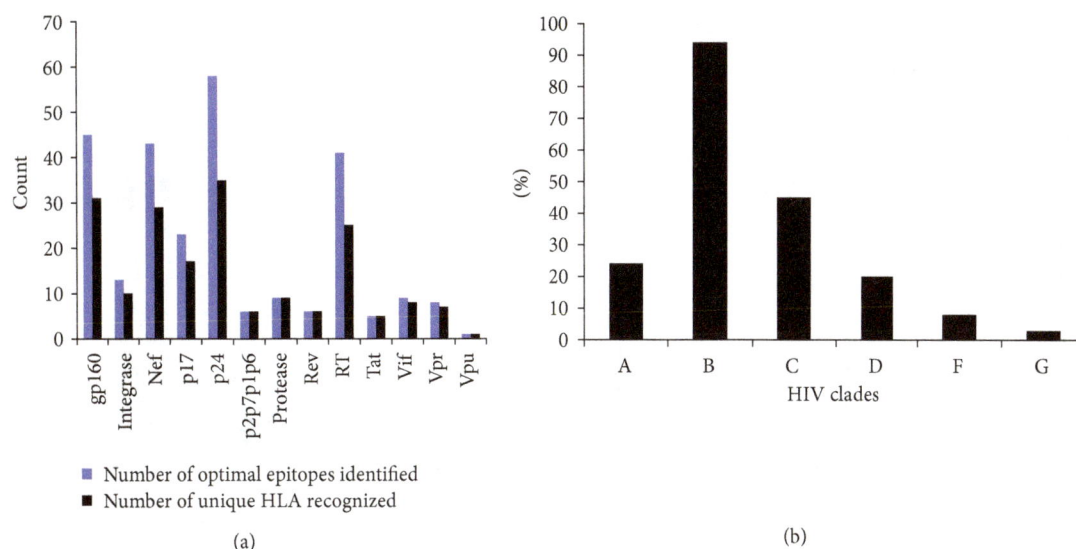

FIGURE 1: Number of optimal CTL epitopes and unique HLA recognized (a) and the percentage of clades (b) to which the optimal epitopes belong for whole HIV proteome were shown.

infected subjects [19]. Studies aided with powerful experimental as well as computational methods are now being conducted with the aim to construct a fine CTL epitope map for the whole HIV-1 proteome. With the advancement of new sophisticated computational and statistical methods, it is now possible to identify the epitope clusters computationally. One significant achievement in computational immunology is the method of identification of immunoproteasome cleavage sites within the query proteins by using different algorithms such as artificial neural network (ANN) which enables the rapid identification of a wide range of potential epitopes that can be analyzed both computationally and experimentally for their affinity to bind with particular HLA class I molecules. Studies, dedicated to identify the CTL epitope clusters by means of computational methods, are now showing some success as far as the identification of new epitope clusters is concerned, as some novel epitope containing clusters were identified. However, more developments in the algorithms are required to construct more realistic models of epitope and cleavage site prediction, so that the predicted proteasomal cleavage events observed in calculation may better mimic the actual processing of viral antigens in the natural environment. In this study, the analysis of the topological arrangement of the 269 experimentally validated optimal epitopes in the HIV proteins listed in Table 2 allowed the identification of epitope clusters in the individual HIV proteins. Among the 13 different HIV proteins (listed in Table 2), epitope clustering was performed for 5 proteins (gp160, Nef, p17, p24, and RT) because for these proteins a relatively higher number of epitopes were identified (Figure 2 and Table 3). The aim of the cluster analysis is to identify the epitope dense regions or "hot spots" in the HIV-1 proteome. To cancel the possibility of random matching, the clusters containing more than 5 overlapping epitopes were only considered. For gp160, 2 major epitope clusters can be observed where the first (amino acid position: 31 to 69) and second (amino acid position: 770–838) cluster harboured 9 epitopes each. Like the gp160 protein 2 major clearly defined clusters were also identified for the Nef and p17 protein. In case of Nef, one spans from 68 to 100 and the second one lies between 105 and 145 amino acid positions. In previous study Penciolelli et al. [20] identified 4 clusters in the Nef protein which falls within the epitope cluster range for Nef observed in this study. For p17 protein, two clusters were similar in epitope composition and length. First p17 cluster with 34 amino acids was found to contain 10 epitopes whereas 2nd p17 cluster with 12 epitopes was composed of 30 amino acids. p24 was found to contain maximum numbers (4) of major epitope clusters. For RT only 1 major cluster was identified which contained 9 overlapping epitopes, whereas the rest of epitopes were found to be distributed randomly in protein.

Data from Table 3 suggest that epitopes in both the RT and gp160 proteins did not exhibit significant clustering properties compared to other HIV proteins. Only 40% and 22% of the epitopes were identified as part of the major cluster in gp160 and RT, respectively, which indicated that the majority of the epitopes were distributed randomly in respective proteins. Epitopes from other three proteins Nef, p17, and p24 showed significant clustering pattern as evident by both Figure 4 and Table 3. Most of the epitopes in these proteins were found to be a part of cluster or epitope dense region.

4. Are the CTL Epitope Clusters Conserved and Hydrophobic in Nature?

By analyzing the nature of the CTL epitopes and their source proteins, Hughes and Hughes proposed two hypotheses about the nature of the CTL epitopes. First they proposed that the endogenous peptides presented by human leukocyte antigen (HLA) class I molecules are largely derived from

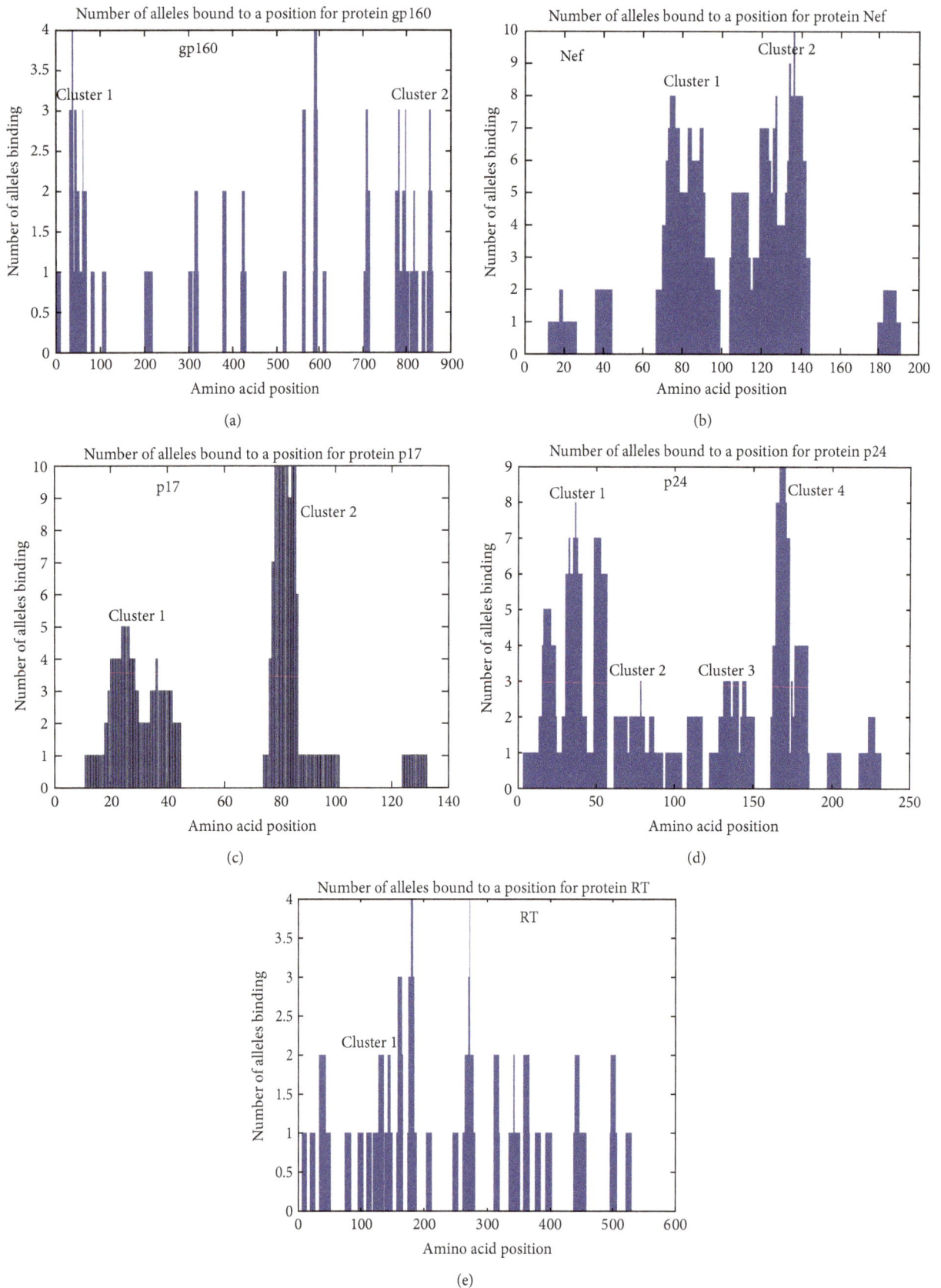

FIGURE 2: Clustering pattern of 5 HIV-1 proteins. The X-axis represents the amino acid position whereas the Y-axis represents the number of allele binding to particular positions.

TABLE 3: Analysis of the identified epitope clusters in HIV-1 proteins.

HIV-1 protein	Total epitope number	Cluster number	Cluster amino acid position	Cluster length (amino acids)	Number of epitopes in cluster	% of clustered epitopes in total epitope pool
Gp160	45	1	31–69	39	9	40
		2	770–838	69	9	
Nef	43	1	68–100	33	18	97
		2	105–145	41	24	
p17	23	1	11–44	34	10	95
		2	72–101	30	12	
p24	54	1	3–56	54	22	98
		2	61–117	57	9	
		3	128–150	43	9	
		4	161–184	24	13	
RT	41	1	107–166	60	9	22

conserved regions of proteins, so in general the CTL epitopes tend to be more conserved than the remainder portion of the source proteins. Secondly they hypothesized that the CTL epitope regions are hydrophobic whereas the source protein may itself be overall hydrophilic in nature [8]. In harmony with these hypotheses, Silva and Hughes showed that the CTL epitopes of HIV-1 Nef protein were derived from the hydrophobic and relatively conserved regions by estimating the relative conservation of CTL epitopes of the Nef protein and relating this to the structure and function of the protein. In another study Lucchiari-Hartz et al. showed that the CTL epitope clusters derived from Nef protein tend to coincide with hydrophobic regions, whereas the noncluster regions are predominantly hydrophilic [8]. Their in vitro analysis of the proteasomal degradation products of HIV-Nef protein demonstrates a differential sensitivity of cluster and noncluster regions to proteasomal processing and the cluster regions are digested by proteasomes with greater preference for hydrophobic P1 residues. But the authors admitted that some cytosolic endoproteases other than proteasomes may also be involved in the production of certain Nef-CTL epitopes in natural condition [8]. In both these studies the primary focus was on one protein (Nef) in the whole HIV proteome. So similar studies on other HIV proteins would certainly be interesting and could reveal some important feature of the HIV-CTL epitopes. In contrast to the notion that HIV CTL epitopes are more conserved and hydrophobic in nature, more recent study revealed that distribution of CTL epitopes in 99% of the HIV-1 protein sequences follows a random pattern and is indistinguishable from the distribution of CTL epitopes in proteins from other proteomes such as hepatitis C virus (HCV), influenza and for three eukaryote proteomes. In this study, the authors opted for the computational approach to predict the large set of CTL epitopes where proteasome cleavage pattern, TAP, and HLA-binding, three most crucial steps in classical endogenous antigen presentation pathway, were predicted by means of computational tools. The use of experimentally validated epitopes instead of computationally predicted epitopes could influence the outcome of the study and may lead the authors to a different conclusion. To shed some light on the contradiction of different studies mentioned above, an investigation involving hydrophobic and conservancy pattern of experimentally validated optimal CTL epitopes (Table 2) from 5 HIV proteins (gp160, Nef, p17, p24, and RT) was conducted in this study. Relative conservancy and hydrophobicity of the five selected HIV proteins were analyzed. 100 proteins sequences of different HIV clades retrieved from the Uniprot database (http://www.uniprot.org/) were used as an input for both conservancy and hydrophobic pattern prediction. To unveil the conservation pattern, multiple sequence alignment (MAS) was constructed using well stabled tool called Clustal W version 2.0 (http://www.ebi.ac.uk/Tools/clustalw2/index.html) developed by European Bioinformatics Institute (EBI). From the MSA the conservancy score for each amino acid position was obtained. To predict the hydrophobicity score Protscale tool of the ExPASy Proteomics Server (http://www.expasy.ch/tools/protscale-ref.html) and algorithm (developed by Abraham and Leo) previously used by Lucchiari-Hartz et al. [8] were employed. Both the hydrophobicity and conservancy scores for each amino acid position within a particular HIV protein were used to calculate the total scores for both these parameters. Figure 3 shows the total hydrophobicity and conservancy scores of 5 individual HIV proteins in agreement with previous study [21]. We found that both the RT and p24 are relatively conserved and more hydrophobic than the rest of the analyzed HIV proteins (Figure 3). To visualize the overlapping pattern and correlation between the hydrophobic pattern and epitope clusters for the five selected proteins, the epitope count/hit and hydrophobicity scores were plotted together for each protein (Figure 4). The epitope hit score for a particular position is the number of alleles binding to that particular position.

To compare the correlation among hydrophobicity, conservancy, and epitope count, correlation coefficient was calculated among them (Appendix 1). For the calculation of correlation coefficient, first the standard deviation of each of

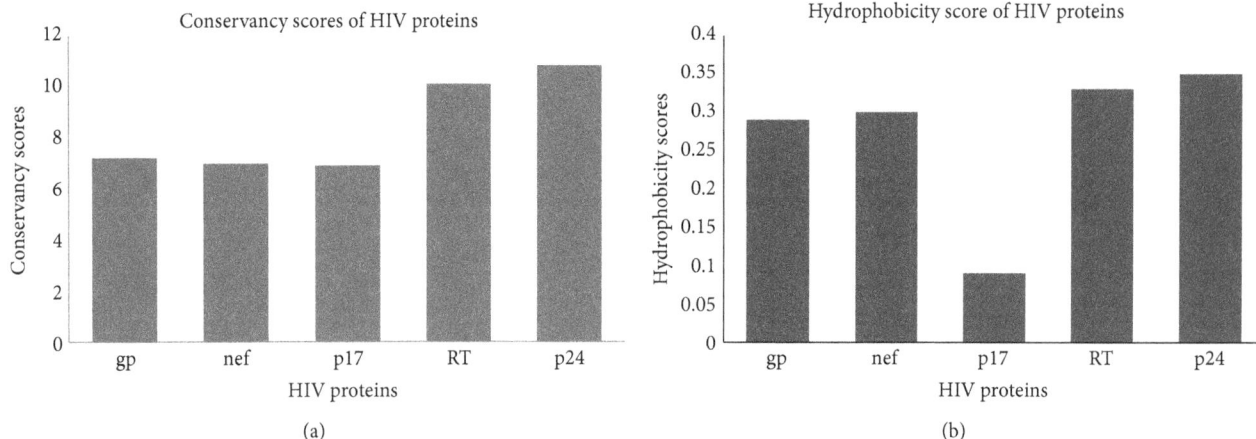

FIGURE 3: Cumulative conservancy (a) and hydrophobicity (b) scores of five individual proteins of HIV-1.

TABLE 4: Correlation coefficient.

Correlation score	HIV-1 proteins				
	gp160	Nef	p17	p24	RT
Epitope hit versus hydrophobicity	0.11	0.27	0.16	0.26	0.12
Epitope hit versus conservancy	0.097	0.21	0.19	0.23	0.047
Hydrophobicity versus conservancy	0.20	0.22	0.23	0.15	0.009

the three score parameters (epitope count, hydrophobicity, and conservancy) was obtained (Appendix 1). Table 4 shows the correlation score values among these three parameters.

From the correlation score it was evident that Nef epitope clusters were strongly correlated with the hydrophobicity and conservancy. p24 protein also showed relatively high correlation between epitope clusters and hydrophobicity and between conservancy and epitope clusters. In contrast, gp160 and RT showed relatively weak yet similar correlation among the parameters. p17 showed strong correlation when epitope hit was compared with conservancy but showed moderate correlation between epitope hit and hydrophobicity. So these findings suggested that not all the epitopes of HIV proteome are derived from conserved and hydrophobic regions of HIV-1 proteome although this hypothesis was found to be valid for two of the five HIV proteins (Nef and p24) as both of these proteins showed a significant correlation among epitope cluster, hydrophobicity, and conservancy. But the very weak correlation obtained for gp160 and RT diminished the general applicability of the hypothesis that all the HIV CTL epitopes were conserved and hydrophobic in nature.

5. The Role of MHC Class I on Immune Control of HIV Infection

Significant variation in the susceptibility to HIV-1 infection and especially in the clinical outcome after infection is observed in HIV infected patients. For instance, variation in the level of circulating virus particles in the plasma during the nonsymptomatic phase is commonly observed among the patients [22]. In addition to this, there is evidence that in

certain cases individuals known as long-term nonprogressor (LTNP), infected with HIV, remain asymptomatic without antiretroviral therapy (ART) in their life time due to the slow or arrested evolution of HIV [23]. The most plausible explanation is that the variation in the susceptibility and outcome of HIV infection is largely due to host factors and viral adaptation to selective pressure. Recently Fellay et al. conducted a whole-genome association study to identify the host factor associated with control of HIV-1. In this study they identified two distinct polymorphisms associated with HLA loci B and C [24]. Surprisingly, almost all HLA class I polymorphisms were found to occur in those residues that belong to peptide-binding groove of these molecules thereby determining the epitopes that bind to each HLA molecule [25]. Among the three MHC class I loci in humans (HLA-A, HLA-B, and HLA-C), HLA-B is the most polymorphic, compared with HLA-A and HLA-C molecules (IMGT/HLA database: http://www.ebi.ac.uk/imgt/hla/). A more direct evidence of the association between HLA polymorphism and disease progression in HIV infected individual came from a previous study where they showed HLA-B*3503 associated with rapid disease progression differs in only one amino acid from HLA-B*3501 for which no such association was observed [26]. The presence of HLA-B*57 allele in a large proportion of LTNPs signifies its role in controlling disease progression and mutations in HLA-B*57-restricted Gag epitopes were frequently present in all viruses from plasma but interestingly inspite of this CTL escape mutations LTNPs can maintain viral suppression [27, 28]. The escape mutation in the HLA-B*57-restricted Gag epitopes can be considered as a consequence of strong evolutionary pressure exerted by the host immune response. Previous studies showed that although mutation in the conserved gag p24 epitope DRFYKTLRAE helps the virus to evade CTL response, it also impairs its ability to replicate because the mutation occurs in a very conserved position which is functionally constrained [29]. Among the three HLA class I molecules, HLA-B is considered as the most important factor for restricting HIV diseases progression and T cells responding to HLA-B-restricted epitopes appear to be immunodominant [30, 31].

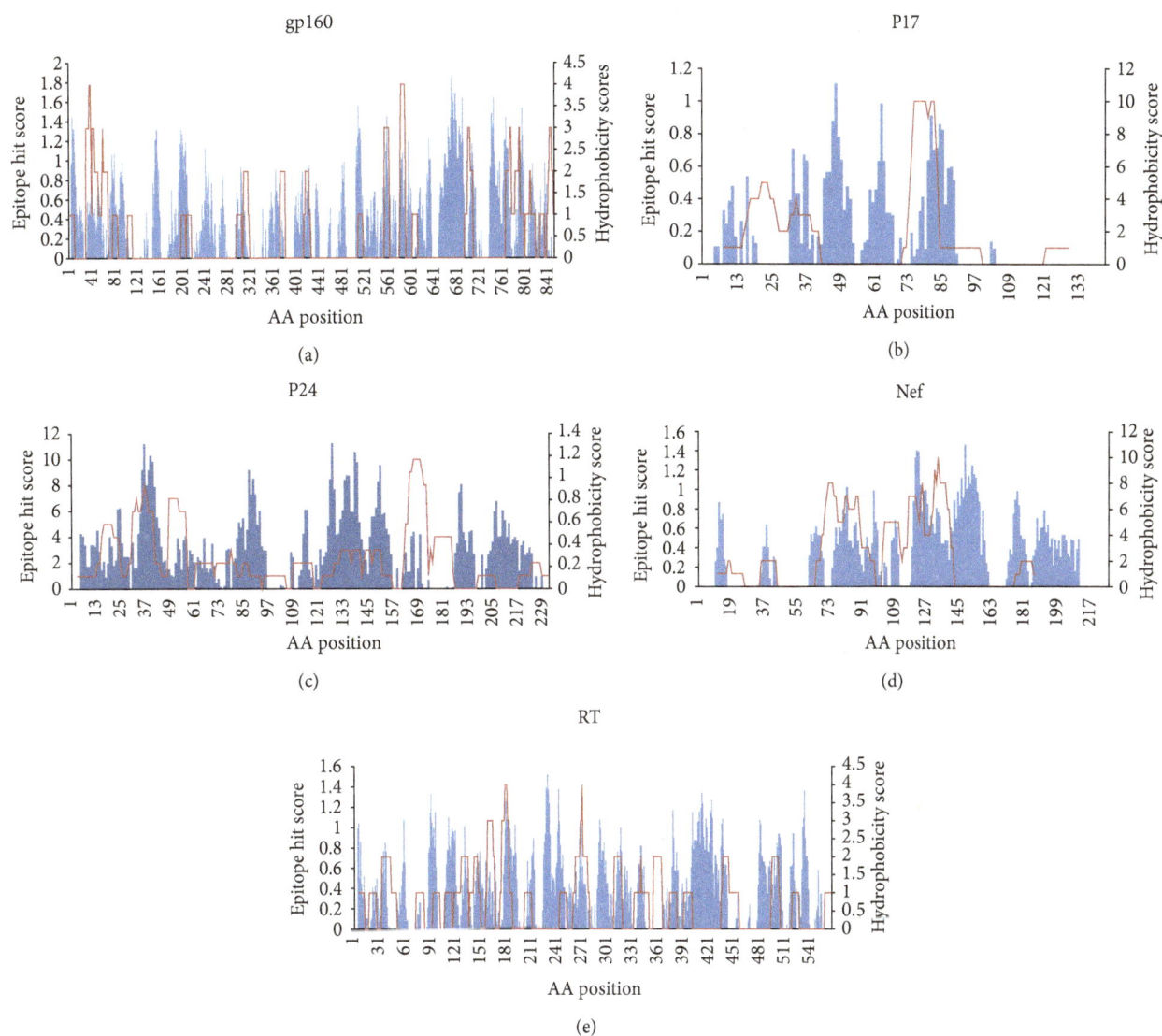

FIGURE 4: Epitope clusters and hydrophobic pattern of five HIV proteins. The X-axis represents the amino acid position. The primary Y-axis (left) shows the hydrophobicity scores whereas the secondary Y-axis (right) represents the epitope hit/count values. The blue bars indicate the hydrophobicity score whereas the red line represents the epitope hit values. For the hydrophobicity scores, the negative values were not shown in the figure, and only the scores greater than 0 were plotted.

Moreover, detailed study of the CTL epitopes in whole HIV-1 proteome revealed that HLA-B-restricted epitopes are more conserved compared to epitopes restricted by HLA-A and C. The same study also showed that although for most of the proteins the fractions of unique HLA-A and B restricted positions are equivalent in the total HIV clade-B proteome, Gag-p24 and Nef seemed to be preferentially targeted by HLA-B alleles as the B-restricted fractions were found to be over threefold higher than the A-restricted residues [32].

In our study, the analysis of different class I HLA alleles that recognize all the listed CTL optimal epitopes revealed some interesting features of HLA restriction patterns of HIV-1 CTL epitopes. Figure 5 shows the number of optimum epitopes recognized by 62 different class I HLA alleles. It was found that HLA-B*57 was the most successful as far as the number of epitope recognition was concerned as it recognizes 22 different optimal epitopes. The other successful HLA alleles were HLA-A*3, A*2, B*7, A*11, and B*35 (Figure 5). Among the total HLA alleles HLA-B contributes to the 50% of the total allele pool whereas HLA-A and C constitute 27% and 23%, respectively (Table 5).

These data also support the previous findings and also signify the role of HLA-B alleles in controlling HIV-1 infection as HLA-B was found to be associated with the 60% of the experimentally validated CTL epitopes. In harmony with the finding of Costaa et al. [32] we also observed a low % (9% of the total optimal epitope pool) of epitopes was recognized by HLA-C alleles. We have also analyzed the % of the epitopes associated with HLA-A and B alleles in individual HIV proteins (gp160, Nef, p24, p17, and RT).

(a)

(b)

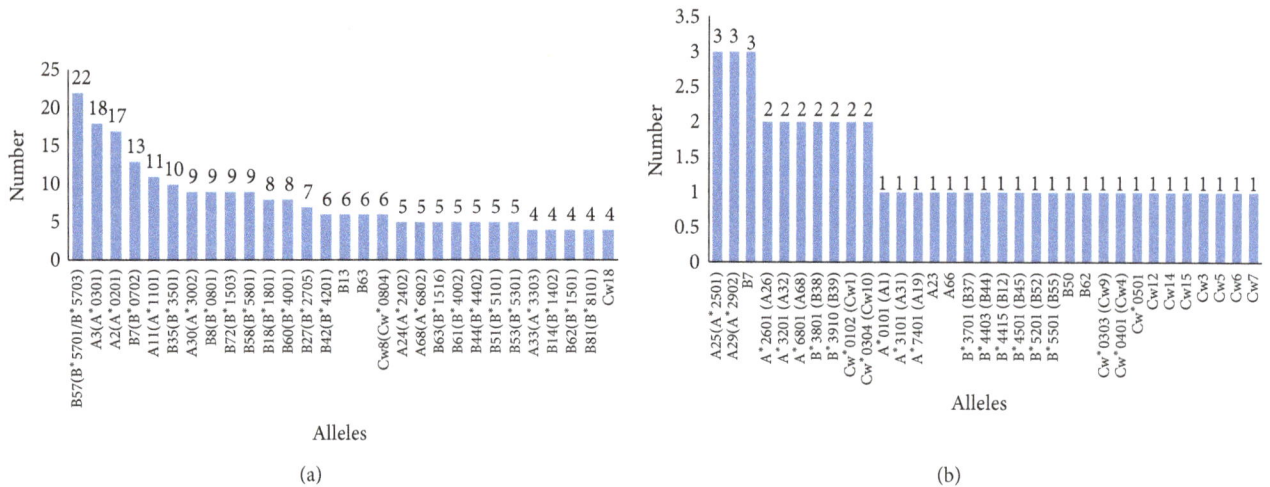

FIGURE 5: The number of unique alleles recognized by the optimal CTL epitopes.

TABLE 5: Analysis of the different HLA alleles.

HLA	Total number and as % of total HLAs	Number of epitope recognized	% of epitope recognized by HLA in total epitope pool
A	17 (27%)	86	31%
B	31 (50%)	165	60%
C	14 (23%)	24	9%

Analysis showed that only the p24 and Nef epitopes were associated with higher numbers of HLA-B alleles than A alleles. In contrast, epitopes derived from gp160, p17, protease, and other proteins (Tat, Rev, Vpu, Vpr, Vif) were recognized by slightly greater number of A alleles compared to B alleles. In case of intergrase, the epitopes were restricted by almost similar number of A and B alleles. There was a significant difference between the HLA*A and HLA*B associated epitopes for RT and p24. For RT the % of HLA*A and HLA*B associated epitopes are 19 and 6.28, respectively. In case of p24 the % of HLA*B associated epitopes are significantly higher (20%) than the HLA*A epitopes (3.5%).

6. Conclusion

As CTL response against HIV infected cells is proved to be crucial in controlling virus population in the host, rationally the CTL based vaccine should have a profound effect on HIV infection. Yusim et al. suggested that epitope clustering methods provide an alternative strategy to design novel multiepitope vaccine. They also suggested that the multiepitope vaccine should not be composed of a string of single epitopes, rather it should be composed of short region containing the epitope clusters and proximal regions flanking the epitope cluster that may be essential for optimal processing of epitopes. These epitope clusters harbor multiple overlapping epitopes which may be recognized by multiple HLA alleles. In conclusion, this study has shown the analysis of the HIV-1 CTL epitopes which revealed that Nef and p24 proteins of HIV-1 can be considered for CTL based multiepitope vaccine design since a significant number of optimal CTL epitopes are derived from Nef and Gag-p24 and almost all these epitopes showed a clustering pattern. A further study is needed to test these proposed vaccine candidates in laboratory animal to test safety and immunity against HIV.

Appendix

Correlation between Different Scores. There are three scores for each of the k-th position of epitope:

(a) epitope hit score: $EHS(k)$,

(b) hydrophobicity score: $HS(k)$,

(c) conservancy score: $CS(k)$.

Let us assume that the position range of epitope, for which each of scores exists, is j to n.

Now

(a) the average epitope hit score $EHS\_AVG = (1/(n - j + 1)) \sum_{k=j}^{n} EHS(k)$;

(b) the average hydrophobicity score $HS\_AVG = (1/(n - j + 1)) \sum_{k=j}^{n} HS(k)$;

(c) the average conservancy score $CS\_AVG = (1/(n - j + 1)) \sum_{k=j}^{n} CS(k)$.

Now

(a) the standard deviation of epitope hit score

$$EHS\_SD = \frac{1}{n - j + 1} \sum_{k=j}^{n} (EHS(k) - EHS\_AVG)^2. \quad (A.1)$$

(b) The standard deviation of hydrophobicity score

$$HS\_SD = \frac{1}{n - j + 1} \sum_{k=j}^{n} (HS(k) - HS\_AVG)^2. \quad (A.2)$$

(c) The standard deviation of conservancy score

$$CS\_SD = \frac{1}{n-j+1} \sum_{k=j}^{n} (CS(k) - CS\_AVG)^2. \quad (A.3)$$

Then the correlation coefficient between the scores are calculated as below.

Correlation between epitope hit count and hydrophobicity score

Corr (EHS, HS)

$$= \frac{\sum_{k=j}^{n} (EHS(k) - EHS\_AVG)(HS(k) - HS\_AVG)}{(n-j) EHS\_SD \times HS\_SD}. \quad (A.4)$$

Correlation between epitope hit count and conservancy score

Corr (EHS, CS)

$$= \frac{\sum_{k=j}^{n} (EHS(k) - EHS\_AVG)(CS(k) - CS\_AVG)}{(n-j) EHS\_SD \times CS\_SD}. \quad (A.5)$$

Correlation between conservancy and hydrophobicity score

Corr (CS, HS)

$$= \frac{\sum_{k=j}^{n} (CS(k) - CS\_AVG)(HS(k) - HS\_AVG)}{(n-j) CS\_SD \times HS\_SD}. \quad (A.6)$$

The calculation of each of these correlations is carried out for each protein group separately.

References

[1] M. D. Moore and W. S. Hu, "HIV-1 RNA dimerization: it takes two to tango," *AIDS Reviews*, vol. 11, no. 2, pp. 91–102, 2009.

[2] United Nations Programme on HIV/AIDS. Global Report fact sheet: the global AIDS epidemic, 2010, http://www .unaids.org/documents/20101123_FS_Global_em_en.pdf.

[3] Report on the Global AIDS Epidemic, UNAIDS, 2013.

[4] J. M. Costin, "Cytopathic mechanisms of HIV-1," *Virology Journal*, vol. 4, article 100, 2007.

[5] J. Hemelaar, E. Gouws, P. D. Ghys, and S. Osmanov, "Global and regional distribution of HIV-1 genetic subtypes and recombinants in 2004," *AIDS*, vol. 20, no. 16, pp. W13–W23, 2006.

[6] M. M. Thomson, L. Pérez-Álvarez, and R. Nájera, "Molecular epidemiology of HIV-1 genetic forms and its significance for vaccine development and therapy," *The Lancet Infectious Diseases*, vol. 2, no. 8, pp. 461–471, 2002.

[7] I. M. Roitt and P. J. Delves, *Roitt's Essential Immunology*, Blackwell Science, London, UK, 10th edition, 2001.

[8] M. Lucchiari-Hartz, V. Lindo, N. Hitziger et al., "Differential proteasomal processing of hydrophobic and hydrophilic protein regions: contribution to cytotoxic T lymphocyte epitope clustering in HIV-1-Nef," *Proceedings of the National Academy of Sciences of the United States of America*, vol. 100, no. 13, pp. 7755–7760, 2003.

[9] M. P. Davenport and J. Petravic, "CD8+ T cell control of HIV—a known unknown," *PLoS Pathogens*, vol. 6, no. 1, Article ID e1000728, 2010.

[10] P. Kunwar, N. Hawkins, W. L. Dinges et al., "Superior control of HIV-1 replication by CD8+ T cells targeting conserved epitopes: implications for HIV vaccine design," *PLoS ONE*, vol. 8, no. 5, Article ID e64405, 2013.

[11] M. E. Severino, N. V. Sipsas, P. T. Nguyen et al., "Inhibition of human immunodeficiency virus type 1 replication in primary CD4+ T lymphocytes, monocytes, and dendritic cells by cytotoxic T lymphocytes," *Journal of Virology*, vol. 74, no. 14, pp. 6695–6699, 2000.

[12] Y. Zhou, R. Bao, N. L. Haigwood, Y. Persidsky, and W. Ho, "SIV infection of rhesus macaques of Chinese origin: a suitable model for HIV infection in humans," *Retrovirology*, vol. 10, no. 89, 2013.

[13] P. J. Goulder and D. I. Watkins, "Impact of MHC class I diversity on immune control of immunodeficiency virus replication," *Nature Reviews Immunology*, vol. 8, no. 8, pp. 619–630, 2008.

[14] M. R. Betts, J. P. Casazza, and R. A. Koup, "Monitoring HIV-specific CD8+ T cell responses by intracellular cytokine production," *Immunology Letters*, vol. 79, no. 1-2, pp. 117–125, 2001.

[15] D. F. Nixon, A. R. M. Townsend, J. G. Elvin, C. R. Rizza, J. Gallwey, and A. J. McMichael, "HIV-1 gag-specific cytotoxic T lymphocytes defined with recombinant vaccinia virus and synthetic peptides," *Nature*, vol. 336, no. 6198, pp. 484–487, 1988.

[16] A. Llanoa, N. Frahm, and C. Brander, "How to optimally define optimal cytotoxic T lymphocyte epitopes in HIV infection?" *HIV Molecular Immunology*, vol. 3, no. 24, 2009.

[17] C. Zhang, J. L. Cornette, J. A. Berzofsky, and C. DeLisi, "The organization of human leucocyte antigen class I epitopes in HIV genome products: implications for HIV evolution and vaccine design," *Vaccine*, vol. 15, no. 12-13, pp. 1291–1302, 1997.

[18] K. Yusim, C. Kesmir, B. Gaschen et al., "Clustering patterns of cytotoxic T-lymphocyte epitopes in human immunodeficiency virus type 1 (HIV-1) proteins reveal imprints of immune evasion on HIV-1 global variation," *Journal of Virology*, vol. 76, no. 17, pp. 8757–8768, 2002.

[19] J. R. Currier, U. Visawapoka, S. Tovanabutra et al., "CTL epitope distribution patterns in the Gag and Nef proteins of HIV-I from subtype A infected subjects in Kenya: use of multiple peptide sets increases the detectable breadth of the CTL response," *BMC Immunology*, vol. 7, article 8, 2006.

[20] B. Culmann-Penciolelli, S. Lamhamedi-Cherradi, I. Couillin et al., "Identification of multirestricted immunodominant regions recognized by cytolytic T lymphocytes in the human immunodeficiency virus type 1 Nef protein," *Journal of Virology*, vol. 68, no. 11, pp. 7336–7343, 1994.

[21] C. L. Kuiken, B. T. Foley, E. Guzman, and B. T. Korber, *Molecular Evolution of HIV*, Johns Hopkins University Press, Baltimore, Md, USA, 1999.

[22] A. Telenti and D. B. Goldstein, "Genomics meets HIV-1," *Nature Reviews Microbiology*, vol. 4, no. 11, pp. 865–873, 2006.

[23] M. Vajpayee, N. Negi, and S. Kurapati, "The enduring tale of T cells in HIV immunopathogenesis," *Indian Journal of Medical Research*, vol. 138, pp. 682–699, 2013.

[24] J. Fellay, K. V. Shianna, D. Ge et al., "A whole-genome association study of major determinants for host control of HIV-1," *Science*, vol. 317, no. 5840, pp. 944–947, 2007.

[25] S. G. E. Marsh, P. Parham, and L. D. Barber, *The HLA Facts Book*, Academic Press, London, UK, 2000.

[26] X. Gao, G. W. Nelson, P. Karacki et al., "Effect of a single amino acid change in MHC class I molecules on the rate of progression to aids," *The New England Journal of Medicine*, vol. 344, no. 22, pp. 1668–1675, 2001.

[27] S. A. Migueles, M. S. Sabbaghian, W. L. Shupert et al., "HLA B∗5701 is highly associated with restriction of virus replication in a subgroup of HIV-infected long term nonprogressors," *Proceedings of the National Academy of Sciences of the United States of America*, vol. 97, no. 6, pp. 2709–2714, 2000.

[28] J. R. Bailey, T. M. Williams, R. F. Siliciano, and J. N. Blankson, "Maintenance of viral suppression in HIV-1-infected HLA-B∗57 + elite suppressors despite CTL escape mutations," *Journal of Experimental Medicine*, vol. 203, no. 5, pp. 1357–1369, 2006.

[29] R. Wagner, B. Leschonsky, E. Harrer et al., "Molecular and functional analysis of a conserved CTL epitope in HIV-1 p24 recognized from a long-term nonprogressor: Constraints on immune escape associated with targeting a sequence essential for viral replication," *Journal of Immunology*, vol. 162, no. 6, pp. 3727–3734, 1999.

[30] P. Klepiela, A. J. Leslie, I. Honeyborne et al., "Dominant influence of HLA-B in mediating the potential co-evolution of HIV and HLA," *Nature*, vol. 432, no. 7018, pp. 769–774, 2004.

[31] F. Bihl, N. Frahm, L. Di Giammarino et al., "Impact of HLA-B alleles, epitope binding affinity, functional avidity, and viral coinfection on the immunodominance of virus-specific CTL responses," *Journal of Immunology*, vol. 176, no. 7, pp. 4094–4101, 2006.

[32] A. I. F. Costaa, X. Rao, E. LeChenadecb, D. V. Baarlea, and C. Keşmirb, "HLA-B molecules target more conserved regions of the HIV-1 proteome," *AIDS*, vol. 24, pp. 211–215, 2010.

17

Evaluating Andrographolide as a Potent Inhibitor of NS3-4A Protease and its Drug-Resistant Mutants using *In Silico* Approaches

Vivek Chandramohan, Anubhav Kaphle, Mamatha Chekuri, Sindhu Gangarudraiah, and Gowrishankar Bychapur Siddaiah

Department of Biotechnology, Siddaganga Institute of Technology, Tumkur, Karnataka 572 103, India

Correspondence should be addressed to Vivek Chandramohan; vivek.bioinf@gmail.com

Academic Editor: Jay C. Brown

Current combination therapy of PEG-INF and ribavirin against the Hepatitis C Virus (HCV) genotype-1 infections is ineffective in maintaining sustained viral response in 50% of the infection cases. New compounds in the form of protease inhibitors can complement the combination therapy. Asunaprevir is new to the drug regiment as the NS3-4A protease inhibitor, but it is susceptible to two mutations, namely, R155K and D168A in the protein. Thus, in our study, we sought to evaluate Andrographolide, a labdane-diterpenoid from the *Andrographis paniculata* plant as an effective compound for inhibiting the NS3-4A protease as well as its concomitant drug-resistant mutants by using molecular docking and dynamic simulations. Our study shows that Andrographolide has best docking scores of −15.0862, −15.2322, and −13.9072 compared to those of Asunaprevir −3.7159, −2.6431, and −5.4149 with wild-type R155K and D168A mutants, respectively. Also, as shown in the MD simulations, the compound was good in binding the target proteins and maintains strong bonds causing very less to negligible perturbation in the protein backbone structures. Our results validate the susceptibility of Asunaprevir to protein variants as seen from our docking studies and trajectory period analysis. Therefore, from our study, we hope to add one more option in the drug regiment to tackle drug resistance in HCV infections.

1. Introduction

More than 25 years after the discovery, Hepatitis C Virus (HCV) is still considered a major global threat to human health. The viral infection is spread over 130–170 million people worldwide [1] and a significant number of people, around 350,000 to 500,000, die each year because of Hepatitis C related liver diseases according to the WHO [2]. Combination of pegylated interferon-α (PEG-INF) with ribavirin has been used as a major treatment for the infection [3, 4]. However, 50% of the HCV genotype-1 infected individuals do not show a sustained virological response (SVR) for the combination, reasons of which have been recently explored by Padmanabhan et al., using systems biology approaches [5]. Several factors have been identified that correlate with these nonresponsive observations, some of which are found to be the genomic differences between individuals, viral genotype, and single-nucleotide polymorphisms (SNPs) in the interferon-λ locus [6, 7]. New drug compounds in the form of protease and polymerase inhibitors are currently in the development as the direct-acting antivirals (DAAs). Studies have shown that, together with the combined therapy of PEG-INF with ribavirin, these antiviral compounds have shown to increase SVR from less than 50% to around 70% in HCV genotype-1 patients [8]. However, the potential of these DAAs has been obscured by high mutation rates and genomic heterogeneity in the virus [9]. Introduction of frequent mutations in the viral genome due to the infidel nature of viral replicase adds plights to drug researchers looking for new antiviral.

Viral proteases are vital for infection and proliferation and hence they can be considered as potential targets for DAAs to intervene viral cycle. In our work, we have selected NS3-4A protease which is responsible for cleaving

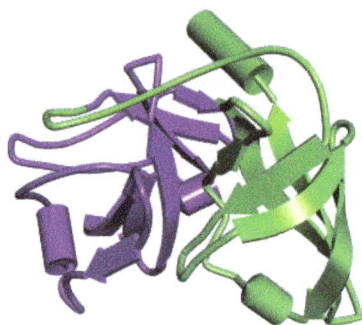

FIGURE 1: The figure depicts the structure of NS3 protease with schematic model. Violet color shows the N-terminal serine protease domain region in the protein, and green color shows the C-terminal domain region that is a member of the DExH/D-box helicase superfamily II with NTPase activity.

single precursor polypeptide, together with NS2-NS3 and NS3 proteinases, of length 3010-3011 aa translated from the long reading frame to yield active proteins [10–12]. Many proteases inhibitors like telaprevir or boceprevir have been approved by the FDA as the potent inhibitors of the protease; however, the mutations in the protein have led to rapid drug inefficacy [13, 14]. Asunaprevir is yet another effective protease inhibitor being developed by Bristol-Myers Squibb and is in its 3rd clinical trial phase. However, the binding capacity of Asunaprevir has been limited by two mutations in the protein structure, namely, R155K and D168A [15]. Crystal structure of the proteases is available publicly on the Protein Data Bank (PDB) website and structure-based drug design approach can be applied to screen plethora of new DAAs that can have maximum binding efficiency against any concomitant mutations in the proteins. Plants are considered as great source of medicinal compounds, and they can be explored to drive drug discovery process fast and smoothly with minimum budget concern [16]. *Andrographis paniculata* Nees is an herbaceous plant in the family Acanthaceae. It has a broad range of pharmacological effects which also include antiviral activity [17–19]. The plant extract contains various phytochemicals majority of which are diterpenoids and flavonoids. Andrographolide, a labdane-diterpenoid, is a major bioactive compound from the plant extract [20].

Thus, our work is directed towards exploring the binding potential of Andrographolide from the plant against the mutations in the protein by molecular docking, dynamics, and comparing its effects with Asunaprevir computationally.

2. Material and Methods

2.1. Protein Preparation. 3D structure of wild-type NS3 protease was retrieved from the Protein Data Bank (PDB) using query ID 4NWL [21]. Cocrystallized ligand, water molecules, and zinc ions were removed from the target structure to obtain clean protein [22]. The protein mutants were prepared by replacing the native residues in the protein with the mutant residues (R155K and D168A) [23] using DS 3.5 "build

mutants" option. The structures thus obtained were optimized classically using CHARMm force field implemented in the DS 3.5, minimized with conjugate gradient energy minimization protocol followed by convergence energy minimization (0.001 kcal/mole), that readied the structures for docking and simulations [24]. Active site residues (*Q41, F43, H57, G58, D81, R109, K136, G137, S138, S139, G140, G141, F154, R155, A156, A157, D168, M485, V524, Q526, and H528*) [25] were selected for both the wild-type protein and mutant structures for molecular docking studies.

2.2. Ligand Preparation. The investigated compounds Andrographolide and Asunaprevir were drawn using Marvin sketch [26]. Ligand optimization was carried out using Chemistry at Harvard Molecular Mechanics (CHARMm) and macro molecular force field (MMF) followed by energy minimization protocol [27]. Several ligand conformations were generated based on bond energy, CHARM energy, dihedral energy, electrostatic energy, initial potential energy, and initial RMS gradient values. The drug likeliness was evaluated using the Lipinski rule of 5 via Lipinski drug filter protocol [28]. The studies on the ADME of aqueous solubility, blood brain barrier level, hepatotoxicity, plasma protein binding levels, and CYP2D6 were carried out [29]. Toxicity profile of the ligand molecules was predicted by using TOPKAT which applies a range of robust, cross validated, and Quantitative Structure-Toxicity Relationship (QSTR) models for assessing specific toxicological endpoints. The toxicity profile also included NTP carcinogenicity, mutagenicity, and developmental toxicity and skin irritation assessment [30]. The studies were performed using Discovery studio 3.5 (Accelrys).

2.3. Molecular Docking and Dynamics. For molecular docking studies, a flexible docking approach was employed using the LeadIT [31] software in which wild NS3 protease and mutants R155K and D168A were considered as receptor proteins. The docking results for receptor-ligand complex comprised intermolecular interaction energies, namely, hydrogen bonding and hydrophobic and electrostatic interaction. Receptor-ligand complex with least binding energy was used to infer the best binding compound. Molecular dynamics (MD) simulations for both proteins and ligands were performed in a flexible manner that allowed binding site to be relaxed around the ligand and directly estimate the effect of explicit water molecules. MD-based computational techniques are available for estimating the binding free energy which includes thermodynamic integration (TI), free energy perturbation (FEP), linear interaction energy (LIE), and molecular mechanics/Poisson-Boltzmann and surface area (MM/PB-SA) methods. Three best receptor-ligand complexes were subjected to molecular dynamics studies based on steepest decent minimization protocol. For dynamics study, the following parameters, heating steps and time steps set as 2000 and 0.001, respectively, equilibration steps and time steps set as 1000 and 0.001, respectively, for the overall production period of 20 ns with time steps as 0.001 and temperature factor of 300 K, were considered. The

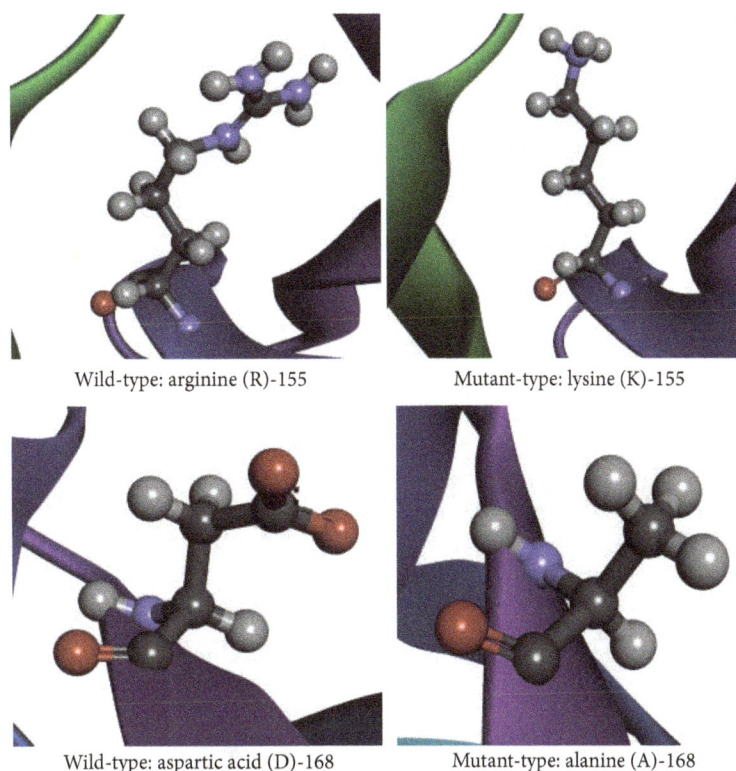

Wild-type: arginine (R)-155   Mutant-type: lysine (K)-155

Wild-type: aspartic acid (D)-168   Mutant-type: alanine (A)-168

FIGURE 2: The figure depicts the native and mutated residues (shown in balls-and-stick model) of the variants R155K and D168A in the structure of NS3 protease (color rainbow). As can be inferred, the change in the residues introduced larger amino acid groups for R155K that will decrease the binding efficiency of drug molecules due to more steric hindrances. Also, the introduction of nonpolar groups for D168A transition will contribute to only weaker molecular interactions and thus reduced binding.

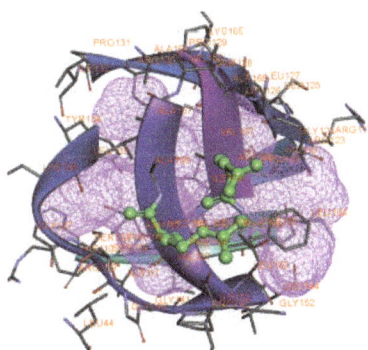

FIGURE 3: Active site of target protein with mutant structure shown in ball-and-stick model of variants R155K and D168A (violet wire mesh indicates binding cavity in the active site).

best conformations were selected based on the least potential energy value [32].

3. Result and Discussions

3.1. Protein Preparation.
The obtained protein structure has a single-chain construct of protease domain of Hepatitis C Virus genotype-1a, with a covalently linked cofactor 4A at the N-terminal [21]. The protease belongs to the hydrolase class in the Enzyme Commission (EC) classification with EC number 3.4.21.98. It is a bifunctional enzyme that has two domains depicted in Figure 1, namely, the N-terminal serine protease domain that locates between −7 and 87 aa and C-terminal domain that is a member of the DExH/D-box helicase superfamily II with NTPase nucleic acid binding and helicase unwinding activities, located between 88 and 182 aa. The "build mutant" option in the DS generated single optimized structures for the mutations R155K and D168A with Discrete Optimized Potential Energy scores [33] (DOPE scores) of −19975.94 and −20031.18, respectively. The change in the amino acids backbone has been compared by keeping the structures side-by-side as shown in Figure 2. The figure clearly shows the difference in the backbone structure and it can be inferred that the change may cause an increase in the steric hindrance for binding of drug molecules. The active site residues have been taken from the PDB records of the structure. Figure 3 shows the structural conformation of the residues in and around the active site. It clearly shows the cavity in the structure where our ligand molecules are expected to fit.

3.2. Ligand Preparation.
Andrographolide is a labdane-diterpenoid compound which is known for its wide range of pharmacological potential. It has been shown to have antiviral, antimalarial activities. Thus, we have considered it as a potent compound for tackling drug resistance in the HCV

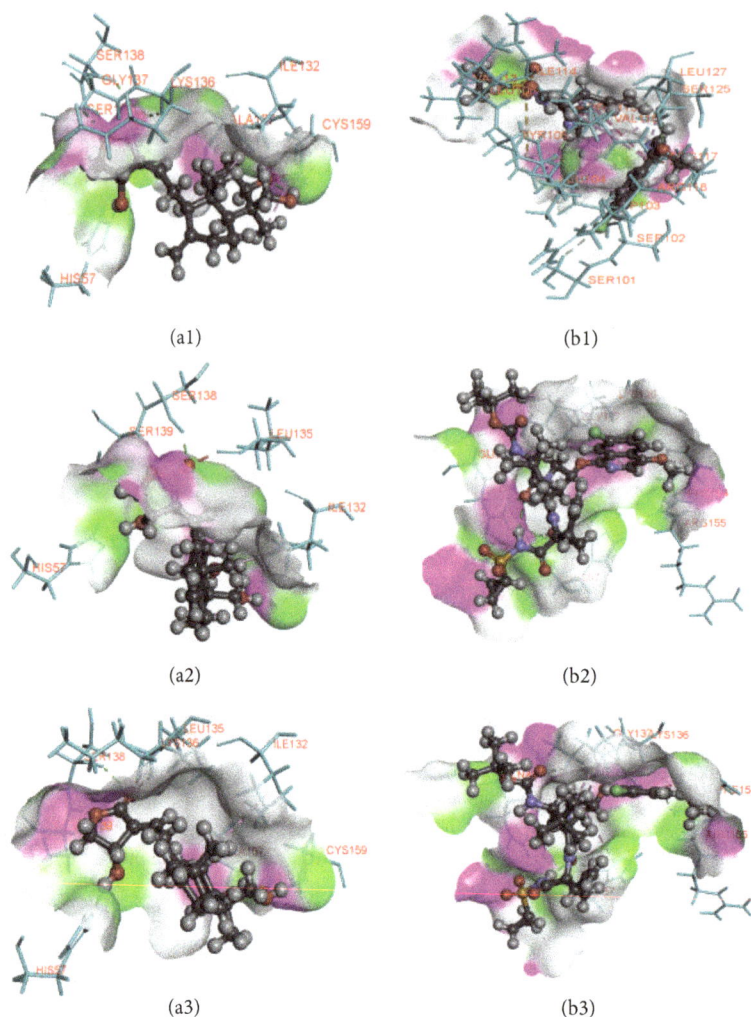

(a1)   (b1)

(a2)   (b2)

(a3)   (b3)

FIGURE 4: Binding poses and atomic interactions between ligands and receptors. (a1, a2, a3) series depicts Andrographolide interactions with the wild-type mutant R155K, and mutant D168A, respectively. (b1, b2, b3) series depicts the same with Asunaprevir and mutants in the same order above (see text for interacting residues). Note: ligands shown in ball-and-stick pattern and interacting residues shown in stick pattern, protein surface. Pink: donor, green: acceptor.

infection and compared its potency against the mutation-sensitive Asunaprevir. The two-dimensional structure and molecular properties of investigated compounds were tabulated in Table 1. The possible 3D conformations generated for Asunaprevir were 1 and for Andrographolide were 16. Out of the generated conformations, the lowest potential energy was selected for further studies. Conformity with ADME and TOPKAT prediction is shown in Tables 2 and 3. Both the compounds are predicted to be safe and show very less toxicity. Asunaprevir has been predicted to be slightly hepatotoxic; however, it should be noted that the predictions are defined based on certain established algorithms and may not be sometimes reliable in the real setup, which is plausible as Asunaprevir has already passed the initial phases of clinical trials (i.e., I and II). The mutagenicity level of both the compounds is also predicted to be low and thus both are predictively nontoxic for any systemic administration.

3.3. Molecular Docking and Dynamics. Molecular docking is an efficient technique to predict the preliminary binding modes of ligand with the protein of solved three-dimensional structure. Studies on binding poses are essential to elucidate key interactions between the small molecules and receptors and they provide helpful data for designing effective inhibitors. In our study, flexible docking method was used, using Biosolve LeadIT to dock compounds into active site of the protein structures. The rationale of using flexible docking is to give compounds enough flexibility to attain all the possible 3D space conformation and not to restrict only certain rigid structures. Docking results showed that Andrographolide occupies binding region of the native protein as well as its structural variants effectively with higher docking score than Asunaprevir. The detailed overview of the binding scores and interacting residues are shown in Table 4. Also the docking poses of ligand-receptor interaction are depicted

TABLE 1: Structure of ligands with their molecular properties.

SN	Compound name	Properties	2D images
1	Andrographolide	Compound ID: 5318517 Molecular weight: 350.4492 (g/mol) Molecular formula: $C_{20}H_{30}O_5$ XLogP3: 2.2 Hydrogen bond donor count: 3 Hydrogen bond acceptor count: 5	
2	Asunaprevir	Compound ID: 16076883 Molecular weight: 748.28584 (g/mol) Molecular formula: $C_{35}H_{46}ClN_5O_9S$ XLogP3: 4.9 Hydrogen bond donor count: 3 Hydrogen bond acceptor count: 10	

TABLE 2: Comparison of the ADME values of ligands.

Name	Solubility level	Blood brain barrier level	Extension CYP2D6	Extension hepatotoxic	Extension PPB
Andrographolide	2	3	−1.54262	−10.8965	16.7621
Asunaprevir	2	4	−9.92277	3.66033	24.0848

Note: solubility: 0–2: highly soluble, BBB: 1: high penetration, 2: medium penetration, and 3: low penetration, CYP2D6: −ve: noninhibitors and +ve: inhibition. HEPATOX: <1: nontoxic, PPB: the greater the value, the greater the binding capacity.

in Figure 4. Lead-IT docking score correlates with the free binding energy. Andrographolide binds the native protein with a Lead-IT score of −15.0862 and interacts with three amino acid residues, namely, SER138, SER139, and HIS57, via hydrogen bonding. In the R155K mutated structure, the compound forms 6 hydrogen bonds with residues SER138, SER139, ALA157, HIS57, LYS136, and GLY137 with docking score of −15.2322. Similarly, the compound has docking score of −13.9072 with the D168A mutated structure and again interacts through 6 hydrogen bonds with amino acid residues, namely, SER138, SER139, ALA157, HIS57, LYS136, and SER139. In all of the protein structures, Asunaprevir has low binding scores, lowest with the R155K mutation with the score of just −2.6431. It was expected because of the high susceptibility towards the mutation as described in various

literatures; the reason is the fact that Asunaprevir makes contacts with R155 residue outside the substrate envelop which is thus stabilized by the D168 residue, and thus any mutation in either of residues will disrupt the interactions between Asunaprevir and the enzyme [15]. Our results thus show that Andrographolide has better binding ability with the protein structures than Asunaprevir.

To compare the structural behavior and flexibility of the wild-type and mutant proteins, both the lead compounds were incorporated in Discovery studio MD simulations run and the studies were performed for 20 ns for each complex with all the parameters as mentioned in Materials and Method. The dynamic simulation runs create a system that tries to mimic physiological environment to check if the ligand is really stable within the cavity of target protein,

TABLE 3: Comparison of the predicted TOPKAT values of ligands.

Name	NTP carcinogenicity call (male mouse) (v3.2) TOPKAT	NTP carcinogenicity call (female mouse) (v3.2) TOPKAT	Developmental toxicity potential (DTP) (v3.1) TOPKAT	Skin irritation (v6.1) TOPKAT	Ames mutagenicity (v3.1) TOPKAT
Andrographolide	0.00	0.00	0.00	1.00	1.00
Asunaprevir	0.00	1.00	0.00	1.00	1.00

Note: 0: negative result, 1: positive result.

TABLE 4: Ligand-protein interaction with docking scores.

Protein	Compound name	Lead-IT (docking)					
		Lead-IT score	H-bond	Amino acid	Amino acid atom	Ligand atom	H-bond length (Å)
Wild-type HCV protease	Andrographolide	−15.0862	3	SER138	HN	O5	1.99421
				SER139	HN	O3	2.17439
				HIS57	NE2	H55	1.719
	Asunaprevir	−3.7159	5	GLY41	HE21	O8	1.6653
				HIS57	HD2	O6	2.12614
				GLY58	HA1	O6	3.07324
				GLY137	HA1	O7	2.76771
				ARG155	O	H92	2.73609
R155K	Andrographolide	−15.2322	6	SER138	HN	O5	2.44552
				SER139	HN	O3	2.29348
				ALA157	O	H53	1.79921
				HIS57	NE2	H55	2.17054
				LYS136	HA	O5	2.5112
				GLY137	HA1	O3	3.09817
	Asunaprevir	−2.6431	6	TYR105	HH	O10	2.87824
				LEU106	HN	O3	1.83884
				SER125	HN	O11	2.00467
				LEU104	O	H95	1.59587
				SER101	HB2	CL1	2.96175
				SER125	O	H80	2.7483
D168A	Andrographolide	−13.9072	6	SER138	HN	O5	2.27611
				SER139	HN	O3	2.20946
				ALA157	O	H53	2.59575
				HIS57	NE2	H55	1.6014
				LYS136	HA	O5	2.20148
				SER139	HB2	O3	3.02411
	Asunaprevir	−5.4149	8	GLN41	HE21	O8	1.65812
				HYS57	O	H97	1.55407
				HYS57	HD2	O6	2.06801
				LYS136	HE21	O10	2.77644
				GLY137	HA1	O7	2.87646
				ARG155	O	H80	3.0789
				ARG155	O	H92	2.76844
				GLY137	HN	O6	2.64645

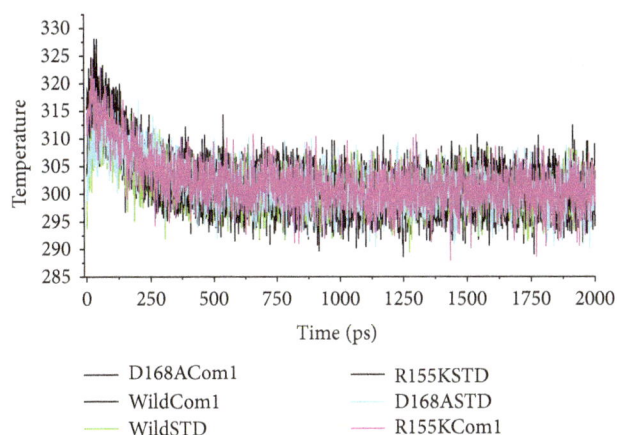

FIGURE 5: Temperature equilibration of the ligand-protein systems. As can be seen from the plots, the systems for all the ligand-protein complexes readily attained the temperature set at 300 K and maintained it throughout the simulation period. (Note: Asunaprevir is mentioned as STD and Andrographolide as Com 1.)

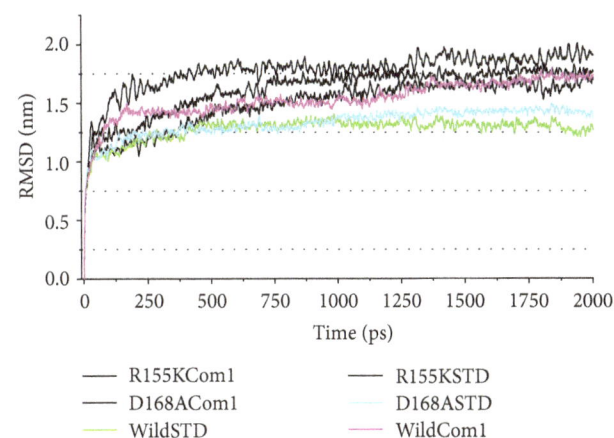

FIGURE 6: Protein backbone RMSD calculation plots for ligand bound complexes. Asunaprevir perturbs backbone of the protein mutant R155K (curve in blue) more than Andrographolide (curve in black). Surprisingly, Andrographolide seems to disturb protein structure for the mutant D168A more than Asunaprevir (check cyan curve for Asunaprevir and red for Andrographolide). (Note: Asunaprevir is mentioned as STD and Andrographolide as Com 1.)

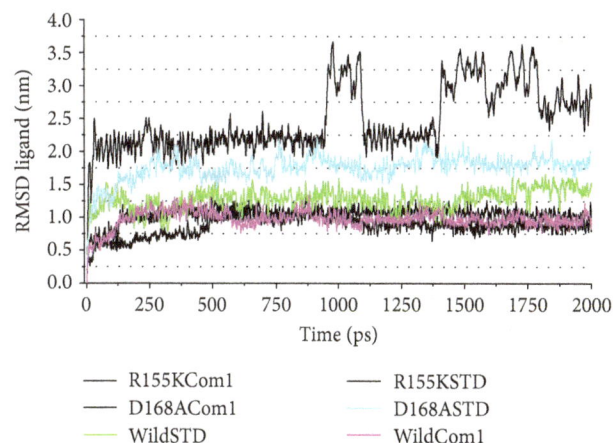

FIGURE 7: Ligand RMSD calculation plots for ligand bound complexes. The curves colored in blue and cyan show the instability of Asunaprevir inside the binding cavity of R155K and D168A mutants, respectively, while good stability is seen in wild type as shown by the green curve. Andrographolide is relatively very stable with very less deviation in the RMSD values for all the complexes (curves colored in black for R155K, red for D168A, and pink for wild-type structures, resp.). (Note: Asunaprevir is mentioned as STD and Andrographolide as Com 1.)

maintain bonds, and be able to inhibit the activity for a certain period of time which will result in therapeutic actions. As can be seen, the ligand-protein systems readily attained the given temperature of 300 K and stayed approximately around it throughout the run (Figure 5). Root mean square deviations (RMSD) [34] of the wild-type and the mutants were calculated against their initial structure in the protein-ligand complexes and graphs were generated to compare the flexibility once the ligand is bound to the structure. Over the simulation period, the backbone of the proteins remained fairly stable, as the graph shows in Figure 6. The binding of Asunaprevir did not disturb protein backbone stability in D168A and wild protein structures. However, in the mutant structure R155K, the binding caused a considerable

perturbation in the backbone with RMSD value eventually deviating by 0.5 nm in the end. Andrographolide did disturb the backbone when compared to Asunaprevir in both wild-type and D168A mutant. However, in case of R155K mutant structure, binding of Andrographolide did not disturb the backbone much as compared to Asunaprevir implying that Andrographolide binds to the mutant stably. This may be because of the small molecular size of Andrographolide that gives it enough freedom in space, whereas Asunaprevir, given its size and flanging chemical moieties, would not have more freedom, and within short simulation period the steric hindrances between the atoms of Asunaprevir and protein start making the system instable. To ensure the binding stability of the drug candidates in the active site of proteins, ligand positional RMSD of each lead molecule were generated and plotted. As can be seen from Figure 7, Asunaprevir showed more fluctuations in noticeable size of 2.0–3.5 nm with the R155K mutant. Also, it was not stably binding with D168A mutant when compared to our ligand molecule; however, the binding stability with the wild type was stable with very low deviations. Andrographolide showed stability in binding to all of the protein structures.

4. Conclusions

Most direct-acting antivirals are directed towards inhibiting proteases and polymerases. NS3-4A serine protease of the HCV is one of the most interesting targets and has a key role in HCV infection and proliferation. A good number of antivirals to inhibit this protease are already in the clinical trial phases, among which Asunaprevir stands in the first line of competitive inhibitors targeting HCV serine protease NS3-4A. However, the resulting side effects and sensitivity of the drug towards the HCV mutants R155K and D168A limit its potential. In this study, we compared the interaction efficiency of Asunaprevir and diterpenoids Andrographolide with the wild-type HCV protease and its mutants. The molecular docking studies using LeadIT revealed that the Asunaprevir binds with docking scores of −3.7159, −2.6431, and −5.4149, and Andrographolide binds with docking scores of −15.0862, −15.2322, and −13.9072, to the wild-type R155K and D168A structures, respectively. It infers that Andrographolide can interact strongly with the protein's active site residues both in the wild type and in mutants with least energy compared to Asunaprevir. The stability of the ligand-protein complexes was evaluated from the molecular dynamic simulations tool in the DS 3.5. Using calculated backbone RMSD data, it was found that Asunaprevir maintains protein stability in both the wild-type and D168A structures and, however, disturbs R155k backbone. Andrographolide did perturb the backbone in both the wild and mutant D168A structures but does not cause much disturbance in the mutant structure R155K when compared to Asunaprevir. We used ligand RMSD calculation data to infer about the binding stability of ligands with the structures. Asunaprevir showed more fluctuations in R155K complex than in others. Andrographolide was binding stably in all the structure types inferring the interactions are strong. Therefore, our study reports that Andrographolide can act as a promising option to target and inhibit NS3-4A along with its drug resistive mutants.

Acknowledgments

The authors wish to thank the Management, Principal, Director, and Head of the Department of Biotechnology, Siddaganga Institute of Technology, Tumkur, Karnataka, India. The authors also appreciate KBITS for their support in providing them with the required computational resources for carrying out this project.

References

[1] D. Lavanchy, "Evolving epidemiology of hepatitis C virus," *Clinical Microbiology and Infection*, vol. 17, no. 2, pp. 107–115, 2011.

[2] M. S. Sulkowski, D. F. Gardiner, M. Rodriguez-Torres et al., "Daclatasvir plus sofosbuvir for previously treated or untreated chronic HCV infection," *The New England Journal of Medicine*, vol. 370, no. 3, pp. 211–221, 2014.

[3] B. Helbling, W. Jochum, I. Stamenic et al., "HCV-related advanced fibrosis/cirrhosis: randomized controlled trial of pegylated interferon α-2a and ribavirin," *Journal of Viral Hepatitis*, vol. 13, no. 11, pp. 762–769, 2006.

[4] E. Voigt, C. Schulz, G. Klausen et al., "Pegylated interferon α-2b plus ribavirin for the treatment of chronic hepatitis C in HIV-coinfected patients," *Journal of Infection*, vol. 53, no. 1, pp. 36–42, 2006.

[5] P. Padmanabhan, U. Garaigorta, and N. M. Dixit, "Emergent properties of the interferon-signalling network may underlie the success of hepatitis C treatment," *Nature Communications*, vol. 5, article 3872, 2014.

[6] C. N. Hayes, M. Imamura, H. Aikata, and K. Chayama, "Genetics of IL28B and HCV-response to infection and treatment," *Nature Reviews Gastroenterology and Hepatology*, vol. 9, no. 7, pp. 406–417, 2012.

[7] A. Kau, J. Vermehren, and C. Sarrazin, "Treatment predictors of a sustained virologic response in hepatitis B and C," *Journal of Hepatology*, vol. 49, no. 4, pp. 634–651, 2008.

[8] T. Asselah and P. Marcellin, "New direct-acting antivirals' combination for the treatment of chronic hepatitis C," *Liver International*, vol. 31, supplement 1, pp. 68–77, 2011.

[9] C. Sarrazin and S. Zeuzem, "Resistance to direct antiviral agents in patients with hepatitis C virus infection," *Gastroenterology*, vol. 138, no. 2, pp. 447–462, 2010.

[10] S. Manabe, I. Fuke, O. Tanishita et al., "Production of nonstructural proteins of hepatitis C virus requires a putative viral protease encoded by NS3," *Virology*, vol. 198, no. 1, pp. 636–644, 1994.

[11] A. Murayama, L. Weng, T. Date et al., "RNA polymerase activity and specific RNA structure are required for efficient HCV replication in cultured cells," *PLoS Pathogens*, vol. 6, no. 4, Article ID e1000885, 2010.

[12] L. Tomei, C. Failla, E. Santolini, R. De Francesco, and N. La Monica, "NS3 is a serine protease required for processing of hepatitis C virus polyprotein," *Journal of Virology*, vol. 67, no. 7, pp. 4017–4026, 1993.

[13] J. J. Kiser, J. R. Burton, P. L. Anderson, and G. T. Everson, "Review and management of drug interactions with boceprevir and telaprevir," *Hepatology*, vol. 55, no. 5, pp. 1620–1628, 2012.

[14] B. L. Pearlman, "Protease inhibitors for the treatment of chronic hepatitis C genotype-1 infection: the new standard of care," *The Lancet Infectious Diseases*, vol. 12, no. 9, pp. 717–728, 2012.

[15] D. I. Soumana, A. Ali, and C. A. Schiffer, "Structural analysis of asunaprevir resistance in HCV NS3/4A protease," *ACS Chemical Biology*, vol. 9, no. 11, pp. 2485–2490, 2014.

[16] S. M. K. Rates, "Plants as source of drugs," *Toxicon*, vol. 39, no. 5, pp. 603–613, 2001.

[17] K. J. Halazun, A. Aldoori, H. Z. Malik et al., "Elevated preoperative neutrophil to lymphocyte ratio predicts survival following hepatic resection for colorectal liver metastases," *European Journal of Surgical Oncology*, vol. 34, no. 1, pp. 55–60, 2008.

[18] S. S. Handa and A. Sharma, "Hepatoprotective activity of andrographolide from *Andrographis paniculata* against carbontetrachloride," *Indian Journal of Medical Research*, vol. 92, pp. 276–283, 1990.

[19] S. P. Thyagarajan, S. Jayaram, V. Gopalakrishnan, R. Hari, P. Jeyakumar, and M. S. Sripathi, "Herbal medicines for liver diseases in India," *Journal of Gastroenterology and Hepatology*, vol. 17, supplement 3, pp. S370–S376, 2002.

[20] S. K. Tewari, A. Niranjan, and A. Lehri, "Variations in yield, quality, and antioxidant potential of Kalmegh (*Andrographis paniculata* nees) with soil alkalinity and season," *Journal of Herbs, Spices and Medicinal Plants*, vol. 16, no. 1, pp. 41–50, 2010.

[21] P. M. Scola, L.-Q. Sun, A. X. Wang et al., "The discovery of asunaprevir (BMS-650032), an orally efficacious NS3 protease inhibitor for the treatment of hepatitis C virus infection," *Journal of Medicinal Chemistry*, vol. 57, no. 5, pp. 1730–1752, 2014.

[22] S. Hwang, S. Thangapandian, and K. W. Lee, "Molecular dynamics simulations of sonic hedgehog-receptor and inhibitor complexes and their applications for potential anticancer agent discovery," *PLoS ONE*, vol. 8, no. 7, Article ID e68271, 2013.

[23] M. Jayakanthan, G. Wadhwa, T. M. Mohan, L. Arul, P. Balasubramanian, and D. Sundar, "Computer-aided drug design for cancer-causing H-Ras p21 mutant protein," *Letters in Drug Design & Discovery*, vol. 6, no. 1, pp. 14–20, 2009.

[24] A.-P. Hynninen and M. F. Crowley, "New faster CHARMM molecular dynamics engine," *Journal of Computational Chemistry*, vol. 35, no. 5, pp. 406–413, 2014.

[25] A. A. Ezat, N. S. El-Bialy, H. I. A. Mostafa, and M. A. Ibrahim, "Molecular docking investigation of the binding interactions of macrocyclic inhibitors with HCV NS3 protease and its mutants (R155K, D168A and A156V)," *The Protein Journal*, vol. 33, no. 1, pp. 32–47, 2014.

[26] K. Wingen, J. S. Schwed, K. Isensee et al., "Benzylpiperidine variations on histamine H_3 receptor ligands for improved drug-likeness," *Bioorganic and Medicinal Chemistry Letters*, vol. 24, no. 10, pp. 2236–2239, 2014.

[27] G. Wu, D. H. Robertson, C. L. Brooks, and M. Vieth, "Detailed analysis of grid-based molecular docking: a case study of CDOCKER—a CHARMm-based MD docking algorithm," *Journal of Computational Chemistry*, vol. 24, no. 13, pp. 1549–1562, 2003.

[28] C. A. Lipinski, "Lead- and drug-like compounds: the rule-of-five revolution," *Drug Discovery Today: Technologies*, vol. 1, no. 4, pp. 337–341, 2004.

[29] T. Hou, J. Wang, W. Zhang, and X. Xu, "ADME evaluation in drug discovery. 6. Can oral bioavailability in humans be effectively predicted by simple molecular property-based rules?" *Journal of Chemical Information and Modeling*, vol. 47, no. 2, pp. 460–463, 2007.

[30] A. Asoodeh, L. Haghighi, J. Chamani, M. A. Ansari-Ogholbeyk, Z. Mojallal-Tabatabaei, and M. Lagzian, "Potential angiotensin I converting enzyme inhibitory peptides from gluten hydrolysate: biochemical characterization and molecular docking study," *Journal of Cereal Science*, vol. 60, no. 1, pp. 92–98, 2014.

[31] M. Dammalli, V. Chandramohan, M. I. Biradar, N. Nagaraju, and B. S. Gangadharappa, "In silico analysis and identification of novel inhibitor for new H1N1 swine influenza virus," *Asian Pacific Journal of Tropical Disease*, vol. 4, no. 2, pp. S635–S640, 2014.

[32] H. R. Naika, V. Krishna, K. Lingaraju et al., "Molecular docking and dynamic studies of bioactive compounds from *Naravelia zeylanica* (L.) DC against glycogen synthase kinase-3β protein," *Journal of Taibah University for Science*, vol. 9, no. 1, pp. 41–49, 2015.

[33] R.-R. Wang, Y.-D. Gao, C.-H. Ma et al., "Mangiferin, an anti-HIV-1 agent targeting protease and effective against resistant strains," *Molecules*, vol. 16, no. 5, pp. 4264–4277, 2011.

[34] S.-Y. Lu, Y.-J. Jiang, J. Lv, T.-X. Wu, Q.-S. Yu, and W.-L. Zhu, "Molecular docking and molecular dynamics simulation studies of GPR40 receptor-agonist interactions," *Journal of Molecular Graphics and Modelling*, vol. 28, no. 8, pp. 766–774, 2010.

18

Serological Survey of Hantavirus in Inhabitants from Tropical and Subtropical Areas of Brazil

Felipe Alves Morais,[1,2] Alexandre Pereira,[3] Aparecida Santo Pietro Pereira,[3,4] Marcos Lazaro Moreli,[5] Luís Marcelo Aranha Camargo,[6,7] Marcello Schiavo Nardi,[8,9] Cristina Farah Tófoli,[8] Jansen Araujo,[1] Lilia Mara Dutra,[1] Tatiana Lopes Ometto,[1] Renata Hurtado,[1] Fábio Carmona de Jesus Maués,[1] Tiene Zingano Hinke,[6] Sati Jaber Mahmud,[6] Monica Correia Lima,[10] Luiz Tadeu Moraes Figueiredo,[11] and Edison Luiz Durigon[1]

[1] *Laboratório de Virologia Clínica e Molecular, Departamento de Microbiologia, Instituto de Ciências Biomédicas, Universidade de São Paulo, 05508-900 São Paulo, SP, Brazil*
[2] *Laboratório de Sinalização Molecular, Departamento de Bioquímica, Instituto de Química, Universidade de São Paulo, 05508-000 São Paulo, SP, Brazil*
[3] *Laboratório de Genética, Instituto Butantan, 05503-900 São Paulo, SP, Brazil*
[4] *Laboratório de Virologia, Instituto Butantan, 05503-900 São Paulo, SP, Brazil*
[5] *Laboratório de Virologia, Universidade Federal de Goiás, 75801-615 Jatai, GO, Brazil*
[6] *Departamento de Parasitologia, Instituto de Ciências Biomédicas, Universidade de São Paulo, 05389-970 São Paulo, SP, Brazil*
[7] *São Lucas Medical School, Rua Alexandre Guimarães 1928, 76000-000 Porto Velho, RO, Brazil*
[8] *Instituto de Pesquisa Ecológica (IPE), 12960-000 Nazaré Paulista, SP, Brazil*
[9] *Divisão Técnica de Medicina Veterinária e Manejo da Fauna Silvestre (DEPAVE-/SVMA), 04030-000 Prefeitura de São Paulo, SP, Brazil*
[10] *Departamento Municipal de Saúde, 11950-000 Prefeitura de Cajati, SP, Brazil*
[11] *Centro de Pesquisa em Virologia da Faculdade de Medicina de Ribeirão Preto da Universidade de São Paulo, 14049-900 Ribeirão Preto, SP, Brazil*

Correspondence should be addressed to Felipe Alves Morais; felipemorais@usp.br

Academic Editor: Subhash C. Verma

Brazil has reported more than 1,600 cases of hantavirus cardiopulmonary syndrome (HPS) since 1993, with a 39% rate of reported fatalities. Using a recombinant nucleocapsid protein of *Araraquara* virus, we performed ELISA to detect IgG antibodies against hantavirus in human sera. The aim of this study was to analyze hantavirus antibody levels in inhabitants from a tropical area (Amazon region) in Rondônia state and a subtropical (Atlantic Rain Forest) region in São Paulo state, Brazil. A total of 1,310 serum samples were obtained between 2003 and 2008 and tested by IgG-ELISA, and 82 samples (6.2%), of which 62 were from the tropical area (5.8%) and 20 from the subtropical area (8.3%), tested positive. Higher levels of hantavirus antibody were observed in inhabitants of the populous subtropical areas compared with those from the tropical areas in Brazil.

1. Introduction

Hantaviruses are emerging pathogens that have gained increasing attention in the last few decades [1]. The genus *Hantavirus* belongs to Bunyaviridae family and is transmitted to human by rodents and possible by other small mammals. More than 40 *Hantavirus* species are currently known and 22 of them are considered pathogenic for humans [2].

The hantaviruses found in Eurasia (e.g., *Hantaan* and *Seoul* virus) are harbored by rodents of the Murinae and Arvicolinae subfamilies and cause hemorrhagic fever with renal syndrome (HFRS) in infected humans. On the other hand the hantaviruses found in the Americas (e.g., *Sin Nombre*, *Juquitiba*, and *Castelo dos Sonhos*) are harbored by rodents of the Sigmodontinae subfamily and cause Hantavirus Pulmonary Syndrome (HPS) in humans [3–5].

As Manigold and Vial [2] wrote, moles, shrews, and bats are also increasingly described as natural hosts of new members of the *Hantavirus* genus (e.g., *Huangpi* virus, *Lianghe* virus, *Longquan* virus, *Yakeshi* virus, and *Seewis* virus). However, the pathogenicity of these viruses for humans is unclear. Also, there are reports of seropositive domestic animals such as dogs and cats, suggesting that these become infected from contact with infected primary hosts. Another interesting study fresh published [5], demonstrates pet rats and whales at United Kingdom. However, there is neither evidence of disease in these species nor of a role as a reservoir for human infection.

In Brazil the HPS cases are mostly caused by five genotypes of hantavirus: *Juquitiba* virus (JUQV), *Araraquara* virus (ARAV), *Laguna Negra-like* virus (LANV-like), *Castelo dos Sonhos* virus (CASV), and *Anajatuba* virus (AJBV) [5]. A significantly higher number of HPS-associated fatalities (50%) were observed in the Midwestern and Southwestern regions compared with other regions of Brazil [6, 7]. Serological evidence of HPS has also been reported in the north and northeast of Brazil where hantavirus genotypes are unknown [5–10].

The first confirmed cases of HPS in North America occurred in 1993, and six months later, it was reported in Brazil [5, 11, 12]. Since then, more than 1600 HPS cases have been reported in Brazil by Brazilian Ministry of Health/SVS, with approximately 39% being fatalities [13].

Serological methods are commonly used for hantavirus diagnosis, including enzyme-linked immunosorbent assays (ELISAs), immunofluorescence assays, and immunoblot assays. Additionally, hantavirus isolation in Vero E6 cell cultures and detection of anti-hantavirus antibodies by plaque reduction neutralization are also used for diagnosis. Nevertheless, both methods require labor-intensive, time-consuming, and biosafety-level-three conditions [6]. On the other hand, molecular biology such as RT-PCR allows molecular characterization (viral genotype) and accurate diagnosis [14].

Here, in an effort to better understand and study the distribution of infections by hantavirus in Brazil, we present the results of a serological survey including individuals living in urban and rural areas near the Amazonian forest and in subtropical areas near rain forests that have degraded environmental conditions. Sera from the participants were tested via an IgG-ELISA [15] that uses a recombinant nucleocapsid protein from ARAV as the antigen [16].

2. Material and Methods

2.1. Sites and Study Population. The design for minimum sample size was performed in accordance with the calculations specified by Luiz and Magnanini [17]. Based on the

presence of wild rodents cohabiting with humans and the occurrence of HPS cases, four study sites were selected for this serological survey between 2003 and 2008. Machado river (from 8°55′57″S/62°03′20″W to 8°10′15″S/62°46′50″W) and Machadinho do Oeste county (09°26′38″S/61°58′53″W) are both in Rondônia state in the Amazon tropical region. In 2003, 435 participants living along the Machado river and working on subsistence farming were enrolled and subjected to blood drawing for the study (Figure 1). In Machadinho do Oeste county, 633 inhabitants were enrolled and subjected to blood drawing in 2005. These participants lived in an urban area surrounded by tropical Amazonian forests (Figure 1). The other two study sites were located more than 2000 km away, near the subtropical rain forests of Sao Paulo state. These regions included Jacupiranga county in the Ribeira Valley, where 65% of the Brazilian Atlantic forest remains (24°54′30″S/048°08′01″W). A total of 157 inhabitants working on banana or orange farms and cattle or fish raising were enrolled in Jacupiranga in 2007 and subjected to blood draws (Figure 1). The fourth site was in Teodoro Sampaio county (22°22′70″S/052°25′66″W) at the mouth of the Paranapanema river, where the land has been subject to disorganized occupation and massive deforestation. Currently, farms at the mouth of the Paranapanema river have intensive agricultural activity (Figure 1). In Teodoro Sampaio, 85 inhabitants were enrolled in the study in 2008 and subjected to blood draws. All clinical samples were transported to the laboratory in nitrogen liquid and stored at −80°C.

2.2. Ethical Considerations. The enrollment of participants in this serological survey was authorized by the Ethics Committee of ICB/USP (670/2005), and the confidentiality of their personal information was ensured. Blood collection of the participants was only performed after signing the informed consent, in compliance with the rules of the ethics committee. After participants signed the consent form, the survey was applied for evaluation of the risk factors, gender, age, education, and another aspects as epidemiological information.

2.3. ELISA. Sera were tested by an indirect IgG enzyme-linked immunosorbent assay (ELISA) using the N recombinant protein (recN) of ARAV as antigen, as described by Figueiredo et al. [15, 16], with some modifications. Briefly, 96-well microplates (PolySorp™, Nunc, USA) were coated with 50 μL of ARAV recN protein or the respective control (0.5 μg/mL) overnight in a wet chamber at 4°C. Both antigens were diluted in carbonate-bicarbonate buffer (0.05 M, pH 9.6). All incubations were conducted at 37°C for 1 h, and the plates were washed six times with wash buffer (phosphate-buffered saline [PBS]-0.1% Tween 20) between each step. In the first step, 50 μL of a blocking solution containing 5% bovine serum albumin in PBS (Sigma, San Francisco, CA, USA) was added to each well, and the plate was incubated for 2 h in a moist chamber at 37°C. Next, 50 μL of the serum sample diluted at 1:100 or control samples diluted to 1:1000 in dilution buffer (PBS/BSA 1%) were added per well. In the third step, the wells were incubated with 50 μL of phosphatase-labeled anti-human IgG antibody (Sigma, San Francisco, USA). Finally, 1 mg/mL of nitrophenyl

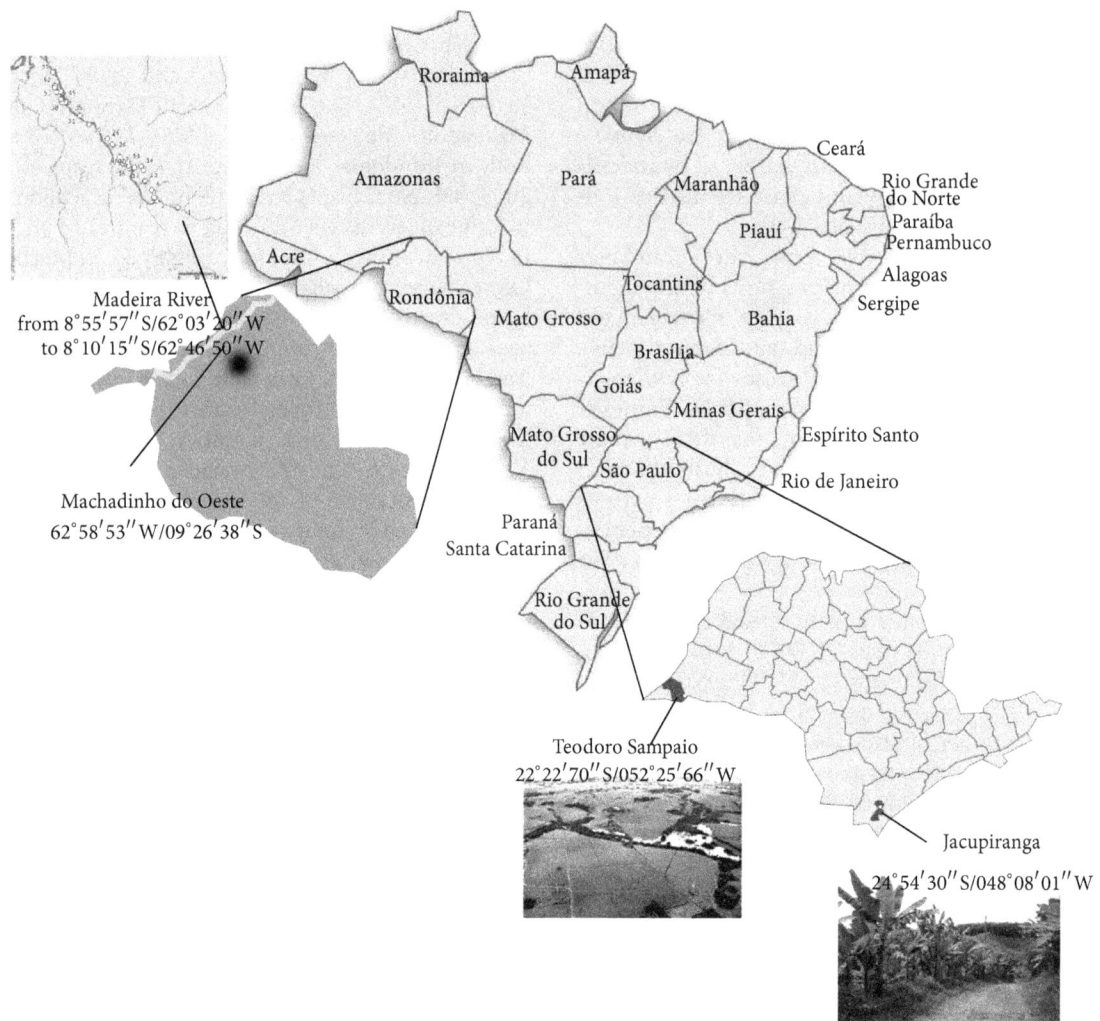

FIGURE 1: Map of Brazil showing the four study sites: the city of Jacupiranga (RV) and Teodoro Sampaio (PP) in São Paulo state in the subtropical region and Machadinho do Oeste city and the Machado river region Rondônia state in the tropical region.

phosphatase substrate (pNPP, Sigma, USA) was added per well, and the reaction was stopped after 20 minutes by adding 50 μL of 3 M NaOH. The absorbance was measured at 405 nm using a Multiskan MS (Labsystems, Helsinki, Finland). The cut-off was established as the mean value + 2 standard deviations control samples and showed a cut-off Optical Density (OD) equal to or greater than 0.500.

The recN protein used as the antigen and the positive/negative control samples were kindly provided by the Laboratories of Arboviruses and Rodent-borne Viruses of the University of São Paulo School of Medicine (Luis Tadeu M. Figueiredo, Ph.D., M.D.).

2.4. Data Analysis. A statistical analysis of the results was performed with Prism version 5.0 (GraphPad Software). Associations of positive serological tests with subtropical areas and demographic and socioeconomic variables were analyzed using the Chi-square test, and $p < 0.05$ was considered significant.

3. Results and Discussion

Sera from all 1,310 participants in this study were tested using the ARAV recN-ELISA, and 82 (6.2%) showed positive results. In the Machado river group (MR), 22 (1.6%) of the 435 participants examined had IgG antibodies to hantavirus. In the Machadinho do Oeste county, 40 (3.0%) among the 633 participants analyzed were previously infected by hantavirus. In Jacupiranga county (RV) in the southeastern part of Brazil, 14 (9.0%) of the 157 participants had IgG antibodies to hantavirus. In Teodoro Sampaio county (PP) in the São Paulo state, 6 (7.0%) of the 85 participants were seropositive for hantavirus, as shown in Figure 2(a).

Participants from Jacupiranga and Teodoro Sampaio counties in the subtropical southeastern region of Brazil were significantly ($p = 0.0008$) more seropositive to hantavirus (20/242, 8.3%) than those from the tropical Amazonian regions of Machado river and Machadinho do Oeste county (62/1068, 5.8%).

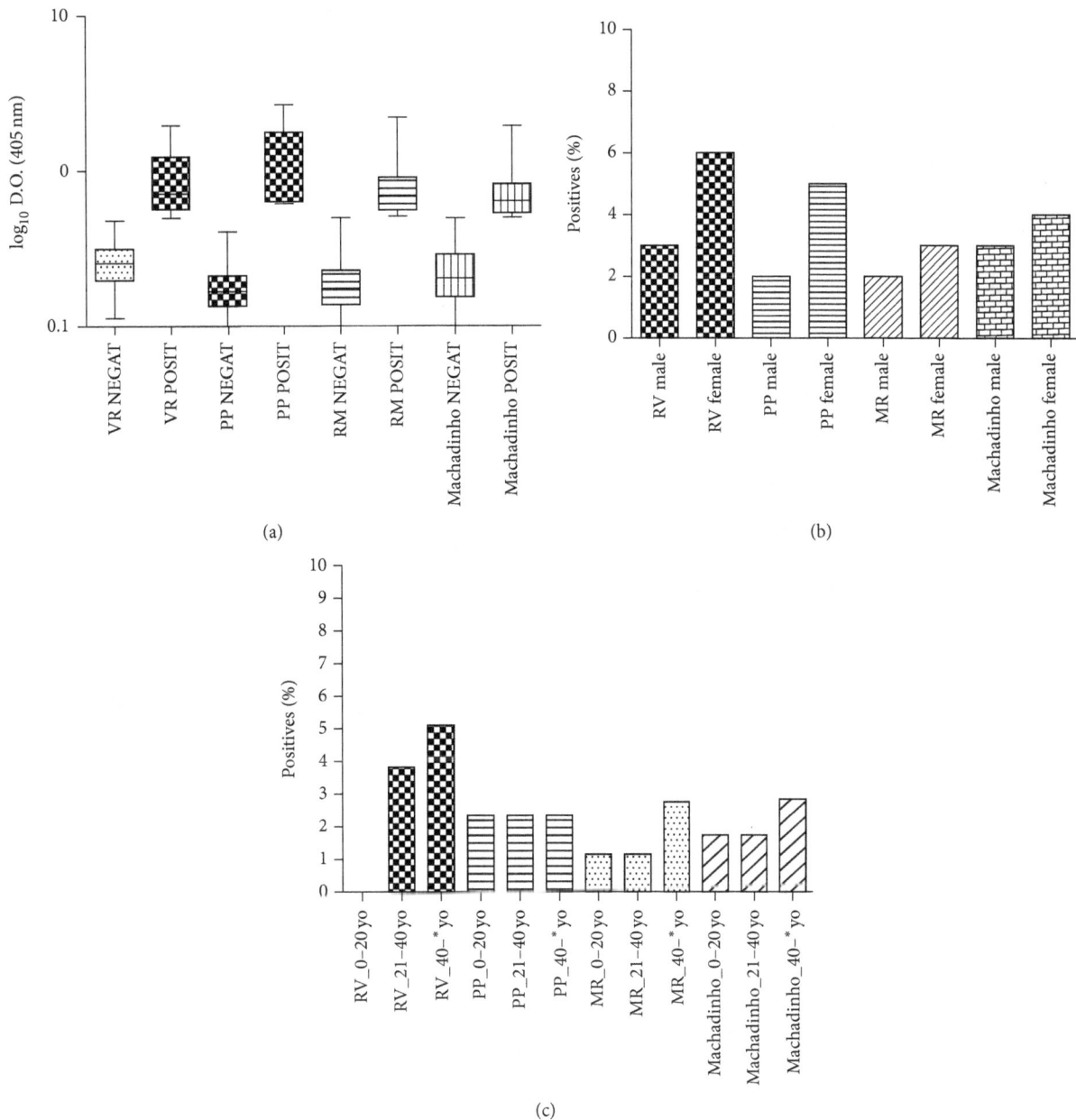

FIGURE 2: (a) demonstrated the absorbance values by ELISA for the total positive (POSIT) and negative (NEGAT) individuals from four different sites (Jacupiranga, SP, in Ribeira Valley, VR; Teodoro Sampaio/SP in Pontal Parapanema, PP; Machado River/RO, RM; and Machadinho do Oeste/RO) for the presence of anti-hantavirus IgG antibodies. (b) The graphic represent results from this four study sites in the tropical and subtropical region (Jacupiranga/SP in Ribeira Valley, VR; Teodoro Sampaio/SP in Pontal Parapanema, PP; Machado River/RO, RM; and Machadinho do Oeste/RO), categorized by gender. (c) The graphics represent results from this four study sites in the tropical and subtropical region (Jacupiranga/SP in Ribeira Valley, VR; Teodoro Sampaio/SP in Pontal Parapanema, PP; Machado River/RO, RM; and Machadinho do Oeste/RO) categorized by age (groups of 20 years). * means "or more."

Among the hantavirus seropositive, 60.9% were female participants, as shown in Figure 2(b), consisting of 877 women and 433 men (50 and 32, resp., were seropositive). In Machadinho do Oeste county, it was observed that 3.63% of females and 2.68% of males were seropositive. In MR it was observed that 3.22% of females and 1.84% of males were seropositive. In RV, it was observed that 5.73% of females and 3.18% of males were seropositive.

The ages of those seropositive to hantavirus ranged from five to 82 years but included mostly young adults greater than 20 years of age, as shown in Figure 2(c).

The seropositivity to hantavirus was not correlated with education level, place of birth, time of residence in the study site, or other risk factors evaluated for epidemiological analyses.

In this study, we evaluated the hantavirus seroprevalence in people living in areas presumed of high risk for this infection. A high proportion of hantavirus seropositive subjects were observed in participants from Jacupiranga county (9%) and Teodoro Sampaio (7%), both of which located at subtropical regions of the state of São Paulo. This value is significantly higher than the 1.6% of seropositive cases reported in a serological survey in the Ribeira Valley performed in 1993 [8, 12].

The seroprevalence of 6% and 8% to hantavirus observed in the present study is not outside standards range observed in the literature, like, for example, the rate of 14% observed in the population of Jardinopolis county in the state of Sao Paulo [18] and of 6.1% in Chile, presumably, with ANDV [19]. Higher seroprevalence was observed, presumably, to the genotype ORNV, in northern Argentina (20%), and, presumably, to the genotype LNV in Paraguayan Chaco, where 12.8% of the urban population and 57% of the indigenous population were infected [20].

In the tropical Amazonian region of Rondônia, the seroprevalence to hantavirus in Machadinho do Oeste county (3%) and Machado river region (2.8%) was similar. Other studies conducted in the Amazon region have yielded different hantavirus seroprevalence rates. A survey of several cities in the Amazonas state between 2007 and 2009 yielded a seroprevalence rate of 0.6% [11], whereas another study performed along a highway that crosses the Amazon region from Cuiabá city to Santarem city reported seroprevalence rates of 2.16% to 9.43% [21].

According description by Nava et al. [22], in the Pontal do Paranapanema, at Morro do Diablo State Park (Teodoro Sampaio, SP), it is a forested region, surrounded by small rural settlements and large private properties. The area is a wedge-shaped region bounded on the north by Rio Paraná and on the south by Rio Paranapanema. As well as the Ribeira Valley (Jacupiranga, SP), the Atlantic Rain Forest in São Paulo state is one of the last remaining significant zones, where only 1.8% of the original natural vegetation remains and is considered a biodiversity hotspot.

A significantly higher proportion of seropositive individuals to hantavirus was observed in the subtropical region of Brazil (São Paulo state) compared to the tropical region (Rondônia state). It is possible that the higher occurrence of hantavirus infections in São Paulo state is associated with the degradation of the local environment. São Paulo, which is the most densely populated state in Brazil, has two main ecosystems: the "cerrado (savannah)" in its western region and neotropical Atlantic Rain Forests along the coast. These ecosystems sustain Sigmodontinae rodents and have been modified, segmented, and damaged by extensive sugarcane, soybean and coffee farming, livestock raising, and rapid and poorly planned urbanization. Such degraded landscapes allow close contact of humans with zoonoses, resulting in enhanced transmission of pathogens to humans. Environmental degradation favors the abundance of opportunistic rodent species (*Necromys lasiurus, Akodon* sp., and *Calomys tener*), which are risk factors for hantavirus infection. In contrast, the environment and landscape in the Amazonian tropical region (Machado river region and Machadinho do Oeste county) is better conserved and has a higher diversity of rodents. Therefore, the participants from these study sites were less frequently infected by hantavirus given that the biodiversity loss would tend to increase pathogen transmission.

One of the limitations of this study is the seroprevalence accuracy of data available at these areas combined with full background from participants of this study. Also, it was impossible to analyze hantavirus infection in rodents from the same studied sites since we did not collect samples. Probably, a comparison of seropositivity to hantavirus among humans and rodents at the different study sites would help to understand why more infections occur in the subtropical area of Brazil (São Paulo state) more than in the tropical area (Rondônia state).

4. Conclusions

In conclusion, our findings highlight a higher seroprevalence rate (IgG) for antibodies against hantavirus in the human population in Brazil, with a higher rate in the subtropical region (Atlantic Rain Forest) than in the tropical region (Amazon Forest). Degraded ecosystems allow close contact of humans with zoonoses, resulting in enhanced transmission of pathogens to humans. Particularly it is relevant because São Paulo state is one of the most densely populated states in Brazil. We are highlighting our findings to provide a better understanding of hantavirus infection and circulation in Brazil, specifically demonstrating hotspots that will require public health action to prevent a possible outbreak.

Authors' Contribution

Felipe Alves Morais and Alexandre Pereira contributed equally to this work.

Acknowledgments

This study received financial support from FAPESP (05/01603-4). The authors would like to extend special thanks to José Maria Lopes (technical specialist at the ICB/USP laboratory), who helped them during sample collection in the field. They also thank Paulo S. da Costa, Fernando A. da Costa, João Ventura, and many other people who helped them in many different ways.

References

[1] M. Zeier, M. Handermann, U. Bahr et al., "New ecological aspects of hantavirus infection: a change of a paradigm and a challenge of prevention—a review," *Virus Genes*, vol. 30, no. 2, pp. 157–180, 2005.

[2] T. Manigold and P. Vial, "Human hantavirus infections: epidemiology, clinical features, pathogenesis and immunology," *Swiss Medical Weekly*, vol. 144, Article ID w13937, 2014.

[3] C. S. Schmaljohn and S. T. Nichol, "Bunyaviridae," in *Fields Virology*, D. M. Knipe and P. M. Howley, Eds., pp. 1741–1789, Wolters Kluwer Health/Lippincott Williams & Wilkins, Philadelphia, Pa, USA, 1997.

[4] S. T. Nichol, J. Arikawa, and Y. Kawaoka, "Emerging viral diseases," *Proceedings of the National Academy of Sciences of the United States of America*, vol. 97, no. 23, pp. 12411–12412, 2000.

[5] L. J. Jameson, S. K. Taori, B. Atkinson et al., "Pet rats as a source of hantavirus in England and Wales, 2013," *Eurosurveillance*, vol. 18, no. 9, 2013.

[6] C. B. Jonsson, L. T. M. Figueiredo, and O. Vapalahti, "A global perspective on hantavirus ecology, epidemiology, and disease," *Clinical Microbiology Reviews*, vol. 23, no. 2, pp. 412–441, 2010.

[7] L. T. M. Figueiredo, M. L. Moreli, G. M. Campos, and R. L. M. Sousa, "Hantaviruses in São Paulo State, Brazil," *Emerging Infectious Diseases*, vol. 9, no. 7, pp. 891–892, 2003.

[8] L. T. M. Figueiredo, M. L. Moreli, R. L. M. De Sousa et al., "Hantavirus pulmonary syndrome, central plateau, southeastern, and southern Brazil," *Emerging Infectious Diseases*, vol. 15, no. 4, pp. 561–567, 2009.

[9] A. Suzuki, I. Bisordi, S. Levis et al., "Identifying rodent hantavirus reservoirs, Brazil," *Emerging Infectious Diseases*, vol. 10, no. 12, pp. 2127–2134, 2004.

[10] W. S. Mendes, A. A. M. da Silva, L. F. C. Aragão et al., "Hantavirus infection in Anajatuba, Maranhão, Brazil," *Emerging Infectious Diseases*, vol. 10, no. 8, pp. 1496–1498, 2004.

[11] J. B. L. Gimaque, M. S. Bastos, W. S. M. Braga et al., "Serological evidence of hantavirus infection in rural and urban regions in the state of Amazonas, Brazil," *Memorias do Instituto Oswaldo Cruz*, vol. 107, no. 1, pp. 135–137, 2012.

[12] A. Plyusnin and S. P. Morzunov, "Virus evolution and genetic diversity of hantaviruses and their rodent hosts," in *Hantaviruses*, vol. 256 of *Current Topics in Microbiology and Immunology*, pp. 47–75, Springer, Berlin, Germany, 2001.

[13] L. B. Iversson, A. P. da Rosa, M. D. Rosa, A. V. Lomar, M. G. Sasaki, and J. W. LeDuc, "Human infection by Hantavirus in southern and southeastern Brazil," *Revista da Associacao Medica Brasileira*, vol. 40, no. 2, pp. 85–92, 1994.

[14] A. Guterres, R. C. de Oliveira, J. Fernandes, C. G. Schrago, and E. R. S. de Lemos, "Detection of different South American hantaviruses," *Virus Research*, vol. 210, pp. 106–113, 2015.

[15] L. T. M. Figueiredo, M. L. Moreli, A. A. Borges et al., "Evaluation of an enzyme-linked immunosorbent assay based on Araraquara virus recombinant nucleocapsid protein," *American Journal of Tropical Medicine and Hygiene*, vol. 81, no. 2, pp. 273–276, 2009.

[16] L. T. M. Figueiredo, M. L. Moreli, A. A. Borges, G. G. Figueiredo, R. L. M. Souza, and V. H. Aquino, "Expression of a hantavirus N protein and its efficacy as antigen in immune assays," *Brazilian Journal of Medical and Biological Research*, vol. 41, no. 7, pp. 596–599, 2008.

[17] R. R. Luiz and M. M. Magnanini, "A lógica da determinação do tamanho da amostra em investigações epidemiológicas," *Cadernos de Saúde Coletiva*, vol. 8, no. 2, pp. 9–28, 2000.

[18] G. M. Campos, R. L. M. De Sousa, S. J. Badra, C. Pane, U. A. Gomes, and L. T. M. Figueiredol, "Serological survey of hantavirus in Jardinopolis County, Brazil," *Journal of Medical Virology*, vol. 71, no. 3, pp. 417–422, 2003.

[19] M. T. Frey, P. C. Vial, C. H. Castillo, P. M. Godoy, B. Hjelle, and M. G. Ferrés, "Hantavirus prevalence in the IX Region of Chile," *Emerging Infectious Diseases*, vol. 9, no. 7, pp. 827–832, 2003.

[20] J. F. Ferrer, C. B. Jonsson, E. Esteban et al., "High prevalence of hantavirus infection in Indian communities of the Paraguayan and Argentinean Gran Chaco," *American Journal of Tropical Medicine and Hygiene*, vol. 59, no. 3, pp. 438–444, 1998.

[21] D. B. A. Medeiros, E. S. T. da Rosa, A. A. R. Marques et al., "Circulation of hantaviruses in the influence area of the Cuiabá-Santarém Highway," *Memorias do Instituto Oswaldo Cruz*, vol. 105, no. 5, pp. 665–671, 2010.

[22] A. F. D. Nava, L. Cullen Jr., D. A. Sana et al., "First evidence of canine distemper in brazilian free-ranging felids," *EcoHealth*, vol. 5, no. 4, pp. 513–518, 2008.

19

Quasispecies Changes with Distinctive Point Mutations in the Hepatitis C Virus Internal Ribosome Entry Site (IRES) Derived from PBMCs and Plasma

Luca Mercuri [iD],[1] **Emma C. Thomson,**[2] **Joseph Hughes,**[2] **and Peter Karayiannis**[3]

[1]*Hepatology Section, Division of Medicine, Faculty of Medicine, Imperial College, London, UK*
[2]*University of Glasgow MRC Centre for Virus Research, Glasgow, UK*
[3]*University of Nicosia Medical School, Nicosia, Cyprus*

Correspondence should be addressed to Luca Mercuri; luca.mercuri@nhs.net

Academic Editor: Amiya K. Banerjee

The 5' untranslated region (UTR) of the hepatitis C virus (HCV) genome contains the internal ribosome entry site (IRES), a highly conserved RNA structure essential for cap-independent translation of the viral polyprotein. HCV, apart from the liver, is thought to be associated with lymphocyte subpopulations of peripheral blood mononuclear cells (PBMCs), in lymph nodes and brain tissue. In this study, RT-PCR, cloning, and sequence analysis were employed to investigate the quasispecies nature of the 5'UTR following extraction of viral RNA from PBMCs and plasma of HCV infected individuals. The nucleotide variation between IRES-derived sequences from PBMCs and plasma indicated the existence of polymorphic sites within the IRES. HCV isolates had divergent variants with unique mutations particularly at positions 107, 204, and 243 of the IRES. Most of the PBMC-derived sequences contained an A-A-A variant at these positions. The mutations associated with the IRESes suggested the presence of unique quasispecies populations in PBMCs compared with plasma.

1. Introduction

Hepatitis C virus (HCV), a member of the *Hepacivirus* genus in the *Flaviviridae* family, is an enveloped virus with an icosahedral capsid that encloses a single-stranded positive sense genomic RNA of approximately 9.5 kb in length [1, 2]. The viral genome consists of a large open reading frame (ORF) flanked by highly conserved untranslated regions (UTR) present at the 5' and 3' termini [3]. The 5'end UTR, which is 341 nt in length, contains an internal ribosome entry site (IRES), a highly conserved region with extensive secondary structure formed by palindromic complementary sequences giving rise to four structural domains [4]. Research into the structure and function of the IRES revealed that its structure is required for the modulation of the cap-independent IRES-mediated translation process by initially recruiting cellular factors and ribosomal subunits [5]. HCV is mainly hepatotropic, but its RNA has been associated with lymph nodes, brain tissue, and peripheral blood mononuclear cells (PBMCs) especially macrophages, B-cells, and

T-cells (CD8+) [6, 7]. HCV replication has been reported in fibroblasts, established human B cell lines (e.g., Rajii and Daudi), T-cell lines (e.g., Molt-4 and Jurkat), and PBMC subpopulations (CD4+, CD8+, and CD19+) from HCV infected patients [8–11]. The presence of HCV RNA in extrahepatic sites has also been detected in HCV infected individuals where HCV negative strands (replicative intermediate) have been associated with PBMCs, CD4+, CD8+, CD19+, T-cells, monocytes, and macrophages [12]. In recent years, the virus has been shown to persist in the PBMCs of patients who had received successful antiviral treatment with Pegylated-IFNα (Peg-IFNα) in combination with ribavirin. In such patients, HCV RNA was associated with PBMCs but not present in the serum [13–15].

HCV RNA derived from PBMCs, liver tissue, and serum samples from chronically infected patients has been shown to display heterogeneity due to the presence of mutations in the 5'UTR [16–18]. The presence of dominant quasispecies with a restricted number of point mutations, distinct from liver and serum, and identical nucleotide substitutions in the IRES region, such as the G107⟶A, C204⟶A and G243⟶A, commonly known as the A-A-A variant detected in patient derived cells have been previously detected also in cell lines such as Daudi B cell lymphoblasts and human T-cell lines [19]. This variant has been associated with lymphoid cell compartment infection [17] including monocyte-derived dendritic cells [20], monocytes/macrophages [12] and PBMCs [16]. Studies on IRES translational efficiency indicated that the A-A-A variant increased translation activity in lymphoid cell lines compared to wild-type G-C-G variants but not in cell lines derived from monocytes, granulocytes, and hepatocytes [21]. IRES translation activity was greatly enhanced in Raji, Bjab, and Molt4 but not in Jurkat cells, hepatocytes (Huh-7), monocytes, and granulocyte cell lines such as HL-60, KG-1, or THP-1 [21]. The A-A-A variant has also been detected in the brain where the A204 and A243 mutations from HCV RNA present in the cerebellum were absent from serum-derived HCV RNA of the same patient suggesting that the CNS was a candidate site of extrahepatic replication [17, 22, 23]. These studies have provided evidence that there are nucleotide substitutions in the HCV IRESes which are associated with lymphoid replication. The aim of this study was to analyze the temporal changes of HCV 5'UTR sequences obtained from PBMCs and compare them with those obtained from plasma in HCV infected patients. HCV RNA extraction, reverse transcription-PCR (RT-PCR), cloning, sequence, and RNA structure analyses were used to determine the presence of the A-A-A variant in PBMCs.

2. Materials and Methods

2.1. Patient Samples. Blood samples were obtained from twenty-four patients (all males, 24-62 years old) positive for anti-HCV and HCV RNA by RT-PCR, who had abnormal liver function tests for the first time since their last scheduled visit, suggesting acute infection with HCV (previously negative). The patients were recruited into the St Mary's acute HCV cohort with inclusion criteria as previously described [25]. None of the patients were coinfected with Hepatitis B virus (HBV) or had other causes of liver disease. The majority of patients were infected with HCV genotype 1a.

All work covering patient material was carried out following informed consent. Moreover, all such work received approval from the Ethics committee at St. Mary's Hospital.

2.2. RNA Extraction and RT-PCR. Blood samples were diluted with phosphate-buffered saline (PBS) (1:1), layered over Ficoll Histopaque 1077 (Sigma-Aldrich), and centrifuged at 400g for 30 min at room temperature. The PBMC layer was removed; the cells were washed twice with PBS and sedimented twice at 100g for 10 min.

Viral RNA was extracted from plasma using a QIAmp Viral RNA MiniKit (Qiagen) in accordance with the manufacturer's instructions while total RNA was extracted from PBMCs using the TRIzol®reagent (Invitrogen). Genomic DNA was removed from total RNA preparations by Turbo DNA-free Dnase treatment (Ambion). Total RNA quality and

quantity were assessed on a NanoDrop ND-1000 spectrophotometer and analyzed on a 1% agarose gel to check size and integrity. RNA was reverse-transcribed into complementary DNA (cDNA) at 42°C for 60 min using 0.5μM of IRES-specific antisense primer (5'-GCACGGTCTACGAGACCT-3') and 50 U of Moloney Murine Leukemia Virus reverse transcriptase (MMLV RT) (RETROscript kit, Ambion) in a volume of 20 μl following the manufacturer's protocol. HCV 5' UTR amplification was performed with the Fast Start High Fidelity PCR System (Roche) to amplify a 243-bp fragment from the HCV 5'UTR region. Primers used were sense primer (5'-GCTTAGCCATGGCGTTAG-3') and antisense primer (5'-GCACGGTCTACGAGACCT-3'). A no template control (NTC) and a no reverse transcriptase control (NRT) were included as negative controls to exclude contamination.

PCR conditions were as follows: preamplification denaturation (one cycle), 95°C for 2 min; amplification (35 cycles) included denaturation at 95°C for 30 sec, annealing at 56°C for 30 sec, and elongation at 72°C for 40sec. A final elongation step (1 cycle) was carried out at 72°C for 7 min. PCR products were separated and visualized on a 2% agarose gel and purified using the QIAquick Gel Extraction kit (Qiagen) following the manufacturer's protocol.

2.3. Cloning. Purified PCR products were ligated into the pGEM-T Easy cloning vector (Promega) using T4 DNA Ligase (3u/μl) and transformed into competent *Escherichia coli* DH5α cells (Invitrogen). A tube with no insert DNA was included as negative control. Recombinant colonies were detected by white/blue selection using 5-bromo-4-chloro-3-indolyl-beta-D-galacto-pyranoside (X-Gal) (Promega) to a final concentration of 40mg/ml and plasmid DNA was extracted using the GenElute Plasmid Miniprep Kit (Sigma) following the manufacturer's protocol. Fifteen clones from each of 15 plasma samples (n=225) and fifteen clones from each of 15 PBMC samples (n=225) were sequenced with the ABI3730xl analyzer (MRC CSC Genomics Core Laboratory, London, United Kingdom).

2.4. Sequence Analysis. Sequence alignments and entropy (Hx) calculations were performed using BioEdit software (version 7.1.3) and the phylogenetic analysis by MEGA software (version 5.1) using the Kimura 2-parameter model using all the sequences obtained from the clones (PBMCs-derived n=225, plasma-derived n=225).

As the majority of patients were infected with HCV subtype 1a the genomic sequences of 438 HCV subtype 1a isolates obtained from the Viral Bioinformatics Resource Centre database and GenBank were aligned with BioEdit software to align many HCV subtype 1a sequences in order to determine the most commonly expressed nucleotide at each position obtaining thus the best consensus sequence, which was used for multiple sequence alignments of PBMC and plasma-derived IRES amplicons. Additionally, the ΔG was computed at 37°C using the Turner model [23, 26] for the RNA parameters with the RNAeval web server (http://rna.tbi.univie.ac.at/cgi-bin/RNAWebSuite/RNAeval.cgi).

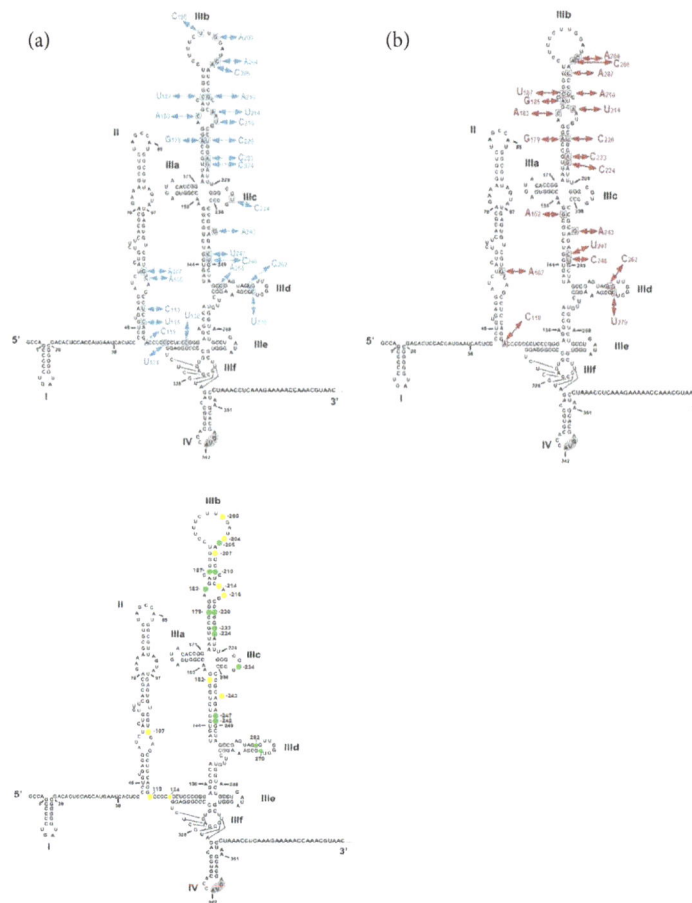

FIGURE 1: Nucleotide differences detected in 5'UTR sequences derived from patient PBMCs and plasma. Nucleotide changes in PBMC-derived IRES sequences (n=225) are indicated in blue while nucleotide changes in plasma-derived IRES sequences (n=225) are indicated in red. Similarities in nucleotide frequency variation between PBMC and plasma-derived samples are represented in green dots while differences are represented in yellow dots. Structural domains are indicated in Roman letters (I-IV). The start codon AUG is highlighted in stem-loop domain IV. Substitution sites are indicated as squares and arrows indicate mutations described in the text. Nucleotide sequence and putative secondary RNA structure of the 5'UTR [nt. 1-383 genotype 1b (GenBank, AJ238799.1)] was adapted from Honda et al. [24].

3. Results

3.1. Sequence Variation between PBMC and Plasma Isolates. The nucleotide differences among all the 225 sequences derived from PBMC and all of the 225 sequences derived from plasma samples are highlighted on the secondary RNA structure of the 5'UTR HCV (nt. 1-383 genotype 1b (GenBank, AJ238799.1) where numbers indicate IRES nucleotide positions relative to strain HCV 1a (GenBank, NC_004102) (Figure 1). As the two-dimensional structure of HCV genotype 1b 5'UTR was used as an outline to position the nucleotide substitutions detected in HCV IRESes, mainly genotype 1a (Figure 1), the HCV IRES genotype 1b was aligned with consensus sequence genotype 1a to pinpoint any differences between the two IRESes. The only differences were substitutions 11U, 12G, 13A, 34G, 35A, 204A and 243A present in genotype 1a (data not included).

PBMC-derived IRES sequences had more nucleotide substitution sites than plasma-derived ones (27 versus 20). Nucleotide substitutions present at positions 108, 113, 115, 124, 130, 198, 200, 216, 234, and 255, were only associated with PBMC-derived IRESes. On the other hand, nucleotide substitutions A152, G185, and A207 were only present in plasma-derived IRESes. Nucleotide substitutions at positions 107, 119, 179, 183, 187, 204, 205, 210, 214, 220, 223, 224, 243, 247, 248, 262, and 270 were associated with both PBMC and plasma-derived IRESes (Figure 1).

Nucleotide frequencies helped to identify the nucleotide position where nucleotide frequency differences occurred between PBMC and plasma-derived HCV IRESes. The majority of the IRES motifs were conserved between PBMC and plasma samples as similar nucleotide frequencies were obtained along the IRES sequence. Nucleotide positions 179, 183, 187, 205, 210, 220, 223, 224, 234, 247, 248, 262, and 270 (Figure 1, highlighted in green) contained two nucleotide frequency changes common in both PBMC and plasma-derived samples. However, positions 107, 119, 124, 152, 200, 204, 207, 214, 216, and 243 had different nucleotide frequencies between PBMC and plasma-derived samples (Figure 1, highlighted in yellow). Overall the majority of changes occurred between nucleotide positions 179 to 187 and 200 to 224 and occasionally from 234 and 270.

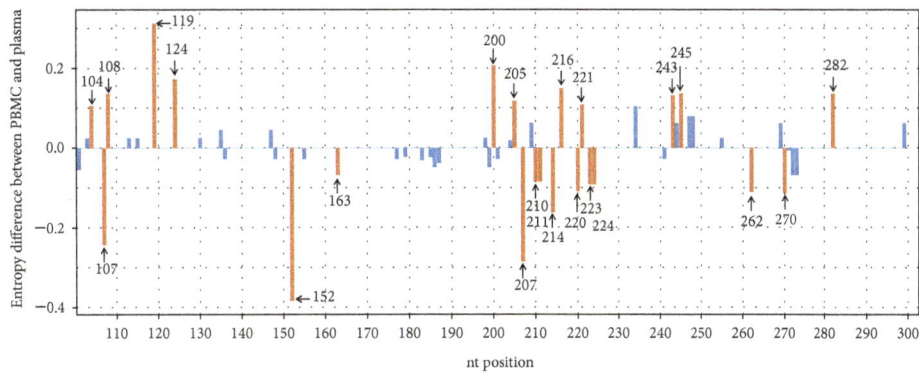

FIGURE 2: Entropy differences between the PBMC and plasma. Entropy differences between sequences (nt 100 to 300 of the HCV IRES) derived from PBMCs (n=225) and plasma (n=225). Each bar represents the entropy difference at a single nucleotide position. Significant differences between the entropy of PBMC and plasma are indicated by orange bars while no significant differences are indicated by blue bars. The results were generated using a Perl script and the figure was produced in R software programming language.

3.2. Entropy Estimations between PBMC and Plasma Sequences.

Entropy differences between PBMC and plasma-derived IRESes were examined to quantify diversity in a single nucleotide position (Figure 2). To calculate the entropy differences between PBMCs- and plasma-derived IRESes, the entropy dataset was randomised with replacement 100 times and the difference in entropy was calculated for each of these random sets. This difference was compared to the difference between PBMC and plasma in the real set to determine whether the difference in entropy was higher than you would expect based on randomly sampling the sequences. A positive value indicated that the entropy was higher in PBMC than in plasma while a negative value indicated that the PBMC had a lower entropy than the plasma. A significant entropy difference according to the randomisation was associated with specific nucleotide positions. High entropy in PBMCs was detected at nucleotide positions 104, 108, 119, 124, 200, 205, 216, 221, 243, 245 and 282, while high entropy in plasma (lower in PBMC), was detected at nucleotide positions 107, 152, 163, 207, 210, 211, 214, 220, 223, 224, 262 and 270 (Figure 2).

Conserved nucleotide sequences were identified among PBMCs and plasma-derived IRES sequences. Two conserved regions were identified in PBMC-derived IRES sequences (131-CTCCCGGGAGAGCCATAGTGGTCTGCGGA ACC-GGTGAGTACACCGGAA-178 and 271-GCGAAAGGC-CTTGTGGTACTGCCTGATAGG-300). In plasma-derived IRES sequences, five conserved regions were detected (120-CCCCCCCTCCCGGGAGAGCCATAGTGGTCTGC-151, 153-GAACCGGTGAGTACACCGGAATTGCC-178, 188-GGGTCCTTTCTTGGAT-203, 225-ATTTGGGCGTGC-CCCCGC-242 and 271-GCGAAAGGCCTTGTGGTACTG-CCTGATAGG-300). The numbering is based on the 1b sequence with accession no. AJ238799.1.

3.3. Distribution of Quasispecies within Patients.

In order to determine in which compartment (PBMCs or plasma) the distribution of each nucleotide combination at position 107, 204 and 243 was more variable, PBMC- and plasma-derived samples were analyzed in each patient (Figure 3).

Plasma-derived samples had more variation in the population of variants than PBMC-derived samples except in patient 3 and 7 where only one specific variant was present in plasma samples. In each patient, the GAA and GAG variants were present in plasma- but not in PBMC-derived samples.

3.4. Phylogenetic Analysis of the HCV IRES Population.

The sequence data obtained from all clones of PBMC- and plasma-derived samples were used to construct an unrooted phylogenetic tree (Figure 4), which revealed lineages containing related PBMC- and plasma-derived HCV strains and one distinctive lineage with isolated plasma-derived strains (Figure 4(a)). A-A-A variants associated with PBMC and plasma suggest possible selection (Figure 4(b)).

The effect of single point mutations on the thermodynamic stability of the RNA was examined on the relevant IRES domains (Figure 5). A decrease in ΔG indicated an increase in stability of the relevant domain while an increase in ΔG indicated a decrease in stability. Mutations G200A, A204C, A214U, U216C, and G243A had no effect on the stability of the RNA secondary structure. However, mutations A107G, A108C in the IRES domain II and mutations A152G, C207A, U262C, and U270C in the IRES domain III increased the stability of the IRES structure. G200A and U216C were only detected in PBMCs while G152A and C207A only in plasma-derived IRESes (Figure 5).

4. Discussion

PBMC-derived samples had a much higher number of mutations in domain II compared to plasma-derived samples (7 versus 2 mutations), particularly between nt 107 and 130 (Figure 1). Recent studies have shown that domain II induces a conformational change of the 40S ribosomal subunit and possibly holds the coding RNA into the decoding centre of the ribosome during the assembly of the translation machinery [27, 28]. Thus, it is possible that nucleotide substitution C108A alone or in combination with G107A detected in domain II might influence initial steps that lead to translation initiation in PBMCs. Mutation C108A, detected only in

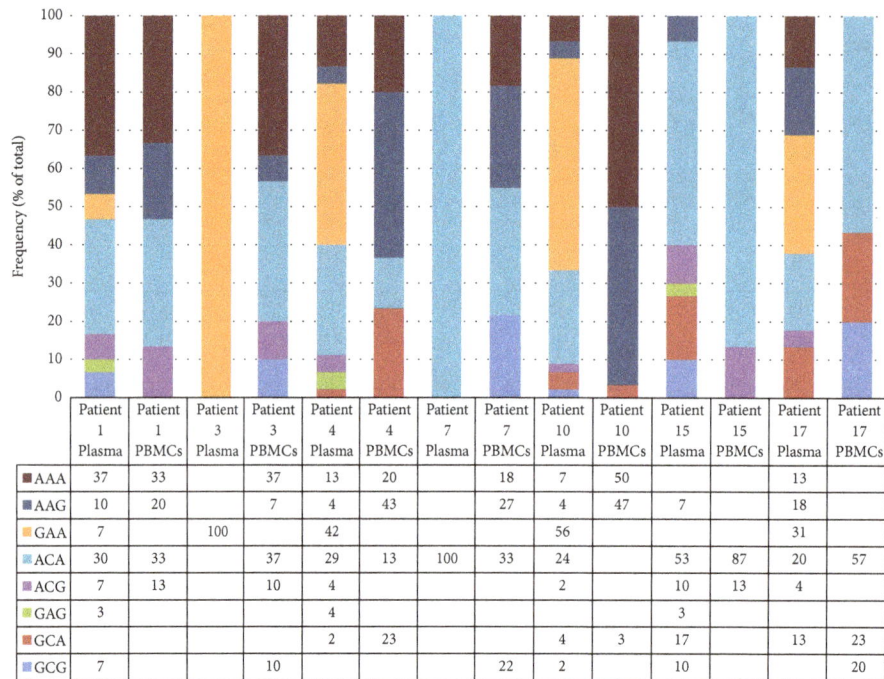

	Patient 1 Plasma	Patient 1 PBMCs	Patient 3 Plasma	Patient 3 PBMCs	Patient 4 Plasma	Patient 4 PBMCs	Patient 7 Plasma	Patient 7 PBMCs	Patient 10 Plasma	Patient 10 PBMCs	Patient 15 Plasma	Patient 15 PBMCs	Patient 17 Plasma	Patient 17 PBMCs
■ AAA	37	33		37	13	20		18	7	50			13	
■ AAG	10	20		7	4	43		27	4	47	7		18	
■ GAA	7		100		42				56				31	
■ ACA	30	33		37	29	13	100	33	24		53	87	20	57
■ ACG	7	13		10	4				2		10	13	4	
■ GAG	3				4						3			
■ GCA					2	23			4	3	17		13	23
■ GCG	7			10				22	2		10			20

FIGURE 3: Distribution of nucleotide combinations at positions 107, 204, and 243 in PBMC- and plasma-derived IRES samples in different patients. Data represent the frequencies in percent of all possible nucleotide combinations at positions 107, 204, and 243 detected in PBMC- and plasma-derived HCV IRES samples from different patients.

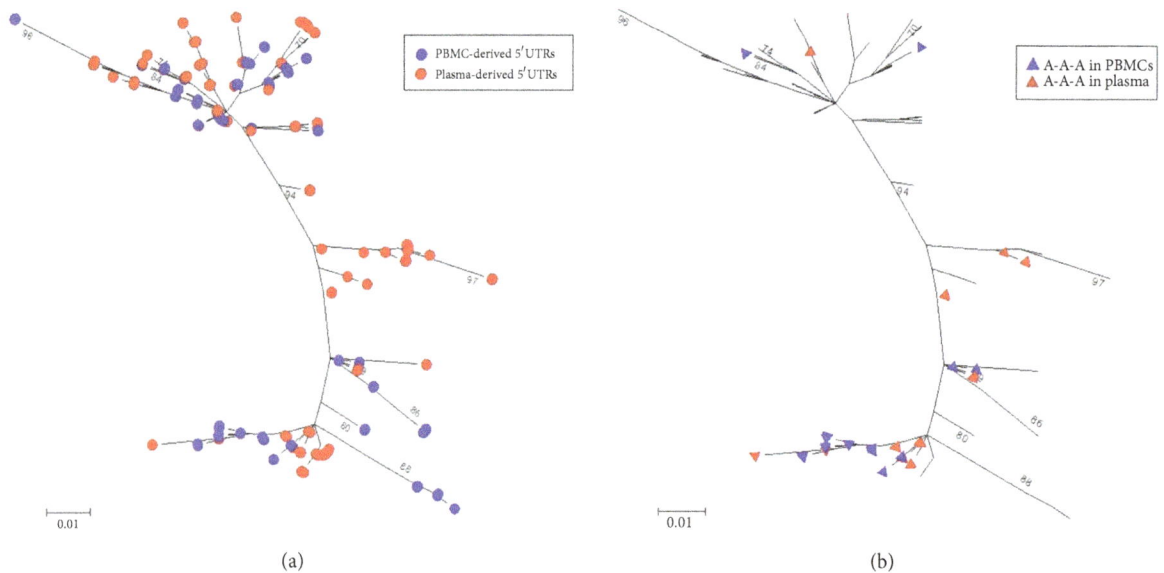

(a)

(b)

FIGURE 4: Phylogenetic tree analysis of all PBMC- and plasma-derived HCV IRES sequences. The evolutionary history was inferred by using the Maximum Likelihood method based on the Kimura 2-parameter model. The tree with the highest log likelihood (-1129.7196) is shown. Initial tree(s) for the heuristic search were obtained by applying the Neighbor-Joining method to a matrix of pairwise distances estimated using the Maximum Composite Likelihood (MCL) approach. The tree is drawn to scale, with branch lengths measured in the number of substitutions per site. The analysis involved 450 nucleotide sequences. PBMC-derived HCV IRES nucleotide sequences (n=225; coloured in blue) and plasma-derived HCV IRES nucleotide sequences (n=225; coloured in red) (a). HCV IRESes with the triple adenine variant (A-A-A) in PBMC (n=36; blue triangles) and in plasma (n=26; red triangles) (b). All positions containing gaps and missing data were eliminated. There were a total of 200 positions (nt 100 to 300) in the final dataset. The bootstrap values along the branches indicate percent confidence of branches. Bootstrap values greater than 70% are shown. The scale bar corresponds to 0.01 substitutions per site. Evolutionary analyses were conducted in MEGA6.

IRES region	Nt. Position	Mutation	ΔG	ΔG	Effect on ΔG (kcal/mol)	Effect on ΔG
Domain II	107	G→A	-23.7	-23.4	-0.3	Decreased
		A→G	-23.4	-23.7	0.3	Increased
	108	C→A	-23.7	-22.7	-1.0	Decreased
		A→C	-22.7	-23.7	1.0	Increased
	113	U→C	-23.7	-17.1	-6.6	Decreased
		C→U	-17.1	-23.7	6.6	Increased
Domain III	152	G→A	-47.5	-41.8	-5.7	Decreased
		A→G	-41.8	-47.5	5.7	Increased
Domain III, Loop IIIb	200	G→A	-47.5	-47.5	0.0	No effect
		A→G	-47.5	-47.5	0.0	No effect
	204	A→C	-47.5	-47.5	0.0	No effect
		C→A	-47.5	-47.5	0.0	No effect
Domain III	207	C→A	-47.5	-48.4	0.9	Increased
		A→C	-48.4	-47.5	-0.9	Decreased
	214	A→U	-47.5	-47.5	0.0	No effect
		U→A	-47.5	-47.5	0.0	No effect
	216	U→C	-47.5	-47.5	0.0	No effect
		C→U	-47.5	-47.5	0.0	No effect
	243	A→G	-47.5	-47.5	0.0	No effect
		G→A	-47.5	-47.5	0.0	No effect
Domain III, Loop IIId	255	C→A	-54.9	-53.6	-1.3	Decreased
		A→C	-53.6	-54.9	1.3	Increased
	259	U→C	-54.9	-54.9	0.0	No effect
		C→U	-54.9	-54.9	0.0	No effect
	262	U→C	-54.9	-56.8	1.9	Increased
		C→U	-56.8	-54.9	-1.9	Decreased
	270	C→U	-54.9	-51.2	-3.7	Decreased
		U→C	-51.2	-54.9	3.7	Increased

FIGURE 5: The effect of single point mutations on the thermodynamic stability of the relevant IRES domains. The thermodynamic stability of the relevant domains was calculated using the RNAeval program with RNA free energy parameters and ΔG set at 37°C. Blue cells indicate mutations detected only in PBMC-derived IRESes while red cells indicate mutations detected only in plasma-derived IRESes.

PBMC-derived IRESes, may help HCV to possibly reduce HCV translation and avoid immune detection.

Sequence variability was more pronounced in domain III, particularly in the apical part, above a 4-way junction (IIIa, b, and c), and between subdomains IIIc and IIId (Figure 1). The apical part of domain III interacts with eIF3, the 40S ribosomal subunit, and some transacting factors [29] and might facilitate the recruitment of the ribosome to the IRES. Sequence polymorphisms between nt 175 to 187 and nt 210 to 224 have also been detected in previous studies [30]. Additionally, the proximity of the 40S subunit to the helical junction IIIabc of domain III might be the cause of the concentration of substitutions in this IRES region, a region that has an important role during the formation of the 48S preinitiation complex and during the recruitment of the 43S subunit and eIF3 [4]. Stem-loops IIIa and IIIc were conserved among PBMCs and plasma-derived IRESes, except for a single mutation in stem-loop IIIc at position 234 detected only in PBMC-derived samples (Figure 1).

The distribution of nucleotide combinations at positions 107, 204, and 243 examined in each patient (Figure 3) indicated that plasma samples were more diverse in the population of variants than PBMC ones. These results may be caused by a reduced exposure of the virus in PBMCs to the immune system, passively evading its response. Thus, the A-A-A variant which appeared to be the natural wild-type for PBMCs might display a preferential tropism for these cells and/or a more functional adaptation for translation in these cells. A previous study has shown that low and high translation variants with sequence changes in the IRES region may be selected by the immune system [31].

The presence of PBMC variants not detected in plasma of the same patient, and therefore derived from his/her liver, may indicate provenance from replication in PBMCs or other extrahepatic sites or may represent stored HCV RNA from past quasispecies no longer present in plasma, or may be caused by selection of viral variants in or associated with PBMCs.

Significant entropy differences according to the randomisation between PBMC- and plasma-derived sequences (Figure 2) indicated high entropy in PBMCs (nt positions 104, 108, 119, 124, 200, 205, 216, 221, 243, 245, and 282) and high entropy in plasma (nt positions 107, 152, 163, 207, 210, 211, 214, 220, 223, 224, 262, and 270). These significant differences between the entropy of the PBMC and the plasma indicate nucleotide diversity at these specific positions suggesting possible selection and virus diversity in each position.

Significant differences with high entropy values were located in domain II in PBMCs and only on the right side of the predicted IRES domain III. On the other hand, the high entropy in plasma concerned only position 107 in IRES domain II and both sides of domain III, in particular domain IIIb and domain IIId, which are involved in the binding of eIF3a and the 40S ribosomal subunit [32]. This suggests selective pressure in these regions and the relation of entropy changes with IRES activity. Regarding nt positions 107, 204, and 243, the entropy was higher only in plasma at positions 107 and 243 while no significant differences occurred at position 204 (Figure 2).

Phylogenetic analysis of all PBMC- and plasma-derived IRES sequences (Figure 4) suggested that, on the whole, PBMC-derived IRESes were the most divergent variants with the longest branch lengths and appeared to have originated from plasma. Additionally, the tree suggested a nonrandom distribution of PBMC- and plasma-derived IRESes. The phylogenetic tree suggested that the A-A-A variant possibly changed and/or derived from plasma. However, more data is required to understand the relationship of the different phylogenetic polytomies.

Future analysis with chemical and enzymatic probing will be critical for validating and interpreting RNA structure or modelling. Overall, the results in this study suggest that the IRES is subject to more modifications in PBMCs than in plasma as a larger number of mutations were identified only in PBMC-derived IRESes. It is possible that some of the different mutations associated with PBMCs and plasma may relate to the origin of the released virus particles, which could be from liver or lymphocytes, and based on the mutations acquired may allow viral replication or inhibition in extrahepatic sites. This hypothesis could also explain the minor variants associated with each population from the cloning study. Previous studies on HCV infection of B-cells indicated that, during chronic infection, B-cells and/or monocytes frequently harboured specific HCV variants [33].

Two conserved IRES regions were associated with PBMC-derived IRESes and five in plasma-derived IRESes. Some of these regions are located in ribosomal binding sites as previously reported [34]. The conserved region between nt 271 to 300 was common in both PBMC- and plasma-derived IRESes as this region is essential for binding to the 40S subunit [35]. This region contains subdomain IIId, an essential component of the HCV IRES, as previous studies have shown that mutations within domain IIId disrupted IRES-mediated initiation caused by the structural reorganization of the HCV IRES leading to reduced IRES activity [32, 36]. PBMC-derived samples had more mutations in domain IIId than in plasma-derived IRESes suggesting a possible reduction in IRES activity by the acquired mutations in domain IIId in PBMCs.

The analysis of the effect of the examined mutations on the thermodynamic stability of the IRES indicated that few mutations in IRES domains II and III (Figure 5) increased or decreased IRES stability. Single point mutations detected only in plasma-derived IRESes at positions 152, 207 and 270 affected the predicted IRES structure and thermodynamic stability (Figure 5). Nucleotide substitution C270U is located in a region (nt 277 to 295) known to be essential for binding to 40S subunits [35]. It is possible that the effect of one mutation on IRES stability might be counteracted by the occurrence of another mutation. For example the effect of A107G and/or A108C might be counteracted by the effect of G107A and/or C108A together or in combination (Figure 5). The IRES may acquire a single point mutation, as C207A was detected only in plasma-derived IRESes to decrease the thermodynamic stability of the IRES and thus possibly reduce viral translation in plasma or acquire mutations such as G107A, G152A and C207A to increase IRES stability and possibly increase viral translation.

The thermodynamic stability values may not predict the affinity of RNA binding to ribosomal subunits but it is possible that the IRES stability might affect HCV replication. Thus, HCV might acquire mutations to decrease the thermodynamic stability of the IRES and to control its replication while present in immunoprivileged sites. This might explain the different disease outcomes among HCV patients. It is possible that the identified mutations may occur during the life cycle of the virus under certain cellular environmental conditions and that the location of the point mutations on the IRES may disrupt IRES function or reduce translation efficiency as seen in previous studies [17, 21, 37]. Additionally, populations with acquired mutations may increase over time resulting in long-term clinical consequences.

5. Conclusions

In conclusion, these results of the study suggest that the quasispecies dynamics are a mechanism through which HCV is able to adapt to its host environment through IRES diversity, possibly to regulate translational activity and tropism in different cells. The mutations and regions of similarities identified in this study may be a consequence of functional, structural, or evolutionary relationships between IRES sequences, which may influence the binding of cellular factors and may reduce or increase translation efficiency accordingly. The results raise the possibility that these strains may compete through the efficiency of the IRES, as previous studies showed plasma IRESes to be more competent than B cell IRESes in hepatocytes indicating a selective pressure on extra-hepatic strains and, therefore, extra-hepatic replication [38]. HCV variants might have a selective advantage in immunoprivileged sites and allow the persistence of the virus in these sites without immune detection as previously suggested [20]. This could explain why nucleotide substitutions occurred more often in PBMCs than in plasma-derived samples.

Acknowledgments

The research was funded by the Liver Research Fund and Dr. Emma Thomson was the recipient of a Welcome trust Fellowship.

References

[1] B. Clarke, "Molecular virology of hepatitis C virus," *Journal of General Virology*, vol. 78, no. 10, pp. 2397–2410, 1997.

[2] P. Karayiannis and M. J. McGarvey, "The GB hepatitis viruses," *Journal of Viral Hepatitis*, vol. 2, no. 5, pp. 221–226, 1995.

[3] M. E. Major and S. M. Feinstone, "The molecular virology of hepatitis C," *Hepatology*, vol. 25, no. 6, pp. 1527–1538, 1997.

[4] J. S. Kieft, K. Zhou, R. Jubin, and J. A. Doudna, "Mechanism of ribosome recruitment by hepatitis C IRES RNA," *RNA*, vol. 7, no. 2, pp. 194–206, 2001.

[5] P. J. Lukavsky, "Structure and function of HCV IRES domains," *Virus Research*, vol. 139, no. 2, pp. 166–171, 2009.

[6] S. Navas, J. Martín, J. A. Quiroga, I. Castillo, and V. Carreño, "Genetic diversity and tissue compartmentalization of the hepatitis C virus genome in blood mononuclear cells, liver, and serum from chronic hepatitis C patients," *Journal of Virology*, vol. 72, no. 2, pp. 1640–1646, 1998.

[7] A. M. Roque Afonso, J. Jiang, F. Penin et al., "Nonrandom distribution of hepatitis C virus quasispecies in plasma and peripheral blood mononuclear cell subsets," *Journal of Virology*, vol. 73, no. 11, pp. 9213–9221, 1999.

[8] P. Baré, I. Massud, C. Parodi et al., "Continuous release of hepatitis C virus (HCV) by peripheral blood mononuclear cells and B-lymphoblastoid cell-line cultures derived from HCV-infected patients," *Journal of General Virology*, vol. 86, no. 6, pp. 1717–1727, 2005.

[9] Y. Kondo, V. M. H. Sung, K. Machida, M. Liu, and M. M. C. Lai, "Hepatitis C virus infects T cells and affects interferon-γ signaling in T cell lines," *Virology*, vol. 361, no. 1, pp. 161–173, 2007.

[10] S. Pal, D. G. Sullivan, S. Kim et al., "Productive Replication of Hepatitis C Virus in Perihepatic Lymph Nodes In Vivo: Implications of HCV Lymphotropism," *Gastroenterology*, vol. 130, no. 4, pp. 1107–1116, 2006.

[11] N. F. Fletcher, J. P. Yang, M. J. Farquhar et al., "Hepatitis C virus infection of neuroepithelioma cell lines," *Gastroenterology*, vol. 139, no. 4, pp. 1365–e2, 2010.

[12] A. H. Lin, *The dynamics of the interactions between solid tumors and lymphocytes: A deterministic model*, ProQuest LLC, Ann Arbor, MI, 2001.

[13] T. N. Q. Pham and T. I. Michalak, "Occult persistence and lymphotropism of hepatitis C virus infection," *World Journal of Gastroenterology*, vol. 14, no. 18, pp. 2789–2793, 2008.

[14] M. Radkowski, J. F. Gallegos-Orozco, J. Jablonska et al., "Persistence of hepatitis C virus in patients successfully treated for chronic hepatitis C," *Hepatology*, vol. 41, no. 1, pp. 106–114, 2005.

[15] D. Januszkiewicz-Lewandowska, J. Wysocki, M. Pernak et al., "Presence of hepatitis C virus (HCV)-RNA in peripheral blood mononuclear cells in HCV serum negative patients during interferon and ribavirin therapy," *Japanese Journal of Infectious Diseases*, vol. 60, no. 1, pp. 29–32, 2007.

[16] T. Laskus, M. Radkowski, L. Wang, M. Nowicki, and J. Rakela, "Uneven distribution of hepatitis C virus quasispecies in tissues from subjects with end-stage liver disease: confounding effect of viral adsorption and mounting evidence for the presence of low-level extrahepatic replication," *Journal of Virology*, vol. 74, no. 2, pp. 1014–1017, 2000.

[17] D. M. Forton, P. Karayiannis, N. Mahmud, S. D. Taylor-Robinson, and H. C. Thomas, "Identification of unique hepatitis C virus quasispecies in the central nervous system and comparative analysis of internal translational efficiency of brain, liver, and serum variants," *Journal of Virology*, vol. 78, no. 10, pp. 5170–5183, 2004.

[18] J. Bukh, R. H. Purcell, and R. H. Miller, "Sequence analysis of the 5' noncoding region of hepatitis C virus," *Proceedings of the National Acadamy of Sciences of the United States of America*, vol. 89, no. 11, pp. 4942–4946, 1992.

[19] N. Nakajima, M. Hijikata, H. Yoshikura, and Y. K. Shimizu, "Characterization of long-term cultures of hepatitis C virus," *Journal of Virology*, vol. 70, no. 5, pp. 3325–3329, 1996.

[20] J. Laporte, C. Bain, P. Maurel, G. Inchauspe, H. Agut, and A. Cahour, "Differential distribution and internal translation efficiency of hepatitis C virus quasispecies present in dendritic and liver cells," *Blood*, vol. 101, no. 1, pp. 52–57, 2003.

[21] H. Lerat, Y. K. Shimizu, and S. M. Lemon, "Cell type-specific enhancement of hepatitis C virus internal ribosome entry site-directed translation due to 5' nontranslated region substitutions selected during passage of virus in lymphoblastoid cells," *Journal of Virology*, vol. 74, no. 15, pp. 7024–7031, 2000.

[22] M. Radkowski, J. Wilkinson, M. Nowicki et al., "Search for hepatitis C virus negative-strand RNA sequences and analysis of viral sequences in the central nervous system: Evidence of replication," *Journal of Virology*, vol. 76, no. 2, pp. 600–608, 2002.

[23] D. H. Turner and D. H. Mathews, "NNDB: The nearest neighbor parameter database for predicting stability of nucleic acid secondary structure," *Nucleic Acids Research*, vol. 38, no. 1, pp. D280–D282, 2009.

[24] M. Honda, R. Rijnbrand, G. Abell, D. Kim, and S. M. Lemon, "Natural variation in translational activities of the 5' nontranslated RNAs of hepatitis C virus genotypes 1a and 1b: Evidence for a long-range RNA- RNA interaction outside of the internal ribosomal entry site," *Journal of Virology*, vol. 73, no. 6, pp. 4941–4951, 1999.

[25] E. C. Thomson, V. M. Fleming, J. Main et al., "Predicting spontaneous clearance of acute hepatitis C virus in a large cohort of HIV-1-infected men," *Gut*, vol. 60, no. 6, pp. 837–845, 2011.

[26] D. H. Mathews, M. D. Disney, J. L. Childs, S. J. Schroeder, M. Zuker, and D. H. Turner, "Incorporating chemical modification constraints into a dynamic programming algorithm for prediction of RNA secondary structure," *Proceedings of the National Acadamy of Sciences of the United States of America*, vol. 101, no. 19, pp. 7287–7292, 2004.

[27] N. Quade, D. Boehringer, M. Leibundgut, J. Van Den Heuvel, and N. Ban, "Cryo-EM structure of Hepatitis C virus IRES bound to the human ribosome at 3.9-Å resolution," *Nature Communications*, vol. 6, article no. 7646, 2015.

[28] H. Yamamoto, M. Collier, J. Loerke et al., "Molecular architecture of the ribosome-bound Hepatitis C Virus internal ribosomal entry site RNA," *EMBO Journal*, vol. 34, no. 24, pp. 3042–3058, 2015.

[29] T. V. Pestova, I. N. Shatsky, S. P. Fletcher, R. J. Jackson, and C. U. T. Hellen, "A prokaryotic-like mode of cytoplasmic eukaryotic ribosome binding to the initiation codon during internal translation initation of hepatitis C and classical swine fever virus RNAs," *Genes & Development*, vol. 12, no. 1, pp. 67–83, 1998.

[30] A. J. Collier, S. Tang, and R. M. Elliott, "Translation efficiencies of the 5' untranslated region from representatives of the six major genotypes of hepatitis C virus using a novel bicistronic reporter assay system," *Journal of General Virology*, vol. 79, no. 10, pp. 2359–2366, 1998.

[31] J. F. Gallegos-Orozco, J. I. Arenas, H. E. Vargas et al., "Selection of different 5' untranslated region hepatitis C virus variants during post-transfusion and post-transplantation infection," *Journal of Viral Hepatitis*, vol. 13, no. 7, pp. 489–498, 2006.

[32] R. Jubin, N. E. Vantuno, J. S. Kieft et al., "Hepatitis C virus internal ribosome entry site (IRES) stem loop IIId contains a phylogenetically conserved GGG triplet essential for translation and IRES folding," *Journal of Virology*, vol. 74, no. 22, pp. 10430–10437, 2000.

[33] D. Ducoulombier, A.-M. Roque-Afonso, G. di Liberto et al., "Frequent compartmentalization of hepatitis C virus variants in circulating B cells and monocytes," *Hepatology*, vol. 39, no. 3, pp. 817–825, 2004.

[34] J. R. Lytle, L. Wu, and H. D. Robertson, "Domains on the hepatitis C virus internal ribosome entry site for 40s subunit binding," *RNA*, vol. 8, no. 8, pp. 1045–1055, 2002.

[35] E. Laletina, D. Graifer, A. Malygin, A. Ivanov, I. Shatsky, and G. Karpova, "Proteins surrounding hairpin IIIe of the hepatitis C virus internal ribosome entry site on the human 40S ribosomal subunit," *Nucleic Acids Research*, vol. 34, no. 7, pp. 2027–2036, 2006.

[36] L. Psaridi, U. Georgopoulou, A. Varaklioti, and P. Mavromara, "Mutational analysis of a conserved tetraloop in the 5' untranslated region of hepatitis C virus identifies a novel RNA element essential for the internal ribosome entry site function," *FEBS Letters*, vol. 453, no. 1-2, pp. 49–53, 1999.

[37] K. I. Kalliampakou, L. Psaridi-Linardaki, and P. Mavromara, "Mutational analysis of the apical region of domain II of the HCV IRES," *FEBS Letters*, vol. 511, no. 1-3, pp. 79–84, 2002.

[38] T. Durand, G. Di Liberto, H. Colman et al., "Occult infection of peripheral B cells by hepatitis C variants which have low translational efficiency in cultured hepatocytes," *Gut*, vol. 59, no. 7, pp. 934–942, 2010.

20

Herpes Simplex Virus Latency: The DNA Repair-Centered Pathway

Jay C. Brown

Department of Microbiology, Immunology, and Cancer Biology, University of Virginia Health System, Charlottesville, VA 22908, USA

Correspondence should be addressed to Jay C. Brown; jcb2g@virginia.edu

Academic Editor: Finn S. Pedersen

Like all herpesviruses, herpes simplex virus 1 (HSV1) is able to produce lytic or latent infections depending on the host cell type. Lytic infections occur in a broad range of cells while latency is highly specific for neurons. Although latency suggests itself as an attractive target for novel anti-HSV1 therapies, progress in their development has been slowed due in part to a lack of agreement about the basic biochemical mechanisms involved. Among the possibilities being considered is a pathway in which DNA repair mechanisms play a central role. Repair is suggested to be involved in both HSV1 entry into latency and reactivation from it. Here I describe the basic features of the DNA repair-centered pathway and discuss some of the experimental evidence supporting it. The pathway is particularly attractive because it is able to account for important features of the latent response, including the specificity for neurons, the specificity for neurons of the peripheral compared to the central nervous system, the high rate of genetic recombination in HSV1-infected cells, and the genetic identity of infecting and reactivated virus.

1. Introduction

All herpesviruses are able to cause both lytic and latent infections. Lytic infection refers to the situation in which the virus replicates in a host cell and causes its lysis, releasing hundreds to thousands of progeny virions. A latent infection is quite different. Here the virus enters into a refractory state in which little or no progeny virus is produced and the cell is not immediately damaged. The virus DNA is present in the latently infected cell nucleus, but there is little DNA replication and only minimal expression of virus-encoded genes. The virus can be reactivated from latency following an appropriate stimulus, however, and reactivation causes lytic virus replication [1–3].

The ability to enter into latency provides herpesviruses with an important survival advantage. In a lytic infection the virus is exposed to components of the immune response that have the potential to clear the virus from the host. In latency, however, infected cells are less readily recognized by the immune system because of the low level of virus gene expression. As a result, the virus can survive an otherwise effective immune response and be reactivated later to spread its infection in a less hostile immunological environment [4].

Herpes simplex virus (HSV1) resembles other herpesviruses in its ability to cause both lytic and latent infections [1, 5]. Lytic infections are produced in epithelial cells of the oral mucosa causing cold sores and other lesions. Progeny virus from this initial infection is able to traffic to sensory neurons in the trigeminal ganglion where a latent infection is produced. Virus reactivated from latently infected neurons migrates back to the site of the initial infection in the oral epithelium producing a second lytic infection. It is usual for a patient to experience many cycles of HSV1 entry into latency and reactivation from it.

The important role of latency in HSV1 pathogenesis has suggested that novel inhibitors targeting latency may be effective as an adjunct to acyclovir for HSV1 therapy. Both entry into latency and reactivation suggest themselves as attractive targets. The pathway to identification of the desired inhibitors would be easier, however, if investigators had a clear understanding of the molecular mechanisms involved in latency. So far, however, despite a large amount of experimental effort and intense interest in the topic, there remains an abundance of viable models for the basic biochemical events involved [1, 5].

Lytic HSV1 replication

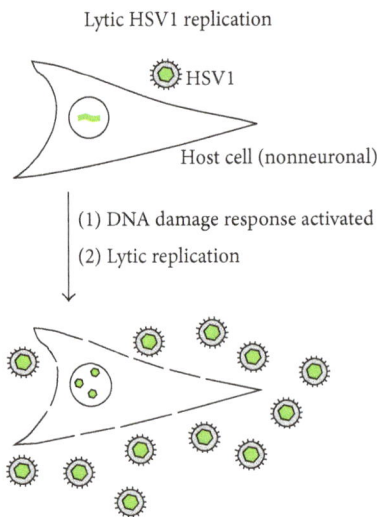

FIGURE 1: Illustration of lytic HSV1 replication as it is observed in nonneuronal cells. Note that host-encoded DNA damage response proteins are activated following DNA entry into nonneuronal cells and DDR proteins actively potentiate lytic virus growth.

Here I describe one of the possibilities, the DNA repair-centered pathway [6]. I begin with a brief summary of HSV1 lytic replication and the basic features of latency. There follows a description of the proposed repair-centered pathway, a summary of the experimental evidence that supports it and an account of how the proposed pathway is compatible with the main features of HSV1 latency and reactivation as they are currently understood.

2. Lytic HSV1 Infection

The most common HSV1 infections begin when extracellular virus binds to epithelial cells surrounding the mouth and present in the oral mucosa [4, 5]. Virus binds to receptors on the host cell surface and there follows a fusion event involving host and virus membranes. Fusion results in deposition of the virus nucleocapsid into the peripheral cytoplasm of the host cell. From there it migrates to the cell nucleus, docks at a nuclear pore, and injects the virus DNA into the nucleoplasm. Only the virus DNA enters the nucleus; the parental capsid remains outside. Once inside the nucleus the virus DNA is replicated primarily by the virus-encoded DNA-dependent DNA polymerase. Virus DNA synthesis is also thought to involve activation of cellular components of the DNA damage response (DDR) as described below (see Figure 1) [7, 8]. At the same time, virus genes are expressed beginning with the synthesis of virus-specific messenger RNAs by the host cell-encoded DNA-dependent RNA polymerase.

Assembly of progeny virus begins after sufficient amounts of virus DNA and proteins have been made. Assembly starts in the nucleus with capsid proteins that have been synthesized in the cytoplasm and imported into the nucleus. Capsids are assembled in the nucleus and filled there with virus DNA using a mechanism in which DNA is injected into a preformed capsid [9, 10].

Further assembly steps take place in the host cell cytoplasm [11]. DNA-filled capsids exit the nucleus and acquire tegument and membrane layers in an engulfment event involving vesicles containing the components of both layers. Mature progeny virions then exit the host cell by direct spreading to adjacent cells or as the host cell is lysed. A single cycle of HSV1 replication can take up to 24 hours.

Cold sore lesions require a few days to develop and they can last for a week or more. Spreading of lesions is eventually controlled by an effective immune response involving components of both the innate and acquired responses. Lesions recede without scarring at the site of infection and cold sores do not ordinarily require medical intervention [12].

3. Latency and Reactivation from Latency

Latency can be thought of as an extension of a lytic infection or perhaps as a diversion from it. Progeny virus produced at the initial site of infection in the oral epithelium spreads in three different ways: (a) it traffics laterally into adjacent epithelial cells to create a cold sore; (b) it is shed from the skin surface to reach contacts of the infected patient; and (c) it spreads internally to infect adjacent sensory neurons. Initiation of latency involves the third of the above pathways, infection of adjacent neurons [13].

Infection of neurons begins in the same manner as infection of other cell types. HSV1 binds to receptors on the cell surface; a membrane fusion event ensues depositing the nucleocapsid into the peripheral cytoplasm; the nucleocapsid traffics to the nucleus and injects the virus DNA. It is at this point that the infection stalls. Virus DNA synthesis and production of progeny virus are blocked completely creating the latent state. Neurons containing latent HSV1 genomes are concentrated in the trigeminal ganglion as these are prominent among the neurons that innervate the oral epithelium. Latency can persist in the trigeminal ganglion for the lifetime of the patient with the possibility for reactivation at any time. Apart from episodes of reactivation, latent infections do not produce symptoms for the patient [5].

For an infecting HSV1 virion, latency would be a terminal event if it were not for the possibility of reactivation. During latency the virus DNA is present in the neuron cell nucleus, but it has no way to be replicated or to create a lytic infection. Reactivation follows after a stimulus that is not well characterized [14, 15]. Most effective stimuli involve stress to the patient. This can be genotoxic stress such as exposure to sunlight; emotional and physical stress can also initiate reactivation. Once reactivation has been triggered, HSV1 replication follows the same pathway found in lytic infections. Replication occurs first in the neuron; it then spreads by way of the neuron to the original site of infection in the oral epithelium. As in the case of a primary lytic infection, progeny virus arising from a reactivated infection can be spread to the patient's contacts. Also, as in primary infections, symptoms arising from reactivated infections are effectively controlled by the immune response. Most affected patients suffer multiple reactivated infections at intervals of weeks to months.

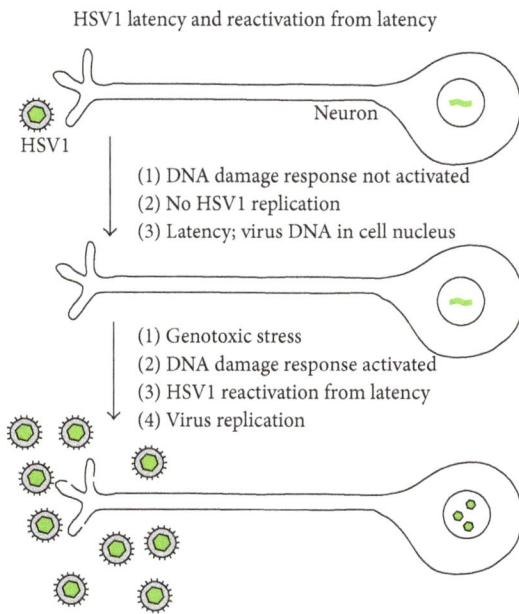

FIGURE 2: Illustration of HSV1 entry into latency and reactivation as proposed in the DNA repair-centered pathway. Note that DNA damage response proteins are not activated following entry of HSV1 into neurons, a condition that permits virus entry into latency. Note also that reactivation occurs following an accumulation of DNA damage in the latently infected cell. Reactivation results in HSV1 replication in the neuron as illustrated here.

4. The DNA Repair-Centered Pathway for HSV1 Latency and Reactivation: Basic Features of the Pathway

The proposed pathway focusses on the observation that lytic HSV1 replication depends on the activity of cell-encoded DNA repair proteins [16–20]. Components of the DNA damage response, for instance, are required and are available because most cell types respond to HSV1 infection by upregulating and activating the DDR. Required repair activities are found among those involved in mismatch repair and homologous recombination-dependent repair. All are readily available in most epithelial cells due to the cell's continuing need for DNA repair capacity. The same is not true of neurons. HSV1 infection of these cells does not activate the DDR to the same extent found in nonneuronal cells [6, 21, 22]. It is proposed therefore that HSV1 is unable to replicate in mature neurons and enters into latency because DNA repair proteins are not activated in response to infection (see Figure 2).

Reactivation is proposed to be the reverse of entry into latency. It is suggested that an accumulation of damage to both the neuronal and virus DNA eventually reaches a level at which DNA repair pathways are activated. Overall activation includes activation of repair functions required for HSV1 lytic growth, and the neuron becomes permissive for replication. HSV1 replication follows producing progeny virus that is transmitted by way of neurons to the site of the original infection in the oral epithelium (Figure 2). There lesions are produced that follow the same pathway of growth and control by the immune system found in the primary infection.

5. Experimental Observations That Support the Repair-Centered Pathway

The DNA repair-centered pathway is supported by experimental studies documenting the following: (1) activation of DNA repair functions following HSV1 infection of nonneuronal cells; (2) the requirement for activation of DNA repair proteins in lytic HSV1 replication; (3) failure of DNA repair protein activation following HSV1 infection of neurons; and (4) reactivation of latent HSV1 from neurons following excess of DNA damage. Studies supporting each of the four conclusions are described briefly below.

5.1. DNA Repair Proteins Are Activated following HSV1 Infection of Nonneuronal Cells. Most studies of repair protein activation have been performed with cells in culture. Cells are infected with HSV1 and assayed thereafter by western blot using antibodies specific for the activated form of repair proteins. For instance, in a representative study, infected HeLa cells were assayed for activation of proteins involved in double strand break repair by the homologous recombination-dependent repair pathway [6]. Activation by phosphorylation was observed for ATM and NBN without any dramatic increase in the overall amount of either protein present. Similar studies have confirmed the activation of NBN [16] and added observations showing activation of three other proteins involved in DNA repair, RPA2 [16], FANCD2, and FANCI [20].

A second line of investigation has also suggested that host-encoded DNA repair proteins are involved in HSV1 lytic replication. Host repair proteins were found to be present in HSV1-induced nuclear replication compartments where virus DNA synthesis takes place [23]. Using immunofluorescence light microscopy it was demonstrated that replication compartments contain host repair proteins including: ATR, ATRIP, ATM, CHEK2, RPA2, RP1, RAD51, NBN, XRCC5, MSH2, MSH6, FANCD2, and FANCI [6, 16, 18–20, 24, 25]. A thorough proteomic analysis has also demonstrated the presence of multiple host DNA repair proteins in HSV1 replication compartments [26]. In this study, repair proteins were considered to be present in replication compartments if they were coisolated in immunoprecipitates of pUL29 (ICP8), an HSV1-encoded protein enriched in replication compartments.

5.2. Activation of DNA Repair Proteins Potentiates Lytic HSV1 Replication. The enhancing role of DNA repair proteins for HSV1 lytic replication is an essential feature of the DNA repair-centered pathway. Entry into latency is expected to be possible only in cells that are minimally permissive for lytic replication due to a deficiency of repair proteins. Two types of studies have been carried out to test the involvement of repair proteins: (1) lytic replication was measured in mutant cells deficient in the repair protein to be tested. Control infections were performed with the same cells after complementation with a gene encoding the functional protein. Virus replication was expected to be observed in the second condition, but not the first if the protein examined enhances lytic HSV1 replication. (2) Lytic replication was examined in cells where expression of the test protein was suppressed by treatment with specific siRNA. Controls in this case were performed with nonspecific siRNA.

An example of the first type of study was performed with a cell line deficient in FANCA, a protein required for DNA repair and involved in the etiology of Fanconi anemia (FA-A cells; [20]). HSV1 replication was tested in FA-A cells and also in FA-A cells after complementation with a gene encoding wild type FANCA protein. The results demonstrated enhanced virus growth only in the complemented cell line providing evidence that FANCA potentiates HSV1 replication. Similar studies involving deletion and complemented cell lines have demonstrated that efficient HSV1 replication requires DNA repair proteins FANCD2, FANCG, MRE11, and ATM [6, 20].

In studies involving siRNA technology, HSV1 replication is measured in control cells and in cells in which expression of a test DNA repair protein is blocked with a specific siRNA. Such studies require a control in which it is demonstrated that the specific siRNA actually depletes test cells of the target repair protein. In a representative study, HFF-1 cells were depleted of RP1 (RPA70), a protein involved in nucleotide excision repair. HSV1 lytic replication was then measured in the depleted cells and in control cells treated with a nonspecific siRNA [19]. Replication was found to be efficient only in control cells indicating that RP1 favors HSV1 growth. Similar studies have demonstrated that efficient HSV1 lytic replication requires proteins involved in mismatch repair and in ATR repair pathway proteins [18, 19].

5.3. DNA Repair Proteins Are Not Activated following HSV1 Infection of Neurons. The failure of neurons to respond to HSV1 infection by activating the DNA damage response is documented in an elegant experiment from the Weitzman lab [6]. The study was carried out with a pluripotent human embryonic stem cell line (Cyth25) that can be induced to differentiate in culture into neurons [27]. HSV1 replication and DNA repair protein (ATM) activation were compared in both the undifferentiated and fully differentiated neuron forms of Cyth25. The results showed that efficient virus replication and repair protein activation occurred only in the undifferentiated form of Cyth25 cells. Neurons were negative in both tests supporting the view that they are well suited to serve as hosts for latent HSV1 infection.

It is relevant to note here that neurons have an additional property that makes them attractive as hosts for latent HSV1. Even in the absence of virus infection, neurons are found to be depleted in overall DNA repair capacity compared to other cell types and also to undifferentiated neuronal precursor cells. Experimental evidences supporting the above conclusions are found in studies involving tests of the ability of neurons to repair damaged virus DNA introduced into the cell [21] and tests of the ability of neurons to repair oxidative damage to their own DNA [22].

5.4. Excess of DNA Damage Leads to Reactivation of Latent HSV1 from Neurons. Quite diverse lines of evidence support the idea that reactivation involves mobilization of DNA repair

functions. One is the observation described above demonstrating that DNA repair functions are necessary for lytic HSV1 replication. Since reactivation results in cycles of lytic HSV1 replication, it is most reasonable to expect that repair functions will also be required for lytic replication resulting from reactivation. A second line of evidence is the clinical observation familiar to physicians who treat HSV1 infections. Reactivated infections are often found to follow exposure of the patient to sunlight [28]. The ultraviolet component of sunlight has the potential to cause DNA damage that could be the initiating factor. Reactivation could also be driven by natural characteristics of the HSV1 genome that are able to launch DNA repair pathways. Such characteristics include the G quadraplex structures resulting from the high G:C content (68%) of HSV1 DNA [29] and features such as inverted and tandem repeats able to promote recombination-dependent repair [30].

Finally, there are relevant cell culture studies that suggest involvement of repair functions in reactivation. One such study was carried out with a mouse neuroblastoma cell line (C1300; [31]) that replicated HSV1 poorly as expected of a neuronal cell. Virus replication was found to be improved significantly, however, when the cells were treated with agents (such as etoposide and hexamethylene bisacetamide) that cause DNA damage.

6. Other Features of Latency Consistent with the DNA Repair-Centered Pathway

It is an attractive feature of the repair-centered pathway that important aspects are compatible with previously known features of HSV1 latency and reactivation. An example is the specificity for latency in neurons. While activation of repair functions is observed in most cell types able to host HSV1 infections, this is not the case with neurons. Here lytic replication is blocked by the absence of activated repair proteins creating an environment conducive to latency. Neurons of the peripheral compared to the central nervous system are especially well suited for a role in latency. The blood brain barrier is found to be more permeable in the PNS [32], a property that favors the availability of small molecules able to promote virus reactivation. Neurons of the PNS are therefore well suited to provide a way for latent HSV1 to escape its confinement in the latent state.

The repair-centered pathway is also compatible with the observation that a high rate of genetic recombination is characteristic of HSV1 genomes during lytic replication. Active recombination has been identified at most sites in the genome with particularly high activity at the junctions of L and S segments [33–36]. The requirement of repair functions for HSV1 lytic replication provides a potential source for the required recombination events. Homologous recombination, for instance, is an integral feature of important repair pathways including the synthesis-dependent strand-annealing pathway of double strand break repair [37]. The high rate of HSV1 genome recombination can therefore be regarded as a consequence of the requirement for DNA repair pathways.

Finally, the genetic identity of infecting and reactivated HSV1 strains also supports the repair-centered pathway. The

DNA sequence identity of initial and reactivated virus is often overlooked because it is obvious that such identity must be the rule. Without it there would be no genetic continuity in the HSV1 species. Viewed more closely, however, it is clear that genetic identity is quite a remarkable fact [38]. During the period of latency the HSV1 genome is expected to be subject to the same variety of toxic affects found for all cells. These include, for example, ionizing radiation, environmental mutagenic chemicals, and mutations occurring during DNA replication. If not repaired, these would introduce instability into the virus genome as they do for the cell. The situation is more severe in the case of latent neurons because repair functions are not activated.

The proposed repair-centered pathway suggests a solution by postulating that reactivation is caused by mobilization of DNA repair capacity. A consequence of the model is that repair of DNA damaged during latency is expected to occur as a part of the reactivation process.

7. Future Directions

As with most scientific hypotheses, the pathway discussed here for HSV1 latency and reactivation would benefit from future experimental testing. The identity of the repair proteins required for lytic growth is an example. It is clear that not all repair proteins are required [39], but can a minimum subset be defined? The proposed mechanism of reactivation in particular is in need of further evaluation. We need to know more about which repair functions are mobilized as the virus is reactivated. Are they the same as those activated as the virus enters into latency or are there differences? Further knowledge about factors that stimulate reactivation would also be most welcome. The roles of stress and mutagenic effects are now appreciated, but it would be of interest to know more about the specific biochemical signaling agents involved.

Competing Interests

The author declares that there is no conflict of interests regarding the publication of this paper.

References

[1] D. C. Bloom, "Alphaherpesvirus latency: a dynamic state of transcription and reactivation," *Advances in Virus Research*, vol. 94, pp. 53–80, 2016.

[2] J. Sinclair, "Human cytomegalovirus: latency and reactivation in the myeloid lineage," *Journal of Clinical Virology*, vol. 41, no. 3, pp. 180–185, 2008.

[3] D. A. Thorley-Lawson, J. B. Hawkins, S. I. Tracy, and M. Shapiro, "The pathogenesis of Epstein-Barr virus persistent infection," *Current Opinion in Virology*, vol. 3, no. 3, pp. 227–232, 2013.

[4] M. P. Nicoll, J. T. Proença, and S. Efstathiou, "The molecular basis of herpes simplex virus latency," *FEMS Microbiology Reviews*, vol. 36, no. 3, pp. 684–705, 2012.

[5] B. Roizman, D. M. Knipe, and R. J. Whitley, "Herpes simplex viruses," in *Fields Virology*, pp. 2803–2819, Lippincott Williams & Wilkins, Philadelphia, Pa, USA, 2013.

[6] C. E. Lilley, C. T. Carson, A. R. Muotri, F. H. Gage, and M. D. Weitzman, "DNA repair proteins affect the lifecycle of herpes simplex virus 1," *Proceedings of the National Academy of Sciences of the United States of America*, vol. 102, no. 16, pp. 5844–5849, 2005.

[7] M. D. Weitzman and J. B. Weitzman, "What's the damage? the impact of pathogens on pathways that maintain host genome integrity," *Cell Host and Microbe*, vol. 15, no. 3, pp. 283–294, 2014.

[8] S. Smith and S. K. Weller, "HSV-I and the cellular DNA damage response," *Future Virology*, vol. 10, no. 4, pp. 383–397, 2015.

[9] J. C. Brown and W. W. Newcomb, "Herpesvirus capsid assembly: insights from structural analysis," *Current Opinion in Virology*, vol. 1, no. 2, pp. 142–149, 2011.

[10] J. C. Brown, M. A. McVoy, and F. L. Homa, "Packaging DNA into herpesvirus capsids," in *Structure-Function Relationships of Human Pathogenic Viruses*, A. Holzenburg and E. Bogner, Eds., Kluwer Academic/Plenum Publishers, London, UK, 2002.

[11] R. J. Diefenbach, M. Miranda-Saksena, M. W. Douglas, and A. L. Cunningham, "Transport and egress of herpes simplex virus in neurons," *Reviews in Medical Virology*, vol. 18, no. 1, pp. 35–51, 2008.

[12] L. A. Waggoner-Fountain and L. B. Grossman, "Herpes Simplex Virus," *Pediatrics in Review*, vol. 25, no. 3, pp. 86–93, 2004.

[13] S. Efstathiou and C. M. Preston, "Towards an understanding of the molecular basis of herpes simplex virus latency," *Virus Research*, vol. 111, no. 2, pp. 108–119, 2005.

[14] J. D. Kriesel, "Reactivation of herpes simplex virus: the role of cytokines and intracellular factors," *Current Opinion in Infectious Diseases*, vol. 12, no. 3, pp. 235–238, 1999.

[15] K. Held and T. Derfuss, "Control of HSV-1 latency in human trigeminal ganglia—current overview," *Journal of Neurovirology*, vol. 17, no. 6, pp. 518–527, 2011.

[16] D. E. Wilkinson and S. K. Weller, "Recruitment of cellular recombination and repair proteins to sites of herpes simplex virus type 1 DNA replication is dependent on the composition of viral proteins within prereplicative sites and correlates with the induction of the DNA damage response," *Journal of Virology*, vol. 78, no. 9, pp. 4783–4796, 2004.

[17] A. S. Turnell and R. J. Grand, "DNA viruses and the cellular DNA-damage response," *Journal of General Virology*, vol. 93, no. 10, pp. 2076–2097, 2012.

[18] K. N. Mohni, A. S. Mastrocola, P. Bai, S. K. Weller, and C. D. Heinen, "DNA mismatch repair proteins are required for efficient herpes simplex virus 1 replication," *Journal of Virology*, vol. 85, no. 23, pp. 12241–12253, 2011.

[19] K. N. Mohni, A. R. Dee, S. Smith, A. J. Schumacher, and S. K. Weller, "Efficient herpes simplex virus 1 replication requires cellular ATR pathway proteins," *Journal of Virology*, vol. 87, no. 1, pp. 531–542, 2013.

[20] H. Karttunen, J. N. Savas, C. McKinney et al., "Co-opting the Fanconi anemia genomic stability pathway enables herpesvirus DNA synthesis and productive growth," *Molecular Cell*, vol. 55, no. 1, pp. 111–122, 2014.

[21] S. Millhouse, X. Wang, N. W. Fraser, L. Faber, and T. M. Block, "Direct evidence that HSV DNA damaged by ultraviolet (UV) irradiation can be repaired in a cell type-dependent manner," *Journal of NeuroVirology*, vol. 18, no. 3, pp. 231–243, 2012.

[22] P. Sykora, J.-L. Yang, L. K. Ferrarelli et al., "Modulation of DNA base excision repair during neuronal differentiation," *Neurobiology of Aging*, vol. 34, no. 7, pp. 1717–1727, 2013.

[23] A. De Bruyn Kops and D. M. Knipe, "Preexisting nuclear architecture defines the intranuclear location of herpesvirus DNA replication structures," *Journal of Virology*, vol. 68, no. 6, pp. 3512–3526, 1994.

[24] N. Shirata, A. Kudoh, T. Daikoku et al., "Activation of ataxia telangiectasia-mutated DNA damage checkpoint signal transduction elicited by herpes simplex virus infection," *Journal of Biological Chemistry*, vol. 280, no. 34, pp. 30336–30341, 2005.

[25] K. N. Mohni, C. M. Livingston, D. Cortez, and S. K. Weller, "ATR and ATRIP are recruited to herpes simplex virus type 1 replication compartments even though ATR signaling is disabled," *Journal of Virology*, vol. 84, no. 23, pp. 12152–12164, 2010.

[26] T. J. Taylor and D. M. Knipe, "Proteomics of herpes simplex virus replication compartments: association of cellular DNA replication, repair, recombination, and chromatin remodeling proteins with ICP8," *Journal of Virology*, vol. 78, no. 11, pp. 5856–5866, 2004.

[27] H. Kawasaki, H. Suemori, K. Mizuseki et al., "Generation of dopaminergic neurons and pigmented epithelia from primate ES cells by stromal cell-derived inducing activity," *Proceedings of the National Academy of Sciences of the United States of America*, vol. 99, no. 3, pp. 1580–1585, 2002.

[28] M. Ichihashi, H. Nagai, and K. Matsunaga, "Sunlight is an important causative factor of recurrent herpes simplex," *Cutis*, vol. 74, no. 5, pp. 14–18, 2004.

[29] M. L. Bochman, K. Paeschke, and V. A. Zakian, "DNA secondary structures: stability and function of G-quadruplex structures," *Nature Reviews Genetics*, vol. 13, no. 11, pp. 770–780, 2012.

[30] J. C. Brown, "The role of DNA repair in herpesvirus pathogenesis," *Genomics*, vol. 104, no. 4, pp. 287–294, 2014.

[31] K. Volcy and N. W. Fraser, "DNA damage promotes herpes simplex virus-1 protein expression in a neuroblastoma cell line," *Journal of NeuroVirology*, vol. 19, no. 1, pp. 57–64, 2013.

[32] E. W. Englander, "DNA damage response in peripheral nervous system: coping with cancer therapy-induced DNA lesions," *DNA Repair*, vol. 12, no. 8, pp. 685–690, 2013.

[33] G. S. Hayward, R. J. Jacob, S. C. Wadsworth, and B. Roizman, "Anatomy of herpes simplex virus DNA: evidence for four populations of molecules that differ in the relative orientations of their long and short components," *Proceedings of the National Academy of Sciences of the United States of America*, vol. 72, no. 11, pp. 4243–4247, 1975.

[34] H. Delius and J. B. Clements, "A partial denaturation map of Herpes simplex virus type 1 DNA: evidence for inversions of the unique DNA regions," *Journal of General Virology*, vol. 33, no. 1, pp. 125–133, 1976.

[35] R. E. Dutch, V. Bianchi, and I. R. Lehman, "Herpes simplex virus type 1 DNA replication is specifically required for high-frequency homologous recombination between repeated sequences," *Journal of Virology*, vol. 69, no. 5, pp. 3084–3089, 1995.

[36] X. Fu, H. Wang, and X. Zhang, "High-frequency intermolecular homologous recombination during herpes simplex virus-mediated plasmid DNA replication," *Journal of Virology*, vol. 76, no. 12, pp. 5866–5874, 2002.

[37] J.-M. Chen, D. N. Cooper, N. Chuzhanova, C. Férec, and G. P. Patrinos, "Gene conversion: mechanisms, evolution and human disease," *Nature Reviews Genetics*, vol. 8, no. 10, pp. 762–775, 2007.

[38] L. Haarr, A. Nilsen, P. M. Knappskog, and N. Langeland, "Stability of glycoprotein gene sequences of herpes simplex virus type 2 from primary to recurrent human infection, and diversity of the sequences among patients attending an STD clinic," *BMC Infectious Diseases*, vol. 14, no. 1, article 63, 2014.

21

Immunogenicity of RSV F DNA Vaccine in BALB/c Mice

Erdal Eroglu,[1,2] **Ankur Singh,**[3] **Swapnil Bawage,**[1] **Pooja M. Tiwari,**[4] **Komal Vig,**[1] **Shreekumar R. Pillai,**[1] **Vida A. Dennis,**[1] **and Shree R. Singh**[1]

[1]Center for NanoBiotechnology Research, Alabama State University, Montgomery, AL, USA
[2]Faculty of Engineering, Bioengineering Department, Celal Bayar University, Muradiye, Manisa, Turkey
[3]College of Medicine, University of South Alabama, Mobile, AL, USA
[4]Yerkes National Primate Research Center, Emory University, Atlanta, GA, USA

Correspondence should be addressed to Shree R. Singh; ssingh@alasu.edu

Academic Editor: Itabajara da Silva Vaz Jr.

Respiratory syncytial virus (RSV) causes severe acute lower respiratory tract disease leading to numerous hospitalizations and deaths among the infant and elderly populations worldwide. There is no vaccine or a less effective drug available against RSV infections. Natural RSV infection stimulates the Th1 immune response and activates the production of neutralizing antibodies, while earlier vaccine trials that used UV-inactivated RSV exacerbated the disease due to the activation of the allergic Th2 response. With a focus on Th1 immunity, we developed a DNA vaccine containing the native RSV fusion (RSV F) protein and studied its immune response in BALB/c mice. High levels of RSV specific antibodies were induced during subsequent immunizations. The serum antibodies were able to neutralize RSV in vitro. The RSV inhibition by sera was also shown by immunofluorescence analyses. Antibody response of the RSV F DNA vaccine showed a strong Th1 response. Also, sera from RSV F immunized and RSV infected mice reduced the RSV infection by 50% and 80%, respectively. Our data evidently showed that the RSV F DNA vaccine activated the Th1 biased immune response and led to the production of neutralizing antibodies, which is the desired immune response required for protection from RSV infections.

1. Introduction

Respiratory syncytial virus (RSV), a member of genus *Pneumovirus* and classified in the family Paramyxoviridae, is the most common cause of severe disease of the lower respiratory tract in infants and the elderly especially in developing countries [1, 2]. There are also some reports claiming that RSV could lead to severe repeated infections such as recurrent wheezing, pneumonia, or asthma in later childhood [3]. Worldwide, the number of RSV-associated cases is estimated to be 33 million and the number of deaths up to 234,000 in children younger than 5 years old in spite of the fact that those numbers are lower in the USA due to the precautions against RSV [4, 5]. Besides the young children, the hospitalization rate of elderly people above 50 years old may be the same as influenza cases [2]. RSV vaccine development efforts such as inactivated RSV, live-attenuated RSV, or subunit vaccines are underway.

However, despite over five decades of intensive research on developing a RSV vaccine, there is no approved vaccine or drug available [6]. Instead of vaccine, some researchers have been attempting to develop prophylactic antibody therapies targeting RSV F protein [7, 8]. Antiviral drugs such as ribavirin (a nucleoside analog), which targets hepatitis C and other viruses including RSV, ALS-8176 (a new nucleoside analog), and GS5806 (pyrazolo[1,5-a]pyrimidine based RSV fusion inhibitor), and neutralizing monoclonal antibodies such as Palivizumab (Synagis™) and Motavizumab (Numax), are administered to infants at high risk of developing respiratory diseases [9–12]. As an alternative to expensive therapies, a vaccine conferring long lasting immunity is a less expensive and more efficient option against recurrent RSV infections [10]. Due to frequent antigenic variations of RNA viruses (RSV, influenza virus, and rhinovirus), developing a vaccine with complete protection is challenging. The incomplete immunity in response to natural RSV infections is responsible for repeated infections. RSV vaccine studies in the 1960s using formalin inactivated RSV (FI-RSV) consisting of the whole virus exacerbated the disease and even in some cases resulted in deaths because of the elevated T helper type-2 (Th2) mediated immune response [1, 13]. In addition, using a vector expressing RSV antigens is found far safer than subunit or inactivated RSV immunization [14]. With these important immunological responses, a safe and stable vaccine with long lasting immunity is an urgent need for the public.

The outer surface glycoproteins, fusion (F) and attachment (G), of RSV are known antigenic proteins that induce the humoral and cellular immune responses and are targets of antigen presenting cells [15]. The RSV F protein mediates the fusion of the virus particle into the host by merging the virion envelope with the host cell membrane following virion attachment using the G protein. In addition, the F protein facilitates fusion of neighboring normal cells with infected cells, thus creating multinuclear giant cells called syncytia, which characterizes RSV infection [9, 16]. The RSV F protein is highly conserved among the different RSV strains compared to other RSV proteins [16]. On the other hand, the variability of the G amino acid sequence among various RSV strains is high [17]. Furthermore, previous reports demonstrated that RSV F vaccines provide protection against both RSV A and RSV B strains by producing neutralizing antibodies [8, 14, 18, 19], whereas RSV G vaccines prominently induced a Th2 biased immune response, thereby enhancing the severity of the disease in subsequent RSV infections [20, 21].

The helper T lymphocytes activate either B cells, which produce specific antibodies, or cytotoxic T lymphocytes, which are responsible for the clearance of RSV infected cells. Cell-mediated protective immunity is important in the clearance of infected cells. However, cell-mediated immunity on its own is not sufficient to provide complete protection against pathogens. Therefore, it is necessary to have memory B cells activating neutralizing antibodies upon reinfection. Although DNA vaccines are not highly immunogenic as compared to whole pathogen or protein vaccines, DNA vaccines have the advantage of expressing the native form of the antigen produced in vivo and inducing strong T and B cells responses. The changes in the epitope regions of the antigen may shift the immune response leading to unwanted allergic immune reactions as seen in the FI-RSV vaccine trials [6, 22, 23]. A highly immunogenic RSV F protein with conserved sequence would be a desirable DNA vaccine candidate for protection from repeated RSV infections. Our group has previously developed a DNA vaccine containing immunogenic regions of RSV F protein (residues 412–524) and showed that the DNA vaccine provides partial protection in BALB/c mice when combined with cholera toxin (CTA_2B) adjuvant [24].

In this study, we developed a full-length RSV F DNA vaccine that was able to induce predominantly a Th1 type response without using any adjuvant. The antibody response in serum was significantly enhanced with subsequent immunizations. The sera from immunized animals were able to neutralize RSV in vitro. The protection afforded by the DNA vaccine was not complete and thus necessitates design and development of other methods of vaccines. A combinatorial concept that can take advantages of various vaccines such as a prime vaccine (DNA vaccine) followed with booster vaccines (subunit or recombinant protein vaccines) may lead to complete protection from RSV.

2. Materials and Methods

2.1. Materials. Restriction enzymes (RE) NotI and BamHI, T4 DNA ligase, Eagle's minimal essential medium (MEM), Hank's balanced salt solution (HBSS), fetal bovine serum (FBS), L-glutamine (100 mM), antibiotics, TrypLE™, 7-aminoactinomycin D (7-AAD), Nucleofector™ electroporation kit for Cos-7 cells, TaqMan master mix 2x, real time probe, primers, superscript II reverse transcriptase, and RNase later solution were all obtained from Life Technologies™ (Carlsbad, CA, USA). All DNA and RNA isolation kits were purchased from QIAGEN™ (Valencia, CA, USA). MEM was supplemented with 10% FBS (MEM-10), penicillin (45 µg/mL), streptomycin (100 µg/mL), kanamycin (75 µg/mL), and L-glutamine (1 mM). Human epithelial type 2 (HEp-2) and monkey kidney (Vero and Cos-7) cells were obtained from American Type Culture Collection (ATC-CTM, Manassas, VA, USA).

2.2. Animals and RSV Stock Preparation. BALB/c female mice (4–6 weeks old) were purchased from Charles River Laboratories (Wilmington, MA). The animals were housed under standard approved conditions with a cycle of 12 h of light and 12 h of darkness and provided daily with sterile food and water *ad libitum*. For all immunization studies, an approved protocol by the Alabama State University Institutional Animal Care and Use Committee was followed. Human RSV long strain was purchased from the American Type Culture Collection (ATCC, Manassas, VA, ATCC # VR-26) and propagated in HEp-2 cells (ATCC # CCL-23). HEp-2 cells were grown in tissue culture flasks in MEM supplemented with 10% FBS and antibiotics. Human RSV was added to the cell monolayer, and virus adsorption was carried out for 1 h at 37°C in a humidified atmosphere with 5% CO_2. MEM with 2% FBS was added to the flask and infection of cells was observed for an additional 3-4 days. RSV infected cells were centrifuged at 3,000 ×g at 4°C to remove cellular debris, aliquoted, and stored at −80°C until they were used.

2.3. Construction of Recombinant RSV F DNA Vaccine. The RSV F DNA sequence originally published by Collins et al. [25] was full-length RSV F gene synthesized by Epoch labs (Missouri City, TX, USA) and amplified by polymerase chain reaction (PCR) using forward and reverse primers shown in Table 1. Both the RSV F gene and the phCMV1 DNA vector were digested with BamHI and NotI RE enzymes. The purified DNA pieces (using QIAGEN gel extraction kit) were ligated using the T4 DNA ligase enzyme and transformed into competent cells of *Escherichia coli* DH5α. For selection, competent cells were grown on kanamycin supplemented Luria Bertani (LB) agar. Clones containing the RSV F gene in the phCMV1 vector were named PF (Figure 1(a)).

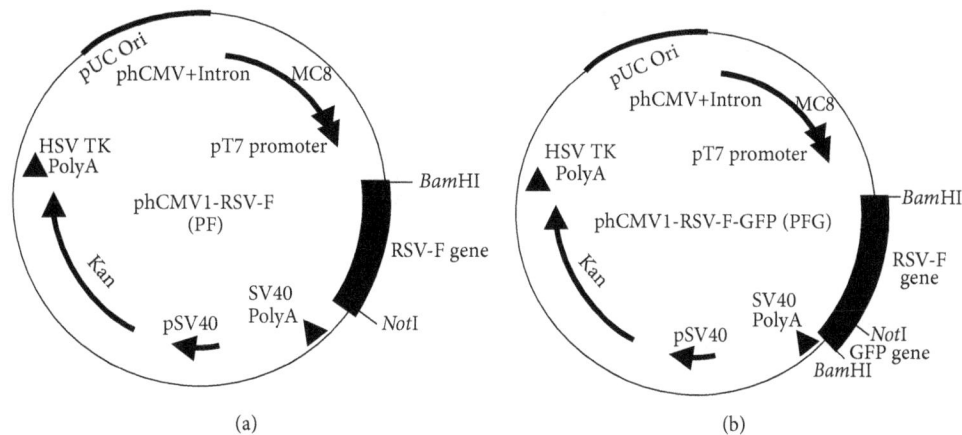

FIGURE 1: Construction of RSV F gene into phCMV1 vector. (a) RSV F gene sequence was amplified with PCR reaction and cloned into *Bam*HI and *Not*I RE sites on phCMV1 vector. (b) GFP tag was amplified with PCR reaction and inserted into *Not*I RE sites at the 3′ end of the RSV F gene.

TABLE 1: The sequences of PCR/qPCR primers and probes.

Names of the primers	Sequences of the primers
RSV F forward primer	GGATCCACCATGATGGTCCTCAAAGCAAATGCAATTACCAC
RSV F reverse primer	CCACCGCGGCCGCTTATCATTGTCGACCAATATTATTTATACCACTC
GFP forward primer	GGATCCACCATGGTGAGCAAGGGCGAGGAGCTGTTCACCGG
GFP reverse primer	CCACGCGGCCGCTCATTACTTGTACAGCTCGTCCATGCCGTGAGTGATCC
RSV F QPCR forward primer	AACAGATGTAAGCAGCTCCGTTATC*
RSV F QPCR reverse primer	CGATTTTTATTGGATGCTGTACATTT*
RSV F QPCR probe	TGCCATAGCATGACACAATGGCTCCT*

*According to the sequences published by Mentel et al. [27]. All sequences are given 5′-3′ direction.

2.4. Construction of phCMV1 Vector Containing GFP Gene. The GFP gene was amplified by PCR using previously published GFP plasmid vector [26] as a template, forward and reverse primers (Table 1) with *Bam*HI and *Not*I restriction sites, respectively. The amplified GFP gene was inserted into the phCMV1 vector following the same protocol and conditions as described above. Clones containing the GFP gene in the phCMV1 vector were named PG. The third vector, containing RSV F and GFP, was also cloned and named PFG (Figure 1(b)). All vectors were purified using the QIAGEN Endofree Giga kit and the purified DNA vector aliquots (100 μg/100 μL) were stored at −80°C until used.

2.5. In Vitro Transfection and Expression of RSV F Protein. Nucleofector™ (Lonza, Germany) electroporation protocol was used for *in vitro* gene transfection following the manufacturer's instructions in Cos-7 cells using the Amaxa™ Nucleofector II electroporation device (Lonza, Germany). The GFP labeled RSV F gene construct was used for immunofluorescence imaging and flow cytometry, whereas the RSV F gene construct was used in RT-PCR analysis. Transfected cells with RSV F DNA were incubated for 3 days at 37°C to allow for protein expression *in vitro*. After the incubation time, images showing protein expression with green color were taken using an immunofluorescence microscope. Also, the cells were used for flow cytometry to detect the green fluorescence of GFP

labeled RSV F protein. Transfection and expression protocol was followed as described above for RSV F DNA to detect the RSV F mRNA by RT-PCR.

2.6. Quantitative PCR (qPCR) Analysis for Detection of RSV F Gene. In order to analyze the transfection efficiency, Cos-7 cells were transfected with the PF construct using the Nucleofector electroporation gene transfection following manufacturer's instructions. Total RNA was isolated from harvested cells and 1 μg of RNA was converted into cDNA using the superscript reverse transcriptase enzyme following manufacturer's protocols. RSV F mRNA specific primers, probe, and experimental protocol for qPCR were adapted from Mentel et al. [27]. The qPCR reaction was carried out with reverse and forward primers (Table 1) using Applied Biosystems ViiA 7 real time PCR (Applied Biosystems International, Foster City, CA, USA). Each qPCR reaction was run in duplicate along with water as a negative control. Dilutions of the RSV F gene amplicon (10^0–10^8 copy numbers) were used to prepare the standard curve. Each experiment was repeated twice from the transfection step in duplicate.

2.7. Immunization of BALB/c Mice and Determination of Antibody Response. Animal studies were performed according to the National Institute of Health (NIH, Bethesda, MD) guidelines following a protocol approved by the Alabama State

University Institutional Animal Care and Use Committee. Animals were housed under standard approved conditions and provided daily with sterile food and water *ad libitum*. Six-to-eight-week-old female BALB/c mice (Charles River Laboratories Inc., Wilmington, Massachusetts) were immunized intramuscularly (i.m.) with PBS (300 μL) and PF DNA (50 μg/300 μL in PBS) to each thigh muscle on days 1, 15, and 29. The RSV control group was immunized intranasally with 2×10^5 plaque forming units (pfu) of live RSV long strain (200 μL) twice on day 1 and day 2 by slow application to the nasal nares. Serum was collected via retro-orbital bleeding from all groups of mice on days 0, 14, 28, and 49 to determine the RSV specific antibody response. Saliva was collected by injecting carbachol (0.25 μg/mouse) intraperitoneally on the same days as sera collections. Serum and saliva samples were stored at $-80°$C until analysis.

Sera and saliva samples collected from the mice were analyzed for antibody response and isotyping. To analyze the anti-RSV F-specific antibody response, ELISA plates were coated with UV-inactivated RSV (10^4 pfu/well) in 100 μL of carbonate buffer (pH 9.2) and incubated overnight at 4°C in a humidified atmosphere. Plates were blocked with 3% milk powder and then incubated with 100 μL of samples at room temperature for 1 hour. Goat anti-mouse HRP-conjugated secondary antibody (100 μL of 1 : 2,000 dilution) specific to isotypes IgA, IgG1, IgG2a, and IgG2b was added to the ELISA plates and incubated at room temperature for 1 hour. The ELISA plates were washed and the enzymatic reaction was developed and absorbance was read at 450 nm using a Tecan ELISA reader (Tecan, Research Triangle Park, NC, USA).

2.8. Viral Neutralization Assay. The viral neutralization assays for the mice sera samples were performed according to the protocols of Singh et al. with slight modifications [18]. Briefly, heat inactivated sera (56°C for 30 min; 25 μL and 12.5 μL per well) from all groups of mice (PBS, RSV, and RSV F DNA) were mixed with 1×10^3 pfu of RSV and incubated at 4°C for 2 hours. Approximately, 1.5×10^3 HEp-2 cells were mixed with sera+RSV mixture in a 96-well plate followed by incubation at 37°C in a CO_2 incubator for 3 days. HEp-2 cells alone and HEp-2 cells infected with RSV (1×10^3 pfu) were used as negative and positive controls, respectively. Cells were washed with 1x PBS (pH 7.0) before fixing the cells by incubating in 80% acetone (v/v) at 4°C for 15 minutes. For the ELISA assay, the plate was blocked with 3% milk and incubated with the primary antibody, goat anti-RSV (1:500 dilution), and then the secondary antibody, rabbit anti-goat IgG-HRP (1:2000 dilution) at room temperature for 1 hour. The plate was washed and the enzymatic reaction was developed with the TMB substrate (KPL, Gaithersburg, Maryland, USA) followed by reading the absorbance at 450 nm in the ELISA reader (TECAN, US Inc., Durham, NC, USA). The same protocol was followed for the immunofluorescence microscopy analysis except for the secondary antibody, rabbit anti-goat IgG-FITC (1:2000 dilution). The cell nuclei were stained with antifade-DAPI, and merged images were taken using the FITC and DAPI channel in the immunofluorescence microscope (Nikon Eclipse Ti, Nikon Instruments Inc., Melville, USA).

2.9. Statistical Analysis. qPCR, ELISA, and virus neutralization assay data are presented as means and standard deviations; statistical analysis of the data was performed using Sigma plot version 11.0 (Systat Software, Inc., Germany). Differences between the means of the four experimental groups were determined using one-way analysis of variance (ANOVA) Tukey's test with the significance level of 1%.

3. Results

3.1. Expression of DNA Vaccine In Vitro. The RSV F gene was cloned into the phCMV1 vector between the *Bam*HI and *Not*I restriction enzymes sites. Positive clones were verified by RE digestion (*Bam*HI and *Not*I) and DNA sequencing. Recombinant clones containing the RSV F gene in the phCMV1 vector were named PF. To test the expression efficiency of the RSV F protein expression *in vitro*, the RSV F gene containing the GFP gene was similarly cloned into the phCMV1 vector generating the PFG clone. The PF and PFG clones were used to transfect Cos-7 cells. Three days after transfection, green fluorescence signals were analyzed (Figures 2(a)–2(c)). The PG clone showed a high level of GFP protein expression (Figure 2(b)). The PFG vector expressed RSV F and GFP proteins (Figure 2(c)) although at a much lower level compared to the PG clones. Transfected cells were trypsinized and protein expression was detected in flow cytometry (Figures 2(d)–2(f)). The transfection efficiency of the PG clone (phCMV1-GFP) was over 90% (Figure 2(e)) while the efficiency of the PFG clone (RSV F-GFP) was over 16% (Figure 2(f)) compared to the negative (phCMV1) clone.

In addition to immunofluorescence and flow cytometry analyses, the transcription efficiencies of the PF and PFG clones were quantified by qPCR analysis in Cos-7 cells using RSV F specific primers (Figure 3). The RSV F mRNA copy numbers for both clones (PF and PFG) were significantly higher (4×10^7) compared to the RSV F mRNA copy number for the negative control (mock transfected cells, $<10^1$). Thus, we confirmed, using three different methods, that PF clones were expressing RSV F protein *in vitro*.

3.2. Analysis of RSV Specific Antibody Response. The humoral immune response induced by immunizing mice with PBS, the RSV F DNA vaccine, or RSV was determined by measuring RSV specific serum and saliva antibody titers using ELISA. Serum and saliva samples were collected from BALB/c mice at 2-week intervals following each immunization. Animals vaccinated with RSV F DNA and RSV showed significantly higher ($P < 0.01$) amount of serum IgG levels compared to the PBS negative control group (Figure 4(a)). Saliva samples from same groups showed no significant RSV specific IgG antibody response except for saliva samples from RSV vaccinated mice collected on day 49 (data not shown). RSV specific IgM antibody was detected only in serum samples (not in saliva) from RSV F immunized mice during all immunization periods (Figure 4(b)). IgM, a basic immunoglobulin produced in B cells, is the first antibody produced in response to an initial exposure to an antigen [28].

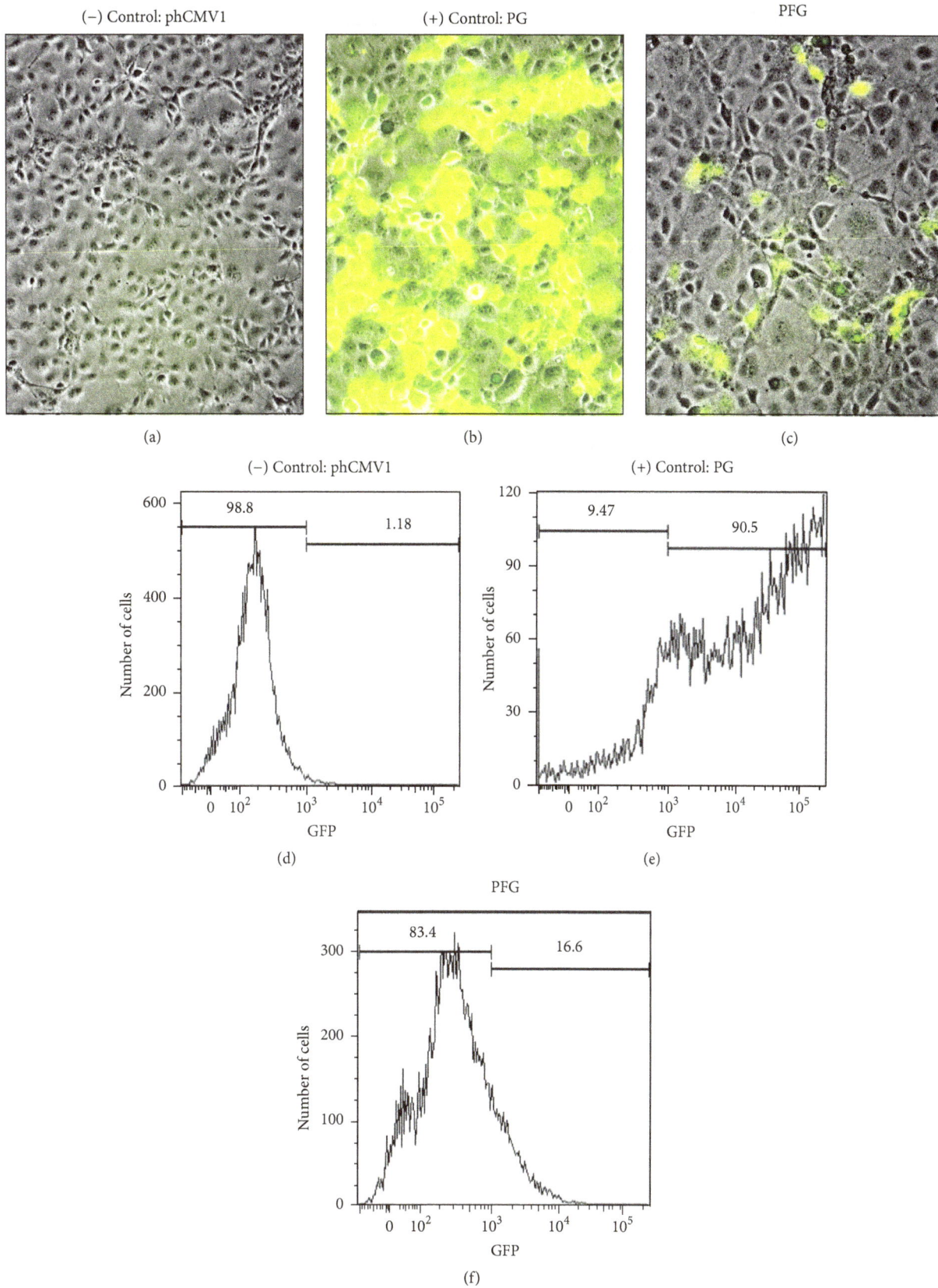

FIGURE 2: Expression of RSV F protein in Cos-7 cells. Visual and quantitative analyses demonstrated that RSV F protein was expressed *in vitro*. Immunofluorescence microscopy of (a) phCMV1 (negative control), (b) phCMV1+GFP (positive control), and (c) PFG (DNA vaccine) transfected cells. Flow cytometric analysis of transfected cells: (d) Cos-7 cells (negative control), (e) phCMV1+GFP (positive control), and (f) PFG (DNA vaccine).

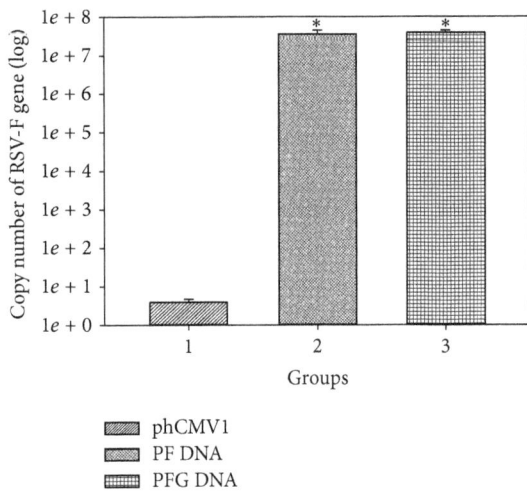

FIGURE 3: qPCR data showing the transcription of RSV F gene in PF/PFG transfected Cos-7 cells. Transcription of RSV F mRNA was >10^7-fold higher than mock transfected cells (negative control, <10^1). *Significantly different ($P < 0.01$).

TABLE 2: Th2/Th1 (IgG1/IgG2a < 1 and IgG2b/IgG2a < 1) antibody ratios. IgG isotypes were detected by ELISA from serum samples of BALB/c mice (RSV-infected and PF DNA-immunized) on day 49. Data is presented as an average of triplicates performed twice.

| | IgG1/IgG2a | | IgG2b/IgG2a | |
	RSV	PF DNA	RSV	PF DNA
Day 14	0.168	0.679	0.911	0.850
Day 28	0.553	0.443	0.827	0.666
Day 49	0.392	0.236	0.767	0.819

3.3. Isotyping of RSV Specific IgG Antibody. Since the Th1 immune response is important in providing protective immunity against RSV infection, we also analyzed and compared the Th1 (IgG2a) and Th2 (IgG1, IgG2b) specific immune responses. Antibody isotyping of serum samples showed significant levels of IgG1 (Figure 5(a)), IgG2b (Figure 5(b)), and IgG2a (Figure 5(c)) antibodies after day 14 of immunization and levels continued to increase on day 28 and day 49 in RSV infected mice. In the RSV F DNA immunized mice, IgG2b and IgG2a production was stimulated only at day 49 while no IgG1 production was detected in all serum samples. The IgG2a isotype antibody response, specific for the Th1 mediated response, was significantly higher than the IgG1 and IgG2b at all immunizations. All IgG1/IgG2a and IgG1/IgG2b ratios were calculated (Table 2) and all of the ratios were constantly lower than 1 clearly demonstrating a Th1-biased response following either RSV infection or RSV F vaccination. A Th1 (IgG1/IgG2a < 1 and IgG2b/IgG2a < 1) response was prominent in both RSV infected and RSV F DNA immunized mice at day 49, the time when the antibody level was highest.

3.4. RSV F DNA Vaccine Stimulates RSV Specific Neutralizing Antibodies. We also tested the efficacy of serum antibodies in neutralizing RSV infection *in vitro* using ELISA. ELISA

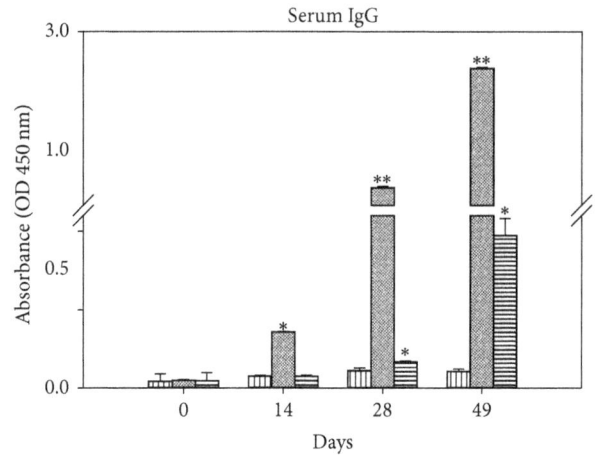

(a)

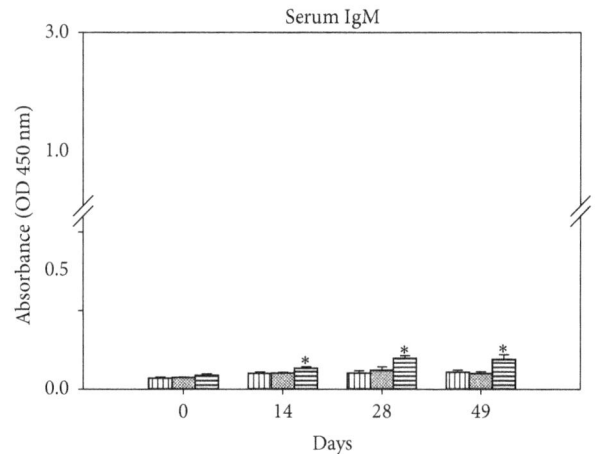

(b)

FIGURE 4: (a) IgG antibody response and (b) IgM antibody response against RSV specific antigens. Serum samples (PBS, RSV infected, and PF DNA-immunized mice) were collected from BALB/c mice on days 0, 14, 28, and 49 and IgG antibody responses were detected by ELISA. Data is presented as an average of triplicates performed twice; error bars represent standard deviations. *Significantly different ($P < 0.01$) from PBS group; **significantly different ($P < 0.01$) from PBS and PF DNA groups. P values ($P < 0.01$) were calculated using ANOVA, Tukey's test.

data indicated that RSV specific neutralizing antibodies from RSV F DNA immunized mice serum reduced the infectivity of RSV by 46% and 30% in 1:8 serum dilution and 1:16 serum dilution, respectively (Figure 6). Consistently, serum from RSV infected mice showed higher RSV reduction with 82% and 76% in 1:8 serum dilution and 1:16 serum dilution, respectively. The data of the neutralization assay was in accordance with the antibody response data. Also,

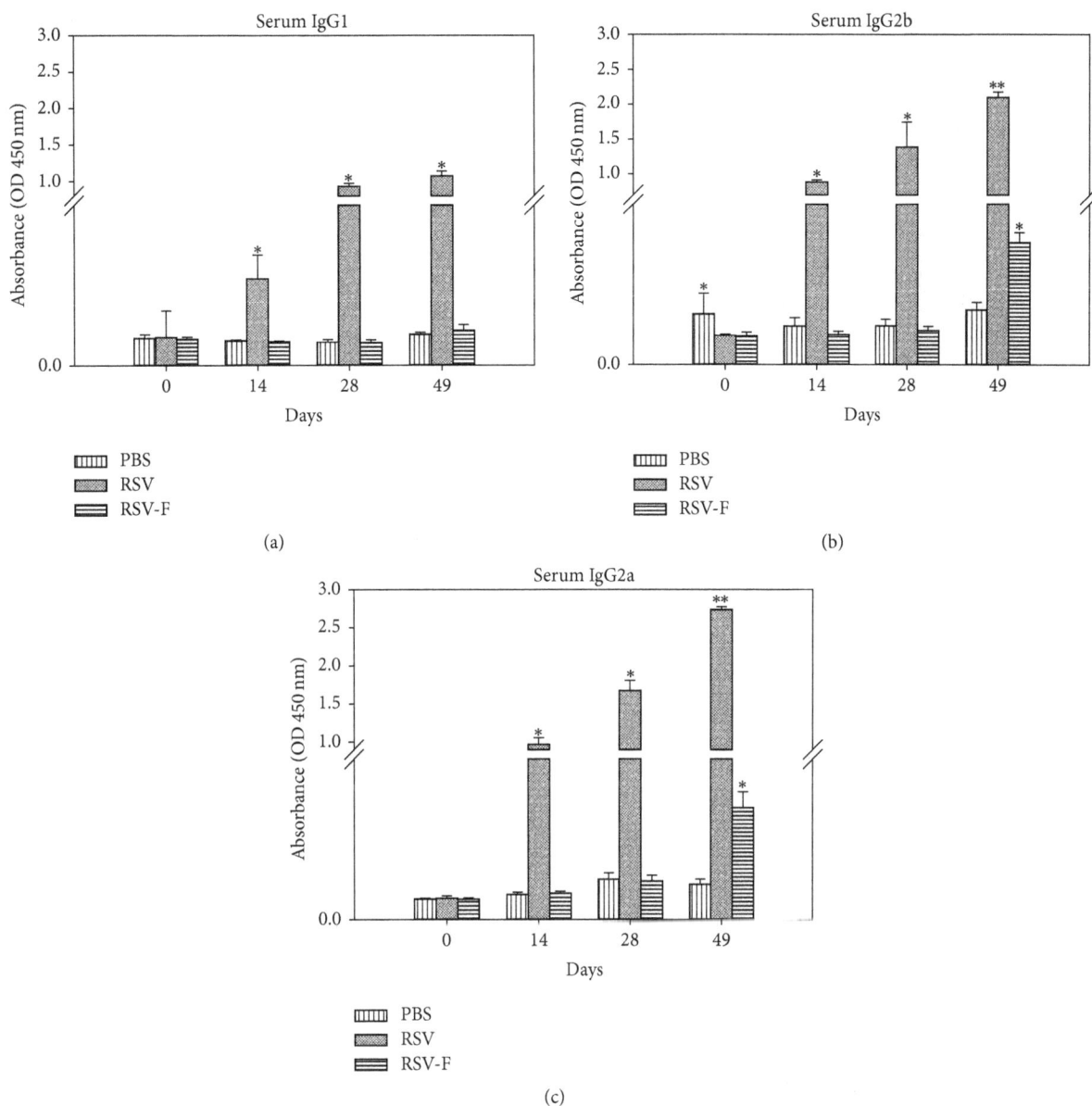

FIGURE 5: IgG isotypes; (a) IgG1, (b) IgG2b, and (c) IgG2a antibody response against RSV specific antigens. Serum samples (PBS, RSV infected, and PF DNA-immunized mice) were collected from BALB/c mice on days 0, 14, 28, and 49 and IgG isotypes were detected by ELISA. Data is presented as an average of triplicates performed twice; error bars represent standard deviations. *Significantly different ($P < 0.01$) from PBS group; **significantly different ($P < 0.01$) from PBS and PF DNA groups. P values ($P < 0.01$) were calculated using ANOVA, Tukey's test.

the ELISA data for RSV neutralization was confirmed with an immunofluorescence assay. The same experiment was repeated under the same conditions and reduction of RSV infection was visualized by a decrease in the FITC signal in immunofluorescence microscopy (Figure 7). RSV infection was visibly observed in the HEp-2 cells incubated with serum from PBS mice compared to untreated Hep-2 cells (Figures 7(a) and 7(b)). The intensity of the FITC signal of the PBS group was detected as strong as the positive control (without serum), whereas no green signal was observed on the cells incubated with RSV+serum from RSV infected mice (Figures 7(c) and 7(d)). On the other hand, RSV immunized serum considerably neutralized RSV infection

in HEp-2 cells and insignificant green signals were detected (Figure 7(e)). Immunofluorescence microscopy observations confirmed the results of the ELISA neutralization assay and the antibody response data.

4. Discussion

As with other pathogenic infections, RSV initially activates the innate response and subsequently develops cellular and humoral immunity. The cellular immunity is needed to clear the infection, whereas the humoral immune response (antibody mediated) is required for protection from initial

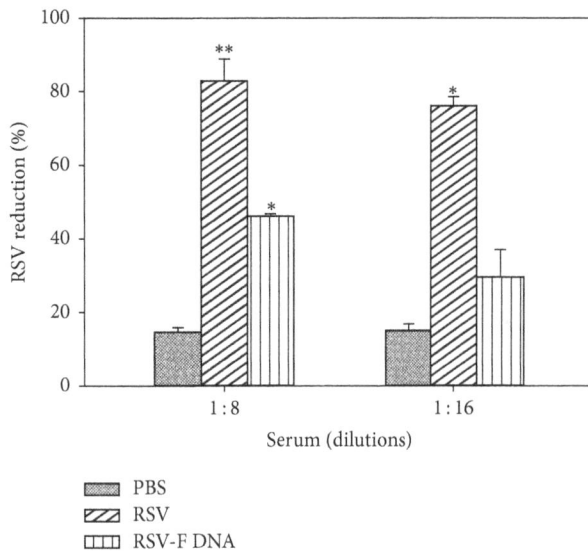

FIGURE 6: Neutralization of RSV on Hep-2 cells, ELISA. The number of RSV mixed with different dilutions of serum (1:16 and 1:8) from mice groups (PBS, RSV, infected and PF DNA-administrated mice) reduced significantly. ELISA was used to detect RSV reduction. Two sera pools from each group of mice were run in duplicate. Bar graphs are represented as means with standard deviations. *Significantly different ($P < 0.01$) from PBS group; **significantly different ($P < 0.01$) from PBS and PF DNA groups. P values ($P < 0.01$) were calculated using ANOVA, Tukey's test.

and subsequent RSV infections. During the 1960s, vaccinations performed with FI-RSV suggested that FI-RSV immunization leads to a predominant Th2 type allergic response. Whereas wild type RSV activates T helper type 1 (Th1) skewed immune providing protection against RSV disease [1]. Thus, the Th1 type immune response is desired for protection against natural RSV infections. To understand the mechanism and type of immune responses for FI-RSV immunizations, different animal models such as monkeys [29], bovine [30], mice [31], and cotton rats [23, 32] were tested. All models challenged with wt RSV following the immunization with FI-RSV stimulated the Th2 type allergic response [23–25, 29–32]. In contrast, animals immunized and challenged with wt RSV developed Th1 type antibody protection against RSV. Likewise, natural RSV infection produces a Th1 mediated immune response against RSV. However, the most desirable immunity against any kind of pathogen is a balanced Th1/Th2 response. Even though the exact mechanism of FI-RSV mediated enhanced disease was not fully understood, Murphy et al. suggested that formalin treatment altered the protective epitopes of F and G proteins and failed to produce neutralizing antibodies against real RSV infections. They also reported that the sera from FI-RSV immunized recipients did not neutralize RSV *in vitro* due to the lack of RSV specific neutralizing antibodies compared to the sera from wt RSV immunized recipients [23]. Consequently, the native form of RSV F is required to produce neutralizing antibodies and provide immunization against RSV infections. DNA vaccines are

thought to be more advantageous due to the processing of antigens in their native forms by eukaryotic cells and due to the efficient presentation of antigens to antigen presenting cells. Thus, antibodies produced against the recombinant antigen expressed in the target host would easily recognize native nondenatured proteins of the pathogen and provide more efficient and specific protection against real pathogens compared to the recombinant protein vaccines expressed in bacteria [33]. In a previous study, we developed a DNA vaccine containing a region of RSV F (412–524 amino acids) conjugated with a modified cholera toxin gene and used to immunize mice which resulted in higher immune response [24].

As mentioned in the FI-RSV vaccine trial, the native form of the RSV F protein is very crucial in stimulating the protective immune response against RSV. Major structural changes in the RSV F protein may lead to disease exacerbation and allergic outcomes. The RSV F DNA vaccine is a preferred immunogen compared to the recombinant RSV F protein produced *ex vivo*. For DNA vaccinations, the intramuscular injection route is the best route that ensures antigen expression and native conformation. Besides the native structure, another advantage of DNA vaccination is that it elicits the Th1 biased immune response due to its endogenous expression and presentation to the immune cells, which is a favorable response for protection from pathogens [13, 34]. The Th1 immune system and the production of neutralizing antibodies are very important for protection from reinfection, which confers long term immunity by recruiting memory B cells. When the host encounters the same pathogen again, memory B cells abruptly produce pathogen specific neutralizing antibodies and immediately inactivate the pathogen before it enters into the host and starts infection [33]. We tested the ability of serum collected from RSV F immunized mice to neutralize RSV *in vitro*. Previous studies have shown that serum from FI-RSV infected mice does not neutralize RSV infection due to the altered structure of the RSV F protein [23].

Based on previous studies, distinct administration routes of the DNA vaccine evoke different types of immune responses. Our study was designed based on the previous DNA vaccine study where intramuscular injection of a DNA vaccine stimulated a moderate T cell response and antibody production compared to the oral administration, which induced a strong T cell response and weak antibody response [35].

RSV vaccine development has been hampered by the failure of previous vaccine trials that led to death of children. The main immunological event responsible of failure of the vaccine was induction of a predominant Th2 response that enhanced RSV disease following natural infection. Our study aimed at developing a safe DNA vaccine that induced a Th1 mediated antibody response. This study provides a basis for future RSV vaccine development that could benefit from DNA vaccine designs and may consider combination of DNA vaccine immunizations followed by traditional recombinant vaccine immunizations for higher protection from RSV infections.

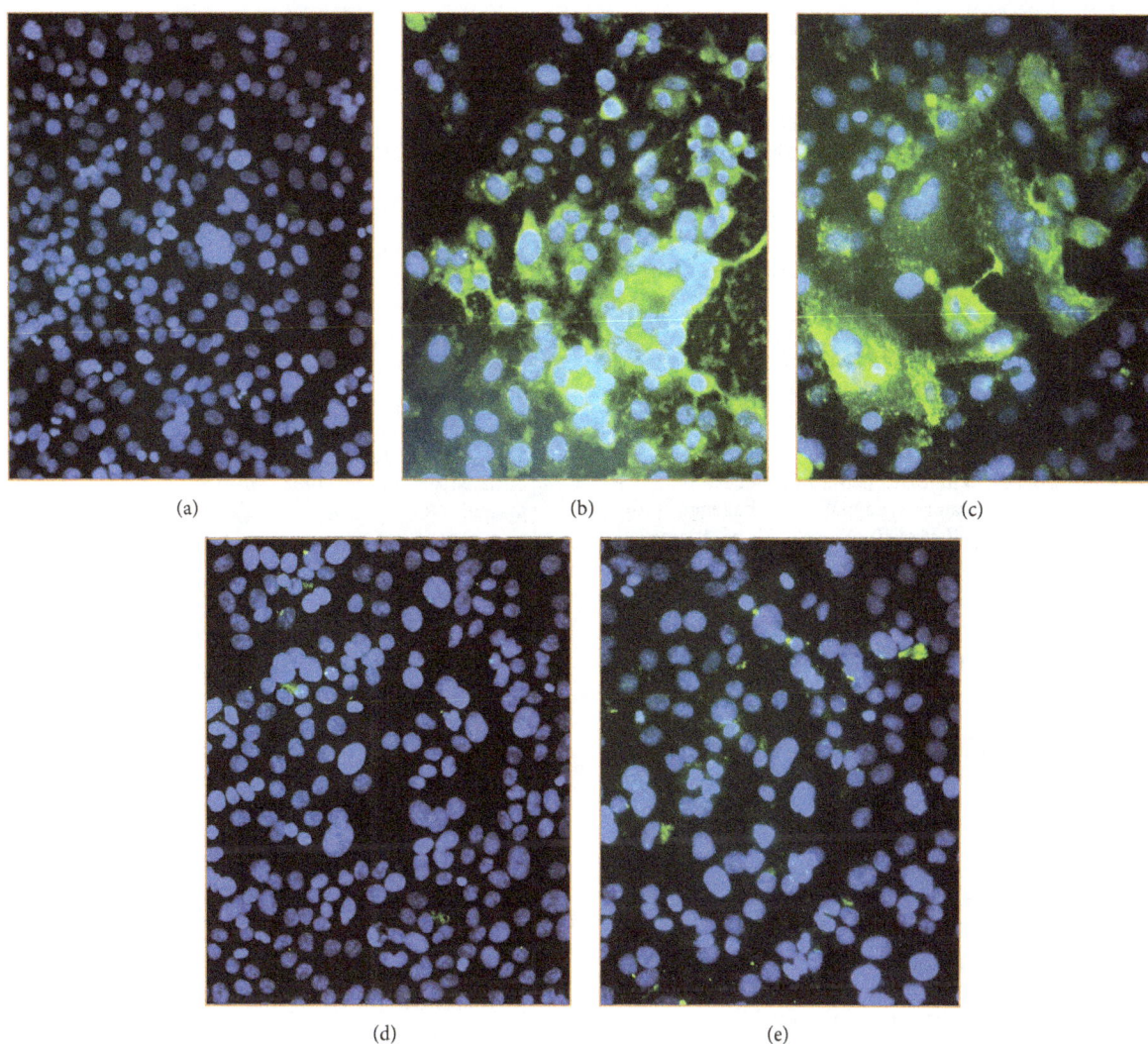

FIGURE 7: Neutralization of RSV on Hep-2 cells, immunofluorescence microscopy. (a) Negative control: uninfected Hep-2 cells, (b) positive control: RSV infected cells only, (c) PBS group: RSV mixed with sera collected from PBS injected mice, (d) RSV sera: RSV mixed with sera collected from RSV infected mice, and (e) PF DNA sera: RSV mixed with sera collected from PF-immunized mice.

Competing Interests

The authors declare there is no conflict of interests in publishing this article.

Acknowledgments

This work was supported by NSF-CREST (HRD-1241701) and NSF-HBCU-UP (HRD-1135863). The authors would like to thank Dr. Saurabh Dixit for his technical support throughout this work.

References

[1] D. Hacking and J. Hull, "Respiratory syncytial virus—viral biology and the host response," *Journal of Infection*, vol. 45, no. 1, pp. 18–24, 2002.

[2] E. D. Clercq, "Chemotherapy of respiratory syncytial virus infections: the final breakthrough," *International Journal of Antimicrobial Agents*, vol. 45, no. 3, pp. 234–237, 2015.

[3] G. F. Langley and L. J. Anderson, "Epidemiology and prevention of respiratory syncytial virus infections among infants and young children," *Pediatric Infectious Disease Journal*, vol. 30, no. 6, pp. 510–517, 2011.

[4] R. Lozano, M. Naghavi, K. Foreman et al., "Global and regional mortality from 235 causes of death for 20 age groups in 1990 and 2010: a systematic analysis for the Global Burden of Disease Study 2010," *The Lancet*, vol. 380, no. 9859, pp. 2095–2128, 2012.

[5] H. Nair, D. J. Nokes, B. D. Gessner et al., "Global burden of acute lower respiratory infections due to respiratory syncytial virus in young children: a systematic review and meta-analysis," *The Lancet*, vol. 375, no. 9725, pp. 1545–1555, 2010.

[6] K. Modjarrad, B. Giersing, D. C. Kaslow, P. G. Smith, and V. S. Moorthy, "WHO consultation on Respiratory Syncytial Virus Vaccine Development Report from a World Health Organization Meeting held on 23-24 March 2015," *Vaccine*, vol. 34, no. 2, pp. 190–197, 2016.

[7] M. S. A. Gilman, S. M. Moin, V. Mas et al., "Characterization of a prefusion-specific antibody that recognizes a quaternary,

cleavage-dependent epitope on the rsv fusion glycoprotein," *PLoS Pathogens*, vol. 11, no. 7, Article ID e1005035, 2015.

[8] J. S. McLellan, M. Chen, M. G. Joyce et al., "Structure-based design of a fusion glycoprotein vaccine for respiratory syncytial virus," *Science*, vol. 342, no. 6158, pp. 592–598, 2013.

[9] K. A. Swanson, E. C. Settembre, C. A. Shaw et al., "Structural basis for immunization with postfusion respiratory syncytial virus fusion F glycoprotein (RSV F) to elicit high neutralizing antibody titers," *Proceedings of the National Academy of Sciences of the United States of America*, vol. 108, no. 23, pp. 9619–9624, 2011.

[10] S. M. Bueno, P. A. González, C. A. Riedel, L. J. Carreño, A. E. Vásquez, and A. M. Kalergis, "Local cytokine response upon respiratory syncytial virus infection," *Immunology Letters*, vol. 136, no. 2, pp. 122–129, 2011.

[11] J. P. DeVincenzo, R. J. Whitley, R. L. Mackman et al., "Oral GS-5806 activity in a respiratory syncytial virus challenge study," *The New England Journal of Medicine*, vol. 371, no. 8, pp. 711–722, 2014.

[12] J. Murray, S. Saxena, and M. Sharland, "Preventing severe respiratory syncytial virus disease: passive, active immunisation and new antivirals," *Archives of Disease in Childhood*, vol. 99, no. 5, pp. 469–473, 2014.

[13] J. S. Lee, Y.-M. Kwon, H. S. Hwang et al., "Baculovirus-expressed virus-like particle vaccine in combination with DNA encoding the fusion protein confers protection against respiratory syncytial virus," *Vaccine*, vol. 32, no. 44, pp. 5866–5874, 2014.

[14] B. Liang, S. Surman, E. Amaro-Carambot et al., "Enhanced neutralizing antibody response induced by respiratory syncytial virus prefusion F protein expressed by a vaccine candidate," *Journal of Virology*, vol. 89, no. 18, pp. 9499–9510, 2015.

[15] H. Mok, S. Lee, T. J. Utley et al., "Venezuelan equine encephalitis virus replicon particles encoding respiratory syncytial virus surface glycoproteins induce protective mucosal responses in mice and cotton rats," *Journal of Virology*, vol. 81, no. 24, pp. 13710–13722, 2007.

[16] J.-F. Valarcher and G. Taylor, "Bovine respiratory syncytial virus infection," *Veterinary Research*, vol. 38, no. 2, pp. 153–180, 2007.

[17] J. P. M. Langedijk, R. H. Meloen, G. Taylor, J. M. Furze, and J. T. Van Oirschot, "Antigenic structure of the central conserved region of protein G of bovine respiratory syncytial virus," *Journal of Virology*, vol. 71, no. 5, pp. 4055–4061, 1997.

[18] S. R. Singh, V. A. Dennis, C. L. Carter et al., "Immunogenicity and efficacy of recombinant RSV-F vaccine in a mouse model," *Vaccine*, vol. 25, no. 33, pp. 6211–6223, 2007.

[19] N. Ternette, B. Tippler, K. Überla, and T. Grunwald, "Immunogenicity and efficacy of codon optimized DNA vaccines encoding the F-protein of respiratory syncytial virus," *Vaccine*, vol. 25, no. 41, pp. 7271–7279, 2007.

[20] T. R. Johnson, R. A. Parker, J. E. Johnson, and B. S. Graham, "IL-13 is sufficient for respiratory syncytial virus G glycoprotein-induced eosinophilia after respiratory syncytial virus challenge," *Journal of Immunology*, vol. 170, no. 4, pp. 2037–2045, 2003.

[21] Y. Huang, S. L. Cyr, D. S. Burt, and R. Anderson, "Murine host responses to respiratory syncytial virus (RSV) following intranasal administration of a Protollin-adjuvanted, epitope-enhanced recombinant G protein vaccine," *Journal of Clinical Virology*, vol. 44, no. 4, pp. 287–291, 2009.

[22] J. L. Harcourt, L. J. Anderson, W. Sullender, and R. A. Tripp, "Pulmonary delivery of respiratory syncytial virus DNA vaccines using macroaggregated albumin particles," *Vaccine*, vol. 22, no. 17-18, pp. 2248–2260, 2004.

[23] B. R. Murphy, A. V. Sotnikov, L. A. Lawrence, S. M. Banks, and G. A. Prince, "Enhanced pulmonary histopathology is observed in cotton rats immunized with formalin-inactivated respiratory syncytial virus (RSV) or purified F glycoprotein and challenged with RSV 3–6 months after immunization," *Vaccine*, vol. 8, no. 5, pp. 497–502, 1990.

[24] H. Wu, V. A. Dennis, S. R. Pillai, and S. R. Singh, "RSV fusion (F) protein DNA vaccine provides partial protection against viral infection," *Virus Research*, vol. 145, no. 1, pp. 39–47, 2009.

[25] P. L. Collins, Y. T. Huang, and G. W. Wertz, "Nucleotide sequence of the gene encoding the fusion (F) glycoprotein of human respiratory syncytial virus," *Proceedings of the National Academy of Sciences of the United States of America*, vol. 81, no. 24, pp. 7683–7687, 1984.

[26] E. Eroglu, P. M. Tiwari, A. B. Waffo et al., "A nonviral pHEMA+chitosan nanosphere-mediated high-efficiency gene delivery system," *International Journal of Nanomedicine*, vol. 8, pp. 1403–1415, 2013.

[27] R. Mentel, U. Wegner, R. Bruns, and L. Gürtler, "Real-time PCR to improve the diagnosis of respiratory syncytial virus infection," *Journal of Medical Microbiology*, vol. 52, no. 10, pp. 893–896, 2003.

[28] R. Racine and G. M. Winslow, "IgM in microbial infections: taken for granted?" *Immunology Letters*, vol. 125, no. 2, pp. 79–85, 2009.

[29] E. Ponnuraj, A. R. Hayward, A. Raj, H. Wilson, and E. A. Simones, "Increased replication of respiratory syncytial virus (RSV) in pulmonary infiltrates is associated with enhanced histopathological disease in bonnet monkeys (*Macac radiata*) pre-immunized with a formaline-inactivated RSV vaccine," *Journal of General Virology*, vol. 82, pp. 2663–2674, 2001.

[30] W. V. Kalina, A. R. Woolums, R. D. Berghaus, and L. J. Gershwin, "Formalin-inactivated bovine RSV vaccine enhances a Th2 mediated immune response in infected cattle," *Vaccine*, vol. 22, no. 11-12, pp. 1465–1474, 2004.

[31] M. E. Waris, C. Tsou, D. D. Erdman, S. R. Zaki, and L. J. Anderson, "Respiratory synctial virus infection in BALB/c mice previously immunized with formalin-inactivated virus induces enhanced pulmonary inflammatory response with a predominant Th2-like cytokine pattern," *Journal of Virology*, vol. 70, no. 5, pp. 2852–2860, 1996.

[32] P. A. Piedra, P. R. Wyde, W. L. Castleman et al., "Enhanced pulmonary pathology associated with the use of formalin-inactivated respiratory syncytial virus vaccine in cotton rats is not a unique viral phenomenon," *Vaccine*, vol. 11, no. 14, pp. 1415–1423, 1993.

[33] G. J. Nabel, "Challenges and opportunities for development of an AIDS vaccine," *Nature*, vol. 410, no. 6831, pp. 1002–1007, 2001.

[34] S. Gurunathan, C.-Y. Wu, B. L. Freidag, and R. A. Seder, "DNA vaccines: a key for inducing long-term cellular immunity," *Current Opinion in Immunology*, vol. 12, no. 4, pp. 442–447, 2000.

[35] H. Hu, X. Huang, L. Tao, Y. Huang, B.-A. Cui, and H. Wang, "Comparative analysis of the immunogenicity of SARS-CoV nucleocapsid DNA vaccine administrated with different routes in mouse model," *Vaccine*, vol. 27, no. 11, pp. 1758–1763, 2009.

22

Alternanthera mosaic potexvirus: Several Features, Properties, and Application

Ekaterina Donchenko, Ekaterina Trifonova ⓘ, Nikolai Nikitin ⓘ, Joseph Atabekov ⓘ, and Olga Karpova ⓘ

Department of Virology, Lomonosov Moscow State University, Moscow 119234, Russia

Correspondence should be addressed to Nikolai Nikitin; nikitin@mail.bio.msu.ru

Academic Editor: Finn S. Pedersen

Alternanthera mosaic virus (AltMV) is a typical member of the *Potexvirus* genus in its morphology and genome structure; still it exhibits a number of unique features. They allow this virus to be considered a promising object for biotechnology. Virions and virus-like particles (VLPs) of AltMV are stable in a wide range of conditions, including sera of laboratory animals. AltMV VLPs can assemble at various pH and ionic strengths. Furthermore, AltMV virions and VLPs demonstrate high immunogenicity, enhancing the immune response to the target antigen thus offering the possibility of being used as potential adjuvants. Recently, for the first time for plant viruses, we showed the structural difference between morphologically similar viral and virus-like particles on AltMV virions and VLPs. In this review, we discuss the features of AltMV virions, AltMV VLP assembly, and their structure and properties, as well as the characteristics of AltMV isolates, host plants, infection symptoms, AltMV isolation and purification, genome structure, viral proteins, and AltMV-based vectors.

1. Introduction

Alternanthera mosaic virus (AltMV) was first described in 1999 as "isolate 451/1" in the state of Queensland, Australia [1]. The new virus was isolated from *Alternanthera pungens* plants of the family *Amaranthaceae*. AltMV belongs to the genus *Potexvirus* and the family *Alphaflexiviridae*. The AltMV genome is a positive-sense single-stranded RNA 6604-6607 nt long depending on the isolates. AltMV virions represent flexible filamentous particles with helical symmetry made up of one type of coat protein (CP) subunits. The mean length of AltMV virions equals 554 nm [1], 536 nm [2], or 570 nm [3]. As reported by Mukhamedzhanova *et al.* [3], the virions are 13 nm in diameter. Recently AltMV virion diameter was corrected to 13.5 nm by means of cryoelectron microscopy [4].

Since plant viruses and virus-like particles (VLPs) are essentially safe for humans they seem promising for technological advances in a broad range of areas from microelectronics to developing candidate vaccines and adjuvants [5, 6]. AltMV virions and VLPs have a considerable number of advantages for successful application in biotechnology [4, 7–9].

2. AltMV Isolates and Their Distribution

Soon after AltMV had been discovered in Australia, other isolates were derived from plants in Europe [10, 11], USA [2, 12–16], Brazil [17], and Asia [18]. Nowadays AltMV is reported to be spread all over the world, and capable of infecting plants of various families including ornamental plants and crops [1, 19]. To date, complete nucleotide sequences are determined for the following AltMV isolates: AltMV-Ac (6604 nt long), AltMV-MU (6606 nt), and AltMV-PA (6607 nt). Four biologically active cDNA of "infectious clones" (3-1, 3-7, 4-1, and 4-7) were derived from AltMV-SP genome; the complete nucleotide sequences of these clones were established (6607 nt). The nucleotide sequences of the other isolates have been only partially determined [2, 14, 20].

The diversity of the obtained isolates both in terms of the host plants being the targets for virulence and geographical distribution implies the existence of phylogenetically

TABLE 1: AltMV isolates.

Virus isolate	Accession no.	Authors	Original host	Origin of infected plants
Phlox-like isolates				
AltMV-AU	AF080448	Geering and Thomas, 1999	*Alternanthera pungens*	Australia, QLD
AltMV-PA	AY863024	Hammond *et al.*, 2004, 2006a, b	*Phlox stolonifera*	USA, PA
AltMV-SP	AY850931	Hammond *et al.*, 2004, 2006a, b	*Phlox stolonifera* cv. Sherwood Purple	USA, MD
AltMV-BR	AY850928	Hammond *et al.*, 2004, 2006a, b	*Phlox stolonifera* cv. Blue Ridge	USA, MD
AltMV-NAN	GU126686	Tang *et al.*, 2010	*Nandina domestica*	USA, OK
AltMV-BW	JX457329	Hammond and Reinsel, 2015	*Phlox stolonifera* cv. Bruce's White	USA, MD
AltMV-LGB	JX457330	Hammond and Reinsel, 2015	*Phlox divaricata* cv. London Grove Blue	USA, MD
AltMV-PGL	JQ405265	Hammond and Reinsel, 2015	*Phlox carolina angusta*	USA
Portulaca-like isolates				
AltMV-IT	AY566288	Ciuffo and Turina, 2004	*Portulaca grandiflora*	Italy, Liguria, Albenga
AltMV-Po	AY850930	Hammond *et al.*, 2004, 2006a, b	*Portulaca grandiflora*	USA, MD
AltMV-MU	FJ822136	Ivanov *et al.*, 2011	*Portulaca grandiflora*	South-Eastern Europe
AltMV-Port	JQ405269	Hammond and Reinsel, 2015	*Portulaca grandiflora*	USA
AltMV-PLR	JQ405266	Hammond and Reinsel, 2015	*Phlox* hybrid (annual)	USA
AltMV-CIN	JQ405268	Hammond and Reinsel, 2015	*Pericallis* hybrid	USA
Asian isolate				
AltMV-Ac	LC107515	Iwabuchi *et al.*, 2016	*Achyranthes bidentata*	Japan, Tokyo
Unassigned isolates				
AltMV Florida isolate	DQ393785	Baker *et al.*, 2006	*Portulaca* sp. *Scutellaria longifolia Crossandra infundibuliformis*	USA, FL
AltMV-T	FJ232066, FJ232067	Duarte *et al.*, 2008	*Torenia* sp.	Brazil, São Paulo
AltMV angelonia isolate	EU679363	Lockhart and Daughtrey, 2008	*Angelonia angustifolia*	USA, NY
AltMV isolate G10-00982	JQ687034	Vitoreli *et al.*, 2011	*Thunbergia laurifolia*	USA, FL

diverged groups within the AltMV taxon [11, 20]. Based on amino acid sequences of RNA-dependent RNA polymerase (RdRp, replicase) and CP, Ivanov *et al.* [11] have distinguished 2 groups within the AltMV species: phlox-like isolates and portulaca-like isolates. In the study by Hammond and Reinsel [20] this differentiation was confirmed by means of amino acid sequence analysis of the three proteins which are products of the "triple gene block" (TGB). Apparently, Asian AltMV isolate should be regarded as a separate group [18].

AltMV-IT, AltMV-MU, AltMV-Port, AltMV-Po, AltMV-CIN, and AltMV-PLR (Table 1) were classified as portulaca-like isolates [20]. Ivanov *et al.* [11] indicated a close evolutionary relationship among the isolates of this group. The amino acid sequences of AltMV-IT, AltMV-MU, and AltMV-Po CPs differ on substitutions at two sites, methionine 106 for isoleucine and serine 185 for phenylalanine [11]. AltMV-Au, AltMV-PA, AltMV-NAN, and AltMV-SP were classified as phlox-like isolates. Belonging to different groups, the CPs of AltMV-MU and AltMV-PA are different by 12 amino acid residues situated mainly on the N-terminus of CP [11]. Nucleotide sequences determined for AltMV isolates obtained from *Scutellaria longifolia* [12], *Torenia* [17], *Angelonia angustifolia* [15], and *Thunbergia laurifolia* plants [21] are too short for detailed analysis, yet most likely they belong to portulaca-like type [20].

The reported distribution of isolates may be determined by the peculiarities of the host plant cultivation. Annual phlox, cineraria, angelonia, torenia, thunbergia, and portulaca are grown mostly in greenhouses as ornamental plants, while perennial phlox and nandina with vegetative reproduction are cultivated in open ground. Similar conditions of plant cultivation most likely account for portulaca-like AltMV detection in greenhouse plants; moreover, portulaca might have served as an infection source for this type of horticulture. Similarly, the spread of phlox-like AltMV obtained from plants cultivated in open ground was the result of cross-contamination from infected perennial phlox plants [20].

According to serological data as well as nucleotide and amino acid sequence similarity, AltMV is the closest relative of the *papaya mosaic virus* (PapMV) [1, 2, 20]. AltMV is easily confused with PapMV based on serological analysis or PCR analysis in case of an incorrect primer selection [2] which leads to AltMV being misinterpreted as PapMV [12, 22]. Hammond *et al.* [2] argued that AltMV is far more widespread than expected, especially in nurseries and greenhouses.

3. AltMV Host Plants and Infection Symptoms

AltMV has a broad host range and can infect plants from at least 31 taxonomic families including *Aizoaceae, Amaranthaceae, Apiaceae, Asteraceae, Brassicaceae, Caryophyllaceae, Chenopodiaceae, Cucurbitaceae, Fabaceae, Plantaginaceae, Polemoniaceae,* and *Solanaceae* [19]. The virus was detected in several ornamental species including *Portulaca grandiflora, Phlox stolonifera, Scutellaria* spp., *Crossandra infundibuliformis, Angelonia angustifolia, Torenia* spp., *Helichrysum* spp., *Salvia splendens,* and *Zinnia elegans* [2, 12, 15, 17]. In addition to ornamentals, the systemically infected plants were found among various horticultural plants including *Solanum lycopersicum, Vicia faba, Helianthus annuus, Citrullus lanatus, Cucumis sativus,* and *Vigna unguiculata* [1]. The infected plants were mostly collected from commercial nurseries [2, 10, 12, 15, 17, 18, 21].

Upon being infected with AltMV all representatives of the *Amaranthaceae* family including *Alternanthera pungens* exhibited symptoms [1]. On the contrary, plants of the families *Caesalpiniaceae, Caricaceae,* and *Poaceae* are not susceptible to AltMV infection. As papaya *Carica papaya* is the main host for PapMV, its insusceptibility to AltMV confirms that these viruses belong to different species [1].

AltMV is known to manifest a wide range of symptoms including chlorotic spotting of various size, chlorosis, chlorotic local lesions, leaf distortion and curling, mosaic, mottle, necrotic spotting of various size, necrotic ringspots, rugosity, veinal necrosis, and interveinal yellowing (Table 2) [1, 2, 12, 15, 17].

AltMV infection symptoms were shown to depend not only on the host plant species, but also on the duration of infection, virus strain, and environmental factors. For example, symptoms being mostly pronounced in plants grown under high light levels and moderate temperature escaped notice in plants grown under low light levels and high temperature [2]. For AltMV-SP isolates derived from tobacco *Nicotiana benthamiana* the symptom severity was shown to correlate with sequence differences of the replicase (RdRp) and Triple Gene Block protein 1 (TGBp1) [23]. Similarly, in case of AltMV-Po isolate, the level of severity in symptom manifestation was demonstrated to go hand in hand with changes in amino acid sequence of CP [24].

The ability of AltMV to infect a wide host range leading to symptomless infection allows for cross-contamination among various species including the cultivated ones. Taking the aforementioned into account, AltMV may be more widespread than the literature suggests, particularly in nurseries [12, 20].

4. AltMV Propagation and Purification

Several methods were developed for AltMV isolation from a variety of host plants (Table 3). Geering and Thomas [1] followed the procedure previously described by Bancroft *et al.* [25]. The technique employed by Hammond *et al.* [2] was initially introduced for potexviruses and later adapted to potyviruses. In the studies by Mukhamedzhanova *et al.* [3] and Ivanov *et al.* [11] AltMV was isolated according to the protocol developed for another potexvirus, namely, *potato virus X* (PVX), with slight modifications. This technique was further substantially modified by Donchenko *et al.* [4].

Geering and Thomas [1] used *Chenopodium amaranticolor* as a host plant to propagate AltMV resulting in a yield of 23.4 mg of virus per 100 g of infected leaves. Even though the yield was relatively high, this host plant cannot be regarded as optimal for AltMV accumulation. Hammond *et al.* [2] managed to isolate AltMV from *Nicotiana benthamiana* with the yield of 8.6-12.5 mg of the virus per 100 g of green biomass while Mukhamedzhanova *et al.* [3] and Ivanov *et al.* [11] used *Portulaca grandiflora* as a host yielding 3.4 mg of virus per 100 g of infected leaves. Since *P. grandiflora* is hardly susceptible to infection with other viruses and *N. benthamiana* is a commonly used model plant, in the study by Donchenko *et al.* [4] portulaca and tobacco plants were selected as hosts [26]. In order to obtain purified AltMV, the infectious material was first propagated in *P. grandiflora* to prevent coinfection and later transmitted and accumulated in *N. benthamiana*. This allowed the yield to be substantially increased up to 20.0 mg and 57.3 mg of virus per 100 g of infected leaves in case of *P. grandiflora* and *N. benthamiana*, respectively [4].

5. Structure of AltMV Genome

The AltMV genome consists of a sole positive-sense single-stranded RNA having a cap at the 5' terminus and polyA

TABLE 2: Diversity of AltMV host plants and infection symptoms.

Family	Host	Symptoms		Authors
		local	systemic	
Aizoaceae	*Tetragonia expansa*	necrotic ringspot	chlorosis, veinal necrosis, leaf curl	Hammond *et al.*, 2006b
Amaranthaceae	*Amaranthus caudatus*	necrotic local lesions	no infection	Hammond *et al.*, 2006b
	Amaranthus tricolor	chlorotic local lesions	mosaic	Hammond *et al.*, 2006b
	Gomphrena celosiodes	asymptomatic infection	mosaic	Geering and Thomas, 1999
	Gomphrena globosa	necrotic local lesions	asymptomatic infection	Geering and Thomas, 1999; Hammond *et al.*, 2006b
	Alternanthera dentata	necrotic local lesions	necrotic spotting, distortion	Hammond *et. al*, 2006b
Apiaceae	*Apium graveolens*, cv. Crisp Salad	necrotic local lesions	mosaic	Geering and Thomas, 1999
Asteraceae	*Aster novi-belgii*	no infection	no infection	
	Dahlia variabilis	no infection	no infection	
	Helianthus annuus	asymptomatic infection	no infection	Hammond *et al.*, 2006b
	Sanvitalia procumbens	no infection	mosaic, leaf curl, distortion	
	Lactuca sativa, cv. Black velvet	necrotic local lesions	no infection	Geering and Thomas, 1999
	Zinnia elegans	asymptomatic infection	mottle/ asymptomatic infection	Geering and Thomas, 1999/Hammond *et al.*, 2006b
Brassicaceae	*Brassica campestris* var. pekinensis cv. Lin White Spoon	asymptomatic infection	no infection	Geering and Thomas, 1999
	Rhaphanus sativus cv. French Breakfast	no infection	no infection	
Caesalpiniaceae	*Cassia floribunda*	no infection	no infection	Geering and Thomas, 1999
	Cassia occidentalis	no infection	no infection	
Caricaceae	*Carica papaya* cv. Richter Gold	no infection	no infection	Geering and Thomas, 1999
Caryophyllaceae	*Gypsophila repens*	asymptomatic infection	no infection	Hammond *et al.*, 2006b
Chenopodiaceae	*Chenopodium amaranticolor*	chlorotic local lesions	mosaic	Geering and Thomas, 1999
	Chenopodium quinoa	chlorotic local lesions/necrotic local lesions	interveinal yellowing/no infection	Geering and Thomas, 1999/Hammond *et al.*, 2006b
	Spinacia oleracea	chlorotic local lesions/no infection	mosaic/necrotic fleck, leaf curl	
Cucurbitaceae	*Citrullus lanatus* var. *Caffer* cv. Candy Red	asymptomatic infection	mosaic	Geering and Thomas, 1999
	Cucumis sativus cv. Green Gem	asymptomatic infection	asymptomatic infection	
	Cucumis sativus (two cultivars)	no infection	no infection	Hammond *et al.*, 2006b
	Cucurbita pepo cv. Green Buttons	asymptomatic infection	no infection	Geering and Thomas, 1999

TABLE 2: Continued.

Family	Host	Symptoms		Authors
		local	systemic	
Fabaceae	Glycine max cv. Bragg	no infection	no infection	Geering and Thomas, 1999
	Phaseolus vulgaris cv. Bountiful	no infection	no infection	Geering and Thomas, 1999/Hammond et al., 2006b
	Phaseolus vulgaris cv Kerman	no infection	no infection	Geering and Thomas, 1999/Hammond et al., 2006b
	Pisum sativum cv. Greenfeast	no infection	no infection	Geering and Thomas, 1999/Hammond et al., 2006b
	Trifolium pratense cv. Montgomery	asymptomatic infection	no infection	Geering and Thomas, 1999
	Vigna unguiculata cv. Black-eye	asymptomatic infection/no infection	asymptomatic infection/no infection	Geering and Thomas, 1999/Hammond et al., 2006b
	Vicia faba	asymptomatic infection/no infection	mosaic/chlorotic fleck	Geering and Thomas, 1999/Hammond et al., 2006b
Papaveraceae	Papaver orientale	no infection	no infection	Hammond et al., 2006b
Plantaginaceae	Plantago lanceolata	asymptomatic infection	mosaic	Geering and Thomas, 1999
Poaceae	Sorghum halapense cv. Silk	no infection	no infection	Geering and Thomas, 1999
	Zea mays cv. Jubilee	no infection	no infection	Geering and Thomas, 1999
Polemoniaceae	Phlox drummondii	no infection	mild mottle	Hammond et al., 2006b
	Phlox stolonifera	mottle	mottle	Hammond et al., 2006b
Solanaceae	Capsicum annuum cv. Yolo Wonder	no infection	no infection	Geering and Thomas, 1999/ Hammond et al., 2006b
	Datura stramonium	no infection	no infection	Geering and Thomas, 1999
	Solanum lycopersicum cv. Gross Lisse	asymptomatic infection/no infection	mottle/mild mottle, leaf curl	Geering and Thomas, 1999/ Hammond et al., 2006b
	Nicotiana benthamiana	asymptomatic infection/chlorotic local lesions	mosaic, rugosity, epinasty	Geering and Thomas, 1999/ Hammond et al., 2006b
	Nicotiana clevelandii	no infection	no infection/mild chlorosis	Hammond et al., 2006b
	Nicotiana edwardsonii	no infection	no infection	Hammond et al., 2006b
	Nicotiana glutinosa	no infection	no infection	Geering and Thomas, 1999/ Hammond et al., 2006b
	Nicotiana megalosiphon	necrotic local lesions, necrotic ringspot	mosaic, necrotic fleck	Hammond et al., 2006b
	Nicotiana rustica	no infection	no infection	Geering and Thomas, 1999/ Hammond et al., 2006b
	Nicotiana tabacum cv. Turkish	no infection	no infection	Geering and Thomas, 1999/ Hammond et al., 2006b
	Nicotiana tabacum cv. Xanthi	no infection	no infection	Geering and Thomas, 1999/ Hammond et al., 2006b
	Physalis floridana	no infection	no infection	Geering and Thomas, 1999
	Solanum melongena	no infection	faint chlorotic spotting, rugosity	Hammond et al., 2006b
	Solanum tuberosum cv. Sebago	no infection	no infection	Geering and Thomas, 1999

TABLE 3: Comparison of AltMV isolation procedures.

Isolation and purification steps	Authors			
	Geering and Thomas, 1999	Hammond et al., 2006b	Mukhamedzhanova et al., 2011; Ivanov et al., 2011	Donchenko et al., 2017
Yield (mg of virus/100 g of green plant biomass) and host plants used	23.4 / Chenopodium amaranticolor/	8.6-12.4/Nicotiana benthamiana/	3.4 / Portulaca grandiflora/	20.0 /Portulaca grandiflora/ 57.3/Nicotiana benthamiana/
Homogenization buffer	0.02 M Sodium borate buffer, 0.5% Na_2SO_3, pH 8.2 (250 ml buffer per 150 g of leaves)	0.5 M K_2H/KH_2PO_4, 0.5% Na_2SO_3, pH 8.4 (3-5 buffer volumes per weight of leaves)	0.3 M glycine-KOH, 1% Na_2SO_3, pH 7.5 (3 ml buffer per 1 g of leaves)	0.3 M glycine-KOH, 1% Na_2SO_3, pH 7.5 (3 ml buffer per 1 g of leaves)
Virus enrichment from plant tissue	centrifugation 0.5% Triton X-100	centrifugation 2% Triton X-100, 4% PEG M_r 8000, 2% NaCl	centrifugation 1% Triton X-100 5% PEG M_r 6000, 2% NaCl (Personal communication)	centrifugation 1% Triton X-100 Two stages of precipitation: (1) 5% PEG M_r 6000, 2% NaCl (2) 8% PEG M_r 6000
Extraction of virus from pellets	—	Extraction in 0.1 M Sodium borate buffer, 0.1 M KCl, pH 8.0 (BK buffer), 0.75-1.5 hours.	Extraction in 0.05 M Tris-HCl, 0.01 M EDTA, pH 8.0, (Personal communication)	Extraction in 0.05 M Tris-HCl, 0.01 M EDTA, pH 8.0, Two stages for 2-6 hours.
Ultracentrifugation steps	(1) Separation by 10-40% sucrose gradient (0.01 M Tris-HCl, 0.001 M EDTA, pH 8.0) (85 000 g, 4 hours) (2) Virus precipitation (85 000 g, 2.5 hours)	(1) Separation by 30% sucrose cushion (BK buffer) (85 600 g, 2.5 hours) (2) Separation by CsCl gradient, density 1.32 g/cm³ (139 000 g, 16-20 hours) (3) Dialysis against 0.5x BK buffer in 1 l (3 changes)	(1) Virus precipitation (100 000 g, 1.5 hours) (Personal communication) (2) Separation by 30% sucrose cushion (110 000 g, 2.5 hours) (Personal communication)	(1) Virus precipitation (111 000 g, 3 hours) (2) Separation by 30% sucrose cushion (extraction buffer) (111 000 g, 4 hours)
Spectrophotometric analysis ($E_{1cm}, 0.1\%_{260nm}$)	2.84	2.5	2.84	2.84

sequence at the 3' terminus [1]. Nucleotide sequence analysis of AltMV-PA genome [2] revealed two untranslated regions (UTR) at the 5' (1-94 nt) and 3' termini (6481-6607 nt) respectively, as well as 5 open reading frames (ORF). The first ORF encountered at the 5' end of the genome is the longest (95-4720 nt) and is capable of encoding a 1540 amino acid (aa) long protein, namely, a viral replicase (RdRp). The next three ORFs presumably encode the three movement proteins referred to as Triple Gene Block Proteins: ORF2 (4704-5402 nt, encodes a 26 kDa 232 aa long protein); ORF3 (5356-5688 nt, encodes a 12 kDa 110 aa long protein); and ORF4 (5624-5815 nt, encodes a 7 kDa 63 aa long protein). The extreme 3' end of the genome contains ORF5 (5858-6481 nt). Translation of the ORF5 produces a 22-23 kDa 207 aa residues long polypeptide being the CP. In addition, a comparatively short ORF6 was identified within all the presently known ORF5 of AltMV isolates [2, 11]. A sequence search through the Uniprot database failed to reveal any significant resemblance of the protein encoded by ORF6 to any known polypeptide. At present there is no evidence of ORF6 translation *in vivo* [2, 11].

The UTR of the AltMV genome is located in the vicinity of the ORF5 stop codon at the 3' terminus. This 129 nt long genome region is highly conserved in AltMV isolates and the degree of nucleotide homology reaches 98% for some of them. This may be connected with the localization of the replicase recognition sites within this region [20]. Nevertheless, the secondary structure and the functions of the 3' UTR of the AltMV genome are yet to be determined. The same is also true for the 5' UTR as well as for putative regulatory elements of the AltMV genome. Furthermore, the presence of the conserved elements, namely, the octanucleotide and hexanucleotide motifs, suggests that their function in AltMV genome may be similar to that in the PVX one [2, 27]. Notably, no conserved polyadenylation signal similar to *bamboo mosaic virus* and PVX was detected in the 3' UTR of AltMV [28].

6. *Alternanthera Mosaic Virus*-Based Vectors

To estimate the correlation of symptom severity and efficiency of cell-to-cell movement with various mutations in TGBp1 and AltMV replicase sequences Lim et al. [29] constructed a viral vector based on AltMV genome. The vector was further applied to outline the functions of AltMV TGBp3 [23] by means of fluorescent reporter proteins DsRed and GFP. Both works employed the same vector design with the reporter gene sequences being inserted between TGBp3 and CP genes under the control of the additional subgenomic (sg) promoter of AltMV CP [29].

A bipartite vector was also derived from AltMV-SP. Its first fragment contained AltMV replicase gene, the second one comprised TGB and CP gene of AltMV. In order to facilitate the cloning, the construction was divided into two parts: following plant tissue transformation with both fragments the complete AltMV genome was produced through recombination. The second component of the vector system was obtained in two versions. The vector was designed for target protein expression in plants and virus-induced gene

silencing (VIGS) [23, 29]. Using the AltMV VIGS vector suppression of endogenous 4/1 protein of *N. benthamiana* expression and influence of Potato spindle tuber viroid movement in 4/1-silenced plants were demonstrated [30, 31]. Vectors AltMV-L-att and AltMV-P-att were created by insertion of the Gateway cloning cassette into the AltMV multiple cloning site (between the triple gene block and the CP gene) for protein expression and VIGS applications, respectively [32].

Several variants of the deconstructed viral AltMV-MU based vectors enabled heterologous proteins to be expressed in plants. TGB was deleted from the AltMV genome, while the target gene was placed under the control of either the AltMV additional sg promoter 1 (AltMV-single vector) or the two consecutive viral promoters sg promoter 1 and sg promoter 3 (AltMV-double vector) previously described in Lim *et al.* [23]. In comparison with AltMV-single, AltMV-double was demonstrated to produce higher target protein yield due to simultaneous functioning of the two sg promoters. AltMV CP and human granulocyte colony-stimulating factor were expressed as model proteins in the recent study [8]. Although an attempt was made to increase the protein accumulation level by using three sg promoters simultaneously, this approach failed to succeed [33].

7. AltMV Proteins

7.1. AltMV RNA-Dependent RNA Polymerase (RdRp). Two predicted domains (the helicase and the polymerase ones) were identified in AltMV RdRp amino acid sequence by the BLAST algorithm [2]. Before helicase domain methyl transferase and 2-oxoglutarate-Fe(II) oxygenase domains are located [19]. Four variants of infectious clones derived from the AltMV-SP genome caused various symptoms in infected *N. benthamiana* plants and differed from each other in several amino acid substitutions in viral proteins including RdRp [23]. The clones were referred to as 3-1, 3-7, 4-1, and 4-7. None of them induced the infection symptoms similar to those of AltMV-SP isolate. Plant infection and two of the clones (3-7 and 4-7) lead to necrosis and eventual death, while in case of combining four of them milder symptoms were revealed. The replication rate increased at least 4 times at 15°C in all the clones. Plants inoculated with the mixture of 4-7 ('severe') and 3-1 ('mild') isolates developed symptoms similar to the ones caused by AltMV-SP, while the clone ratio was different at 25 and 15°C. The clones causing severe symptoms and high necrosis rate (3-7, 4-7) differed from those causing milder infection by several substitutions in the replicase amino acid sequence. Therefore, severe symptoms and higher necrosis rate were characteristic of P1110/R1121/K1255 replicase variant and milder symptoms, of R1110/K1121/R1255 variant. Notably, all the aforementioned amino acid substitutions were located in the polymerase domain of the protein [23].

The AltMV RdRp comprises 1540 aa with solely 68 varying among the isolates [20]. Although the difference by 45 amino acid substitutions between the RdRp of infectious clones 3-1 and 4-7 accounts for significant changes in replication efficiency [7, 29], only slight distinctions between the phlox and the portulaca isolates were detected through a phylogenetic tree analysis [20].

7.2. AltMV TGBp1.

BLAST analysis performed for amino acid sequence of AltMV TGBp1 predicted the existence of N-terminal helicase domain [2].

Similarly to AltMV replicase, TGBp1 of AltMV-SP various clones manifest differences in their amino acid sequences. Part of the clones has leucine residue (TGBp1L88) and the other part has proline residue at 88 position (TGBp1P88) [29]. Solely TGBp1L88 is able to efficiently suppress posttranscriptional gene silencing in plants. Further investigation into the phenomenon revealed that the virus variant expressing TGBp1P88 has a lower replication rate in comparison with the variant expressing TGBp1L88. At the same time, both of the TGBp1 variants are capable of supporting the cell-to-cell movement although at different rates with TGBp1P88 slowing down the spread of infection. Notably, simultaneous expression of the two TGBp1 variants in plant cells reduces the antisilencing activity of the protein which implies the interaction between the two variants [29]. This interaction was confirmed using a yeast two-hybrid system. Subcellular localization of the TGBp1 variants by means of laser scanning confocal microscopy of N. benthamiana leaves indicated that TGBp1L88 is localized in the nuclear membrane and forms discrete aggregates in the nucleolus, while TGBp1P88 is localized in the nuclear periplasm [23]. Outside the nucleus TGBp1L88 was demonstrated to reside at the cell wall as small punctate aggregates, which suggests its association with plasmodesmata. On the contrary, TGBp1P88 was diffusely distributed throughout the cytoplasm. Since the helicase domain I of PVX TGBp1 has been previously reported to be required for in vitro oligomerization [34, 35], amino acid sequences alignment of AltMV TGBp1 and PVX TGBp1 was carried out in search for AltMV TGBp1 oligomerization sites. As a result, 7 conserved sequence motifs were identified in the helicase domain of AltMV TGBp1. The mutants carrying substitutions G31R and GK33/34RR in domain I of TGBp1 were unable to dimerize in the yeast two-hybrid system. The disrupted interaction was also observed in vivo in N. benthamiana plants by means of bimolecular fluorescence complementation. This argues for a crucial role of the protein domain I in the dimerization. As far as domains II and III are concerned, no mutations altered the dimerization process. The oligomerization of AltMV TGBp1 molecules is essential for silencing suppression [36]. Visualizing subcellular localization of AltMV TGBp1 variants both functional and defective in terms of oligomerization revealed the following pattern: AltMV TGBp1 variants capable of oligomerization were localized at the nucleolus or at the cell wall, while the mutant ones occupied the nucleoplasm instead of the nucleolus and were not detected in the vicinity of the cell wall [36]. These data suggest that TGBp1 oligomerization plays a key role both in cell-to-cell movement and silencing suppression.

AltMV TGBp1 is capable of selectively binding to several cellular proteins [37], namely, mitochondrial ATP synthase delta chain subunit, light-harvesting chlorophyll-protein complex I subunit A4, chlorophyll a/b binding protein, chloroplast IscA-like protein, and chloroplast β-ATPase. The latter was demonstrated to specifically bind solely to AltMV TGBp1L88 variant which is efficient silencing suppressor. At the same time, no interaction between the chloroplast β-ATPase and TGBp1P88 was detected. Since the virus-induced suppression of the protein expression induces severe symptoms in the host plant, the β-ATPase is considered to be involved in host plant immune response. Therefore, the interaction between the β-ATPase and TGBp1P88 appears to inhibit this process [37].

Similarly to PVX TGBp1, TGBp1 of AltMV is capable of interacting with one end of the virion thus activating RNA translation in vitro [3, 38].

7.3. AltMV TGBp2.

To date, little is known about AltMV TGBp2 structure except for a transmembrane domain identified by BLAST amino acid sequence analysis [2]. Despite high sequence similarity with only six out of 110 amino acid residues varying among the isolates, portulaca-like AltMV-Po and AltMV-IT comprise a clade clearly distinct from the five phlox isolates, with the bootstrap value of 100% [20].

7.4. AltMV TGBp3.

The three-dimensional structure of AltMV TGBp3 remains unresolved, and no specific domains have been revealed by BLAST search [2]. Up to now, two papers have addressed this issue [29, 39] and demonstrated AltMV TGBp3 to differ substantially from the homologous PVX one by subcellular as well as tissue localization patterns. In the infected leaves the fluorescently labeled TGBp3 was predominantly localized at the outer chloroplast membrane of mesophyll cells. Interestingly, TGBp3 overexpression induced chloroplast membrane vesiculation and veinal necrosis and contributed to the overall symptom severity. Deletion analysis indicated two amino acid residues (17V18L) of TGBp3 serving as the unique signal of AltMV TGBp3 localization in chloroplast membranes [29]. Moreover, TGBp3 is capable of directly interacting with the PsbO protein of the Photosystem II oxygen-evolving complex [39]. This interaction is governed by N-terminal region of TGBp3 from residue 16 to residue 20. The signal sequence required for AltMV TGBp3 chloroplast surface targeting is also localized within this region [29]. This may provide solid evidence for AltMV TGBp3 targeting chloroplast membrane through PsbO interaction, which in its turn is transported to chloroplasts from the cytoplasm where it is synthesized [39]. Thus, the efficiency of the interaction between PsbO and AltMV TRGp3 correlates with the severity of such symptoms as veinal necrosis and chloroplast membrane vesiculation. In case of impaired TGBp3 expression the virus lost the ability to enter the mesophyll cells and therefore cause systemic infection, which underlines the crucial role of TGBp3 in this process. Herewith, the defective virus demonstrated a comparatively limited ability to spread within epidermis with no systemic movement [29]. Visualizing the subcellular localization of AltMV RNA by means of fluorescence in situ hybridization indicated that the viral RNA as well as TGBp3 primarily accumulates near the surface of the chloroplast membrane. At the same time, the major amount of RNA was detected in mesophyll cells [29]. This drives to the conclusion that the AltMV replication occurs mostly in mesophyll cells, more specifically, at the outer chloroplast membrane. Interestingly, the presence of cellular TGBp2 exhibited no influence on AltMV TGBp3 subcellular localization as opposed to PVX TGBp3 [29, 40].

7.5. AltMV CP. AltMV coat protein (CP) is a 22-23-kDa protein comprising 207 aa. Together with AltMV replicase and TGBp1, AltMV CP determines symptom severity in host plants [24].

Both AltMV and PVX virions were found to be translationally activated. The AltMV genomic RNA is normally encapsidated and completely nontranslatable *in vitro*; however, translation can be activated through the phosphorylation of AltMV CP by protein kinase C or by TGBp1 binding to the viral particle [3].

AltMV CP was shown to assemble into stable extended polymers commonly referred to as VLPs *in vitro* under various conditions. Similarly to PapMV CP [41] AltMV CP formed extended VLPs *in vitro* in the absence of RNA at pH 4.0 and low ionic strength [3]. However, PapMV CP was incapable of forming RNA-free VLPs at pH 8.0, while AltMV CP formed particles morphologically resembling native virions under the same conditions [3]. In contrast to PapMV VLPs, AltMV ones were demonstrated to be highly stable under a wide range of conditions [3, 4]. According to their serological properties [3], virions and VLPs of AltMV are structurally different. Recent findings suggest that despite high morphological similarity, AltMV CPs possess a different fold in virions containing RNA and in RNA-free VLPs. By means of cryoelectron microscopy (CryoEM) the diameter of AltMV VLPs was measured to be 15.2 nm, thus exceeding that of AltMV virions (13.5 nm). Authors suggest that the absence RNA contributes significantly in increasing of VLPs central channel diameter (30 Å) versus virions (20 Å). CryoEM image processing demonstrated that VLPs possessed a larger number of CP subunits per turn (9.55) than AltMV virions (8.75) with the same pitch (35.7 Å) [4]. The authors hypothesize that, despite the similarity of AltMV virions and VLPs in the overall morphology when studied at low magnification, the folding and intersubunit interactions of AltMV CP differ in the presence and absence of RNA.

Tyulkina with colleagues [42] designed the hybrid viral vectors based on PVX genome and AltMV CP gene fused with sequences of influenza virus A M2e epitope. This vector was used for expression of chimeric AltMV CP in plant and VLP assembly. The authors considered this VLP as candidate vaccines [42]. Unlike PapMV there are no other works using AltMV virions or VLP as a platforms for epitopes presentation.

Both AltMV VLPs and virions demonstrated high stability under a wide range of conditions. It was shown that viral particles and VLPs do not change their morphology and size during incubation in distilled water, 0.15 M NaCl, and 0.01 M Tris-HCl, 0.15 M NaCl, and pH 7.5. Particularly worth mentioning is that AltMV virions and VLPs also remained stable after 1 hour incubation in mouse serum. Therefore, the absence of RNA in the VLP and the absence of RNA-protein

interactions did not affect the stability of the protein helix of the AltMV VLPs under the selected conditions. [4]. Moreover, high immunostimulating properties resulting in significant enhancement of immune response to a model antigen in test animals were shown for both types of particles [43]. These data ensure the practical application of virions and VLPs of AltMV as an adjuvant platform for vaccine development. Both types of virus particles have numerous advantages such as assembly conditions and stability of the particles in comparison with the AltMV closest relative, namely, PapMV, that has been already applied most successfully in this field of research [44, 45].

8. Conclusion

Alternanthera mosaic virus (AltMV) is a representative of potexviruses with genome structure and virion morphology typical of the group. Additionally, AltMV has been proved to have various desirable properties in terms of its practical application. Moreover, the protocols for the virus particles production and purification have been elaborated establishing the foundation for their further application. Numerous viral vectors were derived from the AltMV genomes providing a perspective tool for target protein production in plants. Stability under a broad range of conditions as well as the immunostimulating properties make AltMV virions and virus-like particles a powerful tool for a plethora of biomedical applications.

Acknowledgments

This work was supported by the Russian Science Foundation (Grant no. 14-24-00007).

References

[1] A. D. W. Geering and J. E. Thomas, "Characterisation of a virus from Australia that is closely related to papaya mosaic potexvirus," *Archives of Virology*, vol. 144, no. 3, pp. 577–592, 1999.

[2] J. Hammond, M. D. Reinsel, and C. J. Maroon-Lango, "Identification and full sequence of an isolate of Alternanthera mosaic potexvirus infecting Phlox stolonifera," *Archives of Virology*, vol. 151, no. 3, pp. 477–493, 2006.

[3] A. A. Mukhamedzhanova, "Characterization of Alternanthera mosaic virus and its Coat Protein," *The Open Virology Journal*, vol. 5, no. 1, pp. 136–140, 2011.

[4] E. K. Donchenko, E. V. Pechnikova, M. Y. Mishyna et al., "Structure and properties of virions and virus-like particles derived from the coat protein of Alternanthera mosaic virus," *PLoS ONE*, vol. 12, no. 8, Article ID e0183824, 2017.

[5] J. G. Atabekov, N. A. Nikitin, and O. V. Karpova, "New type platforms for in vitro vaccine assembly," *Moscow University Biological Sciences Bulletin*, vol. 70, no. 4, pp. 177–183, 2015.

[6] N. A. Nikitin, E. A. Trifonova, O. V. Karpova, and J. G. Atabekov, "Biosafety of plant viruses for human and animals," *Moscow University Biological Sciences Bulletin*, vol. 71, no. 3, pp. 128–134, 2016.

[7] H.-S. Lim, A. M. Vaira, L. L. Domier, S. C. Lee, H. G. Kim, and J. Hammond, "Efficiency of VIGS and gene expression in a novel bipartite potexvirus vector delivery system as a function of strength of TGB1 silencing suppression," *Virology*, vol. 402, no. 1, pp. 149–163, 2010.

[8] E. V. Putlyaev, A. A. Smirnov, O. V. Karpova, and J. G. Atabekov, "Double subgenomic promoter control for a target gene superexpression by a plant viral vector," *Biochemistry (Moscow)*, vol. 80, no. 8, article no. 132, pp. 1039–1046, 2015.

[9] E. A. Trifonova, V. A. Zenin, N. A. Nikitin et al., "Study of rubella candidate vaccine based on a structurally modified plant virus," *Antiviral Research*, vol. 144, pp. 27–33, 2017.

[10] M. Ciuffo and M. Turina, "A potexvirus related to Papaya mosaic virus isolated from moss rose (Portulaca grandiflora) in Italy," *Plant Pathology*, vol. 53, no. 4, p. 515, 2004.

[11] P. A. Ivanov, A. A. Mukhamedzhanova, A. A. Smirnov, N. P. Rodionova, O. V. Karpova, and J. G. Atabekov, "The complete nucleotide sequence of Alternanthera mosaic virus infecting Portulaca grandiflora represents a new strain distinct from phlox isolates," *Virus Genes*, vol. 42, no. 2, pp. 268–271, 2011.

[12] C. A. Baker, L. Breman, and L. Jones, *Plant Disease*, vol. 90, no. 6, pp. 833-833, 2006.

[13] J. Hammond, M. D. Reinsel, and C. J. Maroon-Lango, "Identification of potexvirus isolates from creeping phlox and trailing portulaca as strains of Alternanthera mosaic virus, and comparison of the 3'-terminal portion of the viral genomes," *Acta Horticulturae*, vol. 722, pp. 71–77, 2004.

[14] J. Hammond, M. D. Reinsel, and C. J. Maroon-Lango, "Identification of potexvirus isolates from creeping phlox and trailing portulaca as strains of Alternanthera mosaic virus, and comparison of the 3'-terminal portion of the viral genomes," *Acta Horticulturae*, vol. 722, pp. 71–77, 2006.

[15] B. E. Lockhart and M. L. Daughtrey, "First report of Alternanthera mosaic virus infection in Angelonia in the United States," *Plant Disease*, vol. 92, no. 10, p. 1473, 2008.

[16] J. Tang, J. D. Olson, F. M. Ochoa-Corona, and G. R. G. Clover, "Nandina domestica, a new host of Apple stem grooving virus and Alternanthera mosaic virus," *Australasian Plant Disease Notes*, vol. 5, no. 1, pp. 25–27, 2010.

[17] L. Maria Lembo Duarte, A. Nóbrega Toscano Maria Amélia Vaz Ale, E. Borges Rivas, and R. Harakava, "Identificação e controle do Alternanthera mosaic virus isolado de Torenia sp. (Scrophulariaceae)." *Revista Brasileira de Horticultura Ornamental*, vol. 14, no. 1, 2008.

[18] N. Iwabuchi, T. Yoshida, A. Yusa et al., "Complete genome sequence of Alternanthera mosaic virus, isolated from Achyranthes bidentata in Asia," *Genome Announcements*, vol. 4, no. 2, 2016.

[19] J. Hammond, I. Kim, and H. Lim, "Alternanthera mosaic virus and alternative model potexvirus of broad relevance," *Korean Journal of Agricultural Science*, vol. 44, pp. 145–180, 2017.

[20] J. Hammond and M. D. Reinsel, "Variability in Alternanthera Mosaic Virus isolates from different hosts," *Acta Horticulturae*, vol. 1072, pp. 47–54, 2015.

[21] A. Vitoreli, C. A. Baker, and C. L. Harmon, "Alternanthera mosaic virus identified in clock vine in Florida," *Phytopathology*, vol. 101, p. 183, 2011.

[22] L. L. Breman, "A strain of papaya mosaic potexvirus in Scutellaria," in *Plant Pathology Circular No. 396*, Florida Department of Agriculture and Consumer Services, Division of Plant Industry, Gainsville, Fla, USA, 1999.

[23] H.-S. Lim, A. M. Vaira, M. D. Reinsel et al., "Pathogenicity of Alternanthera mosaic virus is affected by determinants in RNA-dependent RNA polymerase and by reduced efficacy of silencing suppression in a movement-competent TGB1," *Journal of General Virology*, vol. 91, no. 1, pp. 277–287, 2010.

[24] H.-S. Lim, J. Nam, E.-Y. Seo et al., "The coat protein of Alternanthera mosaic virus is the elicitor of a temperature-sensitive systemic necrosis in Nicotiana benthamiana, and interacts with a host boron transporter protein," *Virology*, vol. 452-453, pp. 264–278, 2014.

[25] J. B. Bancroft, M. Abouhaidar, and J. W. Erickson, "The assembly of clover yellow mosaic virus and its protein," *Virology*, vol. 98, no. 1, pp. 121–130, 1979.

[26] A. A. Brunt, K. Crabtree, M. J. Dallwitz, A. J. Gibbs, L. Watson, and E. J. Zurcher, Eds., *Plant Viruses Online: Descriptions and Lists from the VIDE Database. 16th version*, 1997.

[27] K.-H. Kim and C. L. Hemenway, "Long-distance RNA-RNA interactions and conserved sequence elements affect potato virus X plus-strand RNA accumulation," *RNA*, vol. 5, no. 5, pp. 636–645, 1999.

[28] M.-R. Park, R.-D. Jeong, and K.-H. Kim, "Understanding the intracellular trafficking and intercellular transport of potexviruses in their host plants," *Frontiers in Plant Science*, vol. 5, 2014.

[29] H.-S. Lim, A. M. Vaira, H. Bae et al., "Mutation of a chloroplast-targeting signal in Alternanthera mosaic virus TGB3 impairs cell-to-cell movement and eliminates long-distance virus movement," *Journal of General Virology*, vol. 91, no. 8, pp. 2102–2115, 2010.

[30] S. Von Bargen, K. Salchert, M. Paape, B. Piechulla, and J.-W. Kellmann, "Interactions between the tomato spotted wilt virus movement protein and plant proteins showing homologies to myosin, kinesin and DnaJ-like chaperones," *Plant Physiology and Biochemistry*, vol. 39, no. 12, pp. 1083–1093, 2001.

[31] A. G. Solovyev, S. S. Makarova, M. V. Remizowa et al., "Possible role of the Nt-4/1 protein in macromolecular transport in vascular tissue," *Plant Signaling and Behavior*, vol. 8, no. 10, 2013.

[32] N.-Y. Ko, H.-S. Kim, J.-K. Kim et al., "Developing an Alternanthera Mosaic Virus vector for efficient cloning of whitefly cDNA RNAi to screen gene function," *Journal of the Faculty of Agriculture, Kyushu University*, vol. 60, no. 1, pp. 139–149, 2015.

[33] E. V. Putlyaev, A. A. Smirnov, E. A. Lazareva, G. V. Klink, O. V. Karpova, and J. G. Atabekov, "New phytoviral vector for superexpression of target proteins in plants," *Moscow University Biological Sciences Bulletin*, vol. 68, no. 4, pp. 169–173, 2013.

[34] A. D. Leshchiner, A. G. Solovyev, S. Y. Morozov, and N. O. Kalinina, "A minimal region in the NTPase/helicase domain of the TGBp1 plant virus movement protein is responsible for ATPase activity and cooperative RNA binding," *Journal of General Virology*, vol. 87, no. 10, pp. 3087–3095, 2006.

[35] A. D. Leshchiner, E. A. Minina, D. V. Rakitina et al., "Oligomerization of the potato virus X 25-kD movement protein," *Biochemistry (Moscow)*, vol. 73, no. 1, pp. 50–55, 2008.

[36] J. Nam, M. Nam, H. Bae et al., "AltMV TGB1 nucleolar localization requires homologous interaction and correlates with cell wall localization associated with cell-to-cell movement," *Plant Pathology*, vol. 29, no. 4, pp. 454–459, 2013.

[37] E.-Y. Seo, J. Nam, H.-S. Kim et al., "Selective Interaction between chloroplast β-ATPase and TGB1L88 retards severe symptoms caused by Alternanthera mosaic virus infection," *Plant Pathology*, vol. 30, no. 1, pp. 58–67, 2014.

[38] M. V. Arkhipenko, N. A. Nikitin, E. K. Donchenko, O. V. Karpova, and J. G. Atabekov, "Translational cross-activation of Potexviruses virion genomic RNAs," *Acta Naturae*, vol. 9, pp. 52–57, 2017.

[39] C. Jang, E.-Y. Seo, J. Nam et al., "Insights into Alternanthera mosaic virus TGB3 functions: Interactions with Nicotiana benthamiana PsbO correlate with chloroplast vesiculation and veinal necrosis caused by TGB3 over-expression," *Frontiers in Plant Science*, vol. 4, 2013.

[40] M. V. Schepetilnikov, U. Manske, A. G. Solovyev, A. A. Zamyatnin Jr., J. Schiemann, and S. Y. Morozov, "The hydrophobic segment of Potato virus X TGBp3 is a major determinant of the protein intracellular trafficking," *Journal of General Virology*, vol. 86, no. 8, pp. 2379–2391, 2005.

[41] J. W. Erickson, J. B. Bancroft, and R. W. Horne, "The assembly of papaya mosaic virus protein," *Virology*, vol. 72, no. 2, pp. 514–517, 1976.

[42] L. G. Tyulkina, E. V. Skurat, O. Yu. Frolova, T. V. Komarova, E. M. Karger, and I. G. Atabekov, "New Viral Vector for Superproduction of Epitopes of Vaccine Proteins in Plants," *Acta Naturae*, vol. 3, no. 4, pp. 73–82, 2011.

[43] E. K. Petrova, E. A. Trifonova, N. A. Nikitin, and O. V. Karpova, "Adjuvant properties of Alternanthera mosaic virus virions and virus-like particles," *FEBS J*, vol. 282, p. 134, 2015.

[44] M.-È. Lebel, K. Chartrand, E. Tarrab, P. Savard, D. Leclerc, and A. Lamarre, "Potentiating Cancer Immunotherapy Using Papaya Mosaic Virus-Derived Nanoparticles," *Nano Letters*, vol. 16, no. 3, pp. 1826–1832, 2016.

[45] G. Rioux, D. Carignan, A. Russell et al., "Influence of PapMV nanoparticles on the kinetics of the antibody response to flu vaccine," *Journal of Nanobiotechnology*, vol. 14, no. 1, article no. 43, 2016.

Permissions

All chapters in this book were first published in AV, by Hindawi Publishing Corporation; hereby published with permission under the Creative Commons Attribution License or equivalent. Every chapter published in this book has been scrutinized by our experts. Their significance has been extensively debated. The topics covered herein carry significant findings which will fuel the growth of the discipline. They may even be implemented as practical applications or may be referred to as a beginning point for another development.

The contributors of this book come from diverse backgrounds, making this book a truly international effort. This book will bring forth new frontiers with its revolutionizing research information and detailed analysis of the nascent developments around the world.

We would like to thank all the contributing authors for lending their expertise to make the book truly unique. They have played a crucial role in the development of this book. Without their invaluable contributions this book wouldn't have been possible. They have made vital efforts to compile up to date information on the varied aspects of this subject to make this book a valuable addition to the collection of many professionals and students.

This book was conceptualized with the vision of imparting up-to-date information and advanced data in this field. To ensure the same, a matchless editorial board was set up. Every individual on the board went through rigorous rounds of assessment to prove their worth. After which they invested a large part of their time researching and compiling the most relevant data for our readers.

The editorial board has been involved in producing this book since its inception. They have spent rigorous hours researching and exploring the diverse topics which have resulted in the successful publishing of this book. They have passed on their knowledge of decades through this book. To expedite this challenging task, the publisher supported the team at every step. A small team of assistant editors was also appointed to further simplify the editing procedure and attain best results for the readers.

Apart from the editorial board, the designing team has also invested a significant amount of their time in understanding the subject and creating the most relevant covers. They scrutinized every image to scout for the most suitable representation of the subject and create an appropriate cover for the book.

The publishing team has been an ardent support to the editorial, designing and production team. Their endless efforts to recruit the best for this project, has resulted in the accomplishment of this book. They are a veteran in the field of academics and their pool of knowledge is as vast as their experience in printing. Their expertise and guidance has proved useful at every step. Their uncompromising quality standards have made this book an exceptional effort. Their encouragement from time to time has been an inspiration for everyone.

The publisher and the editorial board hope that this book will prove to be a valuable piece of knowledge for researchers, students, practitioners and scholars across the globe.

List of Contributors

Emily Rumschlag-Booms
Department of Biology, Northeastern Illinois University, Chicago, Chicago, IL 60625, USA

Lijun Rong
Department of Microbiology and Immunology, College of Medicine, University of Illinois at Chicago, IL 60612, USA

Sanjeev K. Anand
Vaccine & Infection Disease Organization-International Vaccine Center (VIDO-InterVac), University of Saskatchewan, 120 Veterinary Road, Saskatoon, SK, Canada S7E 5E3
VeterinaryMicrobiology, University of Saskatchewan, 120 Veterinary Road, Saskatoon, SK, Canada S7E 5E3

SureshK. Tikoo
Vaccine & Infection Disease Organization-International Vaccine Center (VIDO-InterVac), University of Saskatchewan, 120 Veterinary Road, Saskatoon, SK, Canada S7E 5E3
VeterinaryMicrobiology, University of Saskatchewan, 120 Veterinary Road, Saskatoon, SK, Canada S7E 5E3
School of PublicHealth, University of Saskatchewan, 120 Veterinary Road, Saskatoon, SK, Canada S7E 5E3

Suchita Bhattacharyya
University of Mumbai and Department of Atomic Energy-Centre for Excellence in Basic Sciences, Health Centre Building, Vidyanagari, Kalina, Santacruz East, Mumbai 400098, India

Thomas J. Hope
Department of Cell and Molecular Biology, Feinberg School of Medicine, Northwestern University, 303 East Superior Avenue, Chicago, IL 60611, USA

Ayaz Ahmad and Raham Sher Khan
Department of Biotechnology, AbdulWali Khan University Mardan, Mardan 23200, Pakistan

Sajid Ali
Department of Biotechnology, AbdulWali Khan University Mardan, Mardan 23200, Pakistan
Center of Biotechnology & Microbiology, University of Peshawar, Peshawar 25120, Pakistan

Sumera Afzal Khan and Abid Ali Khan
Center of Biotechnology & Microbiology, University of Peshawar, Peshawar 25120, Pakistan

Sanaullah Khan and Ali Hydar Baig
Department of Biotechnology, KUST, Kohat 26000, Pakistan

Muhammad Hamayun
Department of Botany, AbdulWali Khan University Mardan, Mardan 23200, Pakistan

Amjad Iqbal
Department of Agriculture, AbdulWali Khan University Mardan, Mardan 23200, Pakistan

Abdul Wadood
Department of Biochemistry, AbdulWali Khan University Mardan, Mardan 23200, Pakistan

Taj Ur Rahman
Department of Chemistry, AbdulWali Khan University Mardan, Mardan 23200, Pakistan

Serge Barcy
Seattle Children's Research Institute, University of Washington, Seattle, WA 98101, USA

Soren Gantt
Seattle Children's Research Institute, University of Washington, Seattle, WA 98101, USA
Department of Pediatrics, University of Washington, Seattle, WA 98105, USA
Department of Global Health, University of Washington, Seattle, WA 98195, USA

Corey Casper
Department of Global Health, University of Washington, Seattle, WA 98195, USA
Department of Medicine, University of Washington, Seattle, WA 98195, USA
Department of Epidemiology, University of Washington, Seattle, WA 98195, USA
Fred Hutchinson Cancer Research Center, Seattle, WA 98109, USA

Eliora Gachelet and Michael Lagunoff
Department of Microbiology, University of Washington, Seattle, WA 98195, USA

Jacquelyn Carlsson
Department of Medicine, University of Washington, Seattle, WA 98195, USA

Wayne L. Gray
Department of Microbiology and Immunology, University of Arkansas for Medical Sciences, 4301West Markham Street, Little Rock, AR 72205, USA

Yuri Drygin
Belozersky Institute of Physico-Chemical Biology, Lomonosov Moscow State University, Vorobévy gory 1, Building 40, Moscow 119992, Russia

Joseph Atabekov
Belozersky Institute of Physico-Chemical Biology, Lomonosov Moscow State University, Vorobévy gory 1, Building 40, Moscow 119992, Russia
Department of Biology, Lomonosov Moscow State University, Vorobévy gory 1, Building 12, Moscow 119992, Russia

Olga Kondakova
Department of Biology, Lomonosov Moscow State University, Vorobévy gory 1, Building 12, Moscow 119992, Russia

Rashmee Topno, Siraj A. Khan and Jagadish Mahanta
Arbovirology Group, Entomology and Filariasis Division, Regional Medical Research Centre, ICMR, Northeast Region, Dibrugarh, Assam 786001, India

Pritom Chowdhury
Arbovirology Group, Entomology and Filariasis Division, Regional Medical Research Centre, ICMR, Northeast Region, Dibrugarh, Assam 786001, India
Department of Biotechnology, Tocklai Tea Research Institute, TRA, Jorhat, Assam 785008, India

Hiba Sami, Meher Rizvi, Mohd Azam, Indu Shukla and Abida Malik
Department of Microbiology, Jawaharlal Nehru Medical College, Aligarh Muslim University, Aligarh 202002, India

Rathindra M. Mukherjee
Asian Institute of Gastroenterology, Hyderabad 500082, India

M. R. Ajmal
Department of Medicine, Jawaharlal Nehru Medical College, Aligarh Muslim University, Aligarh 202002, India

Carolina Alves, Cristina Branco and Celso Cunha
Medical Microbiology Unit, Center for Malaria and Tropical Diseases, Institute of Hygiene and Tropical Medicine, Nova University, Rua da Junqueira 100, 1349-008 Lisbon, Portugal

Hazilawati Hamzah and Mohamed Mustapha Noordin
Department of Veterinary Pathology and Microbiology, Faculty of Veterinary Medicine, Universiti Putra Malaysia, 43400 Serdang, Selangor,Malaysia

Yusuf Abba
Department of Veterinary Pathology and Microbiology, Faculty of Veterinary Medicine, Universiti Putra Malaysia, 43400 Serdang, Selangor,Malaysia
Department of Veterinary Pathology, Faculty of Veterinary Medicine, University of Maiduguri, PMB 1069, Maiduguri, Borno State, Nigeria

Hasliza Hassim
Department of Veterinary Preclinical Sciences, Faculty of Veterinary Medicine, Universiti Putra Malaysia, 43400 Serdang, Selangor,Malaysia

Nikolai Nikitin, Ekaterina Petrova, Ekaterina Trifonova and Olga Karpova
Nikolai Nikitin, Ekaterina Petrova, Ekaterina Trifonova, and Olga Karpova

Kapila Kumar, Sreejith Rajasekharan, Sahil Gulati, Jyoti Rana, Reema Gabrani, Chakresh K. Jain and Sanjay Gupta
Center for Emerging Diseases, Department of Biotechnology, Jaypee Institute of Information Technology, A-10, Sector 62, Noida, Uttar Pradesh 201 307, India

Amita Gupta
Department of Microbiology, University of Delhi, Benito Juarez Marg, New Delhi 110021, India

Vijay K. Chaudhary
Department of Biochemistry, University of Delhi, Benito Juarez Marg, New Delhi 110021, India

Raquel Hernandez and Dennis T. Brown
Department of Molecular and Structural Biochemistry, North Carolina State University, Raleigh, NC 27695, USA

Angel Paredes
U.S. FDA/National Center for Toxicological Research, Department of Health and Human Services, Jefferson, AR 72079, USA

Margaret K. Glausser
The Johnson Center for Child Health and Development, 1700 Rio Grande Street, Austin, TX 78701, USA

Laura Hewitson
The Johnson Center for Child Health and Development, 1700 Rio Grande Street, Austin, TX 78701, USA

Department of Psychiatry, University of Texas Southwestern, Dallas, TX 75390, USA

James B. Thissen and Crystal J. Jaing
Physical & Life Sciences Directorate, Lawrence Livermore National Laboratory, Livermore, CA 94550, USA

Shea N. Gardner and Kevin S. McLoughlin
Computations Directorate, Lawrence Livermore National Laboratory, Livermore, CA 94550, USA

Sajib Chakraborty and Taibur Rahman
Department of Biochemistry andMolecular Biology, Faculty of Biological Sciences,University of Dhaka,Dhaka 1000, Bangladesh

Rajib Chakravorty
Department of EEE, University of Melbourne, National ICT Australia, Melbourne, VIC 3010, Australia

Vivek Chandramohan, Anubhav Kaphle,Mamatha Chekuri, Sindhu Gangarudraiah and Gowrishankar Bychapur Siddaiah
Department of Biotechnology, Siddaganga Institute of Technology, Tumkur, Karnataka 572 103, India

Alexandre Pereira
Laboratório de Genética, Instituto Butantan, 05503-900 São Paulo, SP, Brazil

Marcos Lazaro Moreli
Laboratório de Virologia, Universidade Federal de Goi´as, 75801-615 Jatai, GO, Brazil

Cristina Farah Tófoli
Instituto de Pesquisa Ecol´ogica (IPE), 12960-000 Nazaré Paulista, SP, Brazil

Tiene Zingano Hinke
Departamento de Parasitologia, Instituto de Ci^encias Biom´edicas, Universidade de São Paulo, 05389-970 São Paulo, SP, Brazil

Luiz Tadeu Moraes Figueiredo
Centro de Pesquisa em Virologia da Faculdade de Medicina de Ribeirão Preto da Universidade de São Paulo, 14049-900 Ribeirão Preto, SP, Brazil

Luca Mercuri
Hepatology Section, Division of Medicine, Faculty of Medicine, Imperial College, London, UK

Emma C. Thomson and Joseph Hughes
University of Glasgow MRC Centre for Virus Research, Glasgow, UK

Peter Karayiannis
University of Nicosia Medical School, Nicosia, Cyprus

Jay C. Brown
Department of Microbiology, Immunology, and Cancer Biology, University of Virginia Health System, Charlottesville, VA 22908, USA

Swapnil Bawage, Komal Vig, Shreekumar R. Pillai, Vida A. Dennis and Shree R. Singh
Center for NanoBiotechnology Research, Alabama State University, Montgomery, AL, USA

Erdal Eroglu
Center for NanoBiotechnology Research, Alabama State University, Montgomery, AL, USA
Faculty of Engineering, Bioengineering Department, Celal Bayar University, Muradiye, Manisa, Turkey

Ankur Singh
College of Medicine, University of South Alabama, Mobile, AL, USA

Pooja M. Tiwari
Yerkes National Primate Research Center, Emory University, Atlanta, GA, USA

Ekaterina Donchenko, Ekaterina Trifonova, Nikolai Nikitin, Joseph Atabekov and Olga Karpova
Department of Virology, Lomonosov Moscow State University, Moscow 119234, Russia

Index